Chapman & Hall
Interdisciplinary Statistics

Capture-Recapture Methods for the Social and Medical Sciences

Edited by
Dankmar Böhning
Peter G.M. van der Heijden
John Bunge

CRC Press
Taylor & Francis Group
Boca Raton London New York

CRC Press is an imprint of the
Taylor & Francis Group, an **informa** business

A CHAPMAN & HALL BOOK

CHAPMAN & HALL/CRC
Interdisciplinary Statistics Series

Series editors: N. Keiding, B.J.T. Morgan, C.K. Wikle, P. van der Heijden

Published titles

AGE-PERIOD-COHORT ANALYSIS: NEW MODELS, METHODS, AND EMPIRICAL APPLICATIONS Y. Yang and K. C. Land

ANALYSIS OF CAPTURE-RECAPTURE DATA R. S. McCrea and B. J.T. Morgan

AN INVARIANT APPROACH TO STATISTICAL ANALYSIS OF SHAPES S. Lele and J. Richtsmeier

ASTROSTATISTICS G. Babu and E. Feigelson

BAYESIAN ANALYSIS FOR POPULATION ECOLOGY R. King, B. J.T. Morgan, O. Gimenez, and S. P. Brooks

BAYESIAN DISEASE MAPPING: HIERARCHICAL MODELING IN SPATIAL EPIDEMIOLOGY, SECOND EDITION A. B. Lawson

BIOEQUIVALENCE AND STATISTICS IN CLINICAL PHARMACOLOGY S. Patterson and B. Jones

CAPTURE-RECAPTURE METHODS FOR THE SOCIAL AND MEDICAL SCIENCES D. Böhning, P. G. M. van der Heijden, and J. Bunge

CLINICAL TRIALS IN ONCOLOGY, THIRD EDITION S. Green, J. Benedetti, A. Smith, and J. Crowley

CLUSTER RANDOMISED TRIALS R.J. Hayes and L.H. Moulton

CORRESPONDENCE ANALYSIS IN PRACTICE, THIRD EDITION M. Greenacre

THE DATA BOOK: COLLECTION AND MANAGEMENT OF RESEARCH DATA M. Zozus

DESIGN AND ANALYSIS OF QUALITY OF LIFE STUDIES IN CLINICAL TRIALS, SECOND EDITION D.L. Fairclough

DYNAMICAL SEARCH L. Pronzato, H. Wynn, and A. Zhigljavsky

FLEXIBLE IMPUTATION OF MISSING DATA S. van Buuren

GENERALIZED LATENT VARIABLE MODELING: MULTILEVEL, LONGITUDINAL, AND STRUCTURAL EQUATION MODELS A. Skrondal and S. Rabe-Hesketh

GRAPHICAL ANALYSIS OF MULTI-RESPONSE DATA K. Basford and J. Tukey

INTRODUCTION TO COMPUTATIONAL BIOLOGY: MAPS, SEQUENCES, AND GENOMES M. Waterman

MARKOV CHAIN MONTE CARLO IN PRACTICE W. Gilks, S. Richardson, and D. Spiegelhalter

Published titles

MEASUREMENT ERROR ANDMISCLASSIFICATION IN STATISTICS AND EPIDE-MIOLOGY: IMPACTS AND BAYESIAN ADJUSTMENTS P. Gustafson

MEASUREMENT ERROR: MODELS, METHODS, AND APPLICATIONS
J. P. Buonaccorsi

MEASUREMENT ERROR: MODELS, METHODS, AND APPLICATIONS
J. P. Buonaccorsi

MENDELIAN RANDOMIZATION: METHODS FOR USING GENETIC VARIANTS IN CAUSAL ESTIMATION S.Burgess and S.G.Thompson

META-ANALYSIS OF BINARY DATA USINGPROFILE LIKELIHOOD D. Böhning, R. Kuhnert, and S. Rattanasiri

MISSING DATA ANALYSIS IN PRACTICE T. Raghunathan

MODERN DIRECTIONAL STATISTICS C. Ley and T. Verdebout

POWER ANALYSIS OF TRIALS WITH MULTILEVEL DATA M. Moerbeek and S. Teerenstra

SPATIAL POINT PATTERNS: METHODOLOGY AND APPLICATIONS WITH R
A. Baddeley, E Rubak, and R. Turner

STATISTICAL ANALYSIS OF GENE EXPRESSION MICROARRAY DATA T. Speed

STATISTICAL ANALYSIS OF QUESTIONNAIRES: A UNIFIED APPROACH BASED ON R AND STATA F. Bartolucci, S. Bacci, and M. Gnaldi

STATISTICAL AND COMPUTATIONAL PHARMACOGENOMICS R. Wu and M. Lin

STATISTICS IN MUSICOLOGY J. Beran

STATISTICS OF MEDICAL IMAGING T. Lei

STATISTICAL CONCEPTS AND APPLICATIONS IN CLINICAL MEDICINE
J. Aitchison, J.W. Kay, and I.J. Lauder

STATISTICAL AND PROBABILISTIC METHODS IN ACTUARIAL SCIENCE
P.J. Boland

STATISTICAL DETECTION AND SURVEILLANCE OF GEOGRAPHIC CLUSTERS
P. Rogerson and I. Yamada

STATISTICS FOR ENVIRONMENTAL BIOLOGY AND TOXICOLOGY A. Bailer and W. Piegorsch

STATISTICS FOR FISSION TRACK ANALYSIS R.F. Galbraith

VISUALIZING DATA PATTERNS WITH MICROMAPS D.B. Carr and L.W. Pickle

CRC Press
Taylor & Francis Group
6000 Broken Sound Parkway NW, Suite 300
Boca Raton, FL 33487-2742

First issued in paperback 2021

© 2018 by Taylor & Francis Group, LLC
CRC Press is an imprint of Taylor & Francis Group, an Informa business

No claim to original U.S. Government works

Version Date: 20170623

ISBN 13: 978-1-03-209669-8 (pbk)
ISBN 13: 978-1-4987-4531-4 (hbk)

**Visit the Taylor & Francis Web site at
http://www.taylorandfrancis.com**

**and the CRC Press Web site at
http://www.crcpress.com**

To our families,
our moms and dads.

Contents

VIII Miscellaneous Topics   387

Foreword by Byron J.T. Morgan

This is a timely, important book on the use of capture-recapture methods for social and medical data. To quote one of the book chapters,

> Capture-recapture methods provide a natural way to estimate the unknown size of a partially observed population ... through samples derived using some identification mechanism (traps, lists, registries, etc.). These methods were introduced in the wildlife setting to estimate animal abundance but have been extended to epidemiology, public health.... However, this technique continues to be underused, despite evidence that it can improve prevalence estimates even for diseases like diabetes that are both common and relatively well identified.

Several books have been written on capture-recapture methods for ecology, over many years, and one focussing on social and medical applications has been long overdue.

Much capture-recapture modelling analyses data have been collected on populations of wild animals, and we have recently celebrated 50 years since three iconic papers laid the foundations of the important Cormack–Jolly–Seber models. Issue no 2 of *Statistical Science* in 2016 presents transcripts of interviews with Cormack and Seber, and a range of papers outlining current research in ecology, social and medical areas. These papers indicate that capture-recapture research is continuing to develop in imaginative ways, partly in response to the demands of data arising from new technology, and also the need to monitor the effects of climate change on the environment and wild animals. This book illustrates the power of appropriate capture-recapture analyses in areas other than ecology. Several of the book chapters describe new methods, and suggest avenues for future research. For example, there is a wealth of material on fitting zero-truncated distributions; here one novelty is the use of empirical probability generating functions to fit distributions that are readily described by their generating functions.

The differing emphases among ecology, social and medical applications arise in part because of the primary consideration of closed populations in this book, whereas ecological applications often also involve open populations, and associated estimation of survival and movement probabilities. Indeed, none of the three papers mentioned above are referenced in this book, and frequently different computer packages are involved for model fitting from those in ecology. However there are many similarities, such as investigating sensitivity to model assumptions, including the effects of heterogeneity. As with ecological applications, the use of covariates can be illuminating, e.g., in determining the characteristics of opiate users rather than, say, evaluating the effect of cold weather on the mortality of grey herons. I would expect this book to facilitate cross-fertilisation of new methods between ecological and non-ecological areas, for instance in the area of spatial capture recapture.

The relevance of the methods described is evident, with applications to studies of the prevalence of scrapie, and estimating numbers of injecting drug users, of immigrants, and of victims of domestic violence, etc. Time and again we see the power of statistics in providing answers to really important questions. An interesting chapter considers alternatives to standard censuses of human populations:

The production of socio-economic statistics is undergoing a paradigm shift. ... A case in focus is the transformation of the population census itself. A number of European countries, including notably all the Scandinavian ones, conducted their last round of population census based entirely on the administrative data sources.

The wheel turns full circle here, in that Laplace's iconic 1802 capture-recapture study estimated the population size of France, using birth registers.

I enjoyed reading this book enormously. A great attraction is the wide range of motivating examples, complete with data, which include several from ecology. The way that methods are regularly illustrated on both real and simulated data is engrossing. Models are clearly described and accessible. The book should be required reading, for years to come, for any university course on applied statistical modeling, as well as being a vital reference for research. I am sure that this book will be much read, and make a major impact.

Byron J. T. Morgan, FLSW

Emeritus Professor and Co-Director of the National Centre for Statistical Ecology, Canterbury, Kent, UK

Preface

Capture-recapture methods have developed more and more interest over recent years. Not only have the methods been extended and new developments added, the areas of application have widened. With this book we try to acknowledge some of these recent developments. Our focus is on applications in the social and medical sciences. This is in contrast to existing books including the classical monograph by Seber [259] or more recent books by Amstrup, McDonald and Manly [7], Borchers, Buckland and Zucchini [49], King, Morgan, Gimenez, Brooks [166] or McCrea and Morgan [202] who all are more directed towards ecological applications. Hence we believe that this book covers a different niche. Of course, we do not forget the impact of Chapter 6 on closed population estimation in the book by Bishop, Fienberg and Holland [32] which introduced log-linear modeling into the closed capture-recapture framework and was also directed towards applications in social science and demography. Some of these developments are taken up in part V of the book at hand.

We have divided the contributions into eight different parts:

    I Introduction

   II Ratio regression models

 III Meta-analysis in capture-recapture

 IV Extensions of single source models

  V Multiple sources

 VI Latent variable models

VII Bayesian approaches

VIII Miscellaneous topics

After the introduction in Part I we focus on ratio regression models in Part II with contributions by Marco Alfó, Irene Rocchetti and DB on the fundamental concept of ratio regression modelling, by Antonello Maruotti and Orasa Anan with focus on the Conway–Maxwell–Poisson distribution, and by Veerasak Punyapornwithaya and DB with focus on estimating the burden of dengue fever in Chiang Mai province using ratio plotting on the basis of the geometric distribution. Carla Azevedo, Mark Arnold, and DB look at a capture-recapture setting wherein a subset of the observed data the missing information is available as well and can be incorporated into the inference using a ratio regression approach although other inference attempts would be possible too. Part III applies concepts of meta-analysis to capture-recapture. JB provides the fundamental framework of meta-analysis for capture-recapture. DB and JB provide an application of meta-analysis to maritime accidents and DB, Mehmet Orman, Timur Köse and JB look at an application of meta-analysis for mark-resight experiments. Part IV considers single source models. JB and Sarah Sernaker look at population size estimation via the empirical probability function, whereas Cécile Durot, Jade Giguelay, Sylvie Huet, Francois Koladjo, and Stéphane Robin use concepts of convex distributions for population size determination. Pedro Puig looks at the construction of lower bounds for the population size by means of the empirical probability generating function. Maarten Cruyff, Thomas Husken, and PvdH extend the truncated Poisson regression model to a time-at-risk model. Alberto Vidal-Diez and DB extend the estimator by Anne Chao for covariate information, and Panicha Kaskasamkul and DB consider the case of population size estimation for one-inflated count data. In Part V, we consider multiple sources. PvdH, Maarten Cruyff, Joe Whittaker, Bart Bakker and Paul Smith look at dual and multiple system estimation with fully and partially observed covariates. Eugene Zwane investigates population size estimation in capture-recapture models with continuous covariates whereas Li-Chun Zhang and John Dunne look at trimmed dual system estimation. An interesting application is provided by Bart Bakker, PvdH and Susanna Gerritse on estimating the size of non-registered residents in the Netherlands. Part VI considers latent variable models and is opened by Elena Stanghellini and Maria Giovanna Ranalli who look at population size estimation using a categorical latent variable. Francesco Bartolucci and Antonio Forcina use quantitative latent variables and Rasch models including their marginal extensions in the context of capture-recapture modelling. Zhiyuan Ma, Chang Xuan Mao and Yitong Yang look at hierarchical log-linear models for a heterogeneous population with three lists whereas Elvira Pelle, David Hessen, and PvdH consider a multidimensional Rasch model for multiple system estimation when the number of lists change over time. Rattana Lerdsuwansri and DB extend the Lincoln–Petersen estimator to the setting when both sources are counts in contrast to being binary which is the conventional case. Part VII contributes two Bayesian approaches: Kathryn Barger and JB look at objective Bayes estimation of population size using Kemp distributions and Danilo Alunni Fegatelli, Alessio Farcomeni, and Luca Tardella investigate Bayesian population size estimation with censored counts. The final Part VIII on miscellaneous topics includes a contribution by DB, JB and PvdH on uncertainty assessment in capture-recapture studies and modelling.

It has been a very interesting experience working on this edited book for more than 2 years. We are most grateful to more than 40 contributors for sharing their research work with us and helping us putting this book together. Without their support the book would have not been possible. Finally, our thanks go to the publisher Chapman & Hall/CRC, in particular Rob Calver and his team, for all support, encouragement and patience over the period during which this book has been developed.

Dankmar Böhning, Southampton
John Bunge, Ithaca
Peter G.M. van der Heijden, Utrecht and Southampton

List of Figures

List of Tables

Contributors

Marco Alfó
Sapienza University
Rome, Italy

Mark Arnold
Animal and Plant Health Agency
United Kingdom

Danilo Alunni Fegatelli
Sapienza University
Rome, Italy

Orasa Anan
Thaksin University
Phathalung, Thailand

Carla Azevedo
University of Southampton
Southampton, United Kingdom

Bart F. M. Bakker
Free University Amsterdam
and Statistics Netherlands
Amsterdam and The Hague,
The Netherlands

Kathryn Barger
Tufts University
Boston, United States of America

Francesco Bartolucci
University of Perugia
Perugia, Italy

Dankmar Böhning
University of Southampton
Southampton, United Kingdom

John Bunge
Cornell University
Ithaca, New York,
United States of America

Maarten J.L.F. Cruyff
Utrecht University
Utrecht, The Netherlands

Cécile Durot
Modal'X, Université Paris Nanterre
Paris, France

John Dunne
Central Statistics Office
and University of Southampton
Cork, Irelandand Southampton, UK

Alessio Farcomeni
Sapienza University
Rome, Italy

Antonio Forcina
University of Perugia
Perugia, Italy

Susanna C. Gerritse
VU University
Amsterdam, The Netherlands

Jade Giguelay
MaIAGE, INRA, Université Paris-Saclay
Paris, France

David J. Hessen
Utrecht University
Utrecht, The Netherlands

Sylvie Huet
MaIAGE, INRA, Université Paris-Saclay
Paris, France

Thomas F. Husken
Utrecht University
Utrecht, The Netherlands

Panicha Kaskasamkul
University of Southampton and Naresuan
  University

Southampton, United Kingdom and
    Phitsanulok, Thailand

Timur Köse
Ege University
Izmir, Turkey

Francois Koladjo
INSERM U1181, Université Paris-Saclay
Paris, France

Rattana Lerdsuwansri
Thammasat University
Pathumthani, Thailand

Zhiyuan Ma
Shanghai University of Finance and
    Economics
Shanghai, China

Changxuan Mao
Shanghai University of Finance and
    Economics
Shanghai, China

Antonello Maruotti
Libera Università Maria Ss. Assunta
    (LUMSA)
Rome, Italy

Mehmet Orman
Ege University
Izmir, Turkey

Elvira Pelle
University of Triest
Triest, Italy

Pedro Puig
Universitat Autònoma de Barcelona
Barcelona, Spain

Veerasak Punyapornwithaya
Chiang Mai University
Chiang Mai, Thailand

Maria Giovanna Ranalli
University of Perugia
Perugia, Italy

Stéphane Robin
UMR518 MIA, AgroParisTech, INRA,
    Université Paris-Saclay
Paris, France

Irene Rocchetti
Institute for Official Statistics of Italy
Rome, Italy

Sarah Sernaker
University of Minnesota
Minnesota, United States of America

Elena Stanghellini
University of Perugia
Perugia, Italy

Luca Tardella
Sapienza University
Rome, Italy

Peter G.M. van der Heijden
Utrecht University and University of
    Southampton
Utrecht, The Netherlands and
    Southampton, United Kingdom

Alberto Vidal-Diez
St George's University of London
London, United Kingdom

Joe Whittaker
University of Lancaster
Lancaster, UK

Yitong Yang
Shanghai University of Finance and
    Economics
Shanghai, China

Li-Chun Zhang
University of Southampton & Statistics
    Norway
Southampton, United Kingdom and Oslo,
    Norway

Eugene Zwane
University of Swaziland
Kwaluseni, Swaziland

Part I

Introduction

1

Basic concepts of capture-recapture

Dankmar Böhning

University of Southampton

John Bunge

Cornell University

Peter G.M. van der Heijden

Universities of Utrecht and Southampton

CONTENTS

1.1 Introduction and background

Let us consider a potentially elusive target population whose size we denote by N; it might be a wildlife population, a population of homeless people or drug addicts, software errors or humans with a specific disease. Often, in such a framework an identification device (a trap, a register, a screening test) can be repeatedly used to register units from the population and we may be interested in estimating the global size N of the target population.

In such a context, we may have binary indicator variables y_{it}, $i = 1, \ldots, N$, $t = 1, \ldots, T$, where $y_{it} = 1$ means that the i-th unit has been identified at the t-th occasion, while $y_{it} = 0$ means that the i-th unit has not been identified at t. The binary indicators y_{it} might be observed or not, but it is assumed that $y_i = \sum_{t=1}^{T} y_{it}$ is observed only if $y_i > 0$, that is if at least one $y_{it} > 0$ for $t = 1, \ldots, T$. When $y_{i1} = y_{i2} = \ldots = y_{iT} = 0$ and, consequently $y_i = 0$, the i-th unit remains *unobserved*. The quantity T, e.g., the number of sampling (identification) sources/occasions, may be known a priori, or it may correspond to the maximum observed count.

Here, *clustering* occurs by repeated identifications of the same unit, since the individual sequence $(y_{i1}, \ldots, y_{iT})$ represents a two-level structure; identification (sampling) occasions define lower-level units nested within individuals which represent upper-level units. By simply re-arranging units indices, we may distinguish between the untruncated population of counts $Y_1, Y_2, \ldots, Y_N$ and the truncated sample of counts $Y_1, Y_2, \ldots, Y_n$ where without limitation of generality, we have assumed that $Y_{n+1} = \cdots = Y_N = 0$. Given these assumptions, the target population can be described by a probability density function (y, p_y), where $y = 0, 1, \cdots$, and p_y denotes the probability of exactly y identifications for a generic unit in the population, under the usual constraints $p_y \geq 0$ and $\sum_{y=0}^{\infty} p_y = 1$. If we denote by f_y the frequency of units with count $Y = y$, that is, units that have been identified exactly y times (> 0), f_y/N (which cannot be computed since N is unknown) is an estimate of p_y, whereas f_x/n (which we can compute since n is known) is an estimate of the zero-truncated probability $p_y/(1 - p_0)$. Partial observation leads to a zero-truncated sample of size $n = \sum_{y \geq 1} f_y$. As a result of the study design, f_0 (the frequency of units that have not been observed) and $N = \sum_{y=0}^{T} f_y$ remain unknown. Starting from the observed zero-truncated distribution, the purpose is to find an estimate of the population size N; as it can be easily noticed, the problem at hand is a special form of the capture-recapture problem (see Bunge and Fitzpatrick [59], Wilson and Collins [299], or Chao [75], for reviews on the topic).

1.2 Data sets

1.2.1 Golf tees in St. Andrews

In this experiment 250 golf tee clusters were placed on a golf course in St. Andrews in an area of $1,680 m^2$. Then the area was surveyed by 8 students with the goal to recover as many golf tee clusters as possible. For details see Borchers et al. [49]. The identification history of 9 golf tee clusters is provided in Table 1.1. The distribution of the marginal counts $Y_i = \sum_{t=1}^{8} Y_{it}$ is provided in Figure 1.1. It is clear that only 162 golf tee clusters could be recovered, while 88 remained undetected.

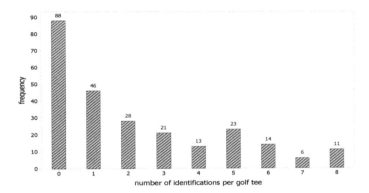

FIGURE 1.1: Frequency distribution of the number of identifications per golf tee for the capture-recapture experiment of recovering 250 golf tees in St. Andrews.

1.2.2 Homeless population of the city of Utrecht

As illustration of the problem we consider the question of estimating the homeless population of Utrecht (NL). The city of Utrecht runs shelters where homeless people are offered to stay overnight. Data are available for a period of 14 nights in 2013 and are shown in Table 1.2. It can be assumed that the shelters cover only the city of Utrecht. The table contains information on how often homeless people stayed in the shelter within this 14-nights period. For example, $f_1 = 36$ people stayed exactly one night, whereas $f_2 = 11$ people stayed exactly two nights, and so forth. In total, 222 different homeless people stayed in the shelters spending a total of $S = \sum_{y=1}^{14} y f_y = 2,009$ nights there. For more details see van der Heijden et al. [284]. Not all homeless people use the shelter at all times or not at all. Hence the register for homeless people based on the shelters is incomplete. The city of Utrecht is interested in the total size of its homeless population. Hence, we are interested to

TABLE 1.1
Entries y_{it} for the first 9 of the 162 golf tee clusters identified by the 8 surveyors; the index i stands for golf tee i, the index t for the observer t

Golf tee i	y_{i1}	y_{i2}	y_{i4}	y_{i4}	y_{i5}	y_{i6}	y_{i7}	y_{i8}	$y_i = \sum_{t=1}^{8} y_{it}$
			Different surveyors						
1	1	0	0	0	0	0	0	0	1
2	0	0	0	0	0	0	0	1	1
3	0	0	0	0	0	0	0	1	1
4	0	0	0	0	0	0	0	1	1
5	0	1	0	0	0	0	0	0	1
6	1	0	0	0	0	0	0	1	2
7	1	0	0	0	1	0	0	0	2
8	1	1	1	1	1	1	0	1	7
9	0	0	0	0	1	0	0	0	1

find an estimate of N, or, equivalently, of f_0, the size of the hidden homeless population. In this case $T = 14$, and it would be tempting to model the distribution of Y with a binomial distribution or a mixed binomial distribution to cope with heterogeneity, and then used the estimated distribution to achieve an estimate of p_0.

1.2.3 McKendrick's Cholera data

McKendrick [205] studied a Cholera epidemic in an Indian village and provided a frequency distribution of Cholera cases per household. Here f_y is the frequency of houses with exactly y cases. There were $f_1 = 32$ households with exactly one case, $f_2 = 16$ households with exactly two cases, $f_3 = 6$ households with exactly three cases, and $f_4 = 1$ household with exactly 4 cases. In total, 55 households had Cholera cases. However, many more households were affected by the Cholera epidemic. The question of interest here is how many households f_0 of those with no Cholera cases are affected by the epidemic (but have no cases)?

1.2.4 Matthews's data on estimating the Dystrophin density in the human muscle

Cullen et al. [92] (see also Matthews and Appleton [201]) attempted to locate dystrophin, a gene product of possible importance in muscular dystrophies, within the muscle fibres of biopsy specimens taken from normal patients. Units (epitopes) of dystrophin cannot be detected by the electron microscope until they have been labelled by a suitable "electron-dense" substance. The technique uses gold-conjugated antibodies which adhere to the dystrophin. However, not all units are succesfully labelled and it is important to account for all labelled and unlabelled units to achieve an unbiased estimate of the dystrophin density. In addition, more than one anti-body molecule may attach to a dystrophin unit. Hence, a count variable Y is observed, counting the number of antibody molecules on each dystrophin unit. As not every epitope is labelled $Y = 0$ is possible and indicates that unit is unlabelled and not observed. Hence Y is a zero-truncated count variable. The associated frequency distribution is $f_1 = 122$, $f_2 = 50$, $f_3 = 18$, $f_4 = 4$, $f_5 = 4$, and in total $n = 198$ labelled units have been observed.

1.2.5 Del Rio Vilas's data on Scrapie surveillance in Great Britain 2005

The occurrence of Scrapie in sheep in the holdings of Great Britain is monitored in the Compulsory Scrapie Flocks Scheme (CSFS) which was established in 2004 and is summarizing abattoir survey, stock survey and the statutory reporting of clinical cases. For more details see Böhning and Del Rio Vilas [39]. The frequency distribution of the count Y of Scrapie cases within each holding for the year 2005 is as follows: $f_1 = 84, f_2 = 15$, $f_3 = 7$, $f_4 = 5$, $f_5 = 2$, $f_6 = 1$, $f_7 = 2$, $f_8 = 2$, with a total of $n = 118$ holdings being observed. The issue here is to estimate the completeness of the surveillance system or to estimate

TABLE 1.2

Frequency distribution of the number of nights y stayed in the shelter per homeless person for the city of Utrecht for a period of 14 nights in 2013

y	1	2	3	4	5	6	7	8	8	10	11	12	13	14
f_y	36	11	6	11	5	7	6	11	3	8	7	12	22	77

the undercount of Scrapie by the surveillance system. For more details on epidemiological capture-recapture studies with illicit drug use application see Hay [140].

1.2.6 Hser's data on estimating hidden intravenous drug users in Los Angeles 1989

Intravenous drug users in LA County were entered into the California Drug Abuse Data System (CAL-DADS). The data in Table 1.3 stem from Hser [149] and refer to the frequency distribution of the episode count per drug user in 1989. Note that drug users with no episode are not entered into the system. The question here is to estimate the size of the hidden drug user group.

1.2.7 Methamphetamine drug use in Bangkok 2001

Drug abuse has become a serious health problem for many countries including Thailand. Surveillance data on drug use are available for 61 health treatment centres in the Bangkok metropolitan region from the Office of the Narcotics Control Board (ONCB). Using this data it was possible to reconstruct the counts of treatment episodes for each patient in the last quarter of 2001. Table 1.4 presents the number of methamphetamine users for each count of treatment episodes (Böhning et al. [35]). Here again, interest is in estimating f_0, the number of hidden methamphetamine users. The maximum observed contact of a drug user was 10.

1.2.8 Chun's data on estimating hidden software errors for the AT&Ts 5ESS switch

Chun [83] presents data from a software error reviewing experiment on issues with the AT&Ts 5ESS switch. Fourty-three faults were detected by at least one of the six reviewers. The details are provided in Table 1.5. The question here is to estimate the number of hidden faults.

TABLE 1.3
Frequency distribution of the episode count per drug user for the year 1989 based on the California Drug Abuse Data System

f_1	f_2	f_3	f_4	f_5	f_6	f_7	f_8	f_9	f_{10}	f_{11}	f_{12}
11,982	3,893	1,959	1,002	575	340	214	90	72	36	21	14

TABLE 1.4
Frequency distribution of the contact count per methamphetamine drug user for a 3-months period in 2001 in Bangkok (Thailand)

f_1	f_2	f_3	f_4	f_5	f_6	f_7	f_8	f_9	f_{10}
3114	163	23	20	9	3	3	3	4	3

1.2.9 Estimating the size of the female grizzly bear population in the Greater Yellowstone Ecosystem

A typical problem in wildlife ecology is estimating the size of a wildlife population. Keating et al. [162] present numbers of sightings of female Grizzle bears with cubs-of-the-year in the Greater Yellowstone Ecosystem. In Table 1.6 we see the frequency f_y of female Grizzle bears that have been observed in the particular year exactly y times. The actual purpose is here to provide a surveillance of the change in the total size of the female Grizzle bear population.

1.2.10 Spinner dolphins around Moorea Island

Oremus [223] estimated the size of a small community of Spinner dolphins around Moorea Island (Tahiti). Observations were done within an 8-month observational period. The following frequencies were reported: $f_1 = 42$, $f_2 = 7$, $f_3 = 2$, in total $n = 52$ different Spinner dolphins have been observed.

1.2.11 Microbial diversity in the Gotland Deep

Microbial ecologists are interested in estimating the number of species N in particular environments. Unlike butterflies, microbial species membership is not clear from visual inspection, so individuals are defined to be members of the same species (or more general taxonomic group) if their DNA sequences (derived from a certain gene) are identical up to some given percentage, 95% in this case. Here the study concerned protistan diversity

TABLE 1.5

Indicator matrix y_{ti} for fault i and reviewer t in a reviewing experiment for the AT&Ts 5ESS switch by six reviewers

Reviewer t	1	2	3	4	5	6	7	8	9	10	11	12	13	14	15
1	1			1				1	1		1	1	1	1	1
2															
3															
4		1			1	1	1				1	1		1	
5			1												1
6		1													

Reviewer t	16	17	18	19	20	21	22	23	24	25	26	27	28	29	30
1	1	1	1			1	1				1			1	1
2		1		1	1										
3													1		
4		1				1	1			1					1
5	1	1							1	1		1			1
6		1	1						1						

Reviewer t	31	32	33	34	35	36	37	38	39	40	41	42	43
1	1	1	1					1	1	1		1	1
2													
3				1	1		1						
4	1												
5											1		
6						1		1					

in the Gotland Deep, a basin in the central Baltic Sea. The sample was collected in May 2005, resulting in the data displayed in Table 1.7. There were $f_1 = 48$ DNA sequences observed exactly once, $f_2 = 9$ were observed exactly twice, up to the maximum observed DNA sequence, which was counted 53 times. The total size of observed different sequences were $n = 84$. For further details see Stock [271].

1.2.12 Illegal immigrants in the Netherlands

As a further social science example, we discuss the estimation of the number of illegal immigrants in four large cities in the Netherlands from police records, analysed with the truncated Poisson regression model by van der Heijden et al. [279, 280] and Böhning and van der Heijden [42]. In their analysis, focus is on those illegal immigrants that, once apprehended, cannot be effectively expelled by the police because, for example, their home country does not cooperate with the organization of deportation. In such cases the police request the individuals to leave the country, but it is unlikely that they will abide by such a request. Hence, they can be apprehended multiple times. The frequency distribution of the apprehension distribution of the illegal immigrant population is as follows: $f_1 = 1645$ were apprehended exactly once, $f_2 = 183$ exactly twice, $f_3 = 37$ exactly three times; the remaining numbers are $f_4 = 13$, $f_5 = 1$ and $f_6 = 1$.

TABLE 1.6

Frequency of sightings of female Grizzle bears with cubs-of-the-year in the Greater Yellowstone Ecosystem for each of the years from 1986 to 2001

Year	S	f_1	f_2	f_3	f_4	f_5	f_6	f_7	f_8	f_9	f_{10}	f_{11}	f_{15}
1986	82	7	5	6	1	1	0	1	2	0	0	0	1
1987	20	7	3	1	1	0	0	0	0	0	0	0	0
1988	36	7	4	4	1	1	0	0	0	0	0	0	0
1989	27	6	5	0	1	0	0	1	0	0	0	0	0
1990	49	7	6	7	1	1	0	0	0	0	0	0	0
1991	62	11	3	3	3	1	2	1	0	0	0	0	0
1992	37	15	5	1	1	0	0	0	0	0	0	0	0
1993	29	7	8	2	0	0	0	0	0	0	0	0	0
1994	29	9	7	2	0	0	0	0	0	0	0	0	0
1995	25	13	2	1	0	1	0	0	0	0	0	0	0
1996	45	15	10	2	1	0	0	0	0	0	0	0	0
1997	65	13	7	4	1	3	0	1	0	0	0	0	0
1998	75	11	13	5	1	1	0	2	0	0	0	0	0
1999	94	9	4	6	2	4	2	0	1	0	0	1	0
2000	72	17	8	1	2	1	0	2	0	1	0	0	0
2001	84	16	12	8	0	1	0	0	1	0	0	0	0

TABLE 1.7

Frequency distribution of the different DNA sequences

f_1	f_2	f_3	f_4	f_5	f_6	f_7	f_8	f_9	f_{10}
48	9	6	2	0	2	0	2	1	1

f_{12}	f_{13}	f_{16}	f_{17}	f_{18}	f_{20}	f_{29}	f_{42}	f_{53}
1	1	1	2	1	1	1	1	1

1.2.13   Shakespeare's unused words

Efron and Thisted [109] tried to answer the question of how many words Shakespeare knew but did not use. They analyzed data collected previously by Spevack [265]. Table 1.8 provides the first part of the frequency distribution of the different word types. According to this, he used $f_1 = 14,376$ words only once, $f_2 = 4,343$ exactly twice and so forth. The complete table can be found in Spevack [265]. According to the analysis Shakespeare knew about 31,500 different words. Efron and Thisted [109] estimated that he knew at least 35,000 more words (but did not use them).

1.3   Estimating population size under homogeneity

Estimates for population size can be achieved using the binomial distribution

$$p_x = P(X = x) = \binom{T}{x}\theta^x(1 - \theta)^{T-x} \tag{1.1}$$

for $x = 0, 1, \cdots, T$. Here T is the number of trapping occasions and $\theta \in (0,1)$ is the probability of capturing a member of the target population at one arbitrary occasion. X is the count of identifications per member of the target population and it is a central underlying assumption that identification occurs independently across occasions and with the same probability θ. The maximum likelihood estimate under the binomial distribution is $\hat{\theta} = \frac{1}{NT}\sum_{x=0}^{T} f_x x$. In our case, the population size N is unknown as is f_0. In other words, we only observe a zero-truncated count of identifications X as members of the target population that have never been identified during the trapping do not occur in the sample.

TABLE 1.8

Frequency distribution f_x of the word types used by Shakespeare exactly x times (only first 100 counts)

f_1	f_2	f_3	f_4	f_5	f_6	f_7	f_8	f_9	f_{10}
14376	4343	2292	1463	1043	837	638	519	430	364
f_{11}	f_{12}	f_{13}	f_{14}	f_{15}	f_{16}	f_{17}	f_{18}	f_{19}	f_{20}
305	259	242	223	187	181	179	130	127	128
f_{21}	f_{22}	f_{23}	f_{24}	f_{25}	f_{26}	f_{27}	f_{28}	f_{29}	f_{30}
104	105	99	112	93	74	83	76	72	63
f_{31}	f_{32}	f_{33}	f_{34}	f_{35}	f_{36}	f_{37}	f_{38}	f_{39}	f_{40}
73	47	56	69	63	45	34	49	45	52
f_{41}	f_{42}	f_{43}	f_{44}	f_{45}	f_{46}	f_{47}	f_{48}	f_{49}	f_{50}
49	41	30	35	37	21	41	30	28	19
f_{51}	f_{52}	f_{53}	f_{54}	f_{55}	f_{56}	f_{57}	f_{58}	f_{59}	f_{60}
25	19	28	27	31	19	19	22	23	14
f_{61}	f_{62}	f_{63}	f_{64}	f_{65}	f_{66}	f_{67}	f_{68}	f_{69}	f_{70}
30	19	21	18	15	10	15	14	11	16
f_{71}	f_{72}	f_{73}	f_{74}	f_{75}	f_{76}	f_{77}	f_{78}	f_{79}	f_{80}
13	12	10	16	19	11	8	15	12	7
f_{81}	f_{82}	f_{83}	f_{84}	f_{85}	f_{86}	f_{87}	f_{88}	f_{89}	f_{90}
13	12	11	8	10	11	7	12	9	8
f_{91}	f_{92}	f_{93}	f_{94}	f_{95}	f_{96}	f_{97}	f_{98}	f_{99}	f_{100}
4	7	6	7	10	10	15	7	7	5

Hence we need to base inference on the zero-truncated binomial distribution

$$p_x^+ = \frac{1}{1-(1-\theta)^T}\binom{T}{x}\theta^x(1-\theta)^{T-x}, \tag{1.2}$$

for $x = 1, \cdots, T$. The maximum likelihood estimate based upon the zero-truncated binomial is not available in closed form, but can easily be constructed by means of the EM algorithm (Dempster et al. [100]). In the *E-step*, given $\hat{\theta}$, we achieve an expected value of f_0 as $\hat{f}_0 = E(f_0|\hat{\theta}, n) = n\frac{(1-\hat{\theta})^T}{1-(1-\hat{\theta})^T}$. In the *M-step*, given $\hat{f}_0$, we find the maximum likelihood estimate for the expected, complete likelihood as $\hat{\theta} = \frac{1}{(n+\hat{f}_0)T}\sum_{x=0}^T f_x x$. The EM algorithm cycles between the E- and M-step until convergence.

An alternative, non-iterative estimator is connected with the name of Alan Turing [131]. The construction of the estimator starts by noting the untruncated count X has expected vale $E(X) = T\theta$ and that the probability for exactly one identification over the period of T possible identifications is $p_1 = T\theta(1-\theta)^{T-1}$. Hence

$$p_0 = (1-\theta)^T = \left(\frac{T\theta(1-\theta)^{T-1}}{T\theta}\right)^{T/(T-1)} = \left(\frac{p_1}{E(X)}\right)^{T/(T-1)}. \tag{1.3}$$

As it is possible to estimate p_1 as f_1/N and $E(X) = S/N$, where $S = \sum_{x=0}^T f_x x$, we can estimate p_0 as $(f_1/S)^{T/(T-1)}$ as the unknown N cancels out. Furthermore, the expected value of the Horvitz–Thompson estimate $n/(1-p_0)$ is N, so that the final estimate of Turing

$$\hat{N} = \frac{n}{1-(f_1/S)^{T/(T-1)}} \tag{1.4}$$

arises. If T becomes large, the simpler version $\hat{N} = \frac{n}{1-(f_1/S)}$ may be used.

We see that in many of our introduced examples, no fixed occasions of trapping or identification are available as these have occurred at some point in time within the observational window. This includes the cholera study in Section 1.2.3, the drug user study in LA County of California in Section 1.2.6, the surveillance study on grizzly bears in Section 1.2.9, the example of protistan diversity in the Baltic Sea in Section 1.2.11, the surveillance study on spinner dolphins in Section 1.2.10, the illegal immigrant study in Section 1.2.12, and the linguistic study on unused words of Shakespeare in Section 1.2.13.

In these cases, a common choice is the Poisson distribution

$$p_x = P(X = x) = \exp(-\lambda)\lambda^x/x! \tag{1.5}$$

for $x = 0, 1, \cdots$ and λ positive. A possible justification of the Poisson assumption is as follows. Suppose the observational window consists of a large number of trapping occasions, each with the same positive capture probability θ. Then, using that $T\theta = \lambda$ remains constant when T becomes large, the binomial distribution (1.2) converges to the Poisson distribution (1.5) with parameter $T\theta = \lambda$. X is again the count of identifications per member of the target population. The only difference is that we do not know what could have been the largest possible count. The underlying assumption remains that identification occurs independently across occasions and with the same probability θ. The maximum likelihood estimate under the Poisson distribution in the untruncated case is $\hat{\lambda} = \frac{1}{N}\sum_{x=0}^T f_x x$. Again, we need to iteratively compute it and the relevant steps in the EM algorithm are the *E-step* as $\hat{f}_0 = E(f_0|\hat{\lambda}, n) = n\frac{\exp(-\hat{\lambda})}{1-\exp(-\hat{\lambda})}$. In the *M-step* we find the maximum likelihood estimate for the expected, complete likelihood as $\hat{\lambda} = \frac{1}{(n+\hat{f}_0)}\sum_{x=0}^m f_x x$, where m is largest count that has occurred in the sample. The EM algorithm cycles between the E- and M-steps until convergence.

In the Poisson case, there is also a Turing estimator possible, in fact, it is the more popular version. The derivation is very similar to the binomial case and leads to

$$\hat{N} = \frac{n}{1 - (f_1/S)}, \tag{1.6}$$

which is the limiting case of the Turing estimator in the binomial case for T becoming large in size. The denominator of the Turing estimator $1 - (f_1/S)$ is an estimate of the *sample coverage* $1 - p_0$, the proportion of the target population covered by the sample. We will give illustrations of these estimators for our examples further below.

1.4   Simple estimates under heterogeneity

In many application studies the assumption of homogeneous catchability or identifiability across members of the target population will not be met. In these case we speak of *heterogeneity*. As it is not observed which member has which parameter, the parameter is considered as a latent variable, a variable whose values are not observed. Consider the Poisson case. In the pair (x, λ), x is a realization of the random variable X, the observed number of captures, and λ is an unobserved realization of the random variable Λ, the parameter of the Poisson density. It follows that the joint density $f(x, \lambda)$ can be written as $f(x|\lambda)g(\lambda)$, where $g(\lambda$ is the marginal distribution of λ w.r.t. $f(x, \lambda)$. As we have not observed the value of λ we consider the margin of $f(x|\lambda)g(\lambda)$ over λ leading to the mixture

$$m_x = \int_\lambda f(x|\lambda)g(\lambda)d\lambda, \tag{1.7}$$

for $x = 0, 1, \cdots$. In the mixture model (1.7), we call $g(\lambda)$ the *mixing distribution* and $f(x|\lambda)$ the *mixture kernel* or simply kernel. In the Poisson case, e.g. when the mixture kernel is a Poisson density, (1.7) becomes $m_x = \int_\lambda \exp(-\lambda)\lambda^x/x! \, g(\lambda)d\lambda$, in the binomial case, (1.7) becomes $m_x = \int_\theta \binom{T}{x}\theta^x(1-\theta)^{T-x} \, g(\theta)d\theta$. Note that we can think of (1.7) as an expected value

$$m_x = E[f(x|\lambda)] = E[\exp(-\Lambda)\Lambda^x/x!], \tag{1.8}$$

where we used the specific case of the Poisson kernel on the RHS of the second equation in (1.8). Chao [72, 73] suggested to use the Cauchy–Schwarz inequality for random variables V, W, which says that $E(VW)^2 \le E(V^2)E(W^2)$, to yield

$$E[\exp(-\Lambda)\Lambda] \le E[\exp(-\Lambda)]E[\exp(-\Lambda)\Lambda^2], \tag{1.9}$$

by choosing $V = \sqrt{\exp(-\Lambda)}$ and $W = \sqrt{\exp(-\Lambda)}\Lambda$. It follows, for the Poisson case, that

$$m_1^2/(2m_2) \le m_0. \tag{1.10}$$

Note that the LHS of (1.10) establishes a lower bound for m_0. Note that this lower bound can easily be estimated as $f_1^2/(2f_2)$ and represents a lower bound estimate for f_0. This leads to Chao's *lower bound estimate* of population size

$$\hat{N}_C = n + f_1^2/(2f_2). \tag{1.11}$$

Chao's estimator of population size is one of the most popular estimators in capture-recapture applications, due to its simplicity and nonparametric character. In the case of

Poisson homogeneity, (1.11) is asymptotically unbiased. For small population sizes, a bias correction should be used. The reason for the bias adjustment is as follows. Ideally, we would like f_1^2/f_2 to be close to $E(f_1)^2/E(f_2)$. However, the estimator f_1^2/f_2 estimates $E[f_1^2/f_2]$, and $E(f_1)^2/E(f_2)$ and $E[f_1^2/f_2]$ are not necessarily close. As it turns out, an excellent bias-corrected estimator is provided by

$$\hat{N}_{CB} = n + f_1(f_1 - 1)/(2f_2 + 2);$$ (1.12)

details are given in Böhning [43] and in Part IV of the book. In a similar manner, we can arrive at the Chao estimator for binomial kernels, namely $\hat{N}_C = n + \frac{T-1}{T}f_1^2/(2f_2)$ and $\hat{N}_{CB} = n + \frac{T-1}{T}f_1(f_1 - 1)/(2f_2 + 2)$, for the bias-corrected version.

Another estimator suggested to cope with heterogeneity is Zelterman's estimator [306]. The basic idea starts with the Horvitz-Thompson estimator $n/(1-p_0) = n/[1-\exp(-\lambda)]$ in the Poisson case, However, instead of replacing the unknown λ by the maximum likelihood estimate, Zelterman observes that

$$\frac{2p_2}{p1} = \frac{2\exp(-\lambda)\lambda^2/2}{\exp(-\lambda)\lambda} = \lambda = E(X)$$ (1.13)

and suggests to estimate λ, using (1.13), as $2f_2/f_1$. This leads not only to a very simple estimate $n/[1 - \exp(-2f_2/f_1)]$ of the population size, the estimate is also not affected by any change in the frequencies $f_3, \cdots, f_m$ where m is the largest observed count. Hence the estimate builds on a *local Poisson* assumption and would not change if contaminations occur. The relationship between Chao's and Zelterman's estimator is investigated in Böhning [43] and a close relationship between the two was found. It is also shown that in some circumstances, Zelterman's estimator can overestimate and does not show the lower bound property of Chao's estimator.

So far we have only talked about *unobserved* heterogeneity, the form of heterogeneity where it is not observed in which way capture probabilities vary across members of the target population. This is in contrast to observed heterogeneity where capture probabilities vary across members of the target population according to observed covariate information such as gender, age or geographic area. Examples of how observed covariate information can be incorporated into Chao and Zelterman estimations are provided in Böhning and van der Heijden [42] and Böhning et al. [45], but more details on this are discussed in Chapter 13, and more general inclusion of covariates into the modeling in Chapter 15 and 16 in this book. The estimators mentioned above are only a selection. Estimators have been developed using a specific heterogeneity distribution for $g(\lambda)$ such as the Chao–Bunge estimator [76] which uses a Γ-distribution for $g(\lambda)$ or an extension of the Chao lower bound estimator to include the first 3 frequencies f_1, f_2, f_3 by Lanumteang and Böhning [173].

1.5 Examples and applications

Let us apply these estimators to some of our data sets presented previously. For computation of these estimators we use the software SPADE developed by Chao et al. [78]. The usage of this R-based software is very simple and easy as only a text file needs to be prepared containing the pairs (x, f_x), in more detail $1f_1\ 2f_2\ 3f_3\ \cdots\ mf_m$ where m is the largest observed count. The estimators provided by SPADE do not incorporate any information on the number of sampling occasions whether they are available or not. Hence maximum likelihood estimation is based on the Poisson density and the versions of Turing's, Chao's,

TABLE 1.9

Application of the estimators to some data sets discussed previously in this chapter

| Sub-section | Estimate of population size with 95%CI | | | |
n	Turing	MLE	Chao	Chao-BC
1.2.3	87.6	88.5	86.6	83.3
55	(71.6 - 119.0)	(72.4 - 119.3)	(68.2 - 130.7)	(67.1 - 124.0)
1.2.4	325.1	315.0	346.4	342.3
198	(286.1 - 381.5)	(280.3 - 364.3)	(290.6 - 435.7)	(288.2 - 428.8)
1.2.5	202.7	170.3	352.0	334.8
118	(168.9 - 259.0)	(150.9 - 201.3)	(239.4 - 569.0)	(232.7 - 527.9)
1.2.6	29,558	26,426	38,637	38,631
20,198	(29,201 - 29.928)	(26,193 - 26,668)	(37,742 - 39,578)	(37,736 - 39,571)
1.2.7	19,395	15,659	33,082	32,892
3,345	(17,043 - 22,151)	(14,350 - 17,124)	(28,461 - 38,554)	(28,320 - 38,299)
1.2.10	158.1	153.4	175.0	156.9
51	(100.5 - 282.7)	(99.6 - 266.8)	(100.4 - 361.9)	(95.5 - 303.1)
1.2.11	118.4	81.8	208.7	193.5
81	(100.2 - 153.9)	(81.1 - 85.9)	(136.0 - 377.1)	(131.3 - 332.8)
1.2.12	7,584	7,059	9,240	9,196
1,877	(6,788 - 8,510)	(6,393 - 7,825)	(8,056 - 10,651)	(8,022 - 10,593)

and Chao's biased-corrected estimators used are (1.6), (1.11), and (1.12), respectively. In the following, we will restrict illustrations to those applications where T is unknown or even might not exist (see also Table 1.9). However, it is reasonable to apply SPADE also for applications with known T if the latter is large as the binomial is then close to the Poisson distribution.

We see that in most applications (except the cholera data of McKendrick 1.2.3) the estimators under homogeneity differ from the estimators under heterogeneity. In fact, it is known that under heterogeneity the maximum likelihood estimator (under homogeneity) as well as Turing's estimator underestimate the true population size. From this perspective, Chao's estimator seems to be the better choice in any case. However, we need to keep in mind that Chao's estimator crucially builds on the frequency of the singletons f_1. If, for some reason, there are more singletons than there should be, for example because units observed twice were not able to be matched, then Chao's estimator is at risk of overestimation. Hence the observation process of how samples are collected should be critically appraised at all times. Note also that Turing's estimator underestimates less than the maximum likelihood estimator with similar confidence interval size. As Turing's estimator is non-iterative, this makes the latter the preferred choice if compared with the maximum likelihood estimator.

1.6   Heterogeneity of sources or occasions

An important estimator for estimating N under heterogeneity of the T sources or occasions goes back to Darroch [95]. Let p_0 be the probability of not identifying a member of the target

TABLE 1.10
Results of the fixed point iteration 1.16

Iteration k	$N^{(k)}$	Iteration k	$N^{(k)}$
1	43	20	66.8186
2	51.6606	—	—
3	56.9503	30	66.8282
4	60.3298	—	—
5	62.5298	38	66.8284
—	—	39	66.8284
10	66.2677	40	66.8284

population during the entire experiment by any source. This probability p_0 can be estimated by $(1-n/N)$. On the other hand, missing a member of the target population during the entire experiment means missing it at the first occasion and at the second occasion up to occasion T. Assuming independence of occasions, the probability for this event is $\prod_{j=1}^{T} p_{j0}$ where p_{j0} is the probability of not identifying a member at the j-th occasion. This probability can be estimated by

$$\prod_{j=1}^{T}(1 - n_j/N,)$$

where n_j is the frequency of members of the target population identified at occasion j. Ideally, these two estimates of p_0 should agree and Darroch's estimator arises when this is the case, or in other words when

$$1 - n/N = \prod_{j=1}^{T}(1 - n_j/N), \tag{1.14}$$

or

$$N = \frac{n}{1 - \prod_{j=1}^{T}(1 - n_j/N)} \tag{1.15}$$

which can viewed as an implicit equation defining $\hat{N}$. In fact, $\hat{N}$ can be found using the fixed point iteration

$$N^{(k+1)} = \frac{n}{1 - \prod_{j=1}^{T}(1 - n_j/N^{(k)})}, \tag{1.16}$$

using some starting value $N^{(1)} = n$, for example, $k = 1, 2, \cdots$. We illustrate Darroch's estimator for the data on software errors of Section 1.2.8. There were $T = 6$ software inspectors (occasions or sources) detecting a total of $n = 43$ errors with $n_1 = 25$, $n_2 = 3$, $n_3 = 4$, $n_4 = 13$, $n_5 = 9$, $n_6 = 6$ individual detections for each of the six inspectors, respectively. In Table 1.10 we see the results of the fixed point iteration (1.16). Convergence is fast with not much of a change any more at iteration $k = 20$. It can be seen that Darroch's estimators suggest that there are 24 additional software errors hidden in the target population.

When sampling is done at T occasions, as it is done here, the setting is called a *Schnabel census*. A full likelihood approach is available for the Schnabel census and we refer for details to McCrea and Morgan [202].

1.6.1 Darroch's estimator and Lincoln–Petersen

Let us consider the situation of only $T = 2$ occasions or sources. Then the characterizing equation (1.6.1) simplifies to $1 - n/N = (1 - n_1/N)(1 - n_2/N)$, or $n = n_1 + n_2 - n_1 n_2/N$,

TABLE 1.11

Setting of a Lincoln–Petersen experiment for two occasions

		Occasion 2		
		1	0	
	1	f_{11}	f_{10}	n_1
Occasion 1				
	0	f_{01}	f_{00}	
		n_2		N

or, ultimately

$$\hat{N} = n_1 n_2 / (n_1 + n_2 - n)$$

which corresponds to the Lincoln–Petersen estimator for two sources.

The situation in which the Lincoln–Petersen estimator arises is described in Table 1.11: two sources or occasions independently identify (a 1 means the occasion identifies, 0 means the occasion does not identify) members of the target population, for example, f_{11} is the frequency of those identified by both sources and f_{00} is the frequency of those remaining unidentified by any of the two sources. As identifying happens independently, the odds ratio of this two-by-two table is one. We can estimate the odds ratio as $\frac{f_{11}f_{00}}{f_{10}f_{01}}$ and equating this estimate to one yields

$$\hat{N} = f_{11} + f_{01} + f_{10} + \frac{f_{10}f_{01}}{f_{11}} = \frac{n_1 n_2}{f_{11}}, \qquad (1.17)$$

which corresponds to $n_1 n_2 / (n_1 + n_2 - n)$ as $f_{11} = n_1 + n_2 - n$.

The benefit of Darroch's estimator is that it allows occasion- or source-specific identification probabilities. However, its restriction is that it assumes independence across occasions or sources. This is in particular exemplified in the case of a two-by-two table for a Lincoln–Petersen experiment as in Table 1.11. The assumption of independence across sources or occasions is often unrealistic, in particular in social and medical science applications. Here it is not possible to relax the independence assumption as this leads to an unidentifiable cross-table. However, given a Schnabel census with $T \geq 3$ occasions it is possible to model associations between occasions or sources. This leads to *log-linear modeling* and will be dealt with in Parts VI and VII of the book.

1.7   Glossary

capture-recapture: process of identifying and re-identifying members of a target population.

catchment area: region of catchment for a specific identifying mechanism. If different mechanisms are in place, ideally all should have the same catchment area.

closed population: a target population that is assumed to remain unchanged within the period of observation: no migration, no births, no deaths.

marking: process of providing an identifier once a member of the target population has been identified: a unique registration id, id card, for animals a numbered tag.

matching: process of establishing that identical members have been identified at different sampling occasions.

trapping: mechanism used to identify members of a target population, for example a registration system, a database, a scanning system or a reviewing system.

Part II

Ratio Regression Models

2

Ratio regression and capture-recapture

Marco Alfó

Sapienza University

Dankmar Böhning

University of Southampton

Irene Rocchetti

Institute for Official Statistics of Italy

CONTENTS

2.1 Introduction

Capture-recapture methods were originally developed in the ecological setting with the aim of assessing the unknown size of an elusive animal population; however, given their wide applicability, they have been gradually extended to epidemiology and public health, see e.g. Böhning et al. [35], to biodiversity studies, see e.g. Bunge et al. [60], and to human studies where the characteristic which defines the population of interest is linked to (possibly deviant) individual behavior, see e.g. van der Heijden et al. [283]. These methods have been also applied to non-animal/human populations, for example in text analysis and library investigations, see e.g. the seminal paper by Efron and Thisted [109], or in software engineering, see e.g. the recent review given by Liu et al. [184], just to mention a few. When we look at specific empirical examples, we may observe that each study has peculiar features which need to be properly taken into account. For example, as far as human populations are concerned, we usually (but not always) observe individual records from *multiple systems* (see also Chapter 15), that is the individual information is given by an indicator variable y_{it} which is equal to one if the i-th unit has been observed by the t-th system (for example, a list corresponding to a disease or to an administrative registry), $i = 1, \ldots, N$, $t = 1, \ldots, T$, and equal to zero otherwise. In this setting, the goal is to estimate the size N of the target population based on several incomplete lists of individuals from that population. Given the study design, the information is available only for those individuals with $y_i = \sum_{t=1}^{T} y_{it} > 0$, that is only for individuals that have been registered by, at least, one source. In this case, being recorded or identified by a list corresponds, in wildlife experiments, to being captured in a sampling occasion, and the probability of being ascertained by a list corresponds to the capture probability. Among the differences between wildlife and human applications, we must also consider that, often, the number of available lists in human studies is lower than the number of trapping/sampling occasions in wildlife studies. Furthermore, animals' response to capture, modifying the probability of being captured in subsequent sampling occasions, is often present and should be adequately taken into account in the analysis. This is a characteristic that we may also find in human studies when, for example, the characteristic of interest is linked to deviant or illegal behaviors, while it can be less common in epidemiology and public health studies.

In this chapter, we will focus on methods specifically designed to estimating the size of a (otherwise unspecific) population; since studies may substantially vary according to data features, available information, assumptions about individual heterogeneity, within and between individuals dependence, let us start from the simplest case of $T = 2$ lists. This setting has gained much interest due to the so-called Lincoln–Petersen (also referred to as the *dual system method*) estimator, Seber [259], which is based on the assumption of individual homogeneity and independence among lists (see also Chapter 1). That is, all individuals have a constant probability to be observed by each list, and the probability may vary across lists. However, being observed in one list does not influence the event of being observed in the other one, and finally, the population size does not vary due to birth or migration events. Obviously, these are simplifying assumptions that are needed to obtain an estimate for the specific empirical situation; however, these can be relaxed when multiple sources/lists or sampling occasions are available. We will assume, throughout the chapter, that the population of interest is closed, an assumption which is clearly linked to the duration of the study, but we will discuss how the assumptions of homogeneity of individual capture probabilities and independence within individuals can be relaxed. For example, dependence or individual heterogeneity in the probability of being observed may be the result of so-called *local dependence*; that is, each individual has a specific probability to be registered by each of the T sources/sampling occasions, and this probability varies across individuals. Since this individual-specific probability, say p_{it}, cannot be observed, it is usually modeled by assuming that it is constant across lists/sampling occasions, i.e., $p_{it} = p_i$, but it varies with individuals according to a distribution $g(\cdot)$. Therefore, for a generic individual, the event to be observed in a given sampling occasion is associated to the event of being observed in a further occasion. In this case, heterogeneity between individuals causes independence within individuals: the ascertainment of the two sources becomes dependent due to the common latent factor that describes how the capture probability varies in the target population. Obviously, we may also extend our treatment to empirical cases where the capture probabilities may vary across lists, which may be characterized by a differential "ability" to observe/capture/register individuals. These two settings should be considered separately as they pose different scientific questions; we will discuss the two cases (individual and aggregated data) in the next paragraph, trying to stress common and specific features.

But, and first of all, let us fix some notation that will be used throughout the chapter. We start from a target population, with the aim of estimating its global size, say N. For this purpose, we use T identification sources/sampling occasions to register units from the population; we may consider, within this setting, both the empirical situations where we repeatedly use the *same* mechanism in T subsequent occasions, and those for which we have T different sources available at the same occasion/time window. The mechanism/s allow to observe only a portion of the population. More precisely, we define the framework by means of the binary indicator variable y_{it}, $i = 1, \ldots, N$, $t = 1, \ldots, T$, where $y_{it} = 1$ if the i-th unit has been identified by the t-th source/sampling occasion. Obviously, we have only knowledge of individuals with $y_i = \sum_{t=1}^{T} y_{it} > 0$, that is, only for individuals with, for at least a specific $t = 1, \ldots, T$, we observe $y_{it} > 0$. For the remaining units, we have $y_i = 0$, and they remain *unobserved*. The number of sampling occasions, T, may be known a priori, as in the case of a finite known number of lists, or when the same mechanism, e.g. a diagnostic test, is repeatedly used, or it may denote the maximum observed count, for example if we look at the number of lesions of a given type in a sample of patients, or at the number of times a word appears in a writer's history. We may re-arrange unit indexes such that we may denote the global population by $Y_1, \ldots, Y_N$ and the observed sample $Y_1, \ldots, Y_n$. In this case, $Y_{n+1} = \cdots = Y_N = 0$.

As it is usual in statistical modeling for discrete responses, the target population can be described by the probability density function (y, p_y), where $y = 0, 1, \cdots$, and p_y denotes the probability that a generic unit from the population is observed exactly y times. Obviously, the usual constraints $p_y \geq 0$ and $\sum_{y=0}^{\infty} p_y = 1$ hold. In the following, we will denote the corresponding empirical distribution by (y, f_y); it should be noted that, due to the adopted sampling design, only units with $y > 0$ are observed. Therefore, the size of the observed sample is $n = \sum_{y>0} f_y$, and the corresponding distribution is truncated at zero, in the sense that units with $y = 0$ are not in the sample.

An obvious estimate for p_y is the relative frequency f_y/N; however, this cannot be computed since N is unknown. At the same time, the empirical relative frequency f_y/n would provide an estimate of the zero-truncated probability $p_y/(1 - p_0)$. Therefore, given the particular features of this study design, f_0 and $N = \sum_{y=0}^{T} f_y$ remain unknown, and finding an estimate of the population size N on the basis of the observed, truncated at zero, distribution represents a special form of the general capture-recapture problem, see Bunge and Fitzpatrick [59], Wilson and Collins [299], and Chao [75].

Conditional on the particular study design and on the specific working assumptions we adopt, we may consider either individual or aggregated data; in the first case, we have information about individual capture histories, that is for any unit in the sample we get a vector $\mathbf{y}_i = (y_{i1}, \ldots, y_{iT})$. When we have no access to individual data, or in those cases where we may suppose that the probability of being observed does not vary across lists/sampling occasions, we may consider the individual-specific synthesis y_i, giving us the information on the count of times the i-th unit has been registered. After a brief paragraph where we will discuss features peculiar to modeling either kind of data, we will proceed, focusing specifically on aggregated data.

2.2 Individual and aggregated data

Let us assume that each individual from a population of interest may have been registered or not by each of the T available sources; presence is denoted by 1 and absence by 0. This means that each individual has a *capture profile* corresponding to his/her registration by the different sources. If we consider $T = 3$ sources, the total number of distinct capture profiles which can be observed is equal to $2^T - 1 = 7$, that is 100, 110, 101, 010, 011, 001, 111, where 100 indicates that the individual has been registered only by the first source.

Obviously the profile 000 corresponds to units that have not been observed; the number of such individuals needs to be estimated by using some appropriate approach. For this purpose, different models could be used; generalized linear models, i.e., multinomial logit models, can be used to estimate the individual probability of having a given profile conditional on some explanatory and list-specific variables. In addition to observed heterogeneity, in the form of covariates information, also latent heterogeneity may be taken into account. Rasch and log-linear models (see also Chapters 19, 20, and 22), for example, lead to estimating the individual probability to be registered by a given source depending on the source capture "ability" (quality) and the individual proneness to be registered (latent factor), see e.g. Agresti [2].

Rasch models are based on the assumption of *unidimensionality*, that is, a unique latent factor reflects the individual proneness to registration; *local independence* states that measures corresponding to the same individual are independent, conditional on the individual specific latent factor. Bartolucci and Forcina [24] relaxed the local independence and the unidimensionality by assuming that individuals are homogeneous within a finite set of latent classes (see also Chapter 20). Furthermore, relaxation of local independence is made also by Stanghellini and van der Heijden [267] and in Chapter 19 of this book. Dependence between lists can also be taken into account; for example, log-linear models for contingency tables, created by aggregating over individual capture histories, could be fitted under both independence or dependence assumptions by introducing suitable interaction terms. The number of missed units can then be estimated by projecting the model over the unobserved cell, see e.g. Cormack [86]. We may also deal with partially observed tables when the available lists refer to different sub-populations with respect to the target population. This may mean that some sources have null probability of capturing a subset of the target population; this obviously leads to heterogeneity in individual capture probabilities and to the adoption of different estimators, see Zwane et al [309].

In many applications, we do not have the information about individual capture histories; rather, we only have the availability of the so-called *frequency of frequency data*. In other cases, the same registration mechanism is used at all occasions and list effects can be therefore assumed to be constant over lists (that is over time). In this empirical setting, a sample of n units is identified by an endogenous mechanism, e.g. a register, from a population with unknown size N. Using this device, each unit may be observed $y = 0, \ldots, T$ times where T denotes the number of sampling occasions. Therefore, the n units have been recorded at least once and n is computed summing up the units identified y times

$$n = f_1 + f_2 + \cdots + f_T = \sum_{y=1}^{T} f_y$$

where (y, f_y) is used to denote the observed distribution of the count $y = 1, \ldots, T$. Starting from the observed frequency distribution, we aim at estimating the total number of unobserved individuals, f_0, or, equivalently, the size of the target population N, where $N = n + f_0$ holds. A common estimation approach is to model the observed distribution through a parametric counting distribution, e.g., a Poisson or a binomial distribution. In this context, let p_y $y = 1, \ldots, T$ denote the probability of observing exactly y identifications, $y = 0, \ldots, T$ for a generic unit drawn from the target population for a given parametric model. Here p_0 denotes the probability of not being registered, and it represents the quantity of interest, since estimating f_0 is equivalent to estimating p_0. Given an estimate $\hat{p}_0$, the conditional maximum likelihood estimator of N is known to be the integer part of the

Horvitz–Thompson estimator:

$$\hat{N} = \frac{n}{1 - \hat{p}_0}.$$

In a parametric context, the distribution p_y is indexed by the natural parameter π, and we may use $p_y(\pi)$ to denote the dependence on the parameter. Based on the observed data, and using the zero-truncated distribution $p_y(\pi)/[1 - p_0(\pi)]$, an estimate $\hat{\pi}$ is calculated and used to derive an estimate of N by means of the Horvitz–Thompson estimator. To illustrate this procedure, let us consider the binomial probability distribution on T sources

$$p_y(\pi) = \Pr(Y = y) = \binom{T}{y} \pi^y (1 - \pi)^{T-y}, \qquad (2.1)$$

$y = 0, \cdots, T$ and $p_y = 0$ for $y > T$. In this case, $p_0(\pi) = (1 - \pi)^T$ and the Horvitz–Thompson estimator is

$$\hat{N} = n/[1 - (1 - \hat{\pi})^T]$$

where $\hat{\pi}$ is estimated fitting a zero-truncated distribution

$$\tilde{p}_y = \binom{T}{y} \pi^y (1 - \pi)^{T-y}/[1 - (1 - \pi)^T], \quad y = 1, \cdots, T$$

to the observed data, usually through an EM-type algorithm, Dempster et al. [100]. In the following, we will focus on aggregated data in the form of a frequency distribution of observed counts. Before discussing some of the proposed estimators for the population size, we introduce two examples which will be dealt with in the following and give some hints on the empirical settings we are talking about.

2.3   Real data examples

Let us consider some real-life data examples. In the first example, f_0 is known, and the study will be used to illustrate how well estimators can recover f_0; in this case, T is fixed and known as frequently happens in diagnostic test examples. In the second example, f_0 is unknown and T is neither fixed nor known.

2.3.1   Fixed number of sources: The bowel cancer data

Human populations are frequently screened for a specific disease to detect its presence early when it is easier to treat and cure. An example could be that of screening for bowel cancer. Bowel cancer can develop without early warning signs and can grow on the inside wall of the bowel for several years before spreading to other parts of the body. Often, very small amounts of blood leak from these growths and pass into the bowel motion before any symptoms are noticed. A test called the fecal occult blood test (FOBT) is frequently used to detect small amounts of blood in the bowel motion, because it is a simple, non-invasive, procedure that can be done in the privacy of your own home. However, it is well known that no screening test can be 100% accurate and a single application of the test might have low sensitivity. Therefore, repeated replications of the diagnostic test over a number of days may help identify most cases of cancer. From 1984 onward, about $50,000$ subjects were screened for bowel cancer at St. Vincent's Hospital in Sydney (Australia). The study design and the observed data are discussed in Lloyd and Frommer [185, 186, 187]. The

TABLE 2.1
Frequency distribution of the number of positive FOBT
tests in populations with verified cancer (primary) and
verified cancer followed by repeated testing, from Lloyd
and Frommer [185, 186]

y	0	1	2	3	4	5	6
f_y cancer (primary)	?	37	22	25	29	34	45
f_y cancer (secondary)	(22)	8	12	16	21	12	31

screening procedure was based on a sequence of binary diagnostic tests, self-administered
on $T = 6$ successive days. On each of these 6 occasions, the absence $y_{it} = 0$ or the presence
$y_{it} = 1$ of blood in feces was recorded. If participants in the screening program have their
true disease status determined, it is said they have been *verified*. Verification was done
by physical examination, sigmoidoscopy and colonoscopy, performed only if at least one
of the six tests was positive. People with all six tests negative were not further assessed
and it remains unknown which is the true disease status. In particular, the frequency n_0 of
persons with all six tests negative in the population of diseased patients remains unknown.
The frequency distribution of the number of positive tests Y is given in Table 2.1 under the
row 'cancer (primary)'. In Lloyd and Frommer [185, 186], it is mentioned that a sample of
122 patients with confirmed cancer status were screened again using the identical screening
procedure. The corresponding frequency distribution of positive out of the $T = 6$ tests is
shown in Table 2.1 in the row 'cancer (secondary)'. The difference between the primary and
the secondary data is that, for the latter, the frequency f_0 of diseased patients with all tests
negative is known. For this reason we will focus on this secondary distribution.

2.3.2   Unknown number of sources: The Shakespeare data

In a historic paper, Efron and Thisted [109] used the work of Shakespeare to discuss the
problem of estimating the (unknown) number of species (see Chapter 1). Following Spevack
[265], Shakespeare's known works comprise 884,647 total words, of which 14,376 appear
just once, 4,343 appear twice, etc. In our notation, Y_i denotes the number of times the i-th
word appears in Shakespeare's work, so that f_y is the number of words appearing exactly
y times.

It We do not report the corresponding observed distribution for the sake of brevity, but
the aim is at estimating the number of words Shakespeare knew but did not use ever in his
writings. As it can be easily noticed, this setting has some peculiar features when compared
to the previous one. While data appear in both cases in the form of frequency of frequencies,
in this case T is not fixed and is unknown. What we may say is just that $T \geq max_i(y_i)$.

2.4   The ratio plot

The main problem with homogeneous models, such as the binomial one, is that they may
not be flexible enough to produce a good fit to the observed (zero-truncated) distribution.
Unobserved heterogeneity may play a substantial role in determining variability in the
probability to be registered; therefore it is important to have a screening tool for *binomiality*
(in general for *homogeneity*). This screening tool may be built on an interesting property

of the binomial distribution, see Hoaglin [142]:

$$\frac{p_{y+1}}{p_y} = \frac{\binom{m}{y+1}\pi^{y+1}(1-\pi)^{T-y-1}}{\binom{T}{y}\pi^y(1-\pi)^{T-y}} = \frac{T-y}{y+1}\frac{\pi}{1-\pi}.$$

By doing a little algebra, we get

$$\frac{y+1}{T-y}\frac{p_{y+1}}{p_y} = a_y\frac{p_{y+1}}{p_y} = \frac{\pi}{1-\pi} = R_y, \tag{2.2}$$

where $a_y = \frac{y+1}{T-y}$. In the binomial distribution, the ratio R_y does not vary with y. The corresponding empirical estimate

$$r_y = a_y\frac{f_{y+1}/N}{f_y/N} = a_y\frac{f_{y+1}}{f_y},$$

where f_y is the observed frequency of count y (frequency of units with exactly y identifications), does not change whether we consider the truncated or the untruncated distribution. The graph $y \to r_y = a_y\frac{f_{y+1}}{f_y}$ is called the *ratio plot*, and it has been developed as a *diagnostic device* for the binomial by Böhning et al. [44]. In a ratio plot, the pattern of a horizontal line can be taken as supporting evidence for a homogeneous distribution (in this case binomial). This is shown in Figure 2.1 below, where 50,000 simulated data from a binomial distribution with index $T = 6$ and parameter $\pi = 0.4$ are reported on the ratio scale (left panel) and as a frequency plot (right panel).

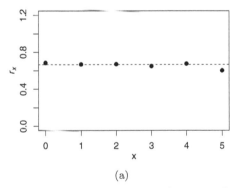 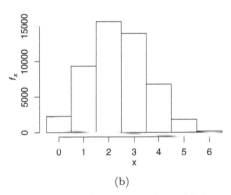

|(a)|(b)|

FIGURE 2.1: Log-ratio plot (left panel), and corresponding frequency chart (right panel), $N = 50,000$ simulated binomial counts, $\pi = 0.4$ and index $T = 6$.

Even in (almost) absence of any random error, the nature of the distribution cannot be recognized (without the use of further statistical techniques) from the frequency plot, while it can be easily evinced by looking at the ratio plot. So, the motivation for the use of the ratio plot can be summarized as follows: it clearly shows whether substantial departures from the homogeneous (binomial) distribution are observed and, in the presence of a large sample size and number of trials, it may help in detecting a discrete mixing. For small sample sizes, random error comes in and the ratio plot may take non-horizontal patterns as can be observed by looking at Figure 2.2.

In Figure 2.2, the ratio plot for a sample drawn from a given population of size N and probability $\pi = 0.4$ is plotted for (clockwise) increasing values of the population size $N = 50, 500, 5000, 50,000$. For small population sizes, sampling error may be present and we need to supplement the ratio plot with error bars. If we apply the ratio plot concept to

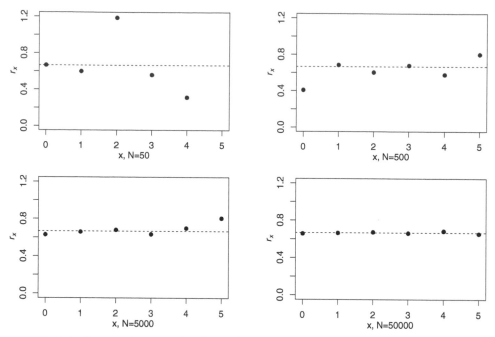

FIGURE 2.2: Log-ratio plots for $N = \{50, 500, 500, 50000\}$ simulated binomial counts, $\pi = 0.4$ and index $m = 6$.

bowel cancer data, there is no evidence of a horizontal line. Instead, we observe a seemingly monotone pattern which might be used as supporting evidence for population heterogeneity. Therefore, the standard, homogeneous, binomial distribution does not seem to adequately fit the observed, zero-truncated distributions. For this purpose, we may consider sources of population heterogeneity, described by a mixing distribution $h(\theta)$ which accounts for individual heterogeneity in the identification probability. The marginal distribution is given by

$$p(y) = \int_0^1 \binom{T}{y} \pi^y (1 - \pi)^{T-y} h(\pi) d\pi. \tag{2.3}$$

The shape of the marginal distribution varies substantially as a function of the mixing distribution, and this last term controls the departure of the marginal distribution in (2.3) from the homogeneous binomial model.

When the mixing distribution is not described by a 1-point mass (leading to the binomial distribution), it can be shown that the ratios R_y are increasing in y. The ratio plots for the benchmark data examples seem to suggest, in both cases, the presence of unobserved population heterogeneity. Parametric choices for $h(\theta)$ such as the beta distribution have been considered which often improve the fit considerably when compared to the binomial model. Discrete mixture models have also been suggested, see e.g. Norris and Pollock [220], Pledger [232], and Böhning and Kuhnert [38]. Besides boundary problems that may arise when the parameter approaches the borders of the segment $(0, 1)$, see Wang and Lindsay [290, 291], identifiability, see Link [180], is an issue of great concern. Given that we only observe the zero-truncated distribution, we are left with the unsolved problem of choosing which mixing is the best, not in terms of the observed fit, but rather in terms of estimating the unknown f_0. While a general solution to the problem does not exist, a sub-optimal solution is to restrict the attention to identifiable, parametric families of distributions.

The question is, how do we achieve alternative families? Could the ratio plot be used to determine the family of interest? Before trying to answer these questions, however, we shift our attention to wider families of distributions to characterize the ratio plot as a diagnostic tool to detect homogeneous versus non-homogeneous distributions in a wider context than the simple binomial setting.

2.4.1 The Katz family

The Katz family of distributions covers a wide spectrum including binomial, negative binomial, and Poisson distributions which arise naturally as models for population size estimation. A major motivation of Katz's [160] work was the problem of discriminating among binomial, negative binomial, and Poisson distributions when data are known to come from one of them.

Let (y, p_y), $y = 0, 1, \ldots$ denote a probability distribution on the non-negative integers. The condition

$$\frac{(y+1)p_{y+1}}{p_y} = \gamma + \delta y, \quad y = 0, 1, 2, \ldots, \tag{2.4}$$

where γ and δ are real constants, characterizes the *Katz family of distributions*, see e.g. Johnson et al. [158]. Katz [160] also suggested another criterion for identifying the discrete distributions

$$\frac{\mu_2 - \mu_1'}{\mu_1'} = \frac{\delta}{1-\delta}, \quad \delta < 1$$

where μ_1' and μ_2 are the mean and variance respectively. The ratio $\frac{\delta}{1-\delta}$ takes the value zero for the Poisson distribution, positive for the negative binomial and negative for the binomial distribution. Condition (2.4) suggests linear regression on the left-hand side upon the right, in some form.

A necessary condition for Equation (2.4) yielding a valid probability distribution is that $\gamma > 0$ and $\delta < 1$. In detail p_y denotes the Poisson distribution if $\delta = 0$, it defines the negative binomial distribution for $\gamma > 0$ and $\delta \in (0, 1)$ while it is the binomial distribution if $\delta < 0$. This last remark says that, in the binomial distribution, the ratio defined by Equation 2.4 is decreasing with y. This result can be linked with the ratio plot which, for the binomial case, has been shown to be constant with y. In detail, let us consider the case $Y \sim Bin(T, \pi)$ and look at the ratio as defined in Equation (2.4)

$$\begin{aligned}
\frac{(y+1)p_{y+1}}{p_y} &= (y+1)\frac{\binom{T}{y+1}\pi^{y+1}(1-\pi)^{T-y-1}}{\binom{T}{y}\pi^y(1-\pi)^{T-y}} \\
&= \frac{\pi}{1-\pi}(T-y) = \frac{\pi}{1-\pi}T - \frac{\pi}{1-\pi}y = \gamma + \delta y,
\end{aligned} \tag{2.5}$$

which leads to $\delta = -\frac{\theta}{1-\theta} < 0$. If we assume a negative binomial model, p_y can be written as

$$p_y = \frac{\Gamma(y+k)}{\Gamma(y+1)\Gamma(k)}p^k(1-p)^y$$

where $k > 0$ is the dispersion parameter and $p \in (0, 1)$ the probability of success in a single trial. Calculating the ratios in Equation (2.4) for the negative binomial we obtain

$$(y+1)p_{y+1}/p_y = (k+y)(1-p) = (1-p)k + (1-p)y.$$

To linearize the right-hand side in k and get an exact approximation for $y = 0$ then, we take the first-order Taylor expansion of $\log(k + y)$ around k, achieving thus

$$\log(k + y) + \log(1 - p) \simeq \log(1 - p) + \log(k) + \frac{y}{k}.$$

Therefore if we write $a_y = (y + 1)$ for the Poisson and the negative binomial and $a_y = \frac{y+1}{T-y}$ for the binomial (with index T) we get that the condition (2.4) may be rewritten as

$$a_y \frac{p_{y+1}}{p_y} = \gamma + \delta y,$$

where $\delta = 0$ for the Poisson and the binomial model and $\delta \in (0, 1)$ for the negative binomial distribution. That is, the condition $\delta \neq 0$ indicates heterogeneity in the observed population. How can this result be related to the ratio plot? To answer this question we need to introduce a wider family of distributions. It is important to know that the non-decreasing monotonicity of the ratio plot with y holds not only for negative binomial distribution but also for any other Poisson mixture. In the next paragraph, we will treat both homogeneous and heterogeneous distributions by introducing the family of power series distributions and look at the properties of the ratio plot in this wider context.

2.4.2 Power series

A number of attempts have been made during the past decades to study power series distributions; Noack [219] was the first to define power series distributions and investigate moment and cumulant properties. Let

$$\eta(\theta) = \sum_{y=0}^{\infty} \alpha_y \theta^y$$

he a power series. Following this definition, we get $\sum_{y=0}^{\infty} \frac{\alpha_y \theta^y}{\eta(\theta)} = 1$. That is, the term

$$p_y(\theta) = \frac{\alpha_y \theta^y}{\eta(\theta)} \quad y = 0, 1, 2, \ldots \quad \theta > 0 \quad \alpha_y > 0 \tag{2.6}$$

defines a power series distribution, where α_y is a known positive coefficient, θ is a positive parameter, $y = 0, 1, \ldots$ ranges over the set of nonnegative integers and $\eta(\theta) = \sum_{y=0}^{\infty} \alpha_y \theta^y$ is the normalizing constant.

Noack [219] established a clear connection between the elements of the power series family and some important discrete distributions on nonnegative integers, such as the binomial, Poisson, negative binomial and logarithmic series distributions. It can be shown that this family is in fact equivalent to the one-parameter discrete exponential family. In detail, the α_y coefficient defines the specific member of the power series, for example $\alpha_y = 1/y!$ defines the Poisson, $\alpha_y = \binom{T}{y}$ for $y = 0, \ldots, T$, defines the binomial ($\alpha_y = 0$ for $y > T$), and $\alpha_y = 1$ gives the geometric. For a known value of the shape parameter $k > 0$, the negative binomial is also part of the power series family and the coefficient α_y is given by $\alpha_y = \frac{\Gamma(y+k)}{\Gamma(y+1)\Gamma(k)}$, for $k = 1$ the negative-binomial becomes the geometric distribution and for $k \to \infty$ the negative-binomial approaches the Poisson distribution. The power series distribution in Equation 2.6 has the following important property. If we consider the ratios of neighboring probabilities multiplied by the inverse ratios of their coefficients

$$R_y = \frac{\alpha_y}{\alpha_{y+1}} \frac{p_{y+1}}{p_y} = a_y \frac{p_{y+1}}{p_y} = \theta$$

That is, the ratio plot is constant with y for densities belonging to the family of power series distributions; this result is worth discussing for the negative binomial distribution. For this distribution, if we use $\alpha_y = \frac{1}{y!}$ as we do for the Poisson distribution, we get a ratio plot which is increasing with y. Rather, if we use $\alpha_y = \frac{\Gamma(y+k)}{\Gamma(y+1)\gamma(k)}$, we get a constant ratio plot. To better explain this duality we need to introduce mixing in the power series family.

2.4.3 Mixed power series

Here, our interest is in connecting unobserved heterogeneity with the concept of the ratio plot. For this purpose, let us define the general mixture model for a distribution in the power series family as

$$P_y = \int_\theta p_y(\theta)g(\theta)d\theta \tag{2.7}$$

where $g(\theta)$ denotes the mixing density for θ and describes unobserved heterogeneity. The ratio plot for mixtures is therefore defined by

$$R_y = \frac{\alpha_y}{\alpha_{y+1}} \frac{P_{y+1}}{P_y}.$$

It can be easily proven that the marginal distribution satisfies the following monotonicity property, Böhning et al. [44]

$$a_0 \frac{p_1}{p_0} \leq a_1 \frac{p_2}{p_1} \leq a_2 \frac{p_3}{p_2} \leq \cdots .$$

where $a_y = \frac{\alpha_y}{\alpha_{y+1}}$.

2.4.4 A specific case: The Beta-binomial distribution

Let us consider a binomial distribution with index T and probability of success π; let us denote the number of successes (number of times an individual has been registered) by $y = 0, 1, 2, \ldots, T$. The binomial distribution can therefore be written as

$$\Pr(Y = y \mid \pi, T) = \binom{T}{y} \pi^y (1-\pi)^{T-y} = \alpha_y \theta^y \mu(\theta),$$

where $\theta = \pi/(1-\pi)$, $\mu(\theta) = (1+\theta)^{-T}$, $\alpha_y = \binom{T}{y}$. If we model unobserved individual specific variation in π through a mixing distribution $g(.)$, the marginal pdf is

$$P_y = \int \alpha_y \theta^y \mu(\theta)g(\theta)d\theta.$$

In the case of mixed power series distribution, it can be proved that

$$\frac{P_y}{\alpha_y} \Big/ \frac{P_{y-1}}{\alpha_{y-1}} \leq \frac{P_{y+1}}{\alpha_{y+1}} \Big/ \frac{P_y}{\alpha_y},$$

that is, the ratio plot increases with y. For $y = 1$, the previous inequality leads to the lower bound for P_0 defined by Chao [73]

$$P_0 \geq \frac{P_1^2(T-1)}{2TP_2}.$$

Replacing the unknown probabilities by the observed frequencies, the Chao estimator is achieved

$$\hat{f}_0 = n + \frac{f_1^2(T-1)}{2f_2 T}.$$

We will briefly focus on the case $\pi \sim beta(a,b)$ as the beta distribution represent a quite general mixing distribution for π that, depending on the values of a and a, may assume very different shapes, see e.g. Gupta and Nadarajah [137]. In this case, the marginal distribution is a beta-binomial

$$P_y = \alpha_y \frac{B(y+a, T-y+b)}{B(a,b)} \quad B(x,z) = \frac{\Gamma(x)\Gamma(z)}{\Gamma(x+z)}$$

where Γ is the ordinary Gamma function. The ratio plot is

$$R_y = \frac{P_{y+1}\alpha_y}{P_y \alpha_{y+1}} = \frac{(y+a)B(y+a, T+b-y)}{(T-y-1+b)B(y+a, T+b-y)} = \frac{y+a}{(T-y-1+b)},$$

which is increasing in y, as this is the case for mixtures of power series distributions.

2.5 The regression approach

Rocchetti et al. [248] considered the following monotone transformation of R_y

$$\left(\frac{R_y}{1+R_y}\right) = \frac{y+a}{(T+a+b-1)}. \tag{2.8}$$

Rewriting $\frac{a}{(T+a+b-1)} = \gamma$ and $\frac{1}{(T+a+b-1)} = \delta$, Equation (2.8) leads to the linear regression model

$$\left(\frac{R_y}{1+R_y}\right) = \gamma + \delta y. \tag{2.9}$$

This result has been used by the authors to define a regression estimator for the frequency of unobserved units and therefore for the population size itself. Let us consider the log transformation

$$\log\left(\frac{R_y}{1+R_y}\right) = \log\left(\frac{y}{T+a+b-1} + \frac{a}{T+a+b-1}\right)$$

$$= -\log(T+a+b-1) + \log(y+a).$$

Applying Taylor expansion around a, we obtain the following approximation for the observed ratios r_y:

$$\log\left(\frac{R_y}{1+R_y}\right) =\simeq \gamma_1 + \delta_1 y$$

leading to the non-linear regression model

$$R_y = \frac{\gamma \exp(\delta_1 y)}{1 - \gamma \exp(\delta_1 y)}, \tag{2.10}$$

where $\gamma_1 = \log\left(\frac{a}{T+a+b-1}\right)$ and $\delta_1 = \frac{\delta}{\gamma}\frac{1}{a}$. In other words, as outlined before, the ratio plot for binomial mixtures is monotone non-decreasing; a similar result is discussed in Hwang and Shen [150]. Estimates $\hat{\gamma}_1$ and δ_1 can be obtained by plugging in the observed frequencies and adopting weighted least squares, as detailed in Rocchetti et al. [248]. Solving for f_0, at $y = 0$, we obtain:

$$\hat{f}_0 = \frac{f_1\,[1 - \exp(\hat{\gamma}_1)]}{\exp(\hat{\gamma}_1)T}, \quad \text{and} \quad \hat{N} = \hat{f}_0^l + n. \tag{2.11}$$

While Rocchetti et al. [248] have used the beta-binomial model as a specific motivation, we must consider the following argument. All that is really needed is that $\log\left[R_y\,(1 + R_y)\right]$ follows, at least approximately, a linear (or a simple) pattern. According to this condition, based on the general monotonicity result for the mixed power series, the specific results for the beta binomial distribution and for the members of the Katz/Kemp families of distributions (see also Chapter 24), we are suggested to explicitly model R_y as a non-decreasing function of y. This *ratio regression* approach can be used to identify an appropriate distributional form without the need to parametrically specify the form of the mixing density $g(\theta)$. Let us assume that there exists an unknown probability distribution $p_1, \dots, p_T$ with $p_y > 0$, $\forall y = 0, \dots, T$, and let us consider the ratios:

$$R_y = \frac{\alpha_y p_{y+1}}{\alpha_{y+1} p_y} = a_y \frac{p_{y+1}}{p_y} \tag{2.12}$$

$y = 0, \dots, T - 1$. The coefficients a_y (respectively α_y) are known constants, determined by the choice of the *reference* distribution we would like to include. The reference distribution represents the homogeneous distribution we obtain when unobserved heterogeneity is not present, which is the conditional distribution in (2.7) to which p_y reduces when $g(\theta)$ is a one-point mass. Therefore, if the upper limit T is known and fixed, the binomial represents the reference distribution and we have $a_y = (y + 1)/(T - y)$. If the range of the counts has no upper limit, we may consider the Poisson as the reference distribution and $a_y = (y + 1)$. The point is that, if the observed count data follow the reference distribution, the associated ratios $R_y = a_y p_{y+1}/p_y$ are constant over $y = 0, \dots, T - 1$. So, any regression model for R_y (or a suitable transformation of it) with only the intercept term represents the reference distribution and a non-null slope estimate implies some unobserved heterogeneity. So, we can link R_y to a known set of predictor functions $z_0(y), \cdots, z_p(y)$, to define the following model:

$$h(R_y) = \beta'\mathbf{z}(y), \tag{2.13}$$

where $y = 0, \cdots, T - 1$, and $h(\cdot)$ is a monotone link function. An example is $\log(R_y) = \beta_0 + \beta_1 y$ with $z_0(y) = 1$ and $z_1(y) = y$, that is $R_y = \exp(\beta_0 + \beta_1 y)$, but other examples, such as the one we have discussed for the beta-binomial case, can be considered as well. A general result can be proven in this context, see Böhning et al. [48].

Theorem 2.1 *Let $R_y > 0$ be given for $y = 0, \cdots, T - 1$, and let a_y, $y = 0, \dots, T - 1$, be known positive coefficients. Then, there exists a unique probability distribution $p_0, \dots, p_T > 0$ such that:*

$$p_{y+1} = R_y p_y/a_y \quad, \forall y = 0, \cdots, T - 1.$$

Furthermore, we have that

$$p_0 = \left[1 + R_0/a_0 + (R_0/a_0)(R_1/a_1) + \cdots + \prod_{y=0}^{T-1} R_y/a_y\right]^{-1}.$$

Proof. Let $R_y > 0$ $y = 0, \cdots, T - 1$. Any probability distribution on $[0, T]$ integers $p_0, \dots, p_T > 0$ will meet the constraint $p_0 + \cdots + p_T = 1$. Given the recurrence relation $p_{y+1} = R_y p_y/a_y$ and the constraint $\sum_y p_y = 1$, we may write

$$1 = p_0 + \cdots + p_T = p_0 + p_0 R_0/a_0 + p_0 R_0/a_0 R_1/a_1 + \cdots + p_0 \prod_{y=0}^{T-1} R_y/a_y$$

$$= p_0(1 + R_0/a_0 + (R_0/a_0)(R_1/a_1) + \cdots + \prod_{y=0}^{T-1} R_y/a_y).$$

Solving for p_0 we obtain

$$p_0 = 1/[1 + R_0/a_0 + (R_0/a_0)(R_1/a_1) + \cdots + \prod_{y=0}^{T-1} R_y/a_y]$$

necessarily, where $0 < p_0 < 1$. The remaining probabilities follow from the recurrence formula, and $p_{y+1} = R_y p_y/a_y$ implies that $0 < p_{y+1} < 1$, $y = 0, \ldots, y-1$. This ends the proof.

According to this theorem, *any* regression model fulfilling the regularity condition $R_y > 0$, $y = 0, \ldots, T-1$ leads to a proper probability distribution, which is obtained by mixing the reference distribution, that is in turn associated to coefficients a_y. The mixing distribution is unknown and need not be parametrically specified; rather, the regression model for R_y and the corresponding link function define a one-to-one mapping from the positive axis into the real line, and guarantees that the regularity conditions $R_y > 0$, $y = 0, \ldots, T-1$ hold. Estimation may be based on the likelihood function, which is defined by the following expression

$$L(\beta) = \prod_{y=1}^{T} \left(\frac{p_y}{1 - p_0} \right)^{f_y}.$$

where p_y is a function of $R_y = g^{-1}(\beta' \mathbf{z}(y))$, and hence of β, via Theorem 2.1. For practical purposes, we suggest using the following procedure. We estimate R_y by its empirical counterpart, $r_y = a_y \frac{f_{y+1}}{f_x}$, and study its dependence from y (for example, by plotting $\log r_y$ versus y). This process could help generate ideas on how to develop an appropriate regression model. Once we have chosen the link function $g(\cdot)$, we fit the model

$$g(r_y) = \beta' \mathbf{z}(y) + \epsilon_y, \tag{2.14}$$

where ϵ_y is such that $\mathrm{E}(\epsilon_y) = \mathbf{0}$, $\mathrm{cov}(\epsilon_y) = \Sigma$, and $\beta = (\beta_0, \cdots, \beta_p)'$ is a $(p+1)$-dimensional vector of unknown fixed parameters, associated to the vector of covariates $\mathbf{z}(y) = (z_0(y), \cdots, z_p(y))'$. Given an estimate $\hat{\Sigma}$, the generalized least-squares estimate of β is known to be

$$\hat{\beta} = (\mathbf{Z}'\hat{\Sigma}^{-1}\mathbf{Z})^{-1}\mathbf{Z}'\hat{\Sigma}^{-1}g(\mathbf{r}),$$

where $g(\hat{\mathbf{r}})$ has elements $g(\hat{r}_x)$, $\mathbf{Z}$ has rows $z_0(y), \ldots, z_p(y)$, $y = 1, \cdots, T-1$, as no observation is available for $y = 0$. Details on how to estimate Σ are discussed in Rocchetti et al. [247]. One of the peculiar features of the ratio regression approach is that the model remains invariant whether the untruncated or the zero-truncated count distribution is considered. In fact, we may observe that:

$$R_y = a_y \frac{p_{y+1}}{p_y} = a_y \frac{p_{y+1}/(1 - p_0)}{p_y/(1 - p_0)}$$

$y = 0, \ldots, T-1$. Clearly, R_0 is defined for the untruncated distribution only. For the zero-count frequency, a regression-based estimator can be derived by:

$$g(\hat{r}_0) = \hat{\beta}' \mathbf{z}(0) \Rightarrow \hat{r}_0 = g^{-1}\left(\hat{\beta}' \mathbf{z}(0) \right).$$

Two estimators can be defined on the basis of the regression model estimated on the empirical ratios. First, we can use the fitted values $\hat{r}_y = g^{-1}(\hat{\beta}'\mathbf{z}(y))$, $y = 0, \cdots, T-1$ to estimate the corresponding probability mass at 0 according to Theorem 2.1:

$$\hat{p}_0 = \left[1 + \hat{r}_0/a_0 + (\hat{r}_0/a_0)(\hat{r}_1/a_1) + \cdots + \prod_{y=0}^{T-1} \hat{r}_y/a_y\right]^{-1}. \tag{2.15}$$

Given this probability mass, the Horvitz–Thompson estimator for the size of the target population can be derived:

$$\hat{N}_{\mathrm{HT}} = \frac{n}{1 - \hat{p}_0} = n + \hat{f}_{0,HT}.$$

Second, once a given regression model has been fitted and corresponding parameters estimated, we may use the recurrence relation $r_y = a_y f_{y+1}/f_y$ and project it onto $y = 0$, to get an estimate of f_0:

$$\hat{f}_0 = a_0 f_1/\hat{r}_0 = a_0 f_1/g^{-1}(\hat{\beta}'\mathbf{z}(0)).$$

The associated population size estimator is obtained summing the estimated number of unrecorded individuals to the size of the observed sample:

$$\hat{N}_{\mathrm{reg}} = n + \hat{f}_0.$$

2.6   Applications

2.6.1   The bowel cancer data

In this case, we have to estimate the frequency of false negative tests f_0 in the secondary data, that is, the number of patients that have been found positive for the disease by further evaluation but have not resulted positive to any of the six tests administered after the evaluation. In this application, $T = 6$ is known and fixed and, therefore, we may use the theory developed for binomial and mixed binomial distributions. As a first step, we report in Figure 2.3 below the ratio plot (left) and the observed distribution (right) corresponding the secondary bowel cancer data. As can be easily noticed, the ratio plot is

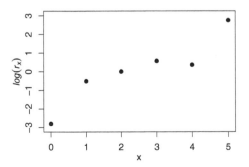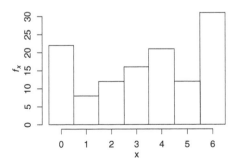

FIGURE 2.3: Sidney secondary bowel cancer data. Log-ratio plot (left panel) and corresponding frequency chart (right panel).

increasing, or at least not constant, with y, and this may point to the presence of unobserved

TABLE 2.2: Secondary bowel cancer data: estimated frequency ($\hat{f}_0$) by several different estimators. True observed frequency $f_0 = 22$. Confidence level $1 - \alpha = 0.95$

	Binom.	Turing	Beta-binom.	Chao	Expon.	Linear
				$\hat{f}_0$		
Point est.	0	1	6	2	8	6
Conf. Int.	—	—	—	—	[2.09, 28.91]	[1.40, 29.42]

individual-specific heterogeneity and, therefore, to the adoption of a mixed binomial model. Table 2.2 reports the observed and the fitted frequencies obtained through the homogeneous binomial, the heterogeneous beta-binomial models, the regression model with two different specifications for the linear predictor and the link function. In particular, the *exponential* estimate corresponds to the model in Eq. (2.10), while the *linear* estimate comes from the model in Eq. (2.9).

The estimates derived by regression models outperform competing estimators and, in particular, the regression model defined by the first-order Taylor approximation of the (log) ratio plot under a beta-binomial model slightly outperforms the results obtained for the beta-binomial distribution when parameters are estimated by ML through an EM algorithm. Obviously, these results may still be enhanced; for example, Böhning et al. [48] report an estimate $\hat{f}_0 = 21$ by using a log link and a logarithmic transform of the count y, that is, $\log(y+1)$ as a covariate in the linear predictor. However, while this particular specification is clearly the best fitting to the complete distribution (remember that in this case f_0 is known), this is not true when the truncated distribution comes at hand. That is, the model that best fits the truncated distribution is not necessarily the model that best fit the *complete* distribution; and the last likely corresponds to the best possible estimate for f_0. This is the motivation for Böhning et al. [48] to explore the wider class of regression models built on fractional polynomials, and to propose averaged estimates for f_0 and N along the lines described by Burnham and Anderson [66].

2.6.2   The Shakespeare data

Efron and Thisted [109] used the work of Shakespeare to illustrate the problem of estimating the number of species. Following Spevack [265], Shakespeare's known works comprise 884,647 total words; out of these, 14,376 are types appearing just once, 4,343 are types appearing twice, and so on. According to the notation we have introduced above, y_i denotes the number of times the i-th word appears in the writings of the author, and f_y denotes the number of words that appear exactly y times. Since the index T denoting the global number of trials is potentially infinite, it seems reasonable to work with a Poisson mixture:

$$P_y = \int_0^\infty \exp(-\lambda)\lambda^y/y! \; g(\lambda)d\lambda, \tag{2.16}$$

where $g(\lambda)$ is some mixing density describing the variability among subjects in the Poisson rate parameter. By re-arranging previous concepts, we set $a_y = y + 1$ so that the Poisson plays here the role of reference distribution, leading to a constant (log) ratio R_y. If we choose a gamma density for λ, a marginal negative binomial distribution arises with event parameter $p \in (0,1)$, and shape $k > 0$. It is interesting to note that the negative binomial is one of the models discussed by Efron and Thisted [109]. If we set $a_y = (y + 1)$, the ratio simplifies to

$$R_y = (1 - p)(y + k),$$

as it has already been established in Section 2.4.1 discussing distributions in the Katz family. That is, if we adopt a negative binomial model, we end up with a straight line defining the ratio plot, with intercept term $\beta_0 = k(1-p)$ and slope $\beta_1 = (1-p)$. This model is seemingly supported when looking at the ratio plot of $r_y = (y+1)f_{y+1}/f_y$ versus y; in fact, it seems to give some evidence of a straight line pattern (see Figure 2.4, left panel).

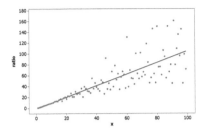 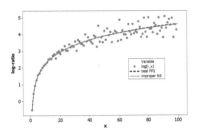

FIGURE 2.4: Ratio plot (left panel), and log-ratio plot (right panel) for the Shakespeare's data.

Taking logs in the equation above, this leads to $\log(R_y) = \log(1-p) + \log(y+k) = \beta_0 + log(y+k)$. However, when fitting the model we get a negative estimate $\hat{k} = -0.3890$, indeed a value which is close to the estimate of Efron and Thisted [109], $\hat{k} = -0.3954$. Although the fit is excellent, as one can derive by looking at Figure 2.4 (right panel), $\hat{r}_0$ is negative and so is $\hat{f}_0$. Constraining solutions to fulfill $k > 0$ diminishes the fit considerably. Alternatively, we can consider a ratio regression approach; for example, we may implement, according to Böhning et al. [48] the following fractional polynomial model:

$$\log r_y = \beta_0 + \beta_1(y+1)^{-2} + \beta_2 \log(y+1) \qquad (2.17)$$

As it is illustrated in Figure 2.4 (right panel), the corresponding fit is coherent with the one from the *improper* negative binomial. The benefit of the ratio regression approach is that a valid count distribution is derived via the result of Theorem 2.1. When the corresponding conditions are valid, an estimate for f_0 can be easily derived, and this may help solve boundary problems.

2.7 Discussion

Some guidelines for the use of the ratio regression approach in practice might be appropriate. The first important choice is the *reference* family as this leads to the coefficients a_y, for $y = 0, 1, \cdots$. It is clear that if there is a finite number of sampling occasions, such as in the application of bowel cancer screening, the natural base family is the binomial and every regression model considered should include an intercept term so that the binomial distribution is included as a special case. In the second application considered here (Shakespeare's word count data) the reference family is less clear as at least the Poisson or the geometric distributions could be considered. Here, ratio plotting might help and the distribution with the least positive trend might be chosen as the base family (and hence determine the coefficients a_y). The choice of link function is usually not a problem as the log-link is typically suitable. Choosing the regression model is clearly important and guidance can be received again from the ratio plot. However, several models might appear equally suitable and model selection criteria such as the Akaike information criterion might be used as suggested in Böhning et al. [48] to select models. The ratio regression approach can be widely applied, clearly also to ecological data. However, it should be mentioned that sample sizes should be at least moderate as the ratios f_{x+1}/f_x need to be constructed on the basis of the frequency distribution of the count of captures Y.

The approach can be extended in several ways. An interesting extension is that validation information can be easily incorporated into the ratio regression modeling (see Chapter 5). In the bowel cancer application we used the secondary data set (in which zero counts were observed) as the benchmark data set to investigate the performance of the methods. Instead, a different way to proceed is to use the primary data set (where no zero counts were observed) jointly with the secondary data set, the validation data, and perform a joint modeling. For example, it is now possible to incorporate the information coming from the validation sample into the modeling as follows:

$$\log[(y + 1)/(T - y)f_{y+1}/f_y] = \beta_1 + \beta_2 y + \beta_3 \mathsf{set} + \epsilon_x. \tag{2.18}$$

Here, set is an indicator variable which is one if y is from the validation sample and zero otherwise. Evidently, this model is a parallel line model and can be extended in various ways. The apparent benefit is that validation data increase efficiency in estimation as well as protect against bias arising from a misspecification of the model. These ideas are also explored in Chapter 5 of the book.

3

The Conway–Maxwell–Poisson distribution and capture-recapture count data

Antonello Maruotti

Libera Università Maria Ss. Assunta

Orasa Anan

Thaksin University

CONTENTS

3.1 Introduction

Capture-recapture models have a long history in (see e.g. McCrea and Morgan [202]; and references therein), starting with the Lincoln–Petersen model (Chao and Huggins [77]), which provided the first method for estimating the unknown population size. Since that time a number of models have been developed to improve inferences from capture-recapture techniques by accounting for heterogeneity, behavioral and time effects. The continued relevance of capture-recapture methodologies has led to a host of recent advances, including those focused on count distributions. While capture-recapture is the most direct way to attain detailed population information, these data can be expansive to collect and challenging to obtain.

In this chapter, we restrict ourselves to a setting in which the identifying mechanism is based upon counting repeated identifications of the same unit within a given time span. This is usually referred to as capture-recapture data in the form of *frequencies of frequencies.*

They consist of the frequency f_1 of units detected exactly once, the frequency f_2 of units detected exactly twice, and so on. The resulting frequency distribution is a zero-truncated distribution as the frequency of undetected units (f_0) is unknown. Several modeling frameworks have been adopted to handle data expressed as counts. Since the number of times a single unit is captured can take only non-negative integer values, the (zero-truncated) Poisson model represents the simplest framework for estimating the frequency f_0. This model relies on several assumptions as it assumes: a unit variance-to-mean ratio (equidispersion assumption); that all the units have the same probability of being captured (homogeneity assumption); and that the probability of being captured is independent of previous captures (independent assumption). Any violation of these assumptions usually leads to overdispersion in the count data distribution and to biased estimates of f_0 (Chao et al. [79]).

Besides very simple models like the Poisson distribution, discrete (Pledger [233]; Bartolucci and Forcina [25]; Morgan and Ridout [211]) and continuous (Dorazio and Royle [102]; Niwitpong et al. [218]; Rocchetti et al. [248]) mixing distributions have been used to mitigate the potential bias in population size estimation due to heterogeneity. See also Chapter 2 and Chapter 21. Nevertheless, the choice of the mixing distribution can influence model inference and may limit its flexibility. A widely used alternative is given by the (zero-truncated) negative binomial model. The negative binomial distribution arises as a mixture of a Poisson distribution with a Gamma distribution, hence it is sometimes also called the Poisson-Gamma model. Due to its enhanced flexibility in fitting count data in comparison to the Poisson, it has been suggested as a more flexible approach in zero-truncated count data modeling. Despite the support, we may look for alternative models because of theoretical as well as practical deficiencies connected to the negative binomial model. These problems do also arise in untruncated count data, but are more pronounced in zero-truncated count data, and thus in the capture-recapture context (Böhning, [46]).

Extending the idea of the Poisson process, we identify that the Conway–Maxwell–Poisson (CMP) can be a strong candidate to handle both under- and overdispersion. In the following, we derive an inferential procedure to infer an unknown population size in the presence of heterogeneity. No other sources of variability such as behavioral and time effects are further considered. Conway and Maxwell [84] originally proposed what is now known as the CMP distribution as a solution to handling queuing systems with state-dependent service rates. Shmueli et al. [262] have provided an excellent summary about the flexible and unique properties of the CMP model both in terms of methodological advancements and applications. Although the CMP distribution is a two-parameter generalization of the Poisson distribution, it has special characteristics that make it especially useful and elegant. Not that the CMP distribution is not the only model capable to describe both under- and over-dispersed data. Competing models include e.g. the generalized Poisson distribution. Nevertheless, the generalized Poisson is not in the exponential family and it is unable to handle different levels of dispersion as well as the CMP distribution, which also belongs to the family of the weighted Poisson distributions (Shmueli et al. [262]).

3.2 The Conway–Maxwell–Poisson distribution and capture-recapture count data

3.2.1 Preliminaries

Throughout the chapter, we consider the following capture-recapture setting. The target population is sampled over a certain number of capture occasions, and for each occasion, captured units are counted only once. Moreover, we consider a closed population, i.e. the unknown population size, say N, is assumed to be constant (with no births/deaths during sampling stages), misclassification is not allowed and all units act independently.

Formally, let Y_i, $i = 1, \ldots, N$ denote the number of times unit i is captured over the S sampling occasions, and let $p_y = \Pr(Y_i = y)$. Also let f_y denote the frequency of units captured exactly y times, $y = 0, 1, \ldots, m$. As $Y_i = 0$ is not observed, the corresponding f_0 is unknown and might be replaced by its expected value $N p_0$. Nevertheless, p_0 is usually unknown too and has to be estimated.

3.2.2 The CMP distribution

In modeling count data, the CMP distribution has recently played an important role. The CMP probability distribution function, $\text{CMP}(\lambda, \nu)$, has the form [262]

$$p_y = \frac{\lambda^y}{(y!)^\nu} \frac{1}{z(\lambda, \nu)}, \quad y = 0, 1, 2, \ldots; \lambda > 0; \nu \geq 0$$

where the normalizing constant

$$z(\lambda, \nu) = \sum_{j=0}^{\infty} \frac{\lambda^j}{(j!)^\nu}$$

is a generalization of well-known infinite sums.

The CMP distribution contains some well-known discrete distributions:

- for $\nu = 1$, $z(\lambda, \nu) = e^\lambda$, and the CMP distribution simply reduces to the ordinary Poisson(λ);

- for $\nu \to \infty$, $z(\lambda, \nu) \to 1 + \lambda$, and the CMP distribution approaches the Bernoulli with parameter $\lambda(1 + \lambda)^{-1}$;

- for $\nu = 0$ and $0 < \lambda < 1$, $z(\lambda, \nu)$ is a geometric sum

$$z(\lambda, \nu) = \sum_{j=0}^{\infty} \lambda^j = \frac{1}{1 - \lambda}$$

and, accordingly, the CMP distribution reduces to the geometric distribution $p_x = \lambda^x(1 - \lambda)$;

- for $\nu = 0$ and $\lambda \geq 1$, $z(\lambda, \nu)$ does not converge, leading to an undefined distribution.

In general, of course, the normalizing constant $z(\lambda, \nu)$ does not permit such a neat, closed-form expression. Asymptotic results are, however, available. Gillispie and Green [129] prove that, for fixed ν,

$$z(\lambda, \nu) \sim \frac{\exp(\nu \lambda^{1/\nu})}{\lambda^{(\nu-1)/2\nu}(2\pi)^{(\nu-1)/2}\sqrt{\nu}}(1 + O(\lambda^{-1/\nu})),$$

as $\lambda \to \infty$, confirming the conjecture made by [262].

Despite its flexibility to accommodate under- and over-dispersion, the mean and the variance of the CMP distribution cannot be expressed in closed forms. Its moments have to be obtained through the following recursion

$$E[Y^{h+1}] = \begin{cases} \lambda E[Y + 1]^{1-\nu}, & h = 0 \\ \lambda \frac{\partial}{\partial \lambda} E[Y^h] + E[Y]E[Y^h], & h > 0 \end{cases}. \tag{3.1}$$

In terms of computation, the infinite sum, which is involved in computing moments and other quantities, might not appear elegant computationally; however, from a practical perspective, it is easily approximated to any level of precision. Advances in computing help to overcome the inconveniences incurred by not having closed form expressions of the mean and variance. For example, the computation can be easily carried out with the aid of existing packages such as `compoisson` and `COMPoissonReg` in R. Accurate approximations are also available by using an asymptotic approximation of $z(\lambda, \nu)$ (see e.g. Nadarajah [214])

$$E(Y) \approx \lambda^{1/\nu} + \frac{1}{2\nu} - \frac{1}{2}$$

$$V(Y) \approx \frac{1}{\nu} \lambda^{1/\nu}.$$

These approximations are accurate when $\nu \leq 1$ or $\lambda \geq 10^\nu$. It is evident that both $E(Y)$ and $Var(Y)$ are increasing functions of λ but decrease with respect to ν.

3.3   Model inference

As widely discussed in Shmueli et al. [262], parameter estimates can be obtained by maximizing the likelihood function by performing constrained numerical maximization techniques. Nevertheless, computational issues may arise as the maximization procedure involves the infinite sum $z(\lambda, \nu)$. Furthermore, in capture-recapture studies, the zero counts are truncated and, hence, the sample frequencies arise from a zero-truncated distribution. Thus, a zero-truncated CMP distribution

$$P_y = \frac{\lambda^{y_i}}{(y_i!)^\nu z(\lambda, \nu)} \Big/ \left(1 - \frac{1}{z(\lambda, \nu)} \right) \tag{3.2}$$

should be considered and this may further complicate model inference.

3.3.1   The ratio plot

Böhning et al. [44] and Böhning [47] recently re-introduces the notion of a ratio plot [225]

$$r_y = (y + 1) \frac{p_{y+1}}{p_y}$$

as a powerful technique for identifying a suitable distribution with which to model a count variable. The authors provide a number of valuable insights into and extensions of the technique and use the approach within the context of Poisson-based distributions for modeling capture-recapture data.

In capture-recapture analyses, the zero counts are truncated and, hence, the observed sample frequencies $f_1, f_2, \dots$ arise from the zero-truncated distribution $\frac{p_y}{1-p_0}$. However, the ratios r_y for the truncated and the untruncated distribution are identical

$$r_y = (y + 1) \frac{p_{y+1}}{p_y} = (y + 1) \frac{p_{y+1}/(1 - p_0)}{p_y/(1 - p_0)}.$$

Accordingly, the ratio for the CMP distribution is

$$r_y = (y+1)\frac{\frac{\lambda^{y+1}}{\{(y+1)!\}^{\nu}}\frac{1}{z(\lambda,\nu)}}{\frac{\lambda^{y}}{(y!)^{\nu}}\frac{1}{z(\lambda,\nu)}} = \lambda(y+1)^{1-\nu} \tag{3.3}$$

and does not depend on the complex normalizing constant term $z(\lambda,\nu)$. Equation (3.3) suggests a non-linear relation between the ratio of successive probabilities and the count y. Clearly,

$$\begin{aligned}
\log(r_y) &= \log\left\{(y+1)\frac{p_{y+1}}{p_y}\right\} = \log\{\lambda(y+1)^{1-\nu}\} \\
&= \log\lambda + (1-\nu)\log(y+1) = \beta_0 + \beta_1\log(y+1). \tag{3.4}
\end{aligned}$$

From (3.4), we have that $\lambda = \exp(\beta_0)$ and $\nu = 1 - \beta_1$; however, due to $\nu \geq 0$ (or, equivalently, $1 - \nu \leq 1$), we must constrain $\beta_1 \leq 1$. There are no restrictions on β_0, $\lambda > 0$ implies $\beta_0 \in (-\infty, +\infty)$. Thus plots of $log(r_y)$ against y are linear with a negative slope if $\nu > 1$, with a positive slope if $0 \leq \nu < 1$ and represent a constant if $\nu = 1$.

In practice, we approximate capture probabilities by relative frequencies, therefore the ratio in (3.3) can be obtained by

$$r_y^* = (y+1)\frac{\hat{p}_{y+1}}{\hat{p}_y} = (y+1)\frac{f_{y+1}/N}{f_y/N} = (y+1)\frac{f_{y+1}}{f_y},$$

and the ratio in (3.4) can be computed as

$$\log(r_y^*) = \log\left\{(y+1)\frac{f_{y+1}}{f_y}\right\},$$

where f_y is the frequency of count y and $N = \sum_{y=0}^{m} f_y$.

A (log-)ratio plot showing a positive slope indicates the presence of overdispersion with respect to the Poisson distribution. On the other hand, in the case of underdispersion, the log-ratio plot displays a straight line with a negative slope. Finally, when the log-ratio plot displays a horizontal line, the equidispersion case is plausible, or, in other words, the Poisson distribution can be used to fit the data.

3.3.2   The ratio regression

The use of the ratio in (3.4) goes beyond a simple graphical technique to check for under/over-dispersion in CR data. Indeed, it can be used as a tool for estimating the model's parameters. Thus, let us consider our basic equation (3.4); we fit the following model

$$\log(r_y^*) = \underbrace{\beta_0 + \beta_1\log(y+1)}_{Systematic} + \underbrace{\epsilon_y}_{Random}, \tag{3.5}$$

where β_0 and β_1 are the intercept and the slope parameters respectively, and ϵ_y is the error term.

Commonly, a least-squares (LS) estimation method is used to provide estimates of β_0 and β_1. However, model (3.5) does not satisfy the classical linear regression assumptions. In the first place, the response is discrete (although log-transformed), so we might consider a generalized linear model. However, this is inadvisable since an appropriate formulation as a generalized linear model leads to an autoregressive equation involving $\log f_x$ as an additional offset term in the linear predictor. These kinds of models experience difficulties in terms of the definition of the likelihood as well as in carrying out inference. Furthermore, CR frequencies often have $f_1 >> f_2 > f_3 > \ldots$, and, additionally, heteroskedasticity might occur in a heterogeneous population due to e.g. unobserved information (see e.g. Rocchetti et al. [248]). All these issues are relevant and should be accounted for. Thus, we address them by using weighted least-squares (WLS) techniques to estimate the regression parameters β_0 and β_1, and accordingly λ and ν. These are obtained by minimizing

$$\sum_{y=1}^{m-1} W_y \left[\log(r_y^*) - \beta_0 - \beta_1 \log(y+1) \right]^2,$$

where W_y denotes the y-th element of an appropriate weight matrix. In other words, we take

$$\begin{pmatrix} \hat{\beta}_0 \\ \hat{\beta}_1 \end{pmatrix} = (\mathbf{X'WX})^{-1} \mathbf{X'WY}, \tag{3.6}$$

where

$$\mathbf{Y} = \begin{pmatrix} \log \frac{2f_2}{f_1} \\ \log \frac{3f_3}{f_2} \\ \vdots \\ \log \frac{m f_m}{f_{m-1}} \end{pmatrix}, \quad \mathbf{X} = \begin{pmatrix} 1 & \log(2) \\ 1 & \log(3) \\ \vdots & \vdots \\ 1 & \log(m) \end{pmatrix}$$

and m is the maximum count used in the estimator.

The application of weighted least squares requires the specification of $\mathbf{W} \approx cov(\mathbf{Y})^{-1}$ to reduce the mean square error. Following Rocchetti et al. [247], covariances between adjacent log-ratios do not play a large role in reducing mean square error, and thus we suggest dropping off-diagonal terms in $cov(\mathbf{Y})$ in approximating $\mathbf{W}$, with little loss of efficiency. Accordingly

$$\mathbf{W} = \begin{bmatrix} \frac{1}{f_1} + \frac{1}{f_2} & 0 & \cdots & 0 \\ 0 & \frac{1}{f_2} + \frac{1}{f_3} & \cdots & 0 \\ \vdots & \vdots & \ddots & \vdots \\ 0 & 0 & 0 & \frac{1}{f_{m-1}} + \frac{1}{f_m} \end{bmatrix}^{-1}. \tag{3.7}$$

To see that (3.7) is the right choice, let $\mathbf{W}_y = \left[Var\{\log(r_y^*)\} \right]^{-1}$; we have

$$\begin{aligned} Var\left\{ \log(r_y^*) \right\} &= Var\left[\log\left\{ (y+1)\frac{\hat{p}_{y+1}}{\hat{p}_y} \right\} \right] \\ &= Var\left\{ \log(y+1) + \log(\hat{p}_{y+1}) - \log(\hat{p}_y) \right\} \\ &= Var\left\{ \log(\hat{p}_{y+1}) \right\} + Var\left\{ \log(\hat{p}_y) \right\} - 2Cov\left\{ \log(\hat{p}_{y+1}), \log(\hat{p}_y) \right\}. \end{aligned}$$

Using the delta method

$$Var\left\{\log(r_y^*)\right\} \approx \frac{1}{p_{y+1}^2}Var(\hat{p}_{y+1}) + \frac{1}{\hat{p}_y^2}Var(\hat{p}_y) - \frac{2Cov(\hat{p}_{y+1},\hat{p}_y)}{\hat{p}_{y+1}\hat{p}_y}$$

$$= \frac{1}{p_{y+1}^2}\left\{\frac{\hat{p}_{y+1}(1-\hat{p}_{y+1})}{n}\right\} + \frac{1}{\hat{p}_y^2}\left\{\frac{\hat{p}_y(1-\hat{p}_y)}{n}\right\} + \frac{\frac{2\hat{p}_{y+1}\hat{p}_y}{n}}{\hat{p}_{y+1}\hat{p}_y}$$

$$= \frac{1-\hat{p}_{y+1}}{n\hat{p}_{y+1}} + \frac{1-\hat{p}_y}{n\hat{p}_y} + \frac{2}{n}$$

$$= \frac{1}{n\hat{p}_{y+1}} - \frac{\hat{p}_{y+1}}{n\hat{p}_{y+1}} + \frac{1}{n\hat{p}_y} - \frac{\hat{p}_y}{n\hat{p}_y} + \frac{2}{n}$$

where n is the number of observations from the target population.

Therefore, the variance of the log-ratio is given by

$$Var\left\{\log(r_y^*)\right\} \approx \frac{1}{n\hat{p}_{y+1}} + \frac{1}{n\hat{p}_y}.$$

In practice, $\hat{p}_{y+1}$ and $\hat{p}_y$ can be estimated by relative observed frequency $\frac{f_{y+1}}{n}$ and $\frac{f_y}{n}$, respectively. Hence

$$\widehat{Var}\left\{\log(r_y^*)\right\} = \frac{1}{n\frac{f_y}{n}} + \frac{1}{n\frac{f_{y+1}}{n}} = \frac{1}{f_y} + \frac{1}{f_{y+1}}.$$

Thus, we get $\hat{\beta}_0$ and $\hat{\beta}_1$ from (3.6), in which $\mathbf{W}$ is given by (3.7). Accordingly, the unknown f_0 can then be estimated by considering that

$$\log\left(\frac{f_1}{f_0}\right) = \hat{\beta}_0$$

$$\frac{f_1}{f_0} = \exp\left(\hat{\beta}_0\right)$$

$$\hat{f}_0 = f_1\exp\left(-\hat{\beta}_0\right),$$

where $\hat{f}_0$ is the unobserved frequency estimator. The linear regression estimator based on the Conway–Maxwell–Poisson distribution (LCMP) of the target population size can be readily achieved as

$$\hat{N}_{LCMP} = n + \hat{f}_0 = n + f_1\exp(-\hat{\beta}_0). \tag{3.8}$$

We also obtain an estimated probability of the count to be zero (unobserved) as

$$\hat{p}_0 = \hat{f}_0/\hat{N}_{LCMP}.$$

For any matrix $\mathbf{W}$, the weighted least-squares estimate in (3.6) is unbiased if $\mathbf{W}$ is non-random, as

$$E\left(\begin{array}{c}\hat{\beta}_0 \\ \hat{\beta}_1\end{array}\right) = (\mathbf{X'WX})^{-1}\mathbf{X'WX}\left(\begin{array}{c}\beta_0 \\ \beta_1\end{array}\right) = \left(\begin{array}{c}\beta_0 \\ \beta_1\end{array}\right).$$

However, an efficient estimator is achieved only if $\mathbf{W} = \mathbf{\Sigma}^{-1}$, where $\mathbf{\Sigma}$ is the true variance-covariance matrix of $\mathbf{Y}$. If an estimator $\hat{\mathbf{\Sigma}}$ of $\mathbf{\Sigma}$ is used (as is often the case in practice and also in our situation), efficiency is usually lost, but not asymptotic unbiasedness. For the latter, only a consistent estimate of $\mathbf{\Sigma}$ is needed. This is the case for our situation. It is shown in Rocchetti et al. [247] that using the weight matrix in (3.7) leads to a gain in efficiency in comparison with the unweighted unbiased estimate

$$\begin{pmatrix} \hat{\beta}_0 \\ \hat{\beta}_1 \end{pmatrix} = (\mathbf{X'X})^{-1}\mathbf{X'Y}.$$

Hence, we prefer to use (3.6) with weight matrix (3.7).

It is clear that some attention has to be paid to the fact that weights are estimated in reality and this is further addressed in the simulation study. We point out here that the Conway-Maxwell-Poisson distribution includes as a special case the geometric ($\nu = 0$) so that an associated weighted least-squares estimator is available for the geometric. It has the simple form $\widehat{\log \lambda} = \left(\sum_{y=1}^{m-1} W_y \log \frac{f_{y+1}}{f_y}\right) / \left(\sum_{y=1}^{m-1} W_y\right)$, where W_y is the y-th diagonal element of (3.7).

3.4 Variance estimation

While the weighted least-squares algorithm provides an efficient means of parameter estimation in the capture-recapture modeling context, the default output does not provide estimates of the uncertainty associated with the parameter estimates. Several approaches can be considered to facilitate the provision of standard errors within this context.

3.4.1 Approaches based upon resample techniques

Examples of variance estimation based on bootstrap methods have been proposed in the literature (Buckland and Garthwaite [58], Norris and Pollock [220], Zwane and van der Heijden [310]). In the following, bootstrap methods to obtain an estimate of the variance associated with the population size estimate are described. For other approaches to assess uncertainty, we refer to Chapter 10 and Chapter 26. Bootstrap methods are straightforward to implement, regardless of the model under consideration. Here, we consider the True Bootstrap (TB), the Imputed Bootstrap (IB) and the Reduced Bootstrap (RB). In our setting, the algorithm for TB, IB and RB variance estimation techniques proceeds as follows

1. Estimate $\hat{N}_{CMP}$ as described in Section 3.3. This provides an estimate of f_0, $\hat{f}_0$.

2. Form R samples comprising observations from the original data as follows:

    (a) Let $\hat{p}_{TB} = \left\{\frac{f_0}{N}, \frac{f_1}{N}, \ldots, \frac{f_m}{N}\right\}$. The true bootstrap can be applied for estimating the variance of the population size estimator of interest only if the population size is known. Accordingly, under the TB approach, each of the R_{TB} samples contains N observations drawn from a multinomial distribution with parameters N and $\hat{p}_{TB}$.

    (b) Let $\hat{p}_{IB} = \left\{\frac{\hat{f}_0}{\hat{N}_{CMP}}, \frac{f_1}{\hat{N}_{CMP}}, \ldots, \frac{f_m}{\hat{N}_{CMP}}\right\}$. Under the IB approach, each of the R_{IB} samples contains $\hat{N}_{CMP}$ observations drawn from a multinomial distribution with parameters $\hat{N}_{CMP}$ and $\hat{p}_{IB}$. This method is particularly attractive if a good estimate for N exists. For each bootstrap replication, $\hat{N}_{CMP}$ capture histories are drawn with probabilities $\hat{p}_{IB}$.

    (c) Let $\hat{p}_{RB} = \left\{\frac{f_1}{n}, \frac{f_2}{n}, \ldots, \frac{f_m}{n}\right\}$. Under the RB approach, each of the R_{RB} samples contains n observations, where the observations are sampled with replacement from the observed data. In other words, this method utilizes sampling with replacement from the capture histories. Specifically, for each replication, n capture histories are drawn, with n constant among replications.

3. For each sample, estimate $\hat{N}_{CMP}$, under the CMP model. In the case of the true bootstrap, it means that the true f_0 is ignored.

4. Estimate the variance of $\hat{N}_{CMP}$ on R bootstrapped samples as follows:

(a) The TB estimate of the variance of $\hat{N}_{CMP}$ is equal to

$$\sigma^2_{TB} = \frac{1}{R-1} \sum_{r=1}^{R} (\hat{N}_{CMP,r} - \bar{\hat{N}}_{CMP,TB}))^2$$

where $\bar{\hat{N}}_{CMP,TB}$ is the TB sample mean.

(b) The IB estimate of the variance of $\hat{N}_{CMP}$ is equal to

$$\sigma^2_{IB} = \frac{1}{R-1} \sum_{r=1}^{R} (\hat{N}_{CMP,r} - \bar{\hat{N}}_{CMP,IB})^2$$

where $\bar{\hat{N}}_{CMP,IB}$ is the IB sample mean.

(c) The RB estimate of the variance of $\hat{N}_{CMP}$ is equal to

$$\sigma^2_{RB} = \frac{1}{R-1} \sum_{r=1}^{R} (\hat{N}_{CMP,r} - \bar{\hat{N}}_{CMP,RB})^2$$

where $\bar{\hat{N}}_{CMP,RB}$ is the RB sample mean.

3.4.2 An approximation-based approach

Another benefit of the ratio regression approach is that variance estimators for f_0 can easily be developed as variance estimators for the estimated regression coefficients are easily available. Let $\hat{N} = \hat{N}_{LCMP} = n + f_1 e^{-\hat{\beta}_0}$ be the population size estimator; then, the variance of $\hat{N}$ arises from two sources; these are influenced by the random variable n and the estimator $\hat{f}_0$. Therefore a simple formula for the variance of the population size estimator is given as

$$Var(\hat{N}) = Var_n\{E(\hat{N}|n\} + E_n\{Var(\hat{N}|n)\}.$$

We apply a technique for computing moments usually referred to as conditioning (see e.g. Böhning [40]) to population size estimation. The technique provides a simple formula for variance computation of population size which can be applied to a general estimator. According to the conditional technique, we have

$$Var(f_1 e^{-\hat{\beta}_0}) = Var_{f_1}\{E(f_1 e^{-\hat{\beta}_0})|f_1\} + E_{f_1}\{Var(f_1 e^{-\hat{\beta}_0})|f_1\},$$

and thus

$$
\begin{aligned}
Var_{f_1}\{E(f_1 e^{-\hat{\beta}_0})|f_1\} &\approx Var(f_1 e^{-\hat{\beta}_0}) = (e^{-\hat{\beta}_0})^2 Var(f_1) \\
&= (e^{-\hat{\beta}_0})^2 N p_1 (1 - p_1) = (e^{-\hat{\beta}_0})^2 f_1 \left(1 - \frac{f_1}{N}\right).
\end{aligned}
$$

Using the delta method, we achieve that $Var(e^{-\hat{\beta}_0}) = (e^{-\hat{\beta}_0})^2 Var(\hat{\beta}_0)$. Hence $E_{f_1}\{Var(f_1 e^{-\hat{\beta}_0})|f_1\} \approx f_1^2 (e^{-\hat{\beta}_0})^2 Var(\hat{\beta}_0)$, where $Var(\hat{\beta}_0)$ comes from the linear regression process. The approximated expression for the variance of the CMP estimator $\hat{N}_{CMP}$ is given as

$$\widehat{Var}(\hat{N}_{LCMP}) = \frac{n f_1 e^{-\hat{\beta}_0}}{n + f_1 e^{-\hat{\beta}_0}} + (e^{-\hat{\beta}_0})^2 f_1 \left(1 - \frac{f_1}{N}\right) + f_1^2 (e^{-\hat{\beta}_0})^2 Var(\hat{\beta}_0).$$

As $1 - \frac{f_1}{N} \leq 1$, a conservative asymptotic variance of $\hat{N}_{CMP}$ is obtained as

$$\hat{\sigma}_{LCMP}^2 = \widehat{Var}(\hat{N}_{LCMP}) = \frac{n f_1 e^{-\hat{\beta}_0}}{n + f_1 e^{-\hat{\beta}_0}} + (e^{-\hat{\beta}_0})^2 f_1 [1 + f_1 Var(\hat{\beta}_0)]. \tag{3.9}$$

3.4.3 Comparing confidence intervals

We then used the approximation-based and bootstrap methods to derive 95% percent quantile confidence intervals for each data set. Using these intervals, we ascertained the coverage proportions for each of the methods. These results shed light on the confidence we can put on the obtained estimates and the related uncertainty. Confidence intervals are computed in different ways. A common procedure is to approximate a 95% confidence interval for the true population size by the interval $\hat{N} \mp z_{0.975}\hat{\sigma}_{LCMP}$, where $\hat{\sigma}_{LCMP}$ is the estimated standard error in (3.9). This is referred to as a symmetric confidence interval (SYM). However, the construction of the symmetric confidence intervals is based on the large-sample normality for population size estimators. Several drawbacks for this method have been highlighted in Chao [72]: the sampling distribution could be skewed, the lower bound of the resulting interval may be less than the number of units captured, and the coverage probabilities may be unsatisfactory. To overcome these issues, coverage of the Burnham confidence interval (BH) $(n + (\hat{N} - n)/c; n + (\hat{N} - n)c)$, where

$$c = \exp\left\{ z_{0.975} \left[\log\left(1 + \frac{\hat{\sigma}_{LCMP}^2}{(\hat{N} - n)^2}\right) \right]^{1/2} \right\}$$

is also evaluated (Tounkara and Rivest [276], Burnham and Overton [65]). We further suggest looking at intervals obtained by using a log-transformation of $\hat{N}$. From the log-normal distribution, it follows that $\log \hat{N}$ has mean $\log N - \frac{1}{2}\log(1 + \sigma_{LCMP}^2/N^2)$ and variance $\log(1 + \sigma_{LCMP}^2/N^2)$. Plugging in estimates for σ_{LCMP}^2 and N leads to a confidence interval for $\log N$ (LOG) given by

$$\log \hat{N} + \frac{1}{2}\log(1 + \hat{\sigma}_{LCMP}^2/\hat{N}^2) \mp z_{0.975}\sqrt{\log(1 + \hat{\sigma}_{LCMP}^2/\hat{N}^2)}.$$

Taking the anti-logs provides the final form of the confidence interval for N. Other approaches can be pursued to get confidence intervals [72].

For the bootstrap methods, considering all estimates $\hat{N}_b, b = 1, \ldots, B$ from B bootstrapped samples results in an empirical distribution around the true value. From this distribution we can compute the standard error $\hat{\sigma}$ of the parameter by taking the sample standard deviation of the resulting distribution. The approximate 95% confidence interval of the population size $\hat{N}$ can be obtained using the percentile method as follows: order $\hat{N}_b$ from the smallest to largest and denote the ordered list by $\hat{N}_{(b)}$; the approximate 95% confidence limits are then given by $\hat{N}_{(B+1)*0.025}$ and $\hat{N}_{(B+1)*0.975}$, both rounded to the nearest integer value.

3.5 Applications

In the following we estimate population sizes through the CMP estimator so far considered in six well-known benchmark datasets. Data are provided in Table 3. Graphical data inspections through the (log) ratio-plot are provided in Figure 3.1. In three cases (the colorectal polyps and the taxicabs data) we know the true population size and, accordingly, the TB approach can be also considered. We would like to provide more insights on the uncertainty of the estimates in real data applications, focusing on implications of using different methods to estimate such an uncertainty. Population size estimates and confidence intervals are reported in Table 3.1, along with the CMP parameter estimates.

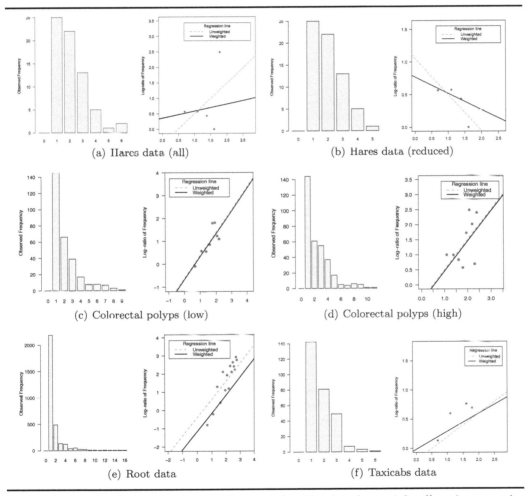

FIGURE 3.1: Applications: Distributions and (weighted and unweighted) ratio regression plots of real data analyzed in Section 3.5.

3.5.1   Snowshoe hares

We revisit the snowshoe hares data (Cormack [86], Agresti [2]), where a sample of $n = 68$ hares was observed at least once on six occasions. For these data, in the literature a strong sensitivity to the dependency structure is recognized, with estimates ranging from $\hat{N} = 70$ to $\hat{N} = 90$ for a set of models with and without heterogeneity. The same data are revisited by Farcomeni and Tardella [115] through models accounting for heterogeneity. A similar sensitivity to the model structure is found, with estimates ranging from $\hat{N} = 76$ to $\hat{N} = 89$. From a graphical inspection through the ratio plot (see Figure 3.1(a)), it is clear that the two animals caught on all occasions create some overdispersion with respect to the Poisson distribution. Therefore, the CMP estimator could be a good candidate to estimate the unknown population size. Parameter estimates are $\hat{\nu} = 0.77$, with $\hat{\lambda} = 1.43$ and the resulting estimated population size is $\hat{N} = 86$, slightly higher than the one estimated in Agresti [2], but in line with aforementioned works. If we remove the 2 hares caught 6 times (see Figure 3.1(b)), as Cormack [86], the situation changes considerably and underdispersion is estimated ($\hat{\lambda} = 2.16$; $\hat{\nu} = 1.25$), with $\hat{N} = 78$, close to the estimate proposed by Farcomeni [114]. It is important to notice that the CMP estimator results are flexible enough to capture even underdispersion.

Similarly, confidence intervals reflect the effect of the 2 hares caught on all occasions, which we have discussed above. They are very large if the complete data are considered, and much smaller if those two hares are left out of the analysis as unrepresentative of the unobserved part of the population. Bootstrap intervals are larger than those obtained by approximating the variance of the sample size estimator, in line with the simulation results. The Burnham- and the log-transformed-based intervals are more in line with the bootstrap ones, confirming that in the underdispersion case, assuming a symmetric confidence interval may lead to unreliable inference.

3.5.2   Colorectal polyps

Colorectal cancer is one of the most common cancer types. Colonoscopy is considered an effective tool for colorectal cancer screening and studies have shown that colonoscopy is associated with a reduction of Colorectal cancer incidence and mortality. In 1990, the Arizona Cancer Center initiated a multicenter trial to determine whether wheat bran fiber can prevent the recurrence of colorectal adenomatous polyps (Alberts et al. [3]). Subjects with a previous history of colorectal adenomatous polyps were recruited and randomly assigned to one of two treatment groups, low fiber and high fiber. From medical research experience it is well recognized that diagnosing adenomatous polyps can be subjected to undercount due to misclassification at colonoscopy. In the following, we evaluate the recurrence of colorectal adenomatous polyps. Subjects with previous history of colorectal adenomatous polyps are allocated to one of two treatment groups, low fiber and high fiber. For both groups the population size is known in advance: 584 for the low-fiber treatment ($f_0 = 285$) and 722 for high-fiber treatment ($f_0 = 381$) respectively (see Figures 3.1(c)–(d)). We assumed that patients with a positive polyp count were diagnosed correctly, whereas it is unclear how many persons with zero polyps were false-negatively diagnosed. Thus, we approach the data as if zero counts were not observed, and we try to estimate the undercount from the nonzero frequencies.

The CMP-based estimator perform very well for the low-fiber case, as $\hat{N} = 583$. Less enthusiastic results were found in the high-fiber case. In the latter case, we underestimate the population size ($\hat{N} = 589$). However, our estimate is the *best* among its competitors. The Poisson-based estimator provides an estimate of 385 (369–401), while heterogeneous estimators such as Chao and Zelterman estimate a population size of 511 (443–579) and 597 (476–718), respectively. None of the considered alternatives, however, provide confidence intervals that cover the true population size. By approximating the variance as described in Section 3.4.2, we get confidence intervals covering the true value $N = 722$.

3.5.3   Root data

Here we analyze the root data already analyzed in Wang [292] (see Figure 3.1(e)) which represent the count distribution of the expressed genes of *Arabidopsis thaliana* in the root tissue. The interest lies in the estimation of the unknown number of unexpressed genes since data are collected from a cDNA library sample that very likely does not allow a full screening of all expressed genes. Researchers agreed that the arabidopsis thaliana has a relatively small genome with approximatively 27,000 protein coding genes not necessarily all expressed in all tissues. Wang [292] provides a conservative estimate of the total number of expressed genes in the root tissue (slightly less than 9000). Our estimate is considerably higher exceeding the value 10,000. Although in this case the population size is not known in advance, previous works [190] suggest a percentage of expressed genes in root tissue greater than 40% of the 27,000 protein coding genes, which fits well with the recommendation provided by the CMP-based estimator.

3.5.4   Taxicab data in Edinburgh

As a final example, we consider the Taxicab data (see Figure 3.1(f)). Carothers [69] reported that 420 taxicabs were registered in Edinburgh, Scotland during his mark-recapture study. This closed population was sampled for 10 consecutive days with observation points and times varied among days. Sighting a cab was considered a *capture*. No taxis were observed on more than six occasions. These data have been analyzed many times in the literature using different estimators (e.g. Chao [72]). The performance of the CMP estimator is remarkably good ($\hat{N} = 428$), compared to other estimators. In all cases, the true N is contained within the confidence intervals, no matter what procedure has been used to obtain them.

3.6   Discussion

A diversity of estimators in the capture-recapture field exists, being widely applied in many areas of interest. Here, we have discussed a parametric method of estimating the population size under a specific form of heterogeneity based on the Conway-Maxwell-Poisson distribution. The CMP-based estimator is accurate, provides small bias in the homogeneous Poisson case which asymptotically disappears and performs well under different heterogeneous data generation processes (i.e. Geometric, Negative Binomial); hence, it improves existing heterogeneous estimators (e.g. Chao's and Zelterman's estimators), see e.g. Anan et al. [8]. The use of the ratio plot allows us to avoid computational issues related to CMP distribution. Furthermore, by using the ratio plot, formal tests can be conducted on null hypotheses of zero-truncated Poisson, i.e. $H_0 : \beta_1 = 0$, or geometric, i.e. $H_0 : \beta_1 = 1$, data. The proposed LCMP estimator performs as well as the MLE under the Poisson and the geometric distribution, supporting that the use of the ratio plot, instead of computing the MLE under the CMP distribution, does not affect estimates.

We also provided a formula of variance approximation of the new estimator. This variance formula is not only useful to determine the efficiency of estimating, but it can be

TABLE 3.1: Population size estimation and uncertainty assessment in real data examples

(a) : Symmetric; (b) : Burnham; (c) : Logarithm transformation

Name	N	$\hat{N}$	Standard error estimation				95% Confidence Intervals			
			Approx.	σ_{TB}	σ_{IB}	σ_{RB}	Approx.	TB	IB	RB
Hares (all data) ($\hat{\lambda}=1.43$, $\hat{\nu}=0.77$)	n.a.	86	12.01	n.a.	15.10	14.43	$(66-113)^{(a)}$ $(66-113)^{(b)}$ $(66-114)^{(c)}$	n.a.	$(68-126)$	$(71-125)$
Hares (reduced data) ($\hat{\lambda}=2.16$, $\hat{\nu}=1.25$)	n.a.	78	4.58	n.a.	14.08	13.50	$(70-87)^{(a)}$ $(68-126)^{(b)}$ $(71-125)^{(c)}$	n.a.	$(66-121)$	$(69-121)$
Colectoral polyps (low) ($\hat{\lambda}=0.51$, $\hat{\nu}=0.00$)	584	583	83.24	36.08	36.21	34.41	$(420-747)^{(a)}$ $(461-798)^{(b)}$ $(445-777)^{(c)}$	$(513-655)$	$(515-658)$	$(516-651)$
Colectoral polyps (high) ($\hat{\lambda}=0.58$, $\hat{\nu}=0.00$)	722	589	120.34	28.99	28.79	27.20	$(354-825)^{(a)}$ $(441-951)^{(b)}$ $(404-893)^{(c)}$	$(535-648)$	$(537-650)$	$(538-644)$
Root data ($\hat{\lambda}=0.31$, $\hat{\nu}=0.00$)	n.a.	10227	241.02	n.a	258.70	262.97	$(5504-14951)^{(a)}$ $(6843-16690)^{(b)}$ $(6661-16572)^{(c)}$	n.a.	$(9723-10842)$	$(9752-10782)$
Taxicabs A ($\hat{\lambda}=0.98$, $\hat{\nu}=0.69$)	420	428	91.28	65.75	65.35	64.12	$(250-607)^{(a)}$ $(284-648)^{(b)}$ $(290-662)^{(c)}$	$(348-600)$	$(348-600)$	$(353-597)$

also used to construct confidence intervals. In this respect, we provided several insights on the behavior of bootstrap methods for variance estimation. Here, three bootstrap methods have been considered: the true bootstrap, the reduced bootstrap, and the imputed bootstrap. What works and what does not? It is very clear that the reduced bootstrap may suffer of underestimating the *true* variance. This is independent of whether the model holds or not. This result indicates that current practice (using reduced bootstrap method in capture-recapture) should be discontinued. The true bootstrap works, if the model holds or not, but it cannot be used in practice. This leaves the imputed bootstrap, which seems to work like the true bootstrap. Hence it behaves similar to the parametric bootstrap. The results are encouraging to investigate the imputed bootstrap in further capture-recapture models and truncated data modeling.

4

The geometric distribution, the ratio plot under the null and the burden of dengue fever in Chiang Mai province

Dankmar Böhning

University of Southampton

Veerasak Punyapornwithaya

Chiang Mai University

CONTENTS

4.1 Introduction

In this application we study counts of dengue fever cases per village. Let y_i denote the count of dengue fever cases in village i, each of which is affected by dengue fever. There are n villages observed with dengue fever, in other words for these villages $y_i > 0$, $i = 1, \cdots, n$. Due to the nature of dengue fever. not every case is detected and hence there will be a number of villages affected by dengue fever which remain unreported to the surveillance system. Let f_0 denote the number of villages affected by dengue fever that remain unreported. This means there are $N = n + f_0$ villages affected by dengue fever. We are interested in estimating f_0 or, equivalently, N, and this is the purpose of this work. Before we go into details of the modeling approach we provide some background on the study data.

4.2 The case study on dengue fever

Dengue fever is a disease caused by a family of viruses transmitted by mosquitoes. The disease causes illness in infants, children and adults. The clinical signs of dengue fever includes headache, exhaustion, fever, muscle pain and rash. A more severe type is dengue hemorrhagic fever characterized by fever, abdominal pain, persistent vomiting, bleeding and breathing difficulty and involves a potentially lethal complication. A high incidence of dengue fever can be found in many tropical countries.

The data considered here were cases of laboratory confirmed dengue fever patients in the year 2013 in Chiang Mai province. Chiang Mai province includes the city of Chiang Mai as well as a large number of surrounding villages. In total there are 2,066 villages which belong to the Chiang Mai region. It is located in Northern Thailand bordering Myanmar, China and Laos. Chiang Mai is the second largest city in Thailand (after Bangkok), and enjoys a large population of foreign (mainly Western tourists. farang in Thai) who come here for short vacation visits or have chosen Chiang Mai as their permanent residence. Chiang Mai's attractions include a lovely country-side and a historic city center. The climate is quite comfortable due to its mild temperatures throughout the year as it is slightly elevated and surrounded by mountains covered vastly by forests. We point out these facts to illustrate that the population affected by dengue fever might include foreign visitors who might be short-term visitors or permanent residents.

The data on dengue fever were collected by the Chiang Mai Provincial Public Health Office. Data consists of patient ID, address, date of sickness, diagnostic date, hospital name where patients were diagnosed and patient gender. The address includes the name of village, sub-district and district. Hence we are analyzing data on the finest available administrative level.

We see from Table 4.1 that 1357 villages were affected by dengue fever. There were 285 villages with exactly one case, 205 villages with two cases, and so forth. In total there were 11,048 cases of dengue fever in the Chiang Mai province in 2013.

4.3 Geometric distribution

A popular distribution for count data is the geometric distribution having probability density function

$$p_y = (1 - \theta)^y \theta \tag{4.1}$$

for $y = 0, 1, \cdots$ and θ being the event parameter. As we have no zeros observed, we consider the zero-truncated geometric distribution

$$\frac{p_y}{1 - p_0} = (1 - \theta)^{y-1}\theta \tag{4.2}$$

for $y = 1, 2, \cdots$ which is again a geometric distribution using the transformed variable $z = y - 1$. Given a sample with frequencies $f_1, f_2, \cdots, f_m$ where m is the largest observed

TABLE 4.1

Frequencies f_y of the number of cases y of dengue fever per village in Chiang Mai province in the year 2013

y	1	2	3	4	5	6	7	8	8	10	11	12	13	14	15
f_y	285	205	157	115	83	68	55	45	47	38	29	21	22	13	15
y	16	17	18	19	20	21	22	23	24	25	26	27	28	29	30
f_y	17	10	10	11	7	8	7	9	7	2	5	3	6	6	5
y	31	33	34	35	36	37	38	39	40	41	43	44	46	50	52
f_y	3	2	4	7	4	2	1	1	1	1	2	1	1	2	1
y	53	57	59	61	62	68	69	85	106	125	1248				
f_y	1	1	1	2	1	2	1	1	1	1	1				

count, maximum likelihood estimation is standard and the maximum likelihood estimator provided in Niwitpong et al. [218] as

$$\hat{N}_{\text{MLE}} = \frac{nS}{S-n},$$ (4.3)

where $S = \sum_{y=1}^{m} y f_y$. However, the major question remains whether the geometric distribution is an appropriate model for the data at hand. This will be investigated in the next section.

4.4   Ratio plot

The geometric distribution is potentially a suitable candidate for case-count distributions as it catches naturally some heterogeneity present in the population. To be more precise, consider the Poisson distribution for the count of cases conditional on some parameter λ. If this parameter has an exponential distribution $g(\lambda)$, then as a marginal distribution the geometric arises:

$$(1-\theta)^y \theta = \int_0^\infty \exp(-\lambda)\lambda^y/y! \times g(\lambda)d\lambda.$$ (4.4)

Hence, in some sense the geometric distribution is more suitable than the Poisson as a case-count distribution. Nevertheless, the geometric distribution needs to be investigated to determine if it is appropriate for our case data.

In Böhning et al. [44] a diagnostic device was suggested to investigate a count data set for a specific distribution. See also Chapter 2, Section 2.4. For related ideas see also Rivest and Baillargeon [246]. The diagnostic device, suggested in Böhning et al. [44] and called the *ratio plot*, is built on the observation that the ratios of neighboring probabilities are constant:

$$r_y = \frac{p_{y+1}}{p_y} = (1-\theta)$$ (4.5)

for $y = 0, 1, \cdots$. Note that these ratios are not dependent on whether untruncated or truncated distributions are considered. A natural estimate of r_x occurs when replacing the unknown probabilities by the estimate f_y/N:

$$\hat{r}_y = \frac{f_{y+1}}{f_y},$$ (4.6)

as the unknown denominator N cancels out. In Figure 4.1 we see a geometric ratio plot on log-scale for the dengue fever data of Chiang Mai. A major difficulty with interpreting this plot is the qualitative judgment on constancy across the count range. This does not improve substantially when standard error bars are added as it does not help to judge constancy across the range of y. See Figure 4.2. Standard errors can be derived using the δ-method and lead to

$$\text{var}(\log \hat{r}_y) \approx \frac{1}{np_{y+1}} + \frac{1}{np_y},$$ (4.7)

which are easily estimated as $1/f_{y+1} + 1/f_y$. To help judge the appropriateness of a distribution, we modify the concept of the ratio plot in the next section.

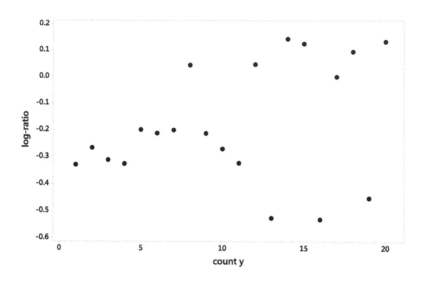

FIGURE 4.1: Ratio plot for a geometric distribution for the dengue fever data of Chiang Mai province.

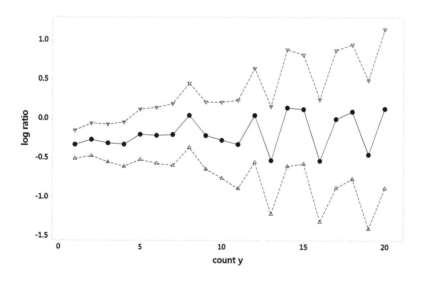

FIGURE 4.2: Ratio plot for a geometric distribution with 95% pointwise confidence intervals (dotted curve) for the dengue fever data of Chiang Mai province.

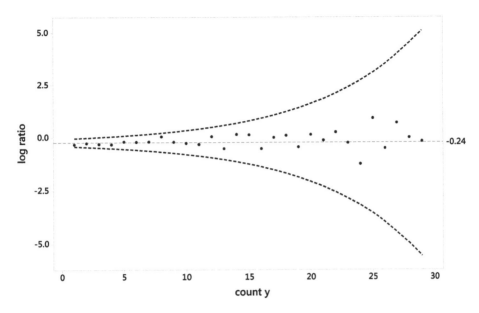

FIGURE 4.3: Ratio plot for a geometric distribution under the null for the dengue fever data of Chiang Mai province.

4.5 Ratio plot under the null

Whereas the ratio plot in the previous sections focused on the idea that an *empirical, nonparametric* estimate of the ratio would follow a straight line, the idea here is to construct a diagnostic device that shows the observed ratio *within limits expected if the data followed a geometric distribution.* This can be accomplished by considering the 95% pointwise error bars

$$\log(1 - \hat{\theta}) \pm 1.96 \times \sqrt{\operatorname{var}(\log \hat{r}_y)}, \tag{4.8}$$

where $y = 1, 2, \cdots$ and $\hat{\theta}$ is the maximum likelihood estimator. Note that under the geometric distribution the ratio is constant across the range of y: $r_y = (1 - \theta)$ for all y. In contrast, to the conventional ratio plot we use parametric variance estimates under the geometric distribution

$$\operatorname{var}(\log \hat{r}_y) \approx \frac{1}{n(1 - \theta)^{y+1}\theta} + \frac{1}{n(1 - \theta)^y\theta}, \tag{4.9}$$

with appropriate estimates for θ as above. As (4.9) is completely specified by the geometric distribution we call this plot the *ratio plot under the null.*

Now we are able to examine more easily if the observed ratios lie in the specified region determined by (4.9). This is evidently the case for the dengue fever counts distribution of cases. Hence this supports using the geometric distribution in this case study.

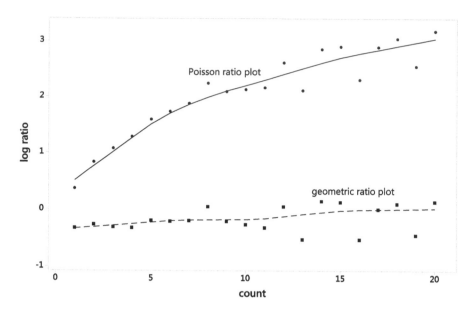

FIGURE 4.4: Poisson and geometric ratio plots for the dengue fever data of Chiang Mai province (with embedded LOWESS smoother as dashed curve).

4.6   Application to estimate the burden of dengue fever

We now apply the concept of the geometric distribution. We find an estimate of the population size of $\hat{N}_{\mathrm{MLE}} = \frac{nS}{S-n} = 1547.02$, so, in other words, we are estimating 200 additional villages to be affected by the disease. The associated variance is provided in Niwitpong et al. [218] as $(Sn)^2/(S-n)^3$, which leads to the approximate 95% confidence interval (1516.22–1577.82).

Another question that arises relates to the fact that we could have used another popular distribution for modeling count data: the Poisson model. The Poisson distribution is given as $p_y = \exp(-\lambda)\lambda^y/y!$ so that $r_y = (y+1)p_{y+1}/p_y = \lambda$ and we expect that $\hat{r}_y = (y+1)f_{y+1}/f_y$ shows a horizontal line pattern. Figure 4.4 shows both ratio plots in comparison and there is clear evidence that there is a positive trend in the Poisson ratio plot. Consequently, we argue here that the geometric distribution is more appropriate in this case study.

5

A ratio regression approach to estimate the size of the Salmonella-infected flock population using validation information

Carla Azevedo

University of Southampton, United Kingdom

Dankmar Böhning

University of Southampton, United Kingdom

Mark Arnold

Animal and Plant Health Agency, United Kingdom

CONTENTS

5.1 Introduction and background

Capture-recapture methods are used to estimate the global size N of a target population of interest when it is incomplete. Many times, in real applications, due to a deficient identification/registration mechanism, only a portion of the population is observed — the positive counts — and we might be able to predict the number of unobserved units of the target population. Estimating the size N of a specific population is of crucial importance in many areas. For example, in ecological applications it is relevant to estimate the size of a wildlife population. In medicine, it is essential to estimate the number of people with a specific disease when a screening test is not totally accurate and we frequently get false negatives.

Let us assume that the members of the population are identified at m observational occasions where m is considered fixed in this work. For each member i the count of identifications X_i for a generic unit returns a count in $0, 1, ..., m$ and i takes values from 1 to N. It is assumed that X_i is available if unit i has been identified for at least one occasion. We have then that X_i is observed and let $X_1, ..., X_n$ denote the observed counts with n representing the total number of recorded individuals. We assume w.l.o.g. that $X_{n+1} = \ldots = X_N = 0$. Hence, units $n + 1$ to N remain unobserved. Let f_x be the frequency of units with count $X = x$. The associated population density function can be described by a probability density function $p_x(\theta)$ and denotes the probability of exactly x identifications for a generic unit where $p_x \geq 0$ and $\sum_{x=0}^{m} p_x = 1$. See also Chapter 1, Section 1.2.

Let us illustrate this theory with a very common example. In the medical field, several screening tests are applied in human populations to detect specific diseases in their early stage when they are easier to be treated and cured. Due to a low sensitivity of the test or even human error, any screening test is not 100% accurate. Moreover, it is possible to find people with a negative test but who actually have the disease. However, people are usually not further assessed when the test shows a negative result, so it remains unknown which disease status they currently have. In other words, we want to investigate how many false negatives we have adopting the described procedure or a similar one.

Using the same notation as above, let us assume we are analyzing for example a clinical disease whose status can be measured at m occasions, where the count x denotes the number of times the screening test is positive (number of captures) as shown in Table 5.1:

TABLE 5.1
Frequency of each status
of a certain disease

x	0	1	2	...	m
f_x	?	f_1	f_2	...	f_m

Here, f_x represents the number of individuals captured exactly x times during this screening test. If the test is negative at all m times, the true status of the person is unknown. Hence, we intend to estimate f_0 using zero-truncated count information to provide an estimate of the total size of the diseased population N.

This is just a simple generic example of an application of capture-recapture methodology. However, this methodology can also be applied in other areas such as epidemiology, ecology or social sciences, see e.g. Böhning et al. [37], van der Heijden [283], also Chapter 1, section 1.2, or Chapter 2, section 2.3. In the case of the example, each capture happens in a fixed period of time and it is assumed that each individual has equal probability of being captured during the study period. Therefore, it is assumed that the population is closed which means that it is kept constant during that period, i.e., there are no births, deaths or migration.

Proceeding with the modelling, we can denote the population density function as $p_x(\theta)$. In the simplest case, a binomial distribution can be chosen with probability distribution as follows:

$$p_x(\theta) = P(X = x) = \binom{m}{x} \theta^x (1 - \theta)^{m-x} \tag{5.1}$$

where X is a random variable of the realizations of a binomial distribution, $x = 0, \ldots, m$.

Naturally, we have that p_0 is the probability for a zero-count (a unit remaining unobserved). In the binomial case, this is equal to $p_0(\theta) = (1-\theta)^m$. Consequently, the probability that an individual is observed is $1 - p_0$ and the total size of the population N can be described by $N = N(1 - p_0) + Np_0$. As $N(1 - p_0)$ corresponds to the observed part of the population, we can estimate $N(1 - p_0)$ by n and set up an estimating equation as $N = n + p_0 N$ from which the Horvitz–Thompson estimator $\hat{N} = \frac{n}{(1-p_0)}$ follows.

Note that the Horvitz–Thompson estimator requires knowledge of p_0. In the following, we will derive a methodology which will allow direct estimation of p_0.

Note also that the population size N can be written as

$$N = f_0 + n \qquad (5.2)$$

where $n = \sum_{x=1}^{m} f_x = f_1 + ... + f_m$.

Situations of heterogeneity in the population can be detected by means of the ratio plot which works like a diagnostic device for the presence of a particular distribution. We can then extend this theory to a regression approach which will consider ratios of neighbouring count probabilities estimated by ratios of the observed frequencies and fit a proper model to the data. Finally, we use the model to derive an estimate for the frequency of hidden counts, f_0, projecting the model backwards.

Sometimes, additional information on the unobserved units is available through another sub-sample of the target population, called a validation sample. In this secondary sample, usually smaller in size, we do observe zero counts which means that there are no hidden cases, all the counts are observed. It is possible to incorporate the information coming from the validation sample into the modelling and decrease the bias in the estimation process. Let us denote by $g_0, g_1, ..., g_m$ the frequency distribution associated with this sample. Notice that all the results and in particular g_0 are known.

Still considering the example above, let us imagine that another sample of people was chosen and assessed to repeat the same tests. The results are shown in Table 5.2.

TABLE 5.2
Frequency of each status of
a certain disease

y	0	1	2	...	m
g_y	g_0	g_1	g_2	$\cdots$	g_m

The structure of the validation sample is similar to the structure of the positive sample. However, it is important to emphasize that here all the counts are observed and, in particular, we have information on g_0 which is unknown in the positive sample. Again, g_y is the frequency of counts exactly equal to y. The first introduction of capture-recapture modelling using validation information can be found in Böhning [47], where it was mentioned as an extension of a generic ratio regression approach.

Simulation studies were conducted to evaluate the performance of the suggested approach. We were able to conclude that the use of a validation sample not only substantially increases the estimation efficiency but also reduces the bias considerably. A zero-inflated model was also considered due to the suspicion of hidden observations in the data in addition to those predicted by non-inflated models.

Overall, this work focuses on the development of methodology to include validation information in the capture-recapture modelling in order to increase the accuracy and efficiency of the final estimate for the unrecorded cases. We are interested in applying this theory to a public health problem scenario which is related to *Salmonella* infection in commercial egg-laying flocks.

5.2 Case study

This project is a joint work with the Animal and Plant Health Agency (APHA) in the UK; therefore we have access to the following data related to *Salmonella* in commercial egg-laying flocks.

Human salmonellosis is a major public health concern in Europe, with the majority of cases in recent years being caused by *Salmonella* strains *Salmonella enteritidis* and *Salmonella typhimurium*, and the most common source of infection thought to be through the consumption of contaminated eggs produced by infected laying hens, see Gillespie et al. [128], Arnold et al. [14].

To assess the current prevalence of infected commercial egg-laying flocks, a European Union wide baseline survey of *Salmonella* was carried out between October 2004 and September 2005. The results of that survey were used as a basis for setting flock prevalence reduction targets for *Salmonella* national control programmes in each member state of the European Union. The target was set at a 10% reduction per annum in the prevalence of *Salmonella* for the UK, for details see Arnold et al. [13]. As part of the baseline survey in the UK, a randomized sample of 454 commercial layer flock holdings was tested for *Salmonella*.

It is important to achieve effective control of the infection at farm level and monitor *Salmonella* strains, and thus reduce the impact on human health. It is crucial that infected flocks are detected so that measures can be taken to avoid consumption of *Salmonella*-contaminated eggs by the public. There may still be a reservoir of *Salmonella* in some commercial laying farms and this extent was largely unknown in the UK before the reported survey, see Arnold et al. [11, 12], Snow et al. [263].

In order to be able to monitor the progress of the national control programme for *Salmonella*, and demonstrate that there is a reduction over time in the prevalence of *Salmonella* in UK egg-laying farms, it is important to be able to obtain an accurate estimate of the initial prevalence at the time of the EU baseline survey. Therefore, it is vital to adjust the under-count of disease occurrence appropriately. The main goal of the present study is to provide an estimate of the number of undetected cases as accurately as possible, i.e. to estimate the number of farms which had *Salmonella*-infected chickens but for which the result in the survey was negative.

5.2.1 *Salmonella* data

In total, 454 holdings were sampled in the survey. From those, 53 tested positive for *Salmonella* in one or more samples of the survey using a method we will denote as the EU baseline survey method. Briefly, this consists of sampling 5 faeces samples, each composing a representative mix of litter from 1/5th of the poultry house, and 2 dust samples collected from around the poultry house, which would then be cultured for *Salmonella*. The EU baseline survey therefore consists of a total of 7 tests, so each farm could have 0,1,...,7 positives as Table 5.3 shows:

TABLE 5.3
Positive sample of *Salmonella*
data

x	0	1	2	3	4	5	6	7
f_x	?	17	9	5	6	5	5	6

Table 5.3 shows the frequency distribution of the number of positive samples from each farm in the EU baseline survey. There are 17 farms that had one positive sample, while 9 farms had two and so on.

The EU baseline survey data reported a prevalence of 11.7% for *Salmonella* (Snow et al. [263]). The sampling method used in the survey is known not to be 100% sensitive, see Arnold et al. [13]. After analysing the data using Bayesian methods, Arnold et al. [13], indicates a prevalence of 18% (95% credibility interval (CI) 12–25%) of holdings infected with *Salmonella*, which is much higher than the prevalence rate reported in the survey.

The prevalence of infected birds varies between farms, possibly related to biosecurity and hygiene practices within the farm, and also dependent on factors such as farm size, and this will affect the sensitivity of sampling methods for *Salmonella* (Arnold et al. [14]). This difference in the sensitivity of sampling methods to detect *Salmonella* will be translated into heterogeneity among the farms.

The EU baseline survey method was applied to 21 suspected infected farms in a subsequent study, which provided the available validation sample as shown in Table 5.4. In fact, other methods were applied in parallel to these 21 farms: 2 sets of methods that involved sampling faeces and dust, and also the testing of ova and caeca from 300 birds. The additional sampling showed that there were 3 flocks which tested positive but which were negative for the EU baseline survey method. A detailed study of the results and power of detection of each method used in the study is discussed and can be found in Arnold et al. [12].

TABLE 5.4
Validation sample of *Salmonella* data

y	0	1	2	3	4	5	6	7
g_y	3	1	3	2	3	3	4	2

Again, it is important to highlight here that we know $g_0 = 3$. This means in this case that the test failed in only 3 of the 21 farms where *Salmonella* infection was detected, which allows us to deduce that the sensitivity of the test applied in the survey was about 87.5%.

5.3   Ratio plot and ratio regression

We are interested in estimating the size N of an elusive target population. Note that $N = n + f_0$ where f_0 is the frequency of units that were not captured any time causing a reduction in the observable available sample with size $n = \sum_{i=1}^{m} f_i$. To find an estimate for N, we can use, for example, the Horvitz–Thompson estimator $\hat{N} = \frac{n}{1-p_0}$, hence we need p_0. On the other hand, to find an estimate for p_0, we need to find a model $p_x = p_x(\theta)$, thus we need to find an estimate $\hat{\theta}$ for θ so that $\hat{p}_x = p_x(\hat{\theta})$. In particular, $\hat{p}_0 = p_0(\hat{\theta})$.

Since we are dealing with a fixed number of sampling occasions, $m = 7$, a binomial distribution to model the data seems to be a natural starting point to be considered. In addition, we are working with a situation of success/failure of a test to detect *Salmonella* infection, consequently, the binomial distribution seems to be the most appropriate to apply in our case study. Let us consider then the binomial probability distribution:

$$p_x(\theta) = P(X = x) = \binom{m}{x} \theta^x (1 - \theta)^{m-x} \tag{5.3}$$

for $x = 0, 1, ..., m$. Here, θ represents the probability a test is positive for each holding.

We have then to derive an estimate $\hat{\theta}$ for θ and use $\hat{\theta}$ in $p_0(\hat{\theta}) = (1 - \hat{\theta})^m$ to estimate N, where p_0 is the probability of a zero count distribution. An estimate for θ is usually obtained fitting a zero-truncated distribution to the available data usually through the Expectation-Maximization algorithm.

However, as we are working with a simple homogeneous model, the fit may not be adequate to provide a good estimate of the distribution due to a lack of flexibility. Also, the benefit of having a validation sample available is neglected, for example to check if the model is correct for the unobserved part of the population. Thus, the variability associated with the fact that all farms may differ, for example in biosecurity issues, translated into unobserved heterogeneity in the data, may play an important aspect to be covered. Ignoring heterogeneity can lead us to underestimate the true population size.

Eventually, the question arises, why not use $\frac{g_0}{n_1}$ where n_1 is the total size of the validation sample as an estimate for $\frac{f_0}{n}$ from the equation $\frac{g_0}{n_1} = \frac{f_0}{f_0+n}$ from which the solution $\hat{f}_0 = n\frac{g_0}{n_1-g_0}$ can be found. We would obtain a result of 9 unreported farms in the survey. This non-parametric estimate is possible. However, it only uses g_0 (and n_1) but neither the full distribution of the validation sample nor the positive distribution of the positive sample. So, it will suffer under instability and lack of efficiency. The binomial model uses both entire distributions, but it was found not flexible enough to provide a good fit. Therefore, it is necessary to have a diagnostic device to test if the binomial model is suitable. Ultimately, we proceed using a methodology which allows heterogeneity.

The main idea of the following approach is to consider ratios of the observed frequencies to estimate ratios of neighbouring count probabilities. This theory was also developed and explored in Chapter 2 (Sections 2.4 and 2.5), Chapter 3 (Section 3.3), and Chapter 4 (Section 4.4). To illustrate this idea, still working with the binomial distribution, let us consider the ratios as follows:

$$\frac{p_{x+1}}{p_x} = \frac{\binom{m}{x+1}\theta^{x+1}(1-\theta)^{m-x-1}}{\binom{m}{x}\theta^x(1-\theta)^{m-x}} = \frac{m-x}{x+1}\frac{\theta}{1-\theta} \tag{5.4}$$

Using the non-negative coefficients $a_x = \frac{x+1}{m-x}$, we can reparametrise these ratios multiplying the neighbouring probability ratios by the inverse of their coefficients as follows:

$$R_x = \underbrace{\frac{x+1}{m-x}}_{a_x}\frac{p_{x+1}}{p_x} = a_x\frac{p_{x+1}}{p_x} = \frac{\theta}{1-\theta}. \tag{5.5}$$

The result is a constant, the odds for the event, regardless of x. Note that R_x does not change, regardless of whether we consider the truncated or the untruncated distributions since it just depends on the parameter θ. In addition, we emphasize that the coefficients a_x directly depend on the chosen base distribution. In this situation, the base is represented by the homogeneous binomial distribution that we get when there is no unobserved heterogeneity.

Naturally, in the case of the binomial distribution, the ratio R_x is constant over x. Since the quantity p_x is unknown, a non-parametric estimation f_x/N of R_x is given by:

$$r_x = a_x\frac{f_{x+1}/N}{f_x/N} = a_x\frac{f_{x+1}}{f_x} \tag{5.6}$$

where f_x is the observed frequency of counts exactly equal to x.

Figure 5.1 (left panel) is called the ratio plot and it works as a diagnostic device for the binomial distribution (Böhning et al. [48]) and its construction depends directly on the coefficients a_x:

$$x \to r_x = a_x\frac{f_{x+1}}{f_x}. \tag{5.7}$$

Note that the coefficients a_x have a large influence in the interpretation of the observed pattern in the ratio plot which will change according to the base distribution we are working with. Under the binomial, we would expect the plot to show a horizontal line pattern.

Let us now consider the ratio plot for the positive sample together with the validation sample in the *Salmonella* data:

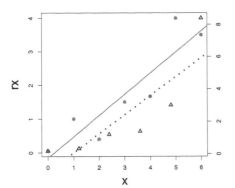 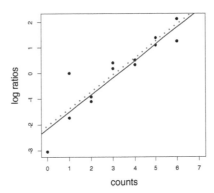

FIGURE 5.1: Left panel: ratio plot for the validation sample (solid points) and for the positive sample (empty triangles) with respective regression lines, continuous for the validation and dashed for the positive sample. Right panel: regression lines of the log ratio on x, continuous for the positive sample, dashed for the validation sample; the estimated regression model is $-2.04 + 0.66x - 0.12S$.

The graphs in Figure 5.1 (left panel) show no evidence of a horizontal line pattern, whether we consider the validation or the positive sample. Instead, it shows substantial departures from the standard binomial distribution as we can see by the monotone increasing trend. This violation of the binomial assumption might be seen as supporting evidence for unobserved population heterogeneity translated in the figure by the non-zero slope. It could be that different farms have different risks for a positive test result, for example, related to biosecurity issues of farm factors as mentioned above.

A closer analysis of the ratio plot shows that there is something in common between the positive and validation sample distributions to be explored, which we can use to improve the inference of f_0. In fact, the regression lines are almost parallel which shows evidence that both samples follow different distributions but having similar shapes. The fit in the case of a standard homogeneous binomial distribution does not seem to be acceptable to the observed, zero-truncated distribution. A chi-square goodness-of-fit test confirms that we can reject the null hypothesis that the data follows a standard binomial distribution. The statistics used was $\chi^2 = \sum_{x=1}^{m-1}(\log \hat{r}_x - \log \hat{r})^2/\widehat{var}(\log \hat{r}_x)$, where $\widehat{var}(\log \hat{r}_x) = \frac{1}{f_{x+1}} + \frac{1}{f_x}$ and for estimating the parameter θ we used the estimate $\hat{r} = \sum_{x=1}^{m-1} a_x \frac{f_{x+1}}{f_x}$. We found the value $\chi^2 = 49.80$ for the positive sample and $\chi^2 = 16.33$ for the validation sample with 6 degrees of freedom. We can definitively reject that the data are consistent with a binomial distribution with a significance level of 0.05.

The ratio plot suggests a regression model taking advantage of the straight line pattern to determine an estimate of f_0. Namely, as $\log(r_x) = \alpha + \beta x + \epsilon_x$, an estimate of f_0 can be found using $\log\left(a_0 \frac{f_1}{f_0}\right) = \hat{\alpha} + \hat{\beta} \times 0$, or, $\hat{f}_0 = a_0 f_1 \exp(-\hat{\alpha})$.

Let us consider a model that allows departures from the homogeneous binomial model by means of allowing population heterogeneity in the form of a distribution on the binomial parameter θ. The marginal distribution of that model is then given by:

$$p_x = \int_0^1 \binom{m}{x} \theta^x (1-\theta)^{m-x} h(\theta) d\theta \qquad (5.8)$$

where $h(\theta)$ is a mixing distribution that controls departures from the homogeneous binomial model. Notice that if $h(\theta)$ is a 1-point distribution putting all the mass at θ, it leads to a binomial distribution with parameter θ.

It is stated in Böhning [47] under general conditions that $R_x = a_x \frac{p_{x+1}}{p_x}$ is monotone if p_x has a mixture model of the type of (5.8). This leads naturally to considering a model with response r_x.

Let us assume that R_x can be linked to a known set of predictor functions $z_0(x), ..., z_p(x)$, so that the following model is defined:

$$g(R_x) = \beta' \mathbf{z(x)} \qquad (5.9)$$

where $x = 0, ..., m-1$ and g is a monotone link function. This link function is essentially used to guarantee that the ratios remain positive, i.e., $r_x > 0, x = 0, ..., m$. If we fit a simple straight line to the ratios r_x, it can lead us to a non-feasible estimate of the unobserved counts since we can get a negative intercept estimate as we can observe in the Figure 5.1. The choice of an appropriate link function avoids this problem. It is also shown in Böhning [47] that any regression model with the form (5.9) corresponds to a proper count distribution.

We are going to use the logarithmic function as the link function, so that our model is given by $log(R_x) = \beta_0 + \beta_1 x$ with $z_0(x) = 1$ and $z_1(x) = x$ and so the ratios are obtained applying the inverse of the link function on both sides of the model equation: $R_x = \exp(\beta_0 + \beta_1 x)$.

The estimation of the parameters β may be based on the likelihood function:

$$L(\beta) = \prod_{x=1}^{m} \left(\frac{p_x}{1 - p_0} \right)^{f_x} \qquad (5.10)$$

where p_x is a function of $R_x = g^{-1}(\beta' z(x))$. However, we follow a simpler approach to find the estimates of β.

In detail, the scheme of this approach towards a proper regression model is firstly to generate the ratio plot by plotting x against the estimates of R_x, $r_x = a_x \frac{f_{x+1}}{f_x}$ and analyse the graph carefully. After an appropriate analysis of the ratio plot and choice the link function g, we may fit the model:

$$g(r_x) = \beta' z(x) + \varepsilon_x \qquad (5.11)$$

where ε_x is such that $E(\varepsilon_x) = 0$ and $cov(\varepsilon_x) = \Sigma$ and $\beta = (\beta_0, ..., \beta_p)'$ represents a $(p+1)$-dimensional vector of unknown fixed parameters, associated to the vector of regression functions $z(x) = (z_0(x), ..., z_p(x))'$. Now we can fit the model (5.11) by regression model techniques.

The primary aim is then to estimate the coefficients β of the regression model. Consequently, the first concern is to estimate Σ, see Rocchetti et al. [247], using the following tridiagonal matrix constructed as follows:

$$
\begin{pmatrix}
\frac{1}{f_1} + \frac{1}{f_2} & \frac{-1}{f_2} & 0 & \cdots & 0 & \cdots & 0 \\
\frac{-1}{f_2} & \frac{1}{f_2} + \frac{1}{f_3} & \frac{-1}{f_3} & 0 & \cdots & \cdots & 0 \\
0 & \ddots & \ddots & \cdots & \cdots & \cdots & \cdots \\
\vdots & & & \ddots & & & \\
0 & 0\cdots & \frac{-1}{f_i} & \frac{1}{f_i} + \frac{1}{f_{i+1}} & \frac{-1}{f_{i+1}} & 0\cdots & 0 \\
\vdots & & & & \ddots & & \\
0 & \cdots & & & 0 & \frac{-1}{f_{m-1}} & \frac{1}{f_{m-1}} + \frac{1}{f_m}
\end{pmatrix}.
\tag{5.12}
$$

It has been indicated in Rocchetti et al. [247] that it is possible to drop the off-diagonal terms of the matrix with a little loss of statistical precision for our purposes; for details see Rocchetti et al. [247]. Thus, we will get an estimate $\hat{\Sigma}$ of Σ determined by just the diagonal elements of the above matrix. Now, $\hat{\Sigma}$ is a diagonal matrix that contains the estimated inverse variances of $Y_1, ..., Y_{m-1}$ given by $\omega_i = (\frac{1}{f_i} + \frac{1}{f_{i+1}})^{-1}$. The generalized weighted least-squares estimate of β is known to be:

$$
\hat{\beta} = (X'\hat{\Sigma}^{-1}X)^{-1}X'\hat{\Sigma}^{-1}Y
\tag{5.13}
$$

where Y has elements $g(\hat{r}_x)$ and X has rows $z_0(x), ..., z_p(x)$, $x = 1, ..., m-1$ since no observation is available for $x = 0$. Note that the estimated covariance matrix of $\hat{\beta}$ is immediately available as $cov(\hat{\beta}) = (X'\hat{\Sigma}^{-1}X)^{-1}$.

In our case, since the link function is the logarithmic function, we have:

$$
Y = \begin{pmatrix} log(\hat{r}_1) \\ \cdots \\ log(\hat{r}_{m-1}) \end{pmatrix} \text{ and } X = \begin{pmatrix} 1 & 1 \\ 1 & 2 \\ \cdots & \cdots \\ 1 & m-1 \end{pmatrix}.
\tag{5.14}
$$

A regression-based estimator in this regression model can be derived for the zero-count frequencies as follows:

$$
g(\hat{r}_0) = \hat{\beta}'z(0) \implies \hat{r}_0 = g^{-1}(\hat{\beta}'z(0)).
\tag{5.15}
$$

Using the recurrence relation $r_x = a_x f_{x+1}/f_x$, we can project it onto $x = 0$ to obtain an estimate of f_0:

$$
\hat{f}_0 = a_0 f_1/\hat{r}_0 = a_0 f_1/g^{-1}(\hat{\beta}'z(0)).
\tag{5.16}
$$

The population size is then the sum of the estimated number of unrecorded individuals and the size of the observed sample:

$$
\hat{N}_{reg} = n + \hat{f}_0.
\tag{5.17}
$$

We see that the estimate for f_0 in the regression model depends directly on f_1 (see (5.16)). In case f_1 suffers from one-inflation, it might be better to base the estimate of f_0 on the entire distribution. Hence, the f_0 using the Horvitz–Thompson estimator could be more appropriate. The Horvitz–Thompson estimator can be calculated as follows. As $N = Np_0 + N(1 - p_0)$, we can get an estimate of N using the moment estimate n for $N(1 - p_0)$. Solving $\hat{N} = \hat{N}p_0 + n$ for $\hat{N}$, the estimate $\frac{n}{1-p_0}$ for N is obtained. Using this estimate and the equality $\hat{N} = n + \hat{f}_0$, we achieve an estimator for f_0 which is given by $\hat{f}_0^{HT} = n\frac{p_0}{1-p_0}$. An estimate for p_0 can be obtained as follows. We are able to estimate the probability mass at 0 using the fitted values $\hat{r}_x = g^{-1}(\hat{\beta}'z(x))$, for R_x, $x = 0, ..., m - 1$, according to the following result from Böhning [47] (see also Chapter 1 Theorem 2.1):

Theorem 5.1 *Let $R_x > 0$ be given for $x = 0, ..., m - 1$, and let a_x, $x = 0, ..., m - 1$, be known positive coefficients. Then, there exists a unique probability distribution $p_0, ..., p_m > 0$ such that:*

$$p_{x+1} = R_x \frac{p_x}{a_x}, \quad \forall x = 0, ..., m - 1. \tag{5.18}$$

Furthermore, we have that:

$$p_0 = \left[1 + R_0/a_0 + (R_0/a_0)(R_1/a_1) + ... + \prod_{x=0}^{m-1} R_x/a_x \right]^{-1}. \tag{5.19}$$

We apply this result now using estimates $\hat{r}_x$ for R_x. This result proves that any valid regression model leads to a proper probability distribution. Notice that the probability density function only depends on the model. This characteristic allows flexible regression modelling.

Using conditioning moment techniques, it is possible to estimate the variance of $\hat{f}_0$ from the variance estimators for the estimated regression coefficients. Böhning [47] demonstrated this for the binomial case:

$$Var(\hat{f}_0) = \frac{1}{m^2} f_1 \exp(-\hat{\beta}_0)^2 (f_1 Var(\hat{\beta}_0) + 1 - f_1/(n + \hat{f}_0)). \tag{5.20}$$

An estimate for $Var(\hat{\beta}_0)$ is available from the result for $cov(\hat{\beta})$ discussed above. Thus, we provide the asymptotic 95% prediction interval for f_0 which is given by

$$\left(\hat{f}_0 - 1.96\sqrt{Var(\hat{f}_0)}, \hat{f}_0 + 1.96\sqrt{Var(\hat{f}_0)} \right). \tag{5.21}$$

Hence, a follow-up prediction interval for N also follows as

$$\left(n + \hat{f}_0 - 1.96\sqrt{Var(\hat{f}_0)}, n + \hat{f}_0 + 1.96\sqrt{Var(\hat{f}_0)} \right). \tag{5.22}$$

Until here, the presented approach covers just the analysis of the positive sample. An interesting extension is to incorporate the validation sample into the modelling. Let us introduce some methods allowing that to be done.

5.4 Ratio regression using validation information

The ratio regression approach can be extended to incorporate the information coming from the validation sample into the ratio regression model. Considering our data, this can be done as follows:

$$log(r_x) = \alpha + \beta x + \delta S + \epsilon_x \tag{5.23}$$

where S represents a dummy variable taking the value of 1 if x is from the positive sample and 0 otherwise. With this approach we allow a regression line for the two samples having the same slope but different intercepts as Figure 5.1 (right panel) shows. The resulting estimate $\hat{f}_0 = f_1 \exp(-\hat{\alpha} - \hat{\delta})$ is 25 undetected farms. Here f_1 is the frequency of ones from the positive sample. Note that if $\delta = 0$, both lines become identical and we allow for a single straight line regression model as Figure 5.2 shows.

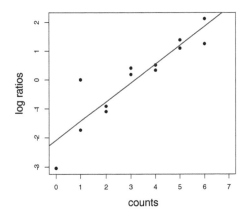

FIGURE 5.2: Single straight line regression model: $-2.30 + 0.70x$.

The use of a validation sample increases the efficiency of our estimation as well as it guarantees that our model provides a reasonable final estimate, see Böhning [47]. We can also consider a model with interaction between the variable S and count x. Here, however, in the case of interaction, the model becomes identical to fitting two separate lines and the benefit of the validation sample diminishes, see Figure 5.3 for illustration.

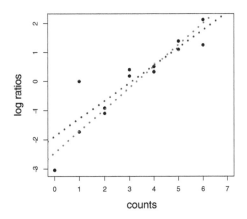

FIGURE 5.3: Separate lines regression model: $-1.85 + 0.60x - 0.63S + 0.15(S \times x)$.

A zero-inflated model was also considered as it appears we have a large number of zeros in addition to those predicted by the non-inflated models. We conducted simulations based on these models and the results show evidence that using the validation sample not only decreases the bias in our estimation, but also leads to more accuracy in the estimation of the population size.

A vast number of choices for regression models are possible once we consider a convenient link function to the ratios of frequencies.

5.4.1   Application to the case study

The three models (single line, parallel lines, separate lines) were applied to the *Salmonella* data and the results are presented in Table 5.5. Note that $n = 53$ for the positive sample and the coefficients a_x were set from the binomial distribution in our analysis.

TABLE 5.5: Estimates of the population size N

Application	$\hat{f}_0$	PI for f_0	$\hat{N}$	PI for N	p-value	AIC	BIC
RR Positive	29	(1.01,56.63)	82	(54.02,109.64)	0.000		
Model 1	24	(3.65,44.90)	77	(56.65,97.90)	0.000	20.53	22.22
Model 2	25	(1.49,48.35)	78	(54.49,101.35)	0.660	22.26	24.52
Model 3	29	(5.98,51.68)	82	(58.98,104.68)	0.316	22.73	25.55

Note: RR denotes the ratio regression approach, PI denotes the prediction interval for the estimate and S the variable indicating type of sample ($S = 1$: positive sample, $S = 0$ otherwise). The model equation for the ratio regression model using just the positive sample is $-2.47 + 0.75x$; for model 1 (single line) we have $-2.30 + 0.70x$; for model 2 (parallel lines) is $-2.21 + 0.70x - 0.12S$ and for model 3 (separate lines) is $-1.85 + 0.60x - 0.63S + 0.15(S \times x)$; column 6 refers to the p-value of the last coefficient of the respective model; column 7 and 8 indicate the AIC and BIC values respectively for each of the three represented models.

We obtained 29 undetected farms using only the positive sample. Model 3 provides exactly the same results as expected. The interaction term is not significant in model 3. The simple regression model (model 1) and the parallel lines model (model 2) produce a very similar result. Model 1 indicates 24 undetected farms while model 2 suggests 25 undetected farms. Table 5.5 includes the estimates for the coefficients of each model as well as prediction intervals for each estimate. As model 2 has a non-significant term for S, we conclude that model 1 is most suitable in our case and the estimate for f_0 is 24 with the shortest prediction interval. When comparing models to the same data, the smaller the AIC or BIC, the better the fit. In this case, AIC and BIC support that model 1 is most appropriate in our case study.

5.5   Simulation study

A question arises as to what is the benefit in using the validation sample in the modelling. A natural way to proceed is to investigate the performance of each model above in the presence and absence of a validation sample through a simulation study. We aim to simulate data with similar properties to our dataset. We generated 1000 sample replications for positive samples in which all the 0 units were considered as missing values and discarded. Another 1000 sample replications were generated for validation samples, each one to pair with each positive sample respectively. Note that all the samples will have a fixed number of 7 trapping occasions.

We will present here only the results for the simulation study based on the single line model. We set $\alpha = -2$ and $\beta = 0.6$ and construct the model $log(r_x) = \alpha + \beta x = -2 + 0.6x$. After that, we can easily find the ratios $r_x = \exp[\alpha + \beta x]$. Using (5.18) from Theorem 5.1, we find p_0 and using the relation $p_x = \frac{r_x}{a_x} p_{x-1}$ for $x = 1, ..., 7$ we find all the probabilities

$p_1, ..., p_7$. These probabilities determine the count distribution $P(X = x) = p_x$ for $x = 0, 1, ..., 7$.

TABLE 5.6
Mean and variance for a positive sample size of 50 and validation sample size of 25 from the simulation study designed based on the single line model

	$\hat{N}$ Positive	$\hat{N}$ HT	$\hat{N}$ SLM
Mean	51.37	51.35	50.42
Variance	74.30	85.04	42.22

Note: the first column represents the estimate of the population size using just the positive sample; the second column represents the same estimate by means of the Horvitz–Thompson estimator (also using only the positive sample) and the last column represents the estimate for the population size using the single line model.

TABLE 5.7
Mean and variance for a positive sample size of 100 and validation sample sizes of 25 and 50 from the simulation study designed based on the single line model

	$\hat{N}$ Positive	$\hat{N}$ HT	$\hat{N}$ SLM
Validation sample size: 25			
Mean	101.40	101.21	100.71
Variance	132.83	144.76	94.94
Validation sample size: 50			
Mean	101.40	101.15	100.41
Variance	130.64	135.43	79.25

Note: the first column represents the estimate of the population size using just the positive sample; the second column represents the same estimate by means of the Horvitz–Thompson estimator (also using only the positive sample) and the last column represents the estimate for the population size using the single line model.

The population size for the positive samples varied among 25, 50, 100, 500 and 1000, as well as for the validation samples. We calculated the population size N using only the positive sample as well as incorporating the validation sample. The estimation of N according to the Horvitz–Thompson estimator was also considered in the study for comparison.

The estimation for N using the simple regression model in the presence of a validation sample is always more accurate than using the ratio regression approach with only the positive sample. It can also be stated that the estimation given by the Horvitz–Thompson estimator is consistently closer to the true value than the estimation using only the positive sample. Also, it is shown that the variance using the model incorporating the validation information is smaller than the other two presented variances using models which just take profit from the positive sample. In fact, the major differences are in terms of efficiency. The gain in efficiency is clear when we work with a validation sample.

TABLE 5.8
Mean and variance for a positive sample size of 500 and validation sample sizes of 25, 50 and 100 from the simulation study designed based on the single line model

	$\hat{N}$ Positive	$\hat{N}$ HT	$\hat{N}$ SLM
Validation sample size: 25			
Mean	501.12	500.99	500.79
Variance	578.15	613.67	536.32
Validation sample size: 50			
Mean	501.30	501.14	501.19
Variance	609.73	642.24	534.70
Validation sample size: 100			
Mean	502.40	502.32	501.46
Variance	593.41	629.47	462.94

Note: the first column represents the estimate of the population size using just the positive sample; the second column represents the same estimate by means of the Horvitz–Thompson estimator (also using only the positive sample) and the last column represents the estimate for the population size using the single line model.

TABLE 5.9
Mean and variance for a positive sample size of 1000 and validation sample sizes of 25, 50, 100 and 1000 from the simulation study designed based on the single line model

	$\hat{N}$ Positive	$\hat{N}$ HT	$\hat{N}$ SLM
Validation sample size: 25			
Mean	1000.43	1000.29	1000.11
Variance	1198.88	1254.86	1146.99
Validation sample size: 50			
Mean	1001.35	1001.10	1001.10
Variance	1105.18	1163.92	1021.33
Validation sample size: 100			
Mean	1002.38	1002.17	1001.66
Variance	1247.03	1326.19	1087.35
Validation sample size: 500			
Mean	1000.79	1000.64	1000.15
Variance	1045.35	1093.94	709.89

Note: the first column represents the estimate of the population size using just the positive sample; the second column represents the same estimate by means of the Horvitz–Thompson estimator (also using only the positive sample) and the last column represents the estimate for the population size using the single line model.

5.6 The inflated model

The previous modelling does not allow for any zero-inflation. Zero-inflation would lead to a first ratio being potentially much lower than the others. To account for zero-inflation, at least in an approximate way, we suggest the model $log(R_x) = \alpha + \beta x + \delta S + \lambda x^2$ estimated as $log(r_x) = -2.47 + 0.94x - 0.13S - 0.04x^2$. This model will allow a bend in the upper straight line corresponding to the positive sample and at the same time taking advantage of the validation sample. A total of 33 undetected farms were obtained employing this model as Table 5.10 shows. In other words, a population size of 86 farms.

The question arises as to whether this kind of approach performs well on our data. As it turns out, the quadratic term is not significant. In fact, the best model for the *Salmonella* data is the single line model. AIC and BIC criteria support that statement, since the values are bigger for this model than for the other three discussed models of Table 5.5.

We conducted simulations that show that the estimation of N using the inflated model, with the validation sample incorporated, produces substantially better results in terms of precision along with an enormous reduction in the bias.

TABLE 5.10: Estimate of the population size N for the zero-inflated model according to the model equation $log(r_x) = -2.47 + 0.94x - 0.13S - 0.04x^2$

Application	f_0	PI for f_0	$\hat{N}$	PI for N	p-value	AIC	BIC
Inflated model	33	(-8.36,73.63)	86	(44.64,126.63)	0.437	23.34	26.16

Note: column 6 refers to the p-value of the last coefficient of the model; PI denotes the prediction interval for the estimate and S denotes the variable S; column 7 and 8 indicate the AIC and BIC values, respectively, for the suggested model.

5.6.1 Simulation study on zero-inflated data

There is no indication that our data suffers zero-inflation, but it could actually happen and we would not know if we did not have a validation sample. We performed a simulation study of a binomial with 50% zero-inflated data and obtained the estimates for f_0 using different models after the analysis of the ratio plot. The simulation work covered the following situations:

- Case 1: Positive sample size: 100 (50 zeros); Validation sample size: 100 (50 zeros).

- Case 2: Positive sample size: 500 (250 zeros); Validation sample size: 500 (250 zeros).

- Case 3: Positive sample size: 1000 (500 zeros); Validation sample size: 1000 (500 zeros).

- Case 4: Positive sample size: 2000 (1000 zeros); Validation sample size: 2000 (1000 zeros).

In Figure 5.4, ratio plots of the simulated frequencies, averaged over the 1000 replications, are shown. Clearly, the effect of the zero-inflation becomes visible.

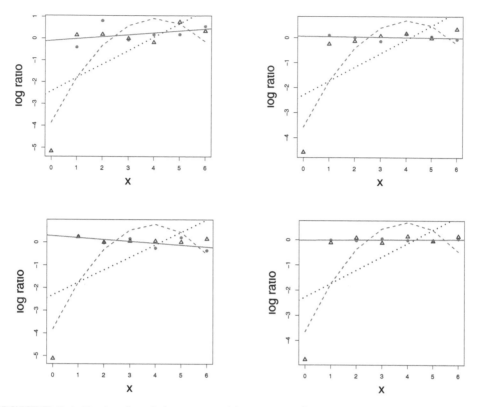

FIGURE 5.4: Ratio plot of the averaged frequencies (case 1, top left panel; case 2, top right panel; case 3, bottom left panel; and case 4, bottom right panel) for the positive samples (solid points) and for the validation samples (empty triangles) with respective regression lines, continuous for the positive samples and dotted for the validation samples. The dashed curve represents the fitted values based on each regression model for the validation sample.

As we observe from Table 5.11, the zero-inflated model is always much closer to the true value in all the analysed situations and it reached the true value in two of the simulation cases (case 2 and 4). The results using just the positive sample are too low to be considered useful. Despite not including it in the table, the Horvitz–Thompson estimate was also calculated and the same values for f_0 were achieved. This can be expected, since we are working with only the positive sample. The single line model and the single quadratic model do not appear to perform well with zero-inflated data.

TABLE 5.11: Estimate of f_0 from the simulation study of a binomial with 50% zero-inflated data for each of the mentioned cases

	Positive	SLM	SQM	Zero-inflation model
$\hat{f}_0$ - case 1	0	1	6	50 (50)
$\hat{f}_0$ - case 2	2	11	50	181 (250)
$\hat{f}_0$ - case 3	3	13	80	500 (500)
$\hat{f}_0$ - case 4	8	40	205	933 (1000)

Note: the first column represents the estimate for f_0 using just the positive sample, the second column using a single line model, the third column uses a single quadratic model (SQM) and the last column uses a zero-inflation model with the true value for f_0 between brackets.

5.7 Discussion and conclusions

The ratio regression approach was discussed and it could be seen how the ratio regression approach for the positive sample could be extended to include information from the validation sample, the untruncated sample including zero counts which are not observed in conventional capture-recapture settings. Including validation samples will reduce bias and increase efficiency. Simulation studies corroborated the role of the validation sample in the estimation process showing that we can rely on the estimate for the population size with more confidence. The identical model might be used for the positive and validation samples, or a partly congruent model such as the parallel lines model, or two separate models such as the separate lines model. In the latter case, there is no gain in efficiency. A zero-inflated model was also considered allowing the first ratio to be particularly lower than the other ratios.

The data used to illustrate the theory of this work was provided by the Animal and Plant Health Agency and it is related with an important public health concern: *Salmonella* infection in poultry. The objective was to adjust the undercount of disease occurrence in UK farms during the period of the EU baseline survey which took place between October 2004 and September 2005. This work focuses essentially on the development of methodology to include validation information in the capture-recapture modelling in order to increase the accuracy and efficiency of the final estimate for the unrecorded cases.

Using the ratio regression approach there are numerous ways of selecting an appropriate model. We have focused here on the Wald-statistic for selecting significant coefficients and model selection criteria were also used, such as AIC and BIC. Another way would be the likelihood ratio statistic. In the case of the *Salmonella* data, on the basis of these criteria, the single line model considering only the counts variable seem to be the most appropriate to explore.

In fact, the number of undetected farms may be much superior to the results we obtained using the various methods discussed in this work. However, a positive detection probability is assumed by the ratio regression approach. If this does not occur, a lower bound for the estimation of unreported farms was determined, which it is necessary to discuss with the responsible authorities for this public health concern.

The EU survey reported a prevalence of *Salmonella* of 11.7% (53 infected farms out of 454 holdings), however, Arnold et al. [13] indicated a prevalence of 18% after analyzing the positive data using Bayesian methods. The results of this work help to confirm that the prevalence was in fact higher than 11.7%. According to the results of the most significant model (single line model), obtained by a ratio regression approach incorporating the validation sample, we report a prevalence of 17% (95% prediction interval (PI) 44.64–126.63).

We see the most important aspect of the use of validation information in the fact that more trust can be developed in the model for the unobserved part.

Part III

Meta-Analysis in Capture-Recapture

6

On meta-analysis in capture-recapture

John Bunge

Cornell University

CONTENTS

6.1   Introduction and background

In this brief note we consider the use of meta-analysis to combine or synthesize the results of several capture-recapture (or equivalently, species richness estimation) studies. Specifically, suppose there are M populations with sizes $N_1, \ldots, N_M$ (conceivably $N_1 = \cdots = N_M = N$), and possibly associated (vector) covariate information $x_1, \ldots, x_M$, also known as "metadata." M studies are performed, producing statistics $\{(\hat{N}_1, \hat{\sigma}_1), \ldots, (\hat{N}_M, \hat{\sigma}_M)\}$, where $\hat{N}_i$ and $\hat{\sigma}_i$ are the estimated population size and associated standard error (respectively) from the ith study. We are interested in using meta-analytic models and procedures to examine the effect of x_i on N_i, or to test various hypotheses about the N_i, or more generally to analyze the parameters of the "super-population" stochastic process that generated the N_i. Here we discuss the existing literature on this problem; we present a meta-analysis of the grizzly bear data from Chapter 1, using a recently proposed method; and we discuss some directions for future research on the topic.

We carried out a comprehensive literature search on this topic in early 2016, which yielded about 40 potential candidate articles, but only one addressed the aforementioned problem directly. There were of course many other interesting papers, and we mention three of these. Koricheva and Gurevitch [168] reviewed and classified meta-analysis studies in plant ecology: in their classification our problem would fall under "combining results of multisite or multiyear experiments," although they do not give any examples where population size is the target parameter. Boulanger et al. [51] carried out a meta-analysis of $M = 7$ grizzly bear capture-recapture studies in British Columbia, but they did not statistically combine the population size estimates $\hat{N}_1, \ldots, \hat{N}_7$. Finally, as an aside we note that Rücker et al. [251] used capture-recapture to estimate the comprehensiveness of literature searches underpinning a meta-analysis, that is, the number of relevant but missing references. While certainly a worthy undertaking, this is a topic for another day.

Our problem as described above is dealt with in Willis et al. [298], and this seems to be its only specific treatment to date. The model proposed there (based partly on the version in Chapter 5 of Demidenko [99]), is as follows. We first suppose that the ith population size is a linear function of k covariates plus a random effect:

$$N_i = \beta_0 + \beta_1 x_{i,1} + \ldots + \beta_k x_{i,k} + U_i,$$

$i = 1, \ldots, M$, where $x_{i,j}$ is the jth covariate measurement for the ith population, β_j is its coefficient, and U_i is a random effect. In matrix terms

$$\vec{N} = \mathbf{X}\vec{\beta} + \vec{U},$$

where $\vec{N} = [N_1, \ldots, N_M]^T$,

$$\mathbf{X} = \begin{bmatrix} 1 & x_{1,1} & x_{1,2} & \cdots & x_{1,k} \\ 1 & x_{2,1} & x_{2,2} & \cdots & x_{2,k} \\ \vdots & \vdots & \vdots & \ddots & \vdots \\ 1 & x_{M,1} & x_{M,2} & \cdots & x_{M,k} \end{bmatrix}$$

$\vec{\beta} = [\beta_0, \beta_1, \ldots, \beta_k]^T$, and $\vec{U} = [U_1, \ldots, U_M]^T$. For now we make the usual assumption that $\vec{U} \sim N(\vec{0}, \sigma_U^2 I_M)$.

Next we suppose that, given $\vec{N}$, we have independent estimates of N_i that are asymptotically normal, with estimable variances σ_i^2. That is,

$$\hat{N}_i | N_i = N_i + \epsilon_i,$$

where $\{\epsilon_1, \ldots, \epsilon_M\}$ are independent Gaussian random errors, independent of $\vec{U}$, with (respective) variances $\{\sigma_1^2, \ldots, \sigma_M^2\}$. Unconditionally, then, we have the final model

$$\hat{N}_i = \beta_0 + \beta_1 x_{i,1} + \ldots + \beta_k x_{i,k} + U_i + \epsilon_i,$$

or in matrix terms

$$\hat{N} = \mathbf{X}\vec{\beta} + \vec{U} + \vec{\epsilon}, \tag{6.1}$$

where $\vec{\hat{N}} = [\hat{N}_1, \ldots, \hat{N}_M]^T$,

$$\vec{\epsilon} = \begin{bmatrix} \epsilon_1 \\ \vdots \\ \epsilon_M \end{bmatrix} \sim N\left(\vec{0}, \begin{bmatrix} \sigma_1^2 & \cdots & 0 \\ \vdots & \ddots & \vdots \\ 0 & \cdots & \sigma_M^2 \end{bmatrix} \right)$$

and $\vec{U}$ and $\vec{\epsilon}$ are independent. We substitute the sample-wise variance estimate $\hat{\sigma}_i^2$ for σ_i^2. The parameters of interest are then $\vec{\beta}$ and σ_U^2: $\vec{\beta}$ represents the (linear) effect of $\mathbf{X}$ on $\vec{N}$, and σ_U^2 represents the amount of variation in the N_i that is not attributable to $\mathbf{X}$. In particular, some hypotheses of interest include:

$$H_0 : \beta_1 = \ldots = \beta_k = \sigma_U^2 = 0, \text{ i.e., } N_1 = \cdots = N_M (= \beta_0);$$
$$H_0 : \beta_1 = \ldots = \beta_k = 0, \text{ i.e. no (linear) effect of } x_i \text{ on } N_i;$$
$$H_0 : \sigma_U^2 = 0, \text{ i.e., all variation in } N_i \text{ is attributable to } x_i.$$

A full statistical analysis of model (6.1), along with a software (R) package called betta, is presented in Willis et al. [298]. That paper addresses (among other things) hypothesis tests, parameter estimates, and goodness-of-fit considerations. Our purpose here is not to reproduce that paper, but to apply its methods to the grizzly bear data from Chapter 1, which is an example of a multiyear study, and to consider some future directions for research in this area.

TABLE 6.1: Observed (n_i) and estimated ($\hat{N}_i$) population sizes, with standard errors ($\hat{\sigma}_i$), for grizzly bear data, by year.

Year i	n_i	$\hat{N}_i$	$\hat{\sigma}_i$
1986	24	33.9	4.5
1987	12	17.8	4.3
1988	17	20.6	2.6
1989	13	25.1	6.7
1990	22	25.9	2.6
1991	24	39.2	6.4
1992	22	70.4	22.3
1993	17	22.4	3.8
1994	18	26.1	5.2
1995	17	53.1	18.8
1996	28	43.4	7.4
1997	29	52.4	8.7
1998	33	58.9	9.1
1999	29	41.9	5.2
2000	32	57.6	9.1
2001	38	75.0	12.0

6.2 Analysis of grizzly bear data

We refer to dataset 1.2.9, Table 1.6, in Chapter 1. This gives the results of a longitudinal study of grizzly bears in Yellowstone Park. For each of $M = 16$ years, from 1986 through 2001, frequency count data is given for the sightings (recaptures) of female grizzly bears. We first calculate sample-wise population size estimates and standard errors, one for each year. We obtained these using CatchAll [60], which implements a mixed Poisson model for the frequency counts, where the mixing distribution is a finite mixture of exponentials. The estimate $\hat{N}$ and its standard error are obtained by maximum likelihood. Table 6.1 summarizes the results. Note: For 1993 and 1994 the data was too sparse for CatchAll to fit any parametric model, so we reverted to Chao's (nonparametric) estimate ACE1 [74] for those years.

Our analysis of this example is intended to illustrate the method of Willis *et al.* [298] in a simple case. We note that the assumption of independence between years is not plausible, and we return to this below. Furthermore, although the $\hat{N}_i$ are known to be asymptotically normal, the sample sizes here are small and small-sample-size normality is not well supported. On the other hand, experience and simulations have shown that the $\hat{\sigma}_i$ represent the standard errors of the $\hat{N}_i$ reasonably well. In this example our only available covariate is time, and we have no substantive theory regarding its effect on N: certainly there does not appear to be a linear trend, for example. We therefore adopt an intercept-only model in this case, i.e., $\vec{\beta} = \beta_0$. Thus our (row-wise) model is

$$\hat{N}_i = \beta_0 + U_i + \epsilon_i,$$

$i = 1, \ldots, M = 16$, with the other assumptions as previously stated.

Running the data in Table 6.1 through the function `betta` in the package `breakaway` [298], we obtain the following results. First, $\hat{\beta}_0 = 37.9$ with SE($\hat{\beta}_0$) = 4.1, and we have $\hat{\sigma}_U^2 = 205.2$ so that $\hat{\sigma}_U = 14.3$, all of which seems reasonable. For $H_0 : \sigma_U^2 = 0$ we have $p < 10^{-4}$ so there is definitely heterogeneity among the N_i. In other words, according to this model, the true number of bears in year i, N_i, varies independently and normally around 37.9 with standard deviation 14.3.

A secondary level of analysis involved *post hoc* estimates — actually predictions — of the random effects U_i. These predictions $\hat{U}_i$ are the Best Linear Unbiased Predictors or BLUPs, which are well-studied in standard mixed-models analysis (Littell *et al.* [182]). That theory also addresses the phenomenon of *shrinkage*, which accounts in particular for the seemingly rather low values of $\hat{\beta}_0$ and σ_U^2 here. We do not discuss BLUPs for this example because they are not yet implemented in the software `betta`, although that is planned for the future.

6.3   Comments and future directions

It is interesting to note that in this setting the *file drawer problem*, which is a nontrivial issue in meta-analysis of published work on a given topic, does not play a major role. This problem arises due to the typical non-publication of non-significant results — they remain in the "file drawer" — which causes the effect sizes observed in the published literature to be biased in favor of statistical significance (Duval and Tweedie [107]). But in our case we suppose that the data $\{(N_i, \hat{\sigma}_i)\}$ arises from a well-defined or controlled collection of experiments or studies, and all the results are readily accessible. We propose three main categories of research in meta-analysis for capture-recapture.

- First, it is desirable to expand and deepen the basic Gaussian mixture model described above. In particular, the question of dependence among observations needs to be addressed. This can be handled in model (6.1) by allowing the covariance matrix of $\vec{U}$ or $\vec{\epsilon}$ (or both) to be non-diagonal. ($\vec{U}$ and $\vec{\epsilon}$ are always assumed independent of one another.) Theory for this exists (Demidenko [99], Littell *et al.* [182]), and it is mainly a matter of computational implementation.

- Second, statistical but non-inferential issues need attention, specifically goodness-of-fit. How well does the above model describe the data? This is a challenging question in general because in a sense the model is imposed by assumption, especially in regard to the normality of the random, latent effects. Some graphical and heuristic methods are given in Willis *et al.* [298] but in general the topic needs clarification. This has only recently begun to be discussed (Chen *et al.* [81]).

- Third, arguably the most important research direction involves expanding the class of "error" distributions for $\vec{U}$ and $\vec{\epsilon}$. It is common to see enormous variation in both the point estimates $\hat{N}_i$ and their standard errors $\hat{\sigma}_i$, even within the context of multiple samples from a single well-planned study. (In such a case, the estimates with *small* standard errors constitute influential points, causing the other estimates to shrink toward them.) The implication is that $\vec{U}$ or $\vec{\epsilon}$ (or both) should sometimes be modeled not as Gaussian random vectors, but with some other distribution that admits greater tail probabilities. Examples (still symmetric distributions) include the double exponential, or stable laws that may not admit some low-order moments. Again research exists in this area for general mixed models (Demidenko [99]) but implementation for meta-analysis in capture-recapture may not be straightforward, and remains to be done.

Acknowledgments

The author thanks Sarah Kimball, Ziyan Liu, Yusi Shao, and Yichi Zhang, who at the time of writing were students in Cornell University's Master of Professional Studies in Applied Statistics program. They carried out a very thorough literature search, and analyzed the grizzly bear data. He thanks his co-author Amy Willis, who worked out the statistical details and computational implementation of the meta-analysis method implemented in the software package `breakaway`.

7

A case study on maritime accidents using meta-analysis in capture-recapture

Dankmar Böhning

University of Southampton

John Bunge

Cornell University

CONTENTS

7.1 Introduction

Here we consider, in a specific case study, the use of meta-analysis to combine or synthesize the results of several capture-recapture studies. For the general layout, notation and theoretical approach on meta-analysis in capture-recapture studies, we refer to Chapter 6. Suppose there are M populations with sizes N_i for $i = 1, \cdots, M$. M capture-recapture studies are performed, producing statistics $\{(\hat{N}_1, \hat{\sigma}_1), \ldots, (\hat{N}_M, \hat{\sigma}_M)\}$, where $\hat{N}_i$ and $\hat{\sigma}_i$ are the estimated population size and associated standard error (respectively) from the ith study. Also, let n_i denote the observed sample size of the i-th study. Also, suppose that two sources are available to identify members of the target population. We denote with

- $n_{11}^{(i)}$ the frequency of members of the target population identified by both sources in study i,

- $n_{10}^{(i)}$ the frequency of members of the target population identified in study i by the first source, but not by the second,

- $n_{01}^{(i)}$ the frequency of members of the target population identified in study i by the second source, but not by the first,

- $n_{00}^{(i)}$ the frequency of members of the target population identified in study i by neither source.

FIGURE 7.1: *Costa Concordia* accident in 2012 in which the vessel hit an underwater rock and capsized partly.

The latter is the target of interest and the associated population size is estimated by the Chapman estimator given as

$$\hat{N}_i = \frac{(n_{1+}^{(i)} + 1)(n_{+1}^{(i)} + 1)}{n_{11}^{(i)} + 1} - 1, \tag{7.1}$$

where $n_{1|}^{(i)} = n_{11}^{(i)} + n_{10}^{(i)}$ and $n_{+1}^{(i)} = n_{11}^{(i)} + n_{01}^{(i)}$. The associated variance estimate is given as

$$\hat{\sigma}_i^2 = \widehat{Var}(\hat{N}_i) = \frac{(n_{1+}^{(i)} + 1)(n_{+1}^{(i)} + 1)(n_{1+}^{(i)} - n_{11})(n_{+1}^{(i)} - n_{11}^{(i)})}{(n_{11}^{(i)} + 1)^2(n_{11}^{(i)} + 2)}. \tag{7.2}$$

For this estimator to be valid, it is assumed that both sources are independent. If they are positively associated, the estimator will provide only a lower bound. More details on the estimator of Chapman are given in Seber [259], Borchers et al. [49] and McCrea and Morgan [202]. Next we are considering applying capture-recapture techniques to maritime accidents on a world-wide scale.

7.2   The case study on maritime accidents

Maritime accidents reach public interest only under unusual, often spectacular circumstances such as in the *Costa Concordia* accident in 2012 (see Figure 7.1). However, in most cases maritime accidents happen without much public notice and, in fact, many even remain unreported.

Hassel et al. [139] consider this serious problem of underreporting the world-wide occurrence of maritime traffic accidents. They say:

Underreporting of maritime accidents is a problem not only for authorities trying to improve maritime safety through legislation, but also to risk management companies and other entities using maritime casualty statistics in risk and accident analysis.

Registered maritime accidents are available for the five years from 2005 to 2009 from two maritime accident registries, called Sea-Web and Flag-State (for details see Hassel et al. [139]). Table 7.1 shows the frequency of registrations by Sea-Web n_{1+}, Flag-State n_{+1} and what they have in common n_{11} by country and year. In the following we are interested in providing a more compact analysis of these types of capture-recapture data using tools from meta-analysis.

TABLE 7.1
Frequency of maritime accidents by register, year, and country according to Hassel et al. [139]

Country	Year	Sea-Web	Flag-State	Common	n
Sweden	2005	18	70	16	72
Sweden	2006	24	54	21	57
Sweden	2007	23	78	17	84
Sweden	2008	21	65	14	72
Sweden	2009	23	66	18	71
Denmark	2005	26	34	384	58
Denmark	2006	50	46	94	83
Denmark	2007	39	40	874	69
Denmark	2008	39	58	120	86
Denmark	2009	35	42	95	68
UK	2005	84	318	42	360
UK	2006	75	297	47	325
UK	2007	81	286	58	309
UK	2008	79	252	44	287
UK	2009	80	274	38	316
US	2005	132	452	28	556
US	2006	128	499	21	606
US	2007	128	447	33	542
US	2008	150	487	32	605
US	2009	94	477	21	550
Canada	2005	146	159	107	198
Canada	2006	117	139	93	163
Canada	2007	118	137	85	170
Canada	2008	115	149	84	180
Canada	2009	112	138	85	165
NL	2005	43	61	6	98
NL	2006	59	70	14	115
NL	2007	78	84	22	140
NL	2008	82	94	21	155
NL	2009	42	33	8	67
Norway	2005	89	105	20	174
Norway	2006	76	93	27	142
Norway	2007	106	132	52	186
Norway	2008	123	115	41	197
Norway	2009	135	151	63	223

7.3 Meta-analysis essentials

Meta-analysis is a statistical methodology for the analysis and integration of results from individual, independent studies. In the last decades, meta-analysis developed a crucial role in many fields of science such as medicine and pharmacy, health science, psychology, and social science, see for example Petitti [231], Schulze et al. [257], Böhning et al. [41], Sutton et al. [273], Egger et al. [110], Borenstein et al. [50], Kulinskaya et al. [169], or Stangl and Berry [266]. Consider the typical set-up in a meta-analysis: effect measure estimates $\hat{\theta}_1$, ..., $\hat{\theta}_M$ are available from M studies with associated known variances σ_1^2, ..., σ_M^2. See also Chapter 6 for more details on the general approach. In our setting of capture-recapture studies we will take $\hat{\theta}_i = \hat{N}_i$ and $\sigma_i^2 = \hat{\sigma}_i^2$ where the latter is given by (7.2) ignoring the uncertainty element and taking it as a known, non-random quantity.

Typically, the random effects model

$$\hat{\theta}_i = \theta + \delta_i + \epsilon_i$$

is employed where $\delta_i \sim N(0, \tau^2)$ is a normal random effect and $\epsilon_i \sim N(0, \sigma_i^2)$ is a normal random error, all being pairwise independent, and $\tau^2 > 0$. Furthermore, let $w_i = 1/\sigma_i^2$ and $W_i = 1/(\sigma_i^2 + \tau^2)$. The heterogeneity statistic Q is defined as

$$Q = \sum_{i=1}^{M} w_i (\hat{\theta}_i - \bar{\theta})^2,$$

where $\bar{\theta} = \sum_{i=1}^{M} w_i \hat{\theta}_i / \sum_{i=1}^{M} w_i$. $\bar{\theta}$ is the mean estimate in the so-called *fixed effects model* and has variance $1/[\sum_i w_i]$. Q is the basis of the DerSimonian-Laird estimator for the heterogeneity variance τ^2 given, in its untruncated form, by

$$\hat{\tau}^2 = \frac{Q - (M-1)}{\sum_{i=1}^{M} w_i - \sum_{i=1}^{M} w_i^2 / [\sum_{i=1}^{M} w_i]}.$$

$\hat{\tau}^2$ is also used in the *random effects model* with overall mean estimate

$$\hat{\theta}_{DL} = \sum_{i=1}^{M} \hat{W}_i \hat{\theta}_i / \sum_{i=1}^{M} \hat{W}_i$$

and associated variance estimate $1/[\sum_i \hat{W}_i]$, $\hat{W}_i = 1/(\sigma_i^2 + \hat{\tau}^2)$. For the fixed and random effects model, confidence intervals are then constructed in the conventional approximate normal way. Q is also the foundation of Higgins's I^2 defined as

$$I^2 = \frac{Q - (M-1)}{Q} \tag{7.3}$$

designed to provide a measure of quantifying the magnitude of heterogeneity involved in the meta-analysis. Indeed, it is a proportion and, if multiplied by 100, a percentage. More precisely, the proportion of total variance due to heterogeneity. This might be not obvious from the definition provided in (7.3) but becomes more evident from the identity

$$I^2 = \frac{\hat{\tau}^2}{\hat{\tau}^2 + s^2}, \tag{7.4}$$

where $s^2 = (M-1) \sum_{i=1}^{M} w_i / [(\sum_{i=1}^{M} w_i)^2 - \sum_{i=1}^{M} w_i^2]$. As s^2 can be viewed as some form of average of the study-specific variances σ_1^2, ..., σ_M^2, I^2 can be validly interpreted, as typically done in variance component models, as the proportion of the total variance (variance due to heterogeneity plus within-study variance) due to heterogeneity. If the meta-analysis shows evidence of strong heterogeneity and if an associated covariate is available, then sub-group analysis is usually performed to explain the heterogeneity.

7.4 Analysis of maritime accident data

We now apply these concepts to our capture-recapture data using $\hat{\theta}_i = \hat{N}_i$ and $\sigma_i^2 = \hat{\sigma}_i^2$ where the latter is given by (7.2). As we said earlier and as it is often done in meta-analytic practice, we take the estimated variances (7.2) as known, non-random quantity. First, we concentrate on population size estimation. Evidently, any meta-analysis of population sizes will need to be restricted to country as the background populations are quite different across countries. For the practical analysis, we use an add-on package of STATA14 called METAN [270]. METAN needs as input the estimated population size with an estimate of the standard error and then is executed in the form

```
metan Nhat seNhat
```

where Nhat and seNhat contain the values of $\hat{N}_i$ and σ_i, respectively.

METAN has a variety of flexible options including the fixed and random effects model as well as sub-group analysis features. Results can also be made available in graphical form as is done in Figure 7.2, which consists of the so-called Forrest plot for the estimated population sizes for Denmark. Besides the fixed effects estimate with 95% confidence interval it also contains the random effects estimate with 95% confidence interval. As there is no evidence of heterogeneity ($I^2 = 0$), the fixed effects and random effects mean population sizes coincide as do their confidence intervals. We see an average population size estimate of 166 accidents over the five years with 95% confidence interval of 133–199. The situation is very different from the UK where substantial heterogeneity is present (see Figure 7.4). Here it is more appropriate to consider a random effects model leading to an average population size estimate of 490 accidents with 95% confidence interval from 415–465. In Table 7.3 we summarize the results for the seven countries involved. In most countries, at least mild forms of heterogeneity exist, indicating some variation in population size of traffic accidents over the five years. The situation changes when we consider completeness of the surveillance system, defined as $n_{ij}/\hat{N}_{ij}$, where i varies over the five years and j over the seven countries. In Figure 7.4, we see a sub-group analysis by country. It is evident that most of the heterogeneity disappears within the sub-group country, so that completeness of identification is very similar over the years, with the exception of Norway, which shows rising levels of completeness and the UK, which shows increasing completeness which falls again after the peak in 2007. The high level of completeness of identification of maritime accidents for Sweden and Canada is also remarkable.

7.5 Comments and future directions

Strictly speaking, the case study provided a meta-analysis of different populations, as the maritime accidents are clearly different across countries and for different years. Still we think

TABLE 7.2: Observed n_i = Sea-Web + Flag-State - Common and estimated $\hat{N}_i$ population sizes for maritime accident data by year and country.

Year	Country	year	Sea-Web	Flag-State	Common	$\hat{N}_i$	$n_i/\hat{N}_i$
1	Sweden	2005	18	70	16	78.35	0.92
2	Sweden	2006	24	54	21	61.50	0.93
3	Sweden	2007	23	78	17	104.33	0.81
4	Sweden	2008	21	65	14	95.80	0.75
5	Sweden	2009	23	66	18	83.63	0.85
6	Denmark	2005	26	34	2	314.00	0.18
7	Denmark	2006	50	46	13	170.21	0.49
8	Denmark	2007	39	40	10	148.09	0.47
9	Denmark	2008	39	58	11	195.67	0.44
10	Denmark	2009	35	42	9	153.80	0.44
11	UK	2005	84	318	42	629.58	0.57
12	UK	2006	75	297	47	470.83	0.69
13	UK	2007	81	286	58	397.88	0.78
14	UK	2008	79	252	44	448.78	0.64
15	UK	2009	80	274	38	570.15	0.55
16	US	2005	132	452	28	2076.55	0.27
17	US	2006	128	499	21	2930.82	0.21
18	US	2007	128	447	33	1698.76	0.32
19	US	2008	150	487	32	2231.97	0.27
20	US	2009	94	477	21	2063.09	0.27
21	Canada	2005	146	159	107	216.78	0.91
22	Canada	2006	117	139	93	174.74	0.93
23	Canada	2007	118	137	85	189.95	0.89
24	Canada	2008	115	149	84	203.71	0.88
25	Canada	2009	112	138	85	181.64	0.91
26	NL	2005	43	61	6	300.71	0.29
27	NL	2006	59	70	14	283.00	0.41
28	NL	2007	78	84	22	290.96	0.48
29	NL	2008	82	94	21	357.41	0.43
30	NL	2009	42	33	8	161.44	0.42
31	Norway	2005	89	105	20	453.29	0.38
32	Norway	2006	76	93	27	257.50	0.55
33	Norway	2007	106	132	52	267.51	0.70
34	Norway	2008	123	115	41	341.48	0.58
35	Norway	2009	135	151	63	322.00	0.69

TABLE 7.3: Average population size estimates with Higgins I^2 estimate of size of heterogeneity by country

Country	I^2 (in %)	$\bar{\theta}$ (95%CI)	$\hat{\theta}_{DL}$ (95%CI)
Sweden	84.7	72 (67–77)	83 (68–98)
Denmark	0.00	166 (133–199)	166 (133–199)
UK	76.8	452 (419–484)	490 (415–565)
US	21.6	2,024 (1,739–2,309)	2,057 (1,727–2,388)
Canada	88.6	190 (185–195)	193 (177–209)
NL	64.3	260 (215–305)	278 (198–359)
Norway	63.2	298 (274–322)	307 (263–351)

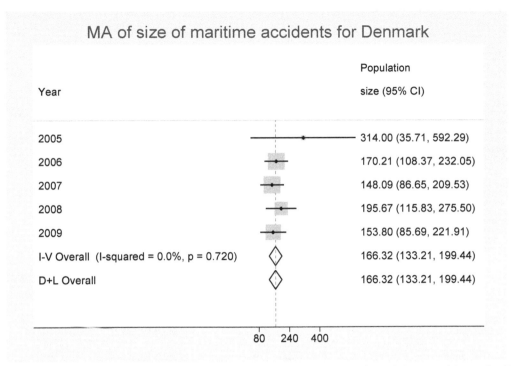

FIGURE 7.2: Meta-analysis of estimated population sizes of maritime accidents for Denmark.

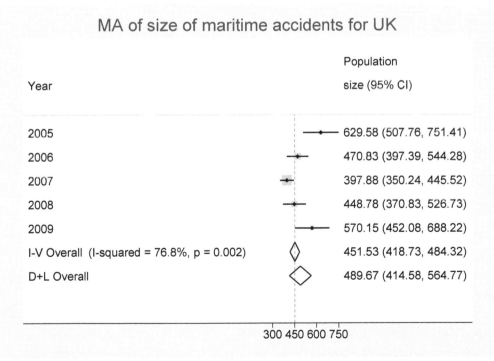

FIGURE 7.3: Meta-analysis of estimated population sizes of maritime accidents for the UK.

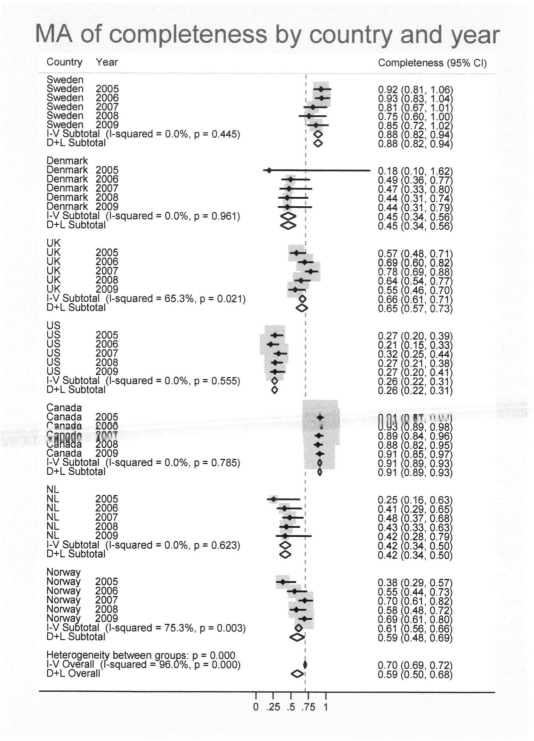

FIGURE 7.4: Meta-analysis of estimated completeness of maritime accidents by year and country.

that a meta-analysis approach is feasible. We could think of a more general population of maritime accidents where location and time are specific characteristics.

An additional, alternative concept would to apply meta-regression. For example, it would be of interest to investigate if there is a year-effect on the completeness. This could be easily accomplished using the add-on package `metareg` of STATA14 [270].

The approach here has used a specific capture-recapture estimator. But there is no need to restrict this to a specific estimator. Any appropriate estimator is possible including potentially different models which vary across studies such as log-linear models with different interaction terms included.

7.6 Software

Here we give some details on how the analysis in this chapter has been accomplished and how the findings have been achieved. We are using the add-on package `metan` of STATA14 [270]. The code given below refers to the following data:

```
+---------------------------------------------------------------+
     | country    year    Nchap    seNchap    NchapL    NchapR |
     |-------------------------------------------------------------|
 1.  | Sweden     2005       78       5.3        68        89 |
 2.  | Sweden     2006       62       3.5        55        68 |
 3.  | Sweden     2007      104        11        84       125 |
 4.  | Sweden     2008       96        12        72       119 |
 5.  | Sweden     2009       84       7.3        69        98 |
     |-------------------------------------------------------------|
 6.  | Denmark    2005      314       142        36       592 |
 7.  | Denmark    2006      170        32       108       232 |
 8.  | Denmark    2007      148        31        87       210 |
 9.  | Denmark    2008      196        41       116       276 |
10.  | Denmark    2009      154        35        86       222 |
     |-------------------------------------------------------------|
11.  |     UK     2005      630        62       508       751 |
12.  |     UK     2006      471        37       397       544 |
13.  |     UK     2007      398        24       350       446 |
14.  |     UK     2008      449        40       371       527 |
15.  |     UK     2009      570        60       452       688 |
     |-------------------------------------------------------------|
16.  |     US     2005     2077       325      1441      2713 |
17.  |     US     2006     2931       544      1864      3998 |
18.  |     US     2007     1699       237      1234      2163 |
19.  |     US     2008     2232       327      1591      2873 |
20.  |     US     2009     2063       368      1341      2785 |
     |-------------------------------------------------------------|
21.  | Canada     2005      217       6.1       205       229 |
22.  | Canada     2006      175       4.7       166       184 |
23.  | Canada     2007      190       6.6       177       203 |
24.  | Canada     2008      204       7.5       189       218 |
25.  | Canada     2009      182       5.9       170       193 |
```

26.	NL	2005	389	119	155	622
27.	NL	2006	283	55	176	390
28.	NL	2007	291	43	207	375
29.	NL	2008	357	56	247	467
30.	NL	2009	161	39	85	238
31.	Norway	2005	453	76	304	602
32.	Norway	2006	258	32	195	320
33.	Norway	2007	268	20	228	307
34.	Norway	2008	341	34	275	408
35.	Norway	2009	322	22	279	365

The following STATA-code (to be best run as a DO-file) refers to the above data and will produce country-specific analysis of estimated sizes of maritime accidents.

```
#
# meta-analysis per country
#
metan Nchap seNchap if country=="Sweden",textsize(250) lcols(year)///
 second(random) nulloff nowt xlabel(40,60,80,100,120,140) ///
title (MA of size of maritime accidents for Sweden) force  effect (population size)
metan Nchap seNchap if country=="Denmark",textsize(200) lcols(year)///
 second(random) nulloff nowt xlabel(80,  240, 400) ///
title (MA of size of maritime accidents for Denmark) effect (population size)
metan Nchap seNchap if country=="UK",textsize(200) lcols(year) ///
second(random) nulloff nowt xlabel(300,450,  600, 750) ///
title (MA of size of maritime accidents for UK) effect (population size)
metan Nchap seNchap if country=="Canada",textsize(200) lcols(year) ///
second(random) nulloff nowt ///
xlabel(150,175,200,225,250) title (MA of size of maritime accidents for Canada)///
 effect (population size) force
metan Nchap seNchap if country=="NL",textsize(200) lcols(year) ///
second(random) nulloff nowt ///
xlabel(50, 200, 350,500, 650) ///
title (MA of size of maritime accidents for NL) ///
effect (population size) force
metan Nchap seNchap if country=="Norway",textsize(200) lcols(year)///
 second(random) nulloff nowt ///
xlabel(50, 200, 350,500, 650) title (MA of size of maritime accidents for Norway)///
 effect (population size) force
metan Nchap seNchap if country=="US",textsize(225) lcols(year)///
 second(random) nulloff nowt ///
xlabel(1000, 2000,3000, 4000) title (MA of size of maritime accidents for US) ///
effect (population size) force
```

For this analysis we have assumed that the normal approximation is justified. An alternative would have been to log-transform the population size estimates first including an appropriate approximation of the estimated standard error of the log-transformed size estimates and then use the `eform` option to display results on the original scale.

For the analysis of completeness, we have used the following data file:

```
+-----------------------------------------------------------------+
     | country   year   completeness   completenessL   completenessR |
     |-----------------------------------------------------------------|
 1.  | Sweden    2005        .92            .81            1.1 |
 2.  | Sweden    2006        .93            .83             1  |
 3.  | Sweden    2007        .81            .67             1  |
 4.  | Sweden    2008        .75            .6              1  |
 5.  | Sweden    2009        .85            .72             1  |
     |-----------------------------------------------------------------|
 6.  | Denmark   2005        .18           .098            1.6 |
 7.  | Denmark   2006        .49            .36            .77 |
 8.  | Denmark   2007        .47            .33            .8  |
 9.  | Denmark   2008        .44            .31            .74 |
10.  | Denmark   2009        .44            .31            .79 |
     |-----------------------------------------------------------------|
11.  |     UK    2005        .57            .48            .71 |
12.  |     UK    2006        .69            .6             .82 |
13.  |     UK    2007        .78            .69            .88 |
14.  |     UK    2008        .64            .54            .77 |
15.  |     UK    2009        .55            .46            .7  |
     |-----------------------------------------------------------------|
16.  |     US    2005        .27            .2             .39 |
17.  |     US    2006        .21            .15            .33 |
18.  |     US    2007        .32            .25            .44 |
19.  |     US    2008        .27            .21            .38 |
20.  |     US    2009        .27            .2             .41 |
     |-----------------------------------------------------------------|
21.  | Canada    2005        .91            .87            .97 |
22.  | Canada    2006        .93            .89            .98 |
23.  | Canada    2007        .89            .84            .96 |
24.  | Canada    2008        .88            .82            .95 |
25.  | Canada    2009        .91            .85            .97 |
     |-----------------------------------------------------------------|
26.  |     NL    2005        .25            .16            .63 |
27.  |     NL    2006        .41            .29            .65 |
28.  |     NL    2007        .48            .37            .68 |
29.  |     NL    2008        .43            .33            .63 |
30.  |     NL    2009        .42            .28            .79 |
     |-----------------------------------------------------------------|
31.  | Norway    2005        .38            .29            .57 |
32.  | Norway    2006        .55            .44            .73 |
33.  | Norway    2007        .7             .61            .82 |
34.  | Norway    2008        .58            .48            .72 |
35.  | Norway    2009        .69            .61            .8  |
     +-----------------------------------------------------------------+
```

The following STATA-code provides a sub-group meta-analysis of completeness of identification of maritime accidents by country.

```
#
# meta-analysis of completeness
#

metan completeness completenessL completenessR, ///
lcols( country year) nulloff nowt effect(completeness) ///
 title(MA of completeness over year and by country) ///
 xlabel(0, .25,.5,.75,1) by(country) textsize(250) second(random)
```

Acknowledgments

D.B. is grateful to the PhD students of statistics at Thammasat University, Faculty of Sciences, Rangsit Campus, Bangkok for their interest and discussions at the occasion of a course on Meta-Analysis on Capture-Recapture Studies which D.B. gave during the period of August 2016.

8

A meta-analytic generalization of the Lincoln–Petersen estimator for mark-and-resight studies

Dankmar Böhning

University of Southampton

Mehmet Orman and Timur Köse

Ege University

John Bunge

Cornell University

CONTENTS

8.1 What are mark-and-resight studies?

In these studies animals are marked (often with a particular color) at the first sampling occasion. This is often done without actually catching them. For example, in stray dog studies, as in the one we discuss in the following, dogs are sprayed with a color from a distance. At follow-up occasions it is only recorded how many are marked and how many are not. The difference to mark-recapture approaches is that the full capture history is not known. For example, it is not known if an animal observed unmarked at occasion three has been observed at occasion two. If there are only two occasions, the capture-recapture and mark-resight approaches are identical.

We introduce some notation. Let n_0 be the number of animals marked at the initial occasion in which marking takes place. We assume there are $M \geq 1$ resighting occasions.

- Let n_i denote the frequency of animals seen at occasion i, $i = 1, \cdots, M$.

- Let m_i denote the frequency of animals seen *marked* at occasion i, $i = 1, \cdots, M$.

With these notations, we are now able to define the Lincoln–Petersen estimator of the population size at occasion i to be

$$\hat{N}_i^{\mathrm{LP}} = \frac{n_0 n_i}{m_i} \tag{8.1}$$

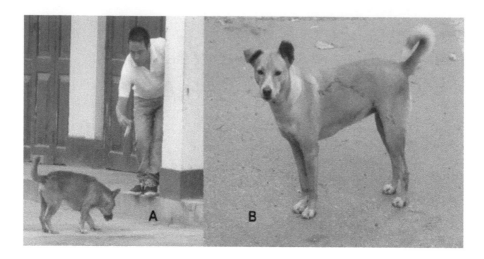

FIGURE 8.1: Mark-and-resight study by Tenzin et al. [275] to determine the size of stray dogs, A: marking of the dogs, B: resighting occasion (copyright permission received from Copyright Clearance Center of Elsevier reference 4006541373616).

and the Chapman-correction as

$$\hat{N}_i = \frac{(n_0 + 1)(n_i + 1)}{m_i + 1} - 1. \tag{8.2}$$

These have been discussed already in Chapter 7.1, but more details on these are given in Sobor [259], Dorchers et al. [49] and McCrea and Morgan [202]. See also Chapter 23 for extensions of the Lincoln–Petersen estimator. Again, we emphasize that for this estimator to be valid, it is assumed that the pairwise sighting occasions are independent. The Chapman estimator has an associated variance estimate of

$$\hat{\sigma}_i^2 = \widehat{Var}(\hat{N}_i) = \frac{(n_0 + 1)(n_i + 1)(n_0 - m_i)(n_i - m_i)}{(m_i + 1)^2(m_i + 2)}. \tag{8.3}$$

8.2   A case study on stray dogs in South Bhutan

To illustrate we consider a case study on stray dogs in South Bhutan [275]. Stray dogs are domestic dogs that are on public areas and not currently under direct control of owners. Tenzin et al. [275] write:

> Many factors are associated with increasing free-roaming dog populations in developing countries including rapid urbanization, increased human population growth, poor waste management, absence of responsible dog ownership and poor management, and cultural tolerance ... Although domestic dogs play an important role in human life, they may also pose significant risks to human health and well-being. The most serious threat to public health is dog bites and as potential sources of infectious diseases including rabies ... Noise pollution, fighting, fecal contamination of the environment, uncontrolled breeding and

spread of rubbish from the bins are some of the additional social problems associated with free-roaming dogs ... In addition, many free-roaming dogs in developing countries suffer from extremely poor welfare as a result of skin diseases such as mange along with secondary bacterial infections, high mortality due to road accidents, malnutrition, starvation and abuse from humans.

Hence it is important to monitor the size of stray dog populations. Tenzin et al. [275] chose two study areas in South Bhutan, namely two urban areas in the south of Bhutan that border India: Gelephu and Phuentsholing. Here we consider the results for one of these areas (Phuentsholing). The results for four resighting occasions are provided in Table 8.1.

It is the purpose of this chapter to provide a more integrative analysis of the various population size estimates using a meta-analytic approach.

8.3   Meta-analysis and mark-resight studies

The meta-analytic essentials have been developed already in Chapter 6 and Chapter 7, and we apply them here once more. We have population size estimates $\hat{N}_1, \cdots, \hat{N}_M$ with associated variances $\sigma_1^2, \cdots, \sigma_M^2$ provided by Equation (8.3). In fact, the variance given by (8.3) is an estimate of the true variance, but we ignore this here for the time being and treat them as non-random. Hence we have the summary estimator

$$\hat{N} = \frac{\sum_i w_i N_i}{\sum_i w_i},\tag{8.4}$$

where $w_i = 1/\sigma_i^2$.

A meta-analysis using the add-on package **metan** of STATA14 [270] provides the Forest-plot in Figure 8.2. There is no evidence of heterogeneity ($I^2 = 0$). The summary estimate is 552 stray dogs in this region with 95% confidence interval of 529–576. Nevertheless, there appears to be a trend visible in Figure 8.2 in the sense that with later sighting occasions there is an increased population size. This question asks for a covariate modeling of occasion. Recall that the basis for the analysis is the random effects model

$$\hat{\theta}_i = \theta + \delta_i + \epsilon_i$$

where $\delta_i \sim N(0, \tau^2)$ is a normal random effect and $\epsilon_i \sim N(0, \sigma_i^2)$ is a normal random error,

TABLE 8.1
Mark-resight study on stray dogs in South Bhutan with 4 resighting occasions: number n_0 of dogs at initial marking occasion, number n_i of dogs seen at sighting occasion i and number m_i of dogs seen marked at occasion i with associated estimates of population size $\hat{N}_i$ given in (8.2)

n_0	n_1	n_2	n_3	n_4
267	244	306	240	312
	m_1	m_2	m_3	m_4
	124	149	114	143
	$\hat{N}_1$	$\hat{N}_2$	$\hat{N}_3$	$\hat{N}_4$
	525	548	562	583

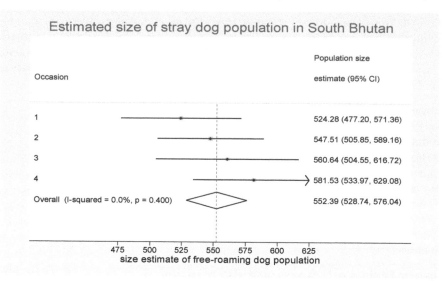

FIGURE 8.2: Forrest plot for mark-and-resight study to determine the size of stray dogs in South Bhutan.

TABLE 8.2: Meta-regression of population size estimated by Chapman's estimator on resighting occasion

Covariate	β	SE	$t = \beta/SE$	$P > |t|$
Occasion	18.48	10.82	1.71	0.230
θ	507.67	28.84	17.60	0.003

all being pairwise independent, and $\tau^2 > 0$. This model is now supplemented with a linear predictor $\mathbf{x}_i'\beta$ where $\mathbf{x}_i$ is a vector of covariates for resighting occasion i with associated parameter vector β. Hence the *meta-regression model* now takes the form

$$\hat{\theta}_i = \theta + \delta_i + \mathbf{x}_i'\beta + \epsilon_i. \tag{8.5}$$

Using another add-on package `metareg` of STATA14 [270], we are able to fit model (8.5) using just one covariate, `occasion`, with values 1,2,3,4. `metareg` uses as fitting criterion the REML likelihood, which is an adjusted normal likelihood with mean structure given by (8.5) and diagonal covariance structure with $\sigma_i^2 + \tau^2$ on the diagonal. The adjustment term corrects for bias in τ^2 estimation caused by estimating the mean structure. The graphical display in Figure 8.3 confirms the previously mentioned trend. However, as Table 8.2 shows, the occasion effect is not significant.

8.4   A Mantel–Haenszel estimator for mark-resight studies

Instead of just reporting M pairwise population size estimators, based upon pairs including occasion 0 with resighting occasion $1, \cdots, M$, the idea is to apply a weighting summary approach to these pairwise estimators and achieve a combined estimator. Let us look at the

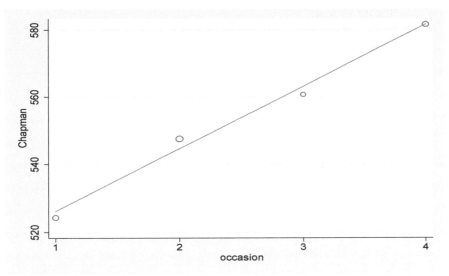

FIGURE 8.3: Meta-regression of population size estimated by the Chapman estimator on the resighting occasion.

weighted Lincoln–Petersen summary estimator

$$\hat{N}_i^{\mathrm{LP}} = \sum_i w_i \frac{n_0 n_i}{m_i} / \sum_i w_i.$$

Instead of using inverse-variance weights, we use the Mantel–Haenszel weights $w_i = m_i$, and the estimator

$$\frac{n_0 \sum_i n_i}{\sum_i m_i}$$

arises. As in general with Mantel–Haenszel estimation, the great benefit is that we achieve a very stable estimator which is not sensitive to sparsity, in particular, small values or zeros for m_i. Note that $\hat{N}^{\mathrm{LP}}$ takes the simple computational form

$$\hat{N}^{\mathrm{LP}} = n_0 n_+ / m_+$$

which is of the form of a Lincoln–Petersen estimator. Note that $n_+ = \sum_i n_i$ and $m_+ = \sum_i m_i$.

In a similar fashion we can proceed for the Chapman estimator and yield

$$\hat{N}^{\mathrm{C}} = \frac{(n_0 + 1) \sum_i (n_i + 1)}{\sum_i (m_i + 1)} - 1 \tag{8.6}$$

using the Mantel–Haenszel weights $w_i = (m_i + 1)$ applied to the Chapman estimator $\hat{N}_i = \frac{(n_0+1)(n_i+1)}{m_i+1} - 1$ at occasion i. $\hat{N}$ has the nice computational form

$$\hat{N}^{\mathrm{C}} = \frac{(n_0 + 1)(n_+ + M)}{(m_+ + M)} - 1$$

where $n_+ = \sum_i n_i$ and $m_+ = \sum_i m_i$ as already defined above. Hence we can think of the meta-analytic generalization of the summary mark-resight estimator as a two-occasion Lincoln–Petersen and Chapman estimator as displayed in Tables 8.3 and 8.4. It should be

TABLE 8.3
Mark-resight with several resighting occasions as a Lincoln–Petersen experiment with two occasions

		Occasion 0		
		1	0	
Other occasions	1	m_+		n_+
	0			
		n_0		N

TABLE 8.4
Mark-resight with several resighting occasions as a Lincoln–Petersen experiment for two occasions with Chapman correction

		Occasion 0		
		1	0	
Other occasions	1	$(m_+ + M - 1) + 1$		$(n_+ + M - 1) + 1$
	0			
		$n_0 + 1$		N

noted though that variance formulae such as (8.3) cannot be applied to Table 8.4, even if appropriately modified in notation such as

$$\frac{(n_0 + 1)(n_+ + M)(n_0 - m_+ - M + 1)(n_+ - m_+)}{(m_+ + M)^2(m_+ + M + 1)}.$$

This is due to various overlap in m_+ and n_+ that will lead to a violation of assumptions underlying (8.3). Formally, this is also seen as $n_0 - m_+ - M + 1$ can easily become negative as in our case study. As an alternative for variance estimation we suggest a parametric bootstrap. In this case, the parametric bootstrap works as follows. Using the original sample data, a combined estimator of the population size $\hat{N}^C$ is developed. This leads to an estimate of the resighting probability. Using this estimate, a marking sample is drawn as well as M independent resighting samples from which a combined estimator $\hat{N}^{C*}$ is created. This process is repeated B times and forms the bootstrap sample from which a variance estimate can be constructed using the bootstrap sample variance. The approach delivers, for the data set considered here, a variance estimate for $\hat{N}^C$ as 142.62 leading to an approximate 95% confidence interval of 530–578 with a value of 554 for $\hat{N}^C$. This is close to the confidence interval of 529–576 given Figure 8.2 using the inverse variance weighted approach.

In our case study, all three estimators are close, but this might be due to the relatively large population size and the large sighting probability of 0.4946. This might be different in situations with smaller sizes and resighting occasions. This will be looked at in more detail in the next section.

8.5 Some simulation work

The purpose of this section is to study how population size N and number of resighting occasions M affect the behavior of the three estimators. We look at the following scenarios:

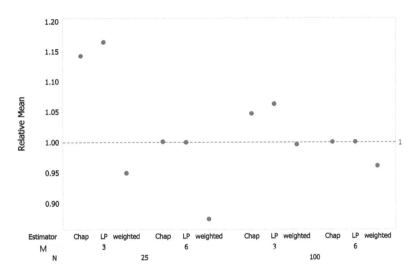

FIGURE 8.4: Relative mean of the Mantel–Haenszel estimator using the Lincoln–Petersen (LP) and Chapman (Chap) estimator as well as the weighted estimator (weighted).

- N takes values 25 and 100,

- M takes values 3 and 6,

- the resighting (and initial sighting) probability is 0.5, and

- the replication size is 1,000.

If population size is varying in a simulation study, we need to choose measures which take into account the variation in N. This is accomplished by choosing the following measures:

- relative mean: $E(\hat{N})/N$,

- relative standard deviation: $\sqrt{var(\hat{N})}/N$.

If an estimator $\hat{N}$ for N is unbiased, then the relative mean is 1; if it is asymptotically unbiased, the limit of the relative mean is 1 for N large.

We see in Figure 8.4 that the weighted estimator underestimates the population size, more for the smaller population size. The other two estimators appear to behave better, also for the small population size, in particular, if the number of resighting occasions are increasing.

Looking at the relative standard deviation, Figure 8.5, we see that there is a clear ranking: the best is the Mantel–Haenszel version of the Chapman estimator, followed by the Mantel–Haenszel version of the Lincoln–Petersen estimator and the inverse variance weighted estimator.

It should be clear that this simulation is quite limited as the range of values is quite narrow. However, it illustrates the potential gain possible using meta-analytic methods.

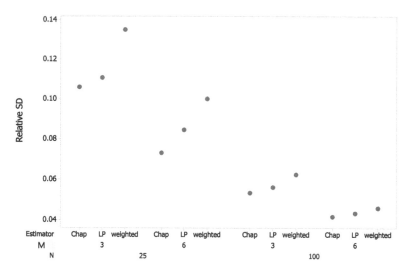

FIGURE 8.5: Relative standard deviation of the Mantel–Haenszel estimator using the Lincoln–Petersen (LP) and Chapman (Chap) estimators as well as the weighted estimator (weighted).

8.6   Concluding remarks

Meta-analytic methods can be very helpful in summarizing information from mark-resight experiments. As was shown in the previous section, combining information in a summary estimator improves the precision of the population size estimator. The more resighting occasions, the higher the precision of the estimator. This relationship can be used to approach the question of how many resighting occasions M should be used.

The technique can also be used to combine M pairwise capture-recapture estimators arising from $M+1$ capture occasions. However, as there is complete identification information, the preferred approach would be based upon log-linear modeling (see also Chapters 19, 20, 21, 22, and 25) as this would allow forms of dependence structure for the capturing process at the $M+1$ occasions.

Part IV

Extensions of Single Source Models

9

Estimating the population size via the empirical probability generating function

John Bunge

Cornell University

Sarah Sernaker

University of Minnesota

CONTENTS

9.1   Introduction and background

We consider the standard model described in Chapter 1: there are N units in the population, labeled $1, \ldots, N$, which produce $Y_1, Y_2, \ldots Y_N$ representatives in the sample (respectively). Unit i is observed only if $Y_i > 0$, and we list the nonzero observations as $Y_1, \ldots, Y_n$; these are summarized by the frequency counts $f_1, f_2, \ldots$ where $f_j := \#\{Y_i : Y_i = j\}, j = 1, 2, \ldots$. We suppose that the untruncated random variables $Y_1, \ldots, Y_N$ are i.i.d. replicates from a distribution p, where $p_y := P(Y_i = y), y = 0, 1, \ldots$. Then the zero-truncated observations are (conditionally) i.i.d. replicates from the corresponding zero-truncated distribution p^+, where $p_y^+ := p_y/(1 - p_0)$. Our inference about N will be based on the frequency count data $f_1, f_2, \ldots$.

Here we are interested in the parametric framework: we assume that p, and hence p^+, depend on a low-dimensional parameter vector θ, and we write $p = p(\theta)$ and $p^+ = p^+(\theta)$. We will use the empirical Horvitz–Thompson estimator of N, namely

$$\hat{N} := \frac{n}{1 - p_0(\hat{\theta})},$$

where $\hat{\theta}$ is an estimate of θ. This estimator has a long history and extensive literature, especially when $\hat{\theta}$ is the maximum likelihood estimate (Bunge et al. [63]). However, in some cases maximum likelihood estimation may not be feasible or tractable. This may happen for various reasons; here we consider a case where the likelihood expression involves the special functions of mathematical physics, and cannot be reduced or approximated without creating further complications. In such a situation one can consider various solutions. In our case the class of distributions of interest has a natural description in terms of the probability generating function (pgf)

$$g(s;\theta) := \sum_{j \geq 0} s^j p_j(\theta),$$

$s \in (-1, 1]$. Note that the pgf of p^+, which is relevant to the observed, zero-truncated data, is then

$$g^+(s;\theta) := \frac{g(s;\theta) - p_0(\theta)}{1 - p_0(\theta)}.$$

One approach, then, is to estimate θ by fitting the pgf to its empirical counterpart, namely the empirical probability generating function or epgf (Nakamura and Perez-Abreu [215]). If the observable random variables are the zero-truncated $Y_1, \ldots, Y_n$ as above, then the epgf is

$$\hat{g}^+(s) := \frac{1}{n} \sum_{j \geq 1} f_j s^j.$$

Generally speaking $\hat{g}^+$ is a (functional) estimate of g^+. In our parametric setting, then, we estimate θ by

$$\hat{\theta} = \arg\min_{\theta \in \Theta} d\left(\hat{g}^+(\cdot), g^+(\cdot;\theta)\right),$$

where Θ is the parameter space for θ and d is a distance measure on a suitable function space. We then use this $\hat{\theta}$ in $\hat{N}$ above to obtain the estimate of N. In this chapter we discuss the empirical pgf method in general, followed by its implementation for a specific family of distributions $p(\theta)$. We demonstrate the performance of the method by simulation, and we apply it to several of the example datasets discussed in the Introduction. Overall we find that, while the empirical pgf method is promising and works well in simulations, it will require extension and refinement (especially computational) for practical application, and we discuss this as a direction for future research.

9.2 Implementation of the empirical pgf method

Several specific questions arise. First, what distance measure d should we use? Second, how should the method be implemented numerically, that is, what is a good search strategy for $\hat{\theta}$? Third, given $\hat{\theta}$, what are suitable error estimates and goodness-of-fit assessments? We consider these general questions first, and then we turn to specific implementation for a particular family of distributions $p(\theta)$.

In recent years a considerable body of research on the empirical pgf method has appeared; we base our development here largely on Ng et al. [216] and references therein. They propose a natural L_2-based distance measure,

$$d_\alpha\left(\hat{g}^+(\cdot), g^+(\cdot;\theta)\right) = \int_0^1 \left| (\hat{g}^+(s))^\alpha - (g^+(s))^\alpha \right|^2 ds,$$

where in general $\alpha > 0$ is continuously variable. Here we specialize to $\alpha = 1/2$ and $\alpha = 1$ for reasons of simplicity, interpretability, and numerical practicality, as do Ng *et al.*, for the same reasons. We denote the corresponding distance measures as $d_{1/2}$ or d_1 respectively.

Given data $f_1, f_2, \ldots$, our objective is to compute

$$\hat{\theta} = \arg\min_{\theta \in \Theta} d_\alpha \left(\hat{g}^+(\cdot), g^+(\cdot; \theta) \right).$$

It is shown in [216] that $\hat{\theta}$ obtained by this empirical pgf method is (strongly) consistent for θ. Our simulations bear this out (as do theirs), but we will see below that consistency is not enough to justify a generally applicable procedure here.

For the integral in d_α we use a standard numerical integration routine that uses global adaptive quadrature and a tolerance level of 10^{-6}. We then use a nonlinear constrained minimization routine on the objective function, constraining the search for the optimal θ over the parameter space. The main issue with this routine is that it requires starting values, possibly in multiple stages, as follows.

9.2.1   Initial values for θ search

Assume that θ is k-dimensional, so that $\theta = [\theta_1, \ldots, \theta_k]^T$. Denote the desired initial (vector) value for the search by $\theta^{(0)} = [\theta_1^{(0)}, \ldots, \theta_k^{(0)}]^T$. We first obtain $\theta^{(0)}$ via the method of moments. Note that, for an arbitrary discrete random variable X with pgf h, we have

$$E\left(\frac{X!}{(X-m)!} \right) = h^{(m)}(1^-), \tag{9.1}$$

$m = 0, 1, 2, \ldots$, where $h^{(m)}(1^-)$ denotes the mth derivative of h evaluated from the left at 1. We apply (9.1) when $k = 3$; the modification for other values of k will be clear. In this case set

$$\left. \frac{\partial}{\partial s} \left(g^+(s; \theta) \right) \right|_{s=1^-} = \frac{1}{n} \sum_{j \geq 1} f_j j$$

$$\left. \frac{\partial^2}{\partial s^2} \left(g^+(s; \theta) \right) \right|_{s=1^-} = \frac{1}{n} \sum_{j \geq 2} f_j j(j-1)$$

$$\left. \frac{\partial^3}{\partial s^3} \left(g^+(s; \theta) \right) \right|_{s=1^-} = \frac{1}{n} \sum_{j \geq 3} f_j j(j-1)(j-2). \tag{9.2}$$

Solving the system of equations (9.2) for $[\theta_1, \theta_2, \theta_3]^T$ yields $[\theta_1^{(0)}, \theta_2^{(0)}, \theta_3^{(0)}]^T$.

Finding the solution above will typically entail a nonlinear search, which will itself require starting values, say $\theta^{(00)} = [\theta_1^{(00)}, \ldots, \theta_k^{(00)}]$. To obtain these values, we substitute $Y - 1$ for the zero-truncated $Y|Y > 0$, and apply the method of moments to the original, non-zero-truncated pgf g. Applying this procedure for $k = 3$, we obtain the equations

$$\left. \frac{\partial}{\partial s} g(s; \theta) \right|_{s=1^-} = \frac{1}{n} \sum_{j \geq 2} f_j(j-1)$$

$$\left. \frac{\partial^2}{\partial s^2} g(s; \theta) \right|_{s=1^-} = \frac{1}{n} \sum_{j \geq 3} f_j(j-1)(j-2)$$

$$\left. \frac{\partial^3}{\partial s^3} g(s; \theta) \right|_{s=1^-} = \frac{1}{n} \sum_{j \geq 4} f_j(j-1)(j-2)(j-3); \tag{9.3}$$

solving the system (9.3) for $[\theta_1, \theta_2, \theta_3]^T$ yields $[\theta_1^{(00)}, \theta_2^{(00)}, \theta_3^{(00)}]^T$. Solving (9.3) probably also entails a nonlinear search which again requires starting values, say $\theta^{(000)} = [\theta_1^{(000)}, \ldots, \theta_k^{(000)}]$. These may be found using a coarse (logarithmic) grid search.

9.2.2   Error estimation for $\hat{N}$

Our objective is to estimate N using $\hat{N} = \hat{N}(\hat{\theta})$, where $\hat{\theta}$ results from the empirical pgf procedure. To accompany $\hat{N}$ we require a variance estimate; the reported standard error for $\hat{N}$ will then be $\text{SE}(\hat{N}) = \sqrt{\widehat{\text{Var}}(\hat{N})}$. Suppose first that θ is known. Since $n \sim$ binomial $(N, 1 - p_0(\theta))$,

$$\text{Var}\left(\frac{n}{1 - p_0(\theta)}\right) = N\frac{p_0(\theta)}{1 - p_0(\theta)}. \tag{9.4}$$

However, θ is (realistically) unknown and must be estimated, and the variance of $\hat{N}(\hat{\theta})$ must account for the variability in the estimate of θ. At present, the covariance of $\hat{\theta}$ (since θ may be multidimensional) resulting from the empirical pgf procedure does not seem to have been analyzed in the literature, and it is not our purpose to do so here. Instead, we wish to find a way to estimate $\text{Var}(\hat{N}(\hat{\theta}))$ that does not require an explicit expression for $\text{Cov}(\hat{\theta})$ (when $\hat{\theta}$ is the empirical pgf-based estimate of θ). We have two strategies for this, each with two versions.

In the first strategy we look for an analytic expression for $\text{Var}(\hat{N}(\hat{\theta}))$ that is based solely on $\hat{\theta}$ and does not require any knowledge of $\text{Cov}(\hat{\theta})$. The empirical version of (9.4), namely

$$\widehat{\text{Var}}_{\text{LB}}(\hat{N}(\hat{\theta})) := \frac{\hat{N}(\hat{\theta})p_0(\hat{\theta})}{1 - p_0(\hat{\theta})}, \tag{9.5}$$

is such an expression, but since it ignores $\text{Cov}(\hat{\theta})$ it can only serve as a lower bound for $\text{Var}(\hat{N}(\hat{\theta}))$. In general we have no idea how much (9.5) underestimates $\text{Var}(\hat{N}(\hat{\theta}))$, but there is one situation where it may be useful: if $\sqrt{\widehat{\text{Var}}_{\text{LB}}(\hat{N}(\hat{\theta}))} = \mathcal{O}(\hat{N}(\hat{\theta}))$, then we can declare $\hat{N}(\hat{\theta})$ to be noninformative about N.

A second expression that omits $\text{Cov}(\hat{\theta})$ is the variance of the (parametric) maximum likelihood estimate of N. This is

$$N \times \left(\frac{1 - p_0(\theta)}{p_0(\theta)} - \frac{1}{p_0^2(\theta)}\nabla_\theta^T(1 - p_0(\theta))\left(\text{Info}^{-1}\right)\nabla_\theta(1 - p_0(\theta))\right)^{-1}$$

(where Info is the Fisher information matrix for the original, untruncated distribution p). Its empirical version, with $\hat{N}$ and $\hat{\theta}$ substituted for N and θ, respectively, will exceed $\widehat{\text{Var}}_{\text{LB}}(\hat{N}(\hat{\theta}))$, and can also serve as a lower bound for $\text{Var}(\hat{N}(\hat{\theta}))$. However, the degree of underestimation of $\text{Var}(\hat{N}(\hat{\theta}))$ is again unknown, and as we will see below, the empirical pgf estimator and the maximum likelihood estimator of θ need not be close in value in finite samples (much less their stochastic behavior), despite the consistency of both. We therefore do not use this formula.

The second and more accurate strategy is to use an expression for $\text{Var}(\hat{N}(\hat{\theta}))$ that does require knowledge of $\text{Cov}(\hat{\theta})$ in some fashion, and to form an empirical version of such an expression by suitable substitutions ($\hat{N}$ for N, $\hat{\theta}$ for θ, etc.). Again there are two possible approaches. In the first, simpler one, we set $\text{Cov}(n, \hat{\theta}) = 0$ as an approximation. Then

$$\text{Var}(\hat{N}(\hat{\theta})) \approx \text{Var}(n)\text{Var}(1 - p_0(\hat{\theta}))^{-1}$$
$$+ \text{Var}(n)E(1 - p_0(\hat{\theta}))^{-2} + E(n^2)\text{Var}(1 - p_0(\hat{\theta}))^{-1}. \tag{9.6}$$

An empirical version of (9.6) is

$$\widehat{\text{Var}}(\hat{N}(\hat{\theta}) := \hat{N}p_0(\hat{\theta})(1 - p_0(\hat{\theta}))\widehat{\text{Var}}\left((1 - p_0(\hat{\theta}))^{-1}\right)$$
$$+ \hat{N}p_0(\hat{\theta})(1 - p_0(\hat{\theta}))^{-1} + n^2\widehat{\text{Var}}\left((1 - p_0(\hat{\theta}))^{-1}\right). \tag{9.7}$$

(Alternatively, $E(n^2)$ could be estimated by $(\hat{N}(1 - p_0(\hat{\theta})))^2$; this is a subject for further investigation.) The question then is how to obtain $\widehat{\mathrm{Var}}\left((1 - p_0(\hat{\theta}))^{-1}\right)$. We use the non-parametric bootstrap from the original frequency count data, calculating the empirical or sample variance of $(1 - p_0(\hat{\theta}))^{-1}$ across B bootstrap replications. One advantage of this approach is that $(1 - p_0(\hat{\theta}))^{-1}$ is one-dimensional regardless of the dimension of θ, and so is relatively simple to compute. Our simulations show that $\widehat{\mathrm{Var}}(\hat{N}(\hat{\theta}))$ in (9.7) is a reasonably good approximation to $\mathrm{Var}(\hat{N}(\hat{\theta}))$, at least in relatively straightforward cases.

Finally, there is an exact expression for $\mathrm{Cov}(\hat{\theta})$ due to Böhning [40]. The empirical version of Böhning's formula is

$$\widehat{\mathrm{Var}}_{\mathrm{exact}}(\hat{N}(\hat{\theta})) = n \frac{p_0(\hat{\theta})}{(1 - p_0(\hat{\theta}))^2}$$

$$+ \left(\frac{n}{(1 - p_0(\hat{\theta}))^2}\right)^2 \nabla_\theta^T (1 - p_0(\hat{\theta})) \mathrm{Cov}(\hat{\theta}) \nabla_\theta (1 - p_0(\hat{\theta})). \tag{9.8}$$

This requires analytical calculation of (or approximation to) the gradient vector of $(1 - p_0(\theta))$, and then substitution of $\hat{\theta}$ into the resulting expression. It also requires the full covariance matrix of $\hat{\theta}$, which may be obtained by bootstrapping from the original frequency count data as above. The computational burden for (9.8) is greater than for (9.7), and we have not yet implemented it, partly due to the complexity of $p(\theta)$ in our example below. We note that for both (9.7) and (9.8) once could consider either the nonparametric or the parametric bootstrap; again this is a topic for further investigation.

9.2.3   Goodness of fit for the empirical pgf procedure

Given the ability to fit a given distribution, that is, estimate its parameters, by the empirical pgf procedure, we can then fit several different distributions to the same dataset and assess their comparative goodness of fit. Note that this is done using the frequency count data and the zero-truncated version of the desired distribution. There are at least three methods for carrying out this assessment.

First we can consider the classical Pearson χ^2 test. This in particular entails computing $p_j(\hat{\theta}), j = 1, 2, \dots$. However, part of the reason for using the empirical pgf procedure for a given distribution is that the likelihood is difficult to compute, and in this case so are the p_j. Furthermore, the domain $\{j = 1, 2, \dots\}$ of the f_j is unbounded, so cells must be concatenated in order to construct a finite sum for the χ^2 statistic, and there are many different ways to do this. In addition, the minimum expected cell count per cell is not settled in the literature, although it is usually taken to be 5. For these reasons we have not yet implemented the χ^2 test for the Kemp distributions in this application.

A second approach has been described recently in several papers, focusing on goodness-of-fit test statistics for the empirical characteristic function. In most of these papers the distribution, and hence critical values, for the test statistic is obtained via some form of bootstrap (Meintanis and Swanepoel [208], Sharifdoust *et al.* [261], Jimenez and Kim [157], Meintanis et al. [209]), and given our long computing times at present (see below) this method is not currently practical for our application. Two approaches that may not require bootstrapping are given in Jimenez et al. [156] and Meintanis et al. [210]. These methods are still not computationally trivial, and at present they are only worked out for the empirical characteristic function, for certain cases (e.g., location-scale families). But the approach could in principle be adapted to our problem; this is a topic for future research.

Finally we consider an information criteria-based approach. We have not found any existing research on information criteria-based model selection for the empirical pgf procedure, and this also remains a topic for future research. For the purposes of this study, we provisionally adopt the following approximate version of the AIC:

$$\text{AIC}^* := 2k + n \ln d_\alpha(\hat{g}^+(\cdot), g^+(\cdot, \hat{\theta})),$$

where k is the number of parameters in the fitted zero-truncated distribution (e.g., $k = 1$ for Poisson, $k = 2$ for negative binomial, etc.), and n is the effective sample size, $n = \sum_{j \geq 1} f_j$.

9.3 The Kemp distributions

We turn now to the family of distributions we wish to fit to capture-recapture frequency counts via the empirical pgf method. These distributions were introduced by A.W. Kemp in 1968 [161], and they have been extensively studied since then (Kemp [164]), though only recently in the capture-recapture context (Willis and Bunge [297]). The Kemp distributions are defined in terms of the generalized hypergeometric function, which is

$$_pF_q(a; b; s) = \sum_{k=0}^{\infty} \frac{(a_1)_k \cdots (a_p)_k}{(b_1)_k \cdots (b_q)_k} \frac{s^k}{k!},$$

where

$$(x)_n = \frac{\Gamma(x + n)}{\Gamma(x)}$$

and $p, q, a_1, \ldots, a_p$ and $b_1, \ldots, b_q$ are parameters. Following the notation of Dacey [93], the general Kemp pgf is then

$$g(s) = g(s; a, b, \lambda) = C_pF_q(a; b; \lambda s),$$

where $a = [a_1, \ldots, a_p], b = [b_1, \ldots, b_1]$, and

$$C^{-1} = {}_pF_q(a; b; \lambda),$$

and $\lambda > 0$ is a (further) parameter. We write $\theta = (a, b, \lambda)$.

We are especially interested in the cases $p = 0, 1; q = 0, 1$. These are:

1. Poisson. $C_0F_0(\cdot; \cdot; \lambda s), \lambda > 0$.

2. Negative binomial. $C_1F_0(a; \cdot; \lambda s), a > 0, \lambda \in (0, 1)$.

3. (no name). $C_0F_1(\cdot; b, \lambda s), b > 0, \lambda > 0$.

4. (no name). $C_1F_1(a; b; \lambda s), a > 0, b > 0, \lambda > 0$.

It is shown in Bunge [64] that the higher-order (in terms of (p, q)) Kemp distributions need not be mixed Poisson: case (3), $_0F_1$, is not mixed Poisson for any parameter values; and case (4), $_1F_1$, is not mixed Poisson when $0 < b < a$. This is important because most approaches (even nonparametric) to fitting distributions to capture-recapture frequency count data are based on mixed Poisson models. The Kemp distributions represent a rare departure from this scenario, and as such offer the possibility of fitting datasets that cannot be accommodated in the classical setting. On the other hand, the general Kemp class includes the Poisson and negative binomial, two of the most commonly used distributions in capture-recapture, and hence this class constitutes a novel direction for generalization of the classical models. In particular these distributions admit a simple representation for the ratios p_{j+1}/p_j, which was exploited in Willis and Bunge [297] to produce a fitting procedure and population-size estimation method based on nonlinear regression. Here we apply the empirical pgf method to estimate the parameters of the Kemp distributions.

9.3.1 Approximate maximum likelihood estimates

We wish to compare our empirical pgf-based results with at least point estimates derived by maximum likelihood (ML), for the datasets we analyze below. The Poisson MLE is implemented in CatchAll [60]. For the negative binomial we use an EM-algorithm-based negative binomial solver due to [94].

For Kemp case 3, $_0F_1$, we use the following procedure. We first fit the zero-truncated version of the $_0F_1$ distribution to the frequency count data $\{f_1, f_2, \ldots\}$ to obtain a (vector) parameter estimate $\hat{\theta}_{\mathrm{ML}}$. Our point estimate of N will then be the "conditional" MLE

$$\hat{N}_{\mathrm{ML}} := \frac{n}{1 - p_0(\hat{\theta}_{\mathrm{ML}})},$$

which is also the ML-based empirical Horvitz–Thompson estimator. (This is well known to be asymptotically equivalent to the global MLE (Sanathanan [252]).) Omitting the combinatorial coefficients, which now become irrelevant, the zero-truncated likelihood is proportional to

$$\left(\frac{p_0(\theta)}{1 - p_0(\theta)} \right)^n \prod_{j \geq 1} (r_j(\theta))^{\sum_{i \geq j} f_j},$$

where $r_j(\theta) := p_j(\theta)/p_{j-1}(\theta), j = 1, 2, \ldots$, and we use the telescoping product representation

$$\prod_{j \geq 1} (p_j(\theta))^{f_j} = (p_0(\theta))^n \prod_{j \geq 1} r_j(\theta)^{\sum_{i \geq j} f_j}.$$

We next apply an analytical approximation to $p_0(\theta)$, based on

$$_0F_1[(); (b); \lambda] =: {}_0F_1|b; \lambda| \approx \left(1 + \frac{\lambda}{b(b+1)} \right)^{b+1}$$

(Spanier and Oldham [264], Chapters 18 and 50), so that

$$p_0(\theta) = (_0F_1[b; \lambda])^{-1} \approx \left(1 + \frac{\lambda}{b(b+1)} \right)^{-b-1} =: \tilde{p}_0.$$

This is fairly accurate outside of a strip in the parameter space where b is small and λ is large: Figure 9.1 shows the ratio $\tilde{p}_0/p_0$, which should be close to 1 if the approximation is good.

We further exploit the simple form that $r_j(\theta)$ takes in the Kemp case. The zero-truncated likelihood is then approximately proportional to

$$\ell_{\mathrm{ZT}}(\theta) := \left(\frac{\left(1 + \frac{\lambda}{b(b+1)} \right)^{-b-1}}{1 - \left(1 + \frac{\lambda}{b(b+1)} \right)^{-b-1}} \right)^n \lambda^{\sum_{j \geq 1} j f_j} \prod_{j \geq 1} \left(\frac{1}{(b+j-1)j} \right)^{\sum_{i \geq j} f_j}.$$

The MLEs are then

$$\hat{\theta}_{\mathrm{ML}} = \begin{bmatrix} \hat{\lambda}_{\mathrm{ML}} \\ \hat{b}_{\mathrm{ML}} \end{bmatrix},$$

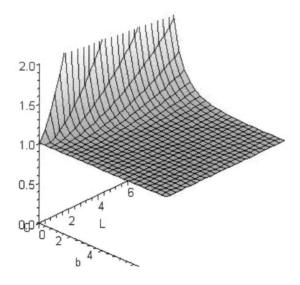

FIGURE 9.1: Ratio $\tilde{p}_0/p_0$ $(L = \lambda)$.

which are found by solving

$$\begin{bmatrix} \frac{\partial}{\partial\lambda} \log \ell(\theta) & = & 0 \\ \frac{\partial}{\partial b} \log \ell(\theta) & = & 0 \end{bmatrix}.$$

To do this we use pre-implemented numerical optimization routines. For simplicity we will call $\hat{\theta}_{\mathrm{ML}}$, and hence $\hat{N}_{\mathrm{ML}}$, the MLEs, although $\hat{\theta}$ actually maximizes an approximation to the conditional likelihood. Despite these approximations and simplifications, though, we are not able to obtain $\hat{N}_{\mathrm{ML}}$ for every dataset. Note: although analogous approximations may be possible for Kemp case 4, ${}_1F'_1$, we have not yet obtained them and consequently do not calculate MLEs for ${}_1F_1$.

9.4   Simulations, data analyses, and discussion

We evaluated the performance of the empirical pgf method on simulated data from the Poisson and negative binomial distributions, computing the estimator $\hat{N}$ and standard error $\sqrt{\widehat{\mathrm{Var}(\hat{N})}}$ (formula (9.7)), with 100 complete replicates at each design point (N, θ). For the variance calculation we used 100 bootstrap replications for the Poisson and 50 for the negative binomial. More replications would yield clearer results but at present we are limited by computational time constraints. We cannot yet simulate the method on the higher-order Kemp distributions, because parameter estimation for those distributions for a single dataset currently takes from minutes to hours on a 3.4-GHz quad core machine, rendering the bootstrap in particular unworkable. We are working on optimizing the code. We also experimented with adding non-distributional noise and outliers to the data, finding little effect on the results (not shown). Overall the results are encouraging, except when the samples are extremely small.

TABLE 9.1: Simulation of empirical pgf procedure in Poisson case

N	θ	$d_{1/2}$ $\bar{\hat{N}}$	SD$(\hat{N})$	$\overline{\text{SE}}$	d_1 $\bar{\hat{N}}$	SD$(\hat{N})$	$\overline{\text{SE}}$
100	0.1	4446.7	4349.6	3399.8	4446.9	4349.4	3399.8
	1	102.2	13.2	14.2	102.2	13.0	14.0
	10	100.0	0.1	0.1	100.0	0.1	0.1
500	0.1	2848.7	9100.3	11269.3	2852.6	9099.5	11267.9
	1	496.7	27.2	27.5	496.8	26.7	27.2
	10	500.0	0.1	0.2	500.0	0.1	0.2
1000	0.1	2695.5	10653.7	9171.8	2699.0	10653.5	9169.1
	1	1002.9	39.8	38.9	1002.7	39.4	38.4
	10	1000.0	0.2	0.2	1000.0	0.2	0.2
5000	0.1	5161.6	1194.2	1182.0	5170.2	1203.2	1192.4
	1	4999.0	83.7	86.8	4998.1	82.3	85.7
	10	5000.0	0.6	0.5	5000.0	0.6	0.5
10000	0.1	10154.2	1625.7	1495.7	10173.7	1636.2	1502.3
	1	10011.7	136.9	122.1	10012.0	136.5	120.6
	10	10000.0	0.6	0.7	10000.0	0.6	0.7

Note: 100 replications/cell. $\bar{\hat{N}}$ and SD$(\hat{N})$ = average and SD of $\hat{N}$ over replications, respectively; $\overline{\text{SE}}$ = average of $\sqrt{\widehat{\text{Var}}(\hat{N})}$ (eqn. (9.7) with 100 bootstrap resamples per replicate), over replications.

The Poisson results are shown in Table 9.1. In particular the standard error formula is an adequate approximation, at least in this simple case, despite the fact that it ignores $\text{Cov}(n,\hat{\theta})$.

Table 9.2 shows the negative binomial results. Here we have used the parametrization $p_j(\theta) = (\Gamma(\theta_1 + j)/(\Gamma(j+1)\Gamma(\theta_1)))\theta_2^{\theta_1}(1-\theta_2)^j, \theta_1 > 0, \theta_2 \in (0,1), j = 0,1,\dots$. As usual for the negative binomial, we find that estimation is challenging when θ is near the (finite) boundaries of the parameter space, i.e., $\theta_1 \approx 0$, $\theta_2 \approx 0$ or 1. In fact, we have omitted results for $\theta_1 = 0.1$, which showed very high variance for both estimates and standard errors. Otherwise the results are reasonable, although less precise than the Poisson owing to one additional parameter. The decrease in accuracy of the standard error formula indicates that for higher-dimensional parameters θ it may be desirable to use Böhning's exact variance formula 9.8, although it requires numerical differentiation which is computationally nontrivial; furthermore as noted above the required bootstrap step is at present too time-consuming. We are currently working on these problems.

We next applied the empirical pgf procedure to ten of the datasets from Chapter 1: golf tees (1.2.1), homeless population (1.2.2), cholera (1.2.3), scrapie (1.2.5), Los Angeles drug users (1.2.6), Bangkok methamphetamine use (1.2.7), dolphins (1.2.10), microbial diversity (1.2.11), Netherlands immigrants (1.2.12), and Shakespeare's words (1.2.13). We applied the empirical pgf procedure to these datasets using all four listed Kemp distributions, and both $d_{1/2}$ and d_1. Due to the time constraints in preparing the present chapter, we present here only point estimates $\hat{N}$. In forthcoming work we examine variance estimation, goodness of fit, and computational algorithms in depth.

We first note that, despite the consistency of both the empirical pgf and maximum likelihood procedures, they may perform quite differently in finite samples. In particular, we compared the minimum distance d_α achieved by the empirical pgf procedure vs. maximum likelihood, finding that the former typically achieved d_α values several orders of magnitude smaller than the latter. Conversely, the MLE maximized the likelihood much more than did the empirical pgf procedure. This is to be expected, but the degree of discrepancy

TABLE 9.2: Simulation of empirical pgf procedure in negative binomial case

N	θ_1	θ_2	$d_{1/2}$ $\bar{\hat{N}}$	SD($\hat{N}$)	$\overline{\text{SE}}$	d_1 $\bar{\hat{N}}$	SD($\hat{N}$)	$\overline{\text{SE}}$
500	1	.1	496.7	19.3	17.1	495.9	18.5	17.8
		.25	481.3	55.3	46.7	493.2	48.1	44.3
		.5	460.8	111.6	342.7	494.0	155.2	335.7
		.75	530.9	824.5	1204.5	409.1	289.6	$\mathcal{O}(10^8)$
		.9	$\mathcal{O}(10^5)$	$\mathcal{O}(10^6)$	$\mathcal{O}(10^8)$	$\mathcal{O}(10^5)$	$\mathcal{O}(10^6)$	$\mathcal{O}(10^6)$
	10	.1	500.0	$\mathcal{O}(10^{-7})$	$\mathcal{O}(10^{-4})$	500.0	$\mathcal{O}(10^{-8})$	$\mathcal{O}(10^{-4})$
		.25	500.0	$\mathcal{O}(10^{-3})$	$\mathcal{O}(10^{-2})$	500.0	$\mathcal{O}(10^{-4})$	$\mathcal{O}(10^{-2})$
		.5	500.2	0.6	0.7	500.1	0.6	0.7
		.75	499.8	7.4	8.1	499.9	6.9	7.7
		.9	517.1	41.4	58.3	517.3	40.8	53.5
1000	1	.1	1002.0	20.5	25.2	997.6	26.6	29.3
		.25	990.3	71.2	72.8	990.9	67.6	69.5
		.5	959.0	218.0	277.1	984.7	143.9	247.7
		.75	779.6	321.5	$\mathcal{O}(10^7)$	969.7	1631.3	930.4
		.9	694.8	307.2	$\mathcal{O}(10^6)$	727.0	339.4	$\mathcal{O}(10^6)$
	10	.1	1000.0	$\mathcal{O}(10^{-7})$	$\mathcal{O}(10^{-4})$	1000.0	$\mathcal{O}(10^{-8})$	$\mathcal{O}(10^{-4})$
		.25	1000.0	$\mathcal{O}(10^{-3})$	$\mathcal{O}(10^{-2})$	1000.0	$\mathcal{O}(10^{-4})$	$\mathcal{O}(10^{-2})$
		.5	1000.0	1.1	1.1	1000.0	1.0	1.0
		.75	1003.9	12.7	11.9	1003.6	12.1	11.2
		.9	1008.1	48.1	57.5	1008.9	47.7	56.2
5000	1	.1	4987.1	74.2	56.9	4985.8	85.4	112.5
		.25	5002.1	192.7	248.7	4922.3	313.1	262.1
		.5	4752.7	644.6	005.0	4605.0	002.4	045.4
		.75	4050.8	917.1	$\mathcal{O}(10^4)$	4324.3	1301.0	1658.5
		.9	3518.7	1714.1	1456.0	3403.7	1026.7	1357.8
	10	.1	5000.0	$\mathcal{O}(10^{-7})$	$\mathcal{O}(10^{-4})$	5000.0	$\mathcal{O}(10^{-7})$	$\mathcal{O}(10^{-4})$
		.25	5000.0	$\mathcal{O}(10^{-3})$	$\mathcal{O}(10^{-2})$	5000.0	$\mathcal{O}(10^{-4})$	$\mathcal{O}(10^{-2})$
		.5	5000.0	2.0	2.4	5000.0	1.8	2.3
		.75	5001.9	20.9	25.2	5001.6	20.7	23.7
		.9	5049.7	127.4	129.4	5052.1	128.3	129.7
10000	1	.1	9963.9	129.2	92.8	9911.5	279.7	209.7
		.25	9892.6	422.1	474.6	9883.0	424.3	534.2
		.5	9656.9	941.8	1128.8	9792.2	773.2	1109.0
		.75	9174.7	1972.0	1638.3	9124.4	1873.8	1694.3
		.9	7101.86	3970.5	8393.6	6930.4	3894.5	2632.4
	10	.1	10000.0	$\mathcal{O}(10^{-7})$	$\mathcal{O}(10^{-2})$	10000.0	$\mathcal{O}(10^{-7})$	$\mathcal{O}(10^{-3})$
		.25	9999.0	0.1	0.1	10000.0	0.1	0.1
		.5	9999.2	3.0	3.3	9999.0	2.9	3.2
		.75	9997.3	27.5	33.8	9998.6	26.6	32.8
		.9	10186.2	208.4	215.6	10192.5	211.1	217.7

Note: 100 replications/cell. $\bar{\hat{N}}$ and SD($\hat{N}$) = average and SD of $\hat{N}$ over replications, respectively; $\overline{\text{SE}}$ = average of $\sqrt{\widehat{\text{Var}}(\hat{N})}$ (Eqn. (9.7) with 50 bootstrap resamples per replicate), over replications.

TABLE 9.3: Point estimates $\hat{N}$ (rounded to nearest integer) for Chapter 1 example datasets

	$_0F_0 =$ Poisson			$_1F_0 =$ negative binomial		
Dataset	$d_{1/2}$	d_1	MLE	$d_{1/2}$	d_1	MLE
1 Golf	175	173	169	219	210	281
2 Homeless	224	222	222	280	253	294
3 Cholera	87	88	88	88	88	1549
5 Scrapie	214	205	170	215	202	96307
6 CA drugs	29946	29234	26426	12876	23713	287962
7 Bangkok	22096	21192	15659	28838	27671	1534640
10 Dolphins	158	157	*	208	208	331
11 Microbial	108	102	82	46565	77298	33067
12 Immigrants	7730	7617	7080	7838	7702	652811
13 Shakespeare	36142	34655	30763	23356430	35552169	6757709

Note: analysis based on 1st and 2nd Kemp models. Two pgf distance metrics and MLE shown. * = computation failed.

TABLE 9.4: Point estimates $\hat{N}$ (rounded to nearest integer) for Chapter 1 example datasets

	$_0F_1$			$_1F_1$	
Dataset	$d_{1/2}$	d_1	MLE	$d_{1/2}$	d_1
1 Golf	174	172	167	276	*
2 Homeless	223	222	222	10900	*
3 Cholera	82	82	*	79	79
5 Scrapie	208	199	146	3932	195254
6 CA drugs	28743	28010	23997	651180	836522
7 Bangkok	20987	19875	7798	38690	37793
10 Dolphins	149	148	142	168	157
11 Microbial	109	100	81	31191	77065
12 Immigrants	7257	7169	5283	7295	7605
13 Shakespeare	35938	34859	30710	18606775	28266404

Note: analysis based on 3rd and 4th Kemp models. Two pgf distance metrics and MLE shown (MLE not computed for 4th Kemp model). * − computation failed.

is somewhat surprising, although the discrepancy did decrease with sample size. This is another topic of our current research.

The population size estimates are shown in Tables 9.3 and 9.4. Without standard errors and goodness-of-fit assessments, only limited interpretations are possible, but we can make the following comments. Overall we find that the third Kemp model $_0F_1$ is "conservative," as is the Poisson model, both returning estimates $\hat{N}$ closer to n than some other models. We know that the Poisson model assumes equal catchability among individuals or species (i.e., a single point-mass abundance model), but $_0F_1$ is known not to be mixed Poisson and so does not admit such an interpretation. It will be interesting to observe the performance of the $_0F_1$ model in the future, and to try to interpret its behavior.

Next, we observe that the negative binomial or second Kemp model not infrequently yields highly divergent results under maximum likelihood optimization. As noted above, we attribute this mainly to the probability parameter estimate's migration toward the boundaries of its range (0 or 1), even when using a reasonably good search algorithm (here, a variant of the EM algorithm). Indeed the empirical pgf procedure appears to outperform maximum likelihood for the negative binomial in some of these cases.

We expected the fourth Kemp model $_1F_1$ to produce higher but not dramatically different estimates compared to the other models, since it is more flexible (with more parameters). This occurred for datasets 1 and 7, but the other results are harder to interpret, either presenting notable differences depending on α (5 and 11), or clearly non-credible estimates (6 and 13). Furthermore the numerical optimization for $_1F_1$ was the most time-consuming, and failed in two cases. We therefore regard this model as requiring further analysis.

While properly computed standard errors will be illuminating when we can compute them efficiently, comparison of point estimates across (parametric) models, with or without standard errors, is of little value without the ability to assess goodness of fit. We believe that the Kemp family is sufficiently rich to provide well-fitting models for a broad range of datasets, especially if we allow larger values of p and q. But to select a good model we require not only a comparative fit measure such as the pseudo-AIC described above, but also an absolute measure such as Pearson's χ^2, and we are presently studying these. Previous work has shown, however, that in many real examples *no* known parametric distribution will fit the entire dataset $\{f_1, f_2, \ldots\}$ (Bunge et al. [63]). For this reason we often use a cutoff τ, carrying out the estimation procedure on the right-truncated dataset $\{f_1, f_2, \ldots, f_\tau\}$ and adding the unused count $\sum_{j>\tau} f_j$ to the final result (essentially treating $\{f_{\tau+1}, f_{\tau+2}, \ldots\}$ as deterministic). This in turn requires selection of τ, which can be done according to goodness-of-fit criteria, but the choice may involve multiple hypothesis tests and consequent adjustment for simultaneous inference. But these problems are amenable to at least practical if not theoretically ideal solution.

In conclusion, we find that the empirical pgf method is promising: it is readily applicable to families of distributions that are best represented via pgfs; it performs well when the distribution of the frequency counts is known or well chosen; it can almost always obtain a result, thanks to its conceptual (if not computational) simplicity; and it admits a straightforward, if time-consuming, bootstrap-assisted standard error calculation. For practical use the procedure will require extension and refinement, especially in terms of model selection, and we are currently working on these topics.

Acknowledgment

The authors are grateful to Shimin Bi and Nan Yao (Cornell University Statistics undergraduate and graduate students, respectively, at the time of writing) for their important contributions to computing the empirical pgf procedure, especially in terms of obtaining the results in Tables 9.3 and 9.4.

10

Convex estimation

Cécile Durot

Modal'X, Université Paris Nanterre

Jade Giguelay, Sylvie Huet

MaIAGE, INRA, Université Paris-Saclay

Francois Koladjo

INSERM U1181, Université Paris-Saclay

Stéphane Robin

UMR518 MIA, AgroParisTech, INRA, Université Paris-Saclay

CONTENTS

10.1 Introduction

10.1.1 Motivation

The estimation of species richness is one of the oldest problems both in statistics and ecology. In this setting, species richness refers to the number of species present in a given place at a given time. The general problem is described in the introduction of this book and traces back to Fisher et al. [123]: based on the number f_1, f_2, f_3... of species for which one, two, three, ... individuals have been observed respectively, one wants to estimate the total number of present species. Obviously, this amounts to evaluating the number f_0 of unobserved species. Although the problem was first raised in the context of ecology, it occurs in a wide variety of domains such as sociology (van der Heijden et al. [280], Example 1.2.2), epidemiology (Cullen et al. [92], Böhning [35, 39], Hser [149], Examples 1.2.3 to 1.2.7), computer sciences (Chun [83], Example 1.2.8) and literature (Efron and Thisted [109], Example 1.2.13).

In this chapter, we use the notation of Chapter 1 (Section 1.1). We denote by y_i the abundance of species i, that is the number of individuals from this species (which is observed only if $y_i > 0$) in the sample. The abundances of the N present species are supposed to be i.i.d. with common distribution $p = (p_y)_{y \geq 0}$. Among the N species, only n of them have a non-zero observed abundance. These species will be referred to as *observed* species. Note that n is random with binomial distribution $\mathcal{B}(N, 1 - p_0)$, whereas N is fixed and to be estimated. Conditional on n, the abundances $Y_1, \ldots, Y_n$ of the n observed species are i.i.d. with distribution $p^+ = (p_y^+)_{y \geq 1}$, where

$$p_y^+ = p_y/(1 - p_0), \tag{10.1}$$

see *e.g.* Durot et al. [104], Lemma 1 of the on line supporting information. Obviously, the estimation of f_0 is related to the estimation of p_0 based on an i.i.d. sample from the truncated distribution p^+. Note that p_0 is not identifiable with no further assumption on the untruncated distribution p.

In the setting we consider, Sanathanan [253] postulated a parametric assumption on p in order to make p_0 identifiable and computed the asymptotic distribution of both the maximum likelihood estimator (MLE) and the so-called conditional MLE. Most authors assumed that f_i is distributed as a Poisson with expectation λ_i, the λ_i's being independent variables from some distribution ω on $(0, \infty)$. Such a setting is called the Poisson mixture setting. It is generally referred to as parametric if a parametric assumption is formulated on ω, and non-parametric otherwise. The parametric Poisson setting is considered in Chao and Bunge [76], Lanumteang and Böhning [173] and implemented in the CatchAll program (Bunge et al. [60]). General results about the MLE in the non-parametric Poisson mixture setting can be found in Laird [171] or Lindsay [178] and the case of Poisson mixtures truncated at zero is considered in Mao and Lindsay [195]. In terms of an inference algorithm, EM-like approaches are considered in Norris and Pollock [220] and Böhning and Schön [37], whereas a penalized log-likelihood procedure is used by Wang and Lindsay [290]. More recently, Wang [292] considered a continuous estimator for ω. Unfortunately, there are no asymptotic results on the aforementioned estimators in the non-parametric Poisson mixture setting. In some sense, Mao and Lindsay [195] proved that no limiting distribution theory could be achievable in this setting because of the discontinuity of the odds $p_0/(1 - p_0)$ with respect to ω. They also proved that asymptotically valid (as $n \to \infty$) confidence intervals for the odds are necessarily one-sided, which means that only lower bounds (for both the odds and N) can be calculated.

10.1.2 Convex abundance distribution

In this chapter we introduce a genuine non-parametric approach that relies on a minimal assumption about the distribution p. We only assume that the abundance distribution p is convex on $\mathbb{N}$, meaning that $(p_{y-1} - p_y) \geq (p_y - p_{y+1})$ for all $y \geq 1$. This implies that the truncated distribution p^+ is also convex on $\mathbb{N}\backslash\{0\}$. Considering the examples from Chapter 1, Section 1.2, it can be seen that most observed empirical abundance distributions $f/n = (f_y/n)_{y \geq 1}$, that estimate an underlying truncated distribution p^+, turn out to be convex, so the assumption is reasonable.

An important property of any discrete convex distribution p on $\mathbb{N}$ is that it admits a representation as a mixture of triangular distributions (see Durot *et al.* [103]). More specifically, denoting by T_j the discrete triangular distribution with support $\{0, 1, \ldots, j-1\}$, that is

$$T_j(y) = \begin{cases} 2(j-y)/[j(j+1)] & \text{if } 0 \le y < j, \\ 0 & \text{otherwise,} \end{cases} \tag{10.2}$$

any discrete convex distribution p on $\mathbb{N}$ admits a representation

$$p_y = \sum_{j \ge 1} \pi_j T_j(y) \tag{10.3}$$

for all $y \ge 0$, where $\pi = (\pi_j)_{j \ge 1}$ is a set of non-negative weights summing up to one. This representation means that sampling a species abundance Y_i according to p is equivalent to first sampling the group j which species i belongs to according to π, and then to sampling Y conditionally to j according to T_j.

Now, note that according to (10.2), T_1 is a Dirac mass at 0. Therefore, the only possible abundance for species from group 1 is zero, which means that such species are actually *absent* species, as none of their individuals can ever be observed. This remark leads to the concept of *convex abundance distributions* defined in Durot et al. [104] as distributions of the form (10.3) with $\pi_1 = 0$. Indeed, if p is supposed to be both convex and to be an abundance distribution (that describes the abundance of *present* species), then the weight π_1 in (10.3) has to be zero, as T_1 describes the abundance of absent species. It will be proved in Section 10.3 that N is identifiable under the assumption that p is a convex abundance distribution. Two different estimators for N are given there.

Examples of convex abundance distributions are given in Durot et al. [104], Section 6.1. For example, geometric distributions are not convex abundance distributions. Also, a Poisson distribution $\mathcal{P}(\lambda)$ is convex if $\lambda \le 2 - \sqrt{2}$ but only $\lambda = 2 - \sqrt{2}$ provides a convex abundance distribution. The classical Fisher's Poisson-gamma model, which assumes that

$$p_y = \frac{\Gamma(\alpha + y)}{\Gamma(\alpha) y!} \left(\frac{\beta}{1+\beta} \right)^\alpha \left(\frac{1}{1+\beta} \right)^y,$$

is a convex abundance distribution provided that $\alpha > 1$ and $\beta = \alpha - 1 + \sqrt{\alpha(\alpha-1)/2}$.

The approach we propose takes advantage of the fact that, as mentioned above, if p is convex on $\mathbb{N}$, then p^+ is convex as well on $\mathbb{N} \setminus \{0\}$. Furthermore, the shifted distribution $q_y = p_{y+1}^+$ is itself convex on $\mathbb{N}$ so its inference (which is obviously equivalent to that of p^+) can be tackled with the methodology developed in Durot *et al.* [103]. This chapter elaborates upon Durot et al. [103, 104] where a similar framework is considered, and provides a theoretical study of the power of the convexity test proposed in the latter and a new derivation (with theoretical justification) of a confidence interval for N.

Section 10.2 is devoted to a theoretical analysis of a test to assess the convexity of the truncated distribution p^+. Section 10.3 is devoted to the estimation of p^+ and N under the constraint of convexity. The standard deviation of the estimator of N, and a confidence interval for N, are given in Section 10.4. The complete procedure is illustrated in Section 10.5 on real-life data. Proofs are given in Section 10.6.

10.2 Testing the convexity of p^+

The aim of this section is to build a statistical test to check whether the convexity assumption of the truncated distribution p^+ is reasonable. The construction of the test, together with theoretical properties, are given in Section 10.6.1. To assess the performance of the testing procedure when N is finite, a simulation study is reported in Section 10.2.2.

10.2.1 The statistical test

Recall that p^+ is convex on $\mathbb{N}\setminus\{0\}$ if and only if $(p_{y-1}^+ - p_y^+) \geq (p_y^+ - p_{y+1}^+)$ for all $y \geq 2$. Let us define the Laplacian of a discrete function $\mu = (\mu_y)_{y\geq 1}$ on $\mathbb{N}\setminus\{0\}$ as

$$\Delta\mu_y = \mu_{y+1} - 2\mu_y + \mu_{y-1} \tag{10.4}$$

for all $y \geq 2$. Then p^+ is convex if and only if $\Delta p_y^+ \geq 0$ for all $y \geq 2$. Hence, we propose to reject the null hypothesis that p^+ is convex on $\mathbb{N}\setminus\{0\}$ if at least one of the empirical estimators $\Delta f_y/n$, for $y \geq 2$, is smaller than some negative threshold. With $Y_{\max} = \max_{i=1,\dots,n} Y_i$ being the maximal point of support of f, we have $\Delta f_y/n > 0$ for $y = Y_{\max}+1$ and $\Delta f_y/n = 0$ for all $y > Y_{\max} + 1$ so we restrict ourselves to $y \leq Y_{\max}$. This means that our test statistic is defined as

$$S_n = \frac{1}{\sqrt{n}} \min_{2\leq y\leq Y_{\max}} \Delta f_y. \tag{10.5}$$

If p^+ is convex, then $S_n \geq \overline{S}_n$ where

$$\overline{S}_n = \frac{1}{\sqrt{n}} \min_{2\leq y\leq Y_{\max}} \left(\Delta f_y - n\Delta p_y^+\right). \tag{10.6}$$

In the following, we assume that p^+ has a finite support. Then, as N goes to infinity, conditional on n, $\overline{S}_n$ converges almost surely to a random variate that depends on p^+ and that we denote by $\overline{S}(p^+)$. Following the classical plug-in procedure, the quantiles of $\overline{S}(p^+)$ are approximated by those of $\overline{S}(f/n)$ and calculated using a simulation procedure. For a chosen $0 < \alpha < 1$, let q_α^* be the α-quantile so calculated. The rejection region for testing that p^+ is convex is then defined as

$$\{S_n \leq q_\alpha^*\}.$$

The procedure for calculating q_α^* is detailed in Section 10.6.1. In that section, it is proved that the test has an asymptotic level smaller than or equal to α as $N \to \infty$ and that triangular distributions are among least favorable distributions. In addition, we prove in Section 10.6.1 that the test is powerful in a certain sense. To be more specific, note that if p^+ is not convex, then there exists $y \geq 2$ such that $\Delta p_y^+ < 0$. We quantify how small Δp_y^+ should be, so that the test rejects the null hypothesis of convexity with high probability: for any small positive β, we can find a positive constant C that depends on β and α, such that if $\sqrt{N(1-p_0)}\Delta p_y^+/2 < -C$ for some $y \geq 2$, then the power of the test is asymptotically greater than $1 - \beta$ (see Theorem 10.2 in Section 10.6.1).

10.2.2 Simulation study

Because in real case studies we do not observe N, we chose to simulate observations conditionnally on n with $n \in \{50, 100, 250, 750, 1500, 3000, 10000, 50000\}$, covering thus a wide range of possible applications. To investigate the behavior of the test, we considered the following convex distributions p^+. In the following, we call the values y knots, such that $\Delta p_y^+ > 0$.

- We chose triangular distributions $p_y^+ = T_j(y-1), y \geq 1$ (see Equation (2.1)) for $j \in \{3, 6, 15, 50\}$. These distributions have only one knot in $y = j+1$ and are known to be least favorable distributions.

TABLE 10.1: Empirical rejection probabilities (in %) for the triangular distributions T_j with $j \in \{3, 6, 15, 50, 100\}$ versus n

n	50	100	250	750	1500	3000	10000	50000
$j = 3$	5.18	5.58	4.92	5.44	4.90	5.32	5.24	5.10
$j = 6$	5.94	5.24	4.54	4.58	5.68	5.84	4.94	5.16
$j = 15$	6.28	5.86	5.92	5.36	5.38	5.18	5.44	5.14
$j = 50$	4.50	5.42	6.02	5.88	6.10	5.18	5.30	5.30
$j = 100$	3.06	3.94	5.74	6.90	5.96	6.04	6.26	5.12

- We considered strictly convex distributions where Δp_y^+ is strictly positive for all y in the support of p^+. Precisely, $p_y^+ = Q_{15}^k(y-1)$ where Q_j^k is the spline of degree k with support $\{0, \dots j-1\}$ defined by

$$Q_j^k(y) = \begin{cases} C_{j-y+k-2}^{k-1} / \sum_{i=0}^{j-1} C_{j-i+k-2}^{k-1} & \text{if } 0 \leq y < j \\ 0 & \text{otherwise.} \end{cases} \quad (10.7)$$

Choosing $k \in \{3, 6, 20\}$, we compared the level of the test for distributions that are more and more hollow (see Figure 10.1).

- Next, we considered mixtures of triangular distributions, $p_y^+ = \sum_{j \geq 1} \pi_j T_j(y-1)$. As will be shown in the following section, the knots of the distribution p^+ are the integers j such that $\pi_j \neq 0$. We chose different types of mixtures in order to consider the effect of the number of knots. Starting from the Q_{15}^3 distribution, we determined a convex distribution that is closed to Q_{15}^3 for different sets of knots. Precisely

$$p_1^+ = 0.24T_3 + 0.59T_7 + 0.17T_{15} \quad (10.8)$$
$$p_2^+ = 0.02T_1 + 0.18T_3 + 0.37T_6 + 0.35T_9 + 0.08T_{15} \quad (10.9)$$
$$p_3^+ = 0.03T_1 + 0.15T_3 + 0.22T_5 + 0.24T_7 +$$
$$0.19T_9 + 0.12T_{11} + 0.05T_{13} + 0.01T_{15} \quad (10.10)$$

- Finally we considered Poisson distributions with parameter λ : for $y \geq 1$, $p_y^+ = \lambda^{y-1} e^{-\lambda}/(y-1)!$, with $\lambda \in \{0.5, 0.55, 2 - \sqrt{2}, 0.6, 0.7, 1\}$. Only the three first values of λ give a convex distribution, $\lambda = 2 - \sqrt{2}$ being the largest value of λ such that the Poisson distribution is convex. The three last values of λ allow us to evaluate the power of the test (see Figure 10.1).

The nominal level was fixed at $\alpha = 0.05$. The level and the power of the test was calculated on the basis of 10000 samples and q_α^* was calculated on the basis of 10000 samples. Note that the standard-error of the estimated level is $\sqrt{0.05 \times 0.95/10000} = 0.002$, so the empirical levels are expected to lie within 0.046 and 0.054 with probability 0.95.

The simulations were carried out with R (R Core Team, 2014).

Level of the test

The simulation study for the triangular distributions T_j (see Table 10.1) shows that the number of observations needed so that the level of the test reaches the asymptotic level α increases with j. Note that triangular distributions T_j are flatter for large values of j, the greatest value being $T_j(0) = 2/(j+1)$. Therefore, for large values of j, the convergence of the empirical distribution to a Gaussian process is very slow.

Considering the Q_j^k's distributions (see Table 10.2), it appears that the empirical level

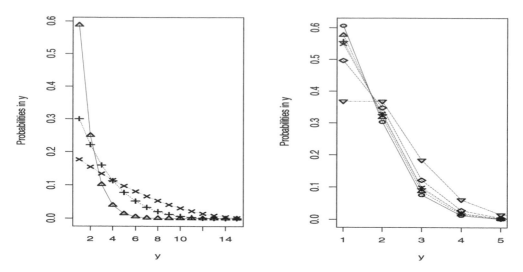

FIGURE 10.1: Simulated distributions. Left: $\times$, $+$, $\triangle$ for the splines of degrees 3, 6, 20 respectively. Right: $\circ$, $\triangle$, $+$, $\times$, $\diamond$, $\triangledown$ for Poisson distributions (represented only on $\{1, .., 5\}$) with parameters 0.5, 0.55, $2 - \sqrt{2}$, 0.6, 0.7, 1 respectively.

TABLE 10.2: Empirical rejection probabilities (in %) for the spline distributions Q_j^k for $j = 15$ and $k \in \{3, 6, 20\}$ versus n

n	50	100	250	750	1500	3000	10000	50000
Q_3	6.20	5.80	5.92	5.02	4.84	3.82	3.72	2.1
Q_6	4.56	4.08	3.84	2.60	1.18	0.50	0.00	0
Q_{20}	0.60	0.18	0.02	0	0	0	0	0

of the test decreases when k increases (that is to say, when the distribution is more convex). Note that when k increases, the values of the Laplacian increase, making it more difficult to reject the null hypothesis.

The results of the simulation study for the Poisson distributions are shown in Table 10.3; the lines corresponding to the values of λ smaller than $2 - \sqrt{2}$ show the level of the test for convex Poisson distributions. The distribution is more convex for smaller values of λ in the sense that when λ decreases, the Laplacian of the distributions calculated in $y = 1$ increases, the smallest value 0 being achieved for $\lambda = 2 - \sqrt{2}$. For this value of λ, the empirical level is close to 5% for all n, whereas it vanishes as n grows for smaller values of λ.

Finally, considering the mixtures of triangular distributions with an increasing number of knots (see Table 10.4), it appears that the level of the test decreases when the number of knots increases. This can be explained as follows: the quantile q_α^* approximates the α-quantile of $\overline{S}_n$ and not the one of S_n, but the higher the number of knots, the higher the difference $\overline{S}_n - S_n$, see Equation (10.6). Note that if the distribution p^+ has no knot on $\{1, \ldots, Y_{\max}\}$, then $\overline{S}_n = S_n$.

TABLE 10.3: Empirical rejection probabilities (in %) for the Poisson distributions with $\lambda \in \{0.5, 0.55, 2 - \sqrt{2}, 0.6, 0.7, 1\}$

n	50	100	250	750	1500	3000	10000	50000
$\lambda = 0.5$	1.12	1.22	0.52	0.06	0	0	0	0
$\lambda = 0.55$	2.52	2.78	2.1	1.18	0.56	1.8	0	0
$\lambda = 2 - \sqrt{2}$	3.88	4.42	4.28	4.56	4.34	4.94	4.56	4.64
$\lambda = 0.6$	4.50	5.44	5.44	7.12	8.28	10.3	48.5	52.4
$\lambda = 0.7$	7.96	11.2	19.3	41.1	66.8	90.1	100	100
$\lambda = 1$	21.0	32.5	62.1	96.6	99.6	100	100	100

TABLE 10.4: Empirical rejection probabilities (in %) for the distributions described by Equations (10.8), (10.9), and (10.10) versus n

n	50	100	250	750	1500	3000	10000	50000
p_1^+	5.25	4.94	4.7	5.12	4.62	4.92	4.42	4.48
p_2^+	5.06	4.28	4.1	3.5	3.72	3.32	3.68	3.22
p_3^+	4.62	3.8	2.94	2.3	1.76	1.86	1.7	1.44
Q_6	4.56	4.08	2.84	2.60	1.46	0.50	0.08	0

Power of the test

The power for non-convex Poisson distributions is given in Table 10.3 for $\lambda \geq 0.6$. As expected, the power tends to 1 as n grows and increases with λ.

10.3 Estimating the number N of species

As detailed in Section 10.1, the number n of observed species has binomial distribution $\mathcal{B}(N, 1 - p_0)$, where N is the total number of species to be estimated and the abundances of the N present species are supposed to be i.i.d. with common distribution $p = (p_y)_{y \geq 0}$. Conditional on n, the abundances $Y_1, \ldots, Y_n$ of the n observed species are i.i.d. with distribution $p^+ = (p_y^+)_{y \geq 1}$ given by (10.1). We assume that p is a *convex abundance distribution*, which means that p admits a representation (10.3) for all $y \geq 0$, where $\pi = (\pi_j)_{j \geq 1}$ is a set of non-negative weights summing up to one with $\pi_1 = 0$.

Under this assumption, N is identifiable (see Section 10.3.1). The precise connection between N and p^+ is given below, so starting with an estimator of p^+, we derive an estimator of N. Two different estimators of p^+ are introduced in Section 10.3.2. The species number problem is addressed afterward.

10.3.1 Identifiability of N

Since p is convex on $\mathbb{N}$, it follows from Theorem 7 in Durot et al. [103] that p has a representation (10.3) for all $y \geq 0$, where the mixing probabilities π_j are determined by

$$\pi_j = \frac{j(j+1)}{2}(p_{j+1} - 2p_j + p_{j-1}) \text{ for all } j \geq 1. \tag{10.11}$$

We have $\pi_1 = 0$ since p is assumed to be a convex abundance distribution, whence $p_0 = 2p_1 - p_2$. Using (10.1), this means that

$$\frac{1}{1 - p_0} = 2p_1^+ - p_2^+ + 1. \qquad (10.12)$$

Note that $p_0 < 1$, otherwise we would have no observation. But p^+ is identifiable since conditional on n, the abundances $Y_1, \ldots, Y_n$ of the n observed species are i.i.d. with distribution p^+, so p_0 also is identifiable. The identifiability of $N = \mathbb{E}(n)/(1 - p_0)$ follows from that of p_0.

10.3.2   Estimating p^+

Conditional on n, the observed abundances $Y_1, \ldots, Y_n$ are i.i.d. with distribution p^+, so $(f_y/n)_{y \geq 1}$ defines an empirical estimator of p^+, where we recall that

$$f_y = \sum_{i=1}^{n} 1_{(Y_i=y)} \qquad (10.13)$$

is the frequency of species for which y individuals have been observed.

The above empirical estimator may be non-convex whereas under our assumptions, p^+ is a convex distribution on $\mathbb{N}\backslash\{0\}$. Hence, in addition to the empirical estimator, we consider an estimator that takes into account the convexity constraint. Precisely, we consider the constrained least-squares estimator (LSE) $\widehat{p}^+$ of p^+ defined as the unique minimizer of

$$\sum_{y \geq 1} (c_y^+ - f_y/n)^2$$

over the set of all convex sequences $c^+ = (c_y^+)_{y>1}$ on $\mathbb{N}\backslash\{0\}$ satisfying $\sum_{y>1}(c_y^+)^2 < \infty$. Existence and uniqueness of $\widehat{p}^+$ follows from Durot et al. [103]. Note that this reference considers convex distributions on $\mathbb{N}$ whereas we are interested here in convex distributions on $\mathbb{N}\backslash\{0\}$, but considering the shifted distribution $q_y = p_{y+1}^+$ for $y \geq 0$, which is convex on $\mathbb{N}$, and the corresponding shifted estimators f_{y+1} and $\widehat{p}_{y+1}^+$ allows to put our framework into that of Durot et al. [103]: we have $\widehat{p}_{y+1}^+ = \widehat{q}_y$ where $\widehat{q} = (\widehat{q}_y)_{y \geq 0}$ is the unique minimizer of

$$\sum_{y \geq 0} (c_y - f_{y+1}/n)^2 \qquad (10.14)$$

over the set of all convex sequences $c = (c_y)_{y \geq 0}$ on $\mathbb{N}$ satisfying $\sum_{y \geq 0} c_y^2 < \infty$. The shifting also allows us to compute the convex LSE of p^+ using the algorithm suggested in Durot et al. [103]. The algorithm is inspired by Groenenboom et al. [136]. It is a support reduction-type algorithm that efficiently computes the least-squares estimator $\widehat{q}$ of $q = (q_y)_{y \geq 0}$ using that similar to (10.3), q admits a representation

$$q_y = \sum_{j \geq 1} \pi_j^q T_j(y) \text{ for all } y \geq 0.$$

Starting with an initial positive measure π^q, the algorithm checks if the gradient of the least squares criterion in (10.14) is non-negative in all directions. This condition is required for a positive measure to be a local minimizer of the least-squares criterion. If the local minimizer is a probability measure, then it is the global minimizer over the set of all possible π^q. The steps of the algorithm and its convergence properties are detailed in Durot et al. [103].

10.3.3 Estimating N

Because $\mathbb{E}(n) = N(1 - p_0)$, it follows from (10.12) that the connection between N and p^+ is that

$$N = E(n)\left(1 + 2p_1^+ - p_2^+\right).$$

We estimate $E(n)$ by n and we plug in an estimator of p^+ to obtain an estimator of N. As we have at hand two different estimators for p^+, we obtain two different estimators for N: the estimator based on empirical frequencies

$$N^f = n\left(1 + 2\frac{f_1}{n} - \frac{f_2}{n}\right) = n + 2f_1 - f_2 \tag{10.15}$$

and the estimator based on the constrained LSE

$$\widehat{N} = n\left(1 + 2\widehat{p}_1^+ - \widehat{p}_2^+\right). \tag{10.16}$$

Both estimators are studied in Durot et al. [104]. The estimator N^f is easy to implement. Its theoretical properties can easily be derived from that of f/n. It has a Gaussian asymptotic distribution with variance in a closed form. On the other hand, implementing $\widehat{N}$ requires a sophisticated algorithm. The asymptotic distribution of $\widehat{N}$ is non-standard and more difficult to approximate. However, the simulation study in Durot et al. [104] shows that it has a smaller prediction error than N^f.

10.4 Confidence intervals and standard errors

We consider the same model and notation as in Section 10.3. This means that we assume that p is a convex abundance distribution. Hence, N is identifiable and we have at hand two different estimators for N. In this section, we investigate the standard deviation of those estimators. Based on these estimators, we build asymptotically valid (as $N \to \infty$) confidence intervals for N.

10.4.1 Estimator based on empirical frequencies

Consider the estimator N^f defined by (10.15). It is proved in Durot et al. [104] that a consistent estimator of the standard error of N^f is given by $\sqrt{6f_1}$ and that $(N^f - N)/\sqrt{6f_1}$ converges to a standard Gaussian distribution as $N \to \infty$. This means that with $\alpha \in (0, 1)$ and $z_{1-\alpha/2}$, the $(1 - \alpha/2)$-quantile of a standard Gaussian distribution,

$$CI^f = \left[N^f - z_{1-\alpha/2}\sqrt{6f_1}\,,\ N^f + z_{1-\alpha/2}\sqrt{6f_1}\right] \tag{10.17}$$

is a confidence interval for N with asymptotic level $1 - \alpha$ as $N \to \infty$.

10.4.2 Estimator based on the constraint LSE

Consider the estimator $\widehat{N}$ defined in (10.16) where we recall that $\widehat{p}^+$ is the constrained LSE of p^+ defined in Section 10.3.2. In the following, we assume that either p^+ has a finite support, or there exists $y \geq 2$ such that both y and $y+1$ are knots of p^+, which means that $\Delta p_y^+ > 0$ and $\Delta p_{y+1}^+ > 0$ with Δ being the discrete Laplacian defined in (10.4).

The limiting distribution of $(\widehat{N} - N)/\sqrt{n}$ is computed in Durot et al. [104] under this assumption, as $N \to \infty$, and confidence intervals for N are proposed based on a plug-in procedure and on bootstrap. However, there is no proof concerning the asymptotic properties of these confidence intervals. For this reason, we introduce a random variable that has the same limiting distribution as $(\widehat{N} - N)/\sqrt{n}$ and which distribution can be estimated using Monte Carlo simulations.

The definition of this random variable is inspired by Balabdaoui et al. [22]. We need to introduce notation to make the definition precise. Denoting by v a positive real number such that $n^{-1/2} \ll v \ll 1$, we define $\hat{k} = \min\{Y_{\max} + 1, \hat{\kappa}\}$ where we recall that $Y_{\max}$ is the greatest point of support of f and $\hat{\kappa}$ is the smallest $y \geq 2$ such that both $n^{-1}\Delta f_y \geq v$ and $n^{-1}\Delta f_{y+1} \geq v$ ($\hat{\kappa}$ being infinite if no such y exists). Now, conditional on $(Y_1, \ldots, Y_n)$, define $\widehat{\Omega}$ and $\widehat{W} = (\widehat{W}_1, \ldots, \widehat{W}_{\hat{k}})$ where $\widehat{W}$ is a centered Gaussian vector in $\mathbb{R}^{\hat{k}}$ with covariance matrix $\widehat{\Gamma}$ defined as

$$\widehat{\Gamma}_{yy'} = \begin{cases} \widehat{p}_y^+ (1 - \widehat{p}_y^+) & \text{if } y = y' \\ -\widehat{p}_y^+ \widehat{p}_{y'}^+ & \text{if } y \neq y' \end{cases}$$

for all $1 \leq y, y' \leq \hat{k}$, and where $\widehat{\Omega}$ is a centered Gaussian variable independent of $\widehat{W}$, with variance $\left(2\widehat{p}_1^+ - \widehat{p}_2^+ + 1\right)\left(\widehat{p}_2^+ - 2\widehat{p}_1^+\right)$. Note that the procedure is still valid if one replaces $\widehat{p}^+$ by f/n in the definitions of $\widehat{\Gamma}_{yy'}$ and $\widehat{\Omega}$. Next, define $\widehat{g}$ as the minimizer of

$$\sum_{y=1}^{\hat{k}} \left(q_y - \widehat{W}_y\right)^2$$

over the set of all sequences $q \in \mathbb{R}^{\hat{k}}$ such that $\Delta q_y \geq 0$ for all $y \in \{2, \ldots, \hat{k} - 1\}$ such that $\Delta \widehat{p}_y^+ < v$, with no constraint at other points y. Note that $\widehat{g}$ can be computed using the Dykstra type algorithm described in Section 4.1 of Balabdaoui et al. [21]. One can simulate a large number of independent replications of $(\widehat{W}, \widehat{\Omega})$ conditional on $(Y_1, \ldots, Y_n)$, and then obtain the corresponding independent replications of $\widehat{g} + \widehat{\Omega}$. The empirical distribution of these independent replications then provides an estimate for the conditional distribution of $\widehat{g} + \widehat{\Omega}$.

It turns out that the conditional distribution of $\widehat{g} + \widehat{\Omega}$ converges to the limiting distribution of $(\widehat{N} - N)/\sqrt{n}$ (see Theorem 10.3 in Section 10.6.2). This means that the empirical standard error of i.i.d. replications of $\widehat{g} + \widehat{\Omega}$ provides an estimate of the standard error of $\widehat{N}$. Moreover, with $\alpha \in (0, 1)$ and $\check{z}_\alpha$, the α-quantile of the conditional distribution of $\widehat{g} + \widehat{\Omega}$,

$$\left[\widehat{N} - \sqrt{n}\check{z}_{1-\alpha/2}, \ \widehat{N} - \sqrt{n}\check{z}_{\alpha/2}\right] \tag{10.18}$$

is a confidence interval of asymptotic level $1 - \alpha$ for N (as $N \to \infty$). This confidence interval depends on v, which controls the estimation of the set of knots of p^+. In order to assess the effect of v on the level of the confidence interval, we carried out a large simulation study (not shown here). The conclusion was that taking for v a very small positive value, was the best choice in order to keep confidence intervals with covering probabilities close to $1 - \alpha$. Although the choice $v = 0$ is not supported by theoretical proofs, it appears to perform very well on simulations. Hence, to keep the calibration as simple as possible, we recommend the choice $v = 10^{-15}$ to take into account the machine precision.

We point out that the validity of the confidence interval follows from the convergences in Theorem 10.3 (Section 10.6.2) provided that the limiting distribution in that theorem is continuous. Although we feel that the limiting distribution is indeed continuous, the proof of that property is out of the scope of the contribution.

10.5 Case studies

The purpose of the present section is to illustrate the proposed methodology on a series of examples.

Analysis of various abundance distributions

We first considered some of the examples introduced in Chapter 1, Section 1.2 of this book. As observed in Table 10.5, the convexity hypothesis is accepted in most cases and is only rejected for the homeless dataset. This rejection is due to the increase of the frequencies f_y for the largest values of y ($y = 11, 12, 13$), as observed in Figure 10.2. When truncated to $y \leq 10$, the test for convexity is accepted with p-value $= 0.217$. Still this truncation yields a smaller estimation of the total abundance: $\widehat{N} = 165$ (se $= 14.7$). So, except for this example where convexity is questionable, the assumption is consistent with most of the empirical datasets.

For all examples from Chapter 1, Section 1.2, the two estimates N^f and $\widehat{N}$ coincide. This comes from the first three frequencies f_1, f_2 and f_3, which display a convex shape in all these cases. As a result, the standard errors are also equal and the bounds of the confidence intervals are also very close (not shown). To emphasize the influence of the first frequencies, we added the Bird abundance dataset introduced in Norris and Pollock [221], in which the first three frequencies are 11, 12 and 10, respectively. As shown in Figure 10.2, this induces a substantial difference in the estimation of p_0 and in the estimation of N: $N^f = 82$ vs $\widehat{N} = 87$. Because $\widehat{N}$ relies on the convexity assumption (which is accepted by the convexity test), its standard-error is smaller and it results in a narrower confidence interval: $[70.0; 95.5]$ vs $[66.1; 97.9]$ for N^f.

TABLE 10.5: Test for convexity and estimated abundance for a series of examples

	$\widehat{T}$	p-value	N^f	SE	$\widehat{N}$	SE
Homeless in Utrecht	-2.66	0.002	206	14.70	206	14.70
Cholera data	0.52	1.000	115	14.70	115	14.70
Scrapie surveillance	-0.18	0.981	271	22.45	271	22.45
Drug use in L.A.	-2.16	0.032	39963	268.13	39963	268.13
Methamphetamine use	-0.14	0.860	9410	136.69	9410	136.69
Dolphins in Moorea	0.42	0.98	128	15.87	128	15.87
Microbial diversity	-0.45	0.811	167	16.97	167	16.97
Shakespeare's words	-0.27	0.988	55097	293.69	55097	293.69
Bird abundance [221]	-0.82	0.585	82	8.12	87	6.57

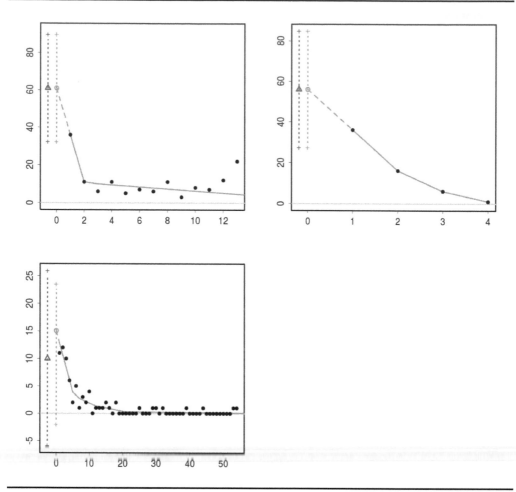

FIGURE 10.2: Abundance prediction for a series of examples. Top: homeless population in Utrecht, McKendrick's cholera data. Bottom: bird abundance from Norris and Pollock [221]. Observed frequencies: dots; fitted convex abundance distribution: solid line; abundance prediction: triangle $= N^f$, circle $= \widehat{N}$; confidence intervals: dotted lines.

Grizzly bear population in Yellowstone

We then studied the population of female grizzly bears in Yellowstone as described in Chapter 1, Section 1.2.9. The dataset consists of the collection of observed frequencies f_j in the years from 1986 to 2001. As shown in Table 10.6, according to the convexity test described in Section 10.2, the convexity hypothesis is never rejected for any of the 16 years under study. The same table compares the estimated population size N under the convexity assumptions using both estimates (10.15) and (10.16). The two estimates are quite close, $\widehat{N}$ being a bit larger. The standard deviation of $\widehat{N}$ is smaller than that of N^f. Figure 10.3 provides N^f, $\widehat{N}$ together with the confidence intervals (10.17) and (10.18). The general trend is that the population has increased during the period under study.

TABLE 10.6: Estimated abundance of grizzlies by year

	p-value	N^f	SE	$\widehat{N}$	SE	n
1986	0.333	33	6.48	33	5.16	24
1987	0.872	23	6.48	23	6.48	12
1988	0.665	27	6.48	27	6.51	17
1989	0.236	20	6.00	23	5.51	13
1990	0.249	30	6.48	31	5.36	22
1991	0.923	43	8.12	43	8.09	24
1992	0.915	47	9.49	47	9.49	22
1993	0.166	23	6.48	28	5.50	17
1994	0.552	29	7.35	31	6.56	18
1995	0.636	41	8.83	41	8.83	17
1996	0.658	48	9.49	50	8.45	28
1997	0.459	48	8.83	48	8.83	29
1998	0.159	42	8.12	50	7.04	33
1999	0.394	43	7.35	42	7.11	29
2000	0.543	58	10.10	58	10.10	32
2001	0.618	58	9.80	60	7.94	38

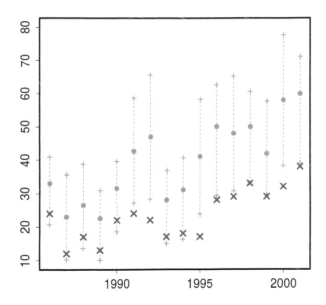

FIGURE 10.3: Estimated abundance of grizzlies from 1986 to 2001. •: convex estimation $\widehat{N}$ (+---+: 95% confidence intervals). ×: total number of observed individuals n.

Comments

As already noticed, in some applications, the two estimators N^f and $\widehat{N}$ may be very close. This is obviously the case when the empirical distribution is convex or nearly convex, see the cholera data for example. More generally a large simulation study has shown that $\widehat{N}$ tends to over-estimate N for small values of N and p_0, while N^f is unbiased. However $\widehat{N}$ has a smaller standard error than N^f leading finally to a smaller prediction error (see Durot et al. [104]).

10.6 Appendix

10.6.1 Testing convexity of a discrete distribution

This section is devoted to a precise description of the test of Section 10.2. We assume here that the truncated distribution p^+ has finite support and we denote by τ the maximal point of support, which means that $p_\tau^+ > 0$ and $p_y^+ = 0$ for all $y \geq \tau + 1$. As described in Section 10.2, we reject the null hypothesis that p^+ is convex if the test statistic defined by (10.5) is smaller than some negative threshold q_α^*. We need notation to precisely define q_α^*.

Let Γ^+ be the $\tau \times \tau$ matrix with components $\Gamma_{yy'}^+ = -p_y^+ p_{y'}^+$ if $y \neq y'$ and $\Gamma_{yy}^+ = p_y^+(1-p_y^+)$ for $1 \leq y, y' \leq \tau$, and let Γ^f be the plug-in estimator of Γ^+ obtained by replacing p_y^+ and τ by f_y/n and $Y_{\max}$, respectively. Let us introduce the $(\tau-1) \times \tau$ matrix A whose lines A_y^T satisfy $\Delta p_y^+ = A_{y-1}^T p^+$ for $y = 2, \ldots, \tau$. We will see below that Γ^+ is a variance matrix. This means that Γ^+, as well as $A\Gamma^+ A^T$, is symmetric and semi-positive definite. Hence, the square-root of $A\Gamma^+ A^T$ is well defined. In the following, we denote by M the square-root, which means that M is a symmetric matrix that satisfies $MM = A\Gamma^+ A^T$. We define M^f as the plug-in estimator of M. Finally, for $\alpha > 0$ we define q_α^* as the α-quantile of the conditional distribution of Z^f given $Y_1, \ldots, Y_n$, where

$$Z^f = \min_{1 \leq y \leq Y_{\max}-1} \sum_{y'=1}^{Y_{\max}-1} M_{yy'}^f Z_{y'}$$

and $Z_1, Z_2, \ldots$ are independent standard Gaussian variates.

The following theorem shows that the test has asymptotic level α and that triangular distributions are among the least favorable distributions.

Theorem 10.1
Assume that p^+ is convex with a finite support. Then, for all $\alpha \in (0, 1)$ we have

$$\lim_{N \to \infty} \mathbb{P}\left(S_n \geq q_\alpha^*\right) \leq \alpha$$

with an equality if p^+ is a triangular distribution.

The following lemma will be used repeatedly in the proofs.

Lemma 10.1 *Assume that p^+ has a finite support. Then with τ the maximal point of the support, we have $Y_{\max} = \tau$ with probability that tends to one. Moreover, with probability one we have*

$$\lim_{N \to \infty} \frac{f}{n} = p^+.$$

Proof of Lemma 10.1. Note that $\tau \geq 1$ is also the maximal point of support of p, and we have $Y_{\max} = \max_{i=1,\ldots,n} Y_i = \max_{i=1,\ldots,N} y_i$. For the sake of clarity, we denote $Y_{\max}$ by $Y_{\max,N}$. Now,

$$\mathbb{P}\left(Y_{\max,N} < \tau\right) = (1 - p_\tau)^N$$

which converges to zero as $N \to \infty$ since $p_\tau > 0$. This proves the first assertion.

To prove the second assertion, note that

$$\mathbb{P}\left(\lim_{N \to \infty} \frac{f}{n} = p^+\right) = E\left[\mathbb{P}\left(\lim_{n \to \infty} \frac{f}{n} = p^+|n\right)\right] + o(1),$$

using again that n/N converges to $1 - p_0 > 0$ with probability one. Conditional on n, $Y_1, \ldots, Y_n$ are i.i.d. with distribution p^+ so the second assertion follows from the strong law of large numbers combined with (10.13), together with the dominated convergence theorem. This completes the proof of Lemma 10.1. $\square$

Proof of Theorem 10.1. In the following, $\mathbb{P}(\cdot|Y)$ refers to the conditional probability given $Y_1, \ldots, Y_n$. We consider the variable Z defined as

$$Z = \min_{1 \le y \le \tau-1} \sum_{y'=1}^{\tau-1} M_{yy'} Z_{y'}.$$

It can be proved using Balabdaoui et al. [22] that Z has a continuous distribution (details are omitted). We denote by q_α the $(1 - \alpha)$-quantile of Z.

First, we prove that the conditional distribution of Z^f given $(Y_1, \ldots, Y_n)$ converges in probability to the distribution of Z. Since the distribution of Z is continuous, this means that for all $\epsilon > 0$ we have

$$\lim_{N \to \infty} \mathbb{P}\left(\sup_{t \in \mathbb{R}} \left|\mathbb{P}(Z^f \le t|Y) - \mathbb{P}(Z \le t)\right| > \epsilon\right) = 0. \tag{10.19}$$

To prove (10.19), we consider the random variable $\overline{Z}^f$, which we define in the same way as Z^f, but with $Y_{\max}$ replaced by τ, the greatest point of support of p^+. By Lemma 10.1, we have $\overline{Z}^f = Z^f$ with probability that tends to one and therefore, the probability in (10.19) is equal to

$$\mathbb{P}\left(\sup_{t \in \mathbb{R}} \left|\mathbb{P}(\overline{Z}^f \le t|Y) - \mathbb{P}(Z \le t)\right| > \epsilon\right) + o(1).$$

But f/n converges to p^+ with probability one (see Lemma 10.1), so (10.19) follows from the continuity of the distribution of Z.

Since the conditional distribution of Z^f given $(Y_1, \ldots, Y_n)$ converges in probability to the distribution of Z, for any subsequence we can extract a further subsequence along which the convergence holds with probability one. Hence, it follows from Lemma 21.2 in van der Vaart [288] that for any subsequence, we can extract a further subsequence along which q_α^* converges to q_α with probability one. This means that q_α^* converges to q_α in probability as $N \to \infty$. As a consequence, for all $\delta > 0$ we have

$$|q_\alpha^* - q_\alpha| \le \delta \text{ with probability that tends to one.} \tag{10.20}$$

Now, we prove that $\overline{S}_n$ converges in distribution to Z as $N \to \infty$. Similar to the above, we consider the random variable $\overline{S}_{n,\tau}$ that is defined in the same way as $\overline{S}_n$ but with $Y_{\max}$ replaced by τ. Using again that n/N converges almost surely to $1 - p_0$, it follows from the central limit theorem that the conditional distribution of the vector $\sqrt{n}A(f/n - p^+)$ given n converges to the distribution of the centered Gaussian vector with variance matrix equal to $A\Gamma^+ A^T$, with probability one as $N \to \infty$. It follows that the conditional distribution of $\overline{S}_{n,\tau}$ given n converges to the distribution of Z with probability one, as $N \to \infty$. Since $\overline{S}_{n,\tau} = \overline{S}_n$ with probability that tends to one, we conclude that the conditional distribution of $\overline{S}_n$ given n converges in probability to the distribution of Z as $N \to \infty$. Hence, it follows from the dominated convergence theorem that $\overline{S}_n$ converges in distribution to Z as $N \to \infty$.

Since p^+ is convex we have $S_n \geq \overline{S}_n$ with $\overline{S}_n$ defined by (10.6). Hence,

$$\begin{aligned} \mathbb{P}\left(S_n \leq q_\alpha^*\right) &\leq \mathbb{P}\left(\overline{S}_n \leq q_\alpha^*\right) \\ &\leq \mathbb{P}\left(\overline{S}_n \leq q_\alpha + \delta\right) + o(1) \end{aligned}$$

using (10.20) with $\delta > 0$ arbitrarily small. Now, $\overline{S}_n$ converges in law to Z so

$$\lim_{N \to \infty} \mathbb{P}\left(S_n \leq q_\alpha^*\right) \leq \mathbb{P}\left(Z \leq q_\alpha + \delta\right)$$

where $\delta > 0$ is arbitrarily small. Letting $\delta \to 0$, we conclude that

$$\lim_{N \to \infty} \mathbb{P}\left(S_n \leq q_\alpha^*\right) \leq \mathbb{P}\left(Z \leq q_\alpha\right) = \alpha.$$

This completes the proof of the first assertion.

To prove the second assertion, we use Lemma 10.1 to conclude that $S_n = \overline{S}_n$ with probability that tends to one if p^+ is a triangular distribution. Hence,

$$\begin{aligned} \mathbb{P}\left(S_n \leq q_\alpha^*\right) &= \mathbb{P}\left(\overline{S}_n \leq q_\alpha^*\right) + o(1) \\ &\geq \mathbb{P}\left(\overline{S}_n \leq q_\alpha - \delta\right) + o(1) \\ &\geq \mathbb{P}\left(Z \leq q_\alpha - \delta\right) + o(1) \end{aligned}$$

using (10.20) and the convergence of $\overline{S}_n$ to Z, where $\delta > 0$ is arbitrarily small. We conclude by letting $\delta \to 0$, using that Z has a continuous distribution. This completes the proof of Theorem 10.1. □

The following theorem describes the power of the test. The power is studied along local alternatives, which means that we consider that the underlying truncated distribution depends on N, is non-convex, and tends to a convex distribution as $N \to \infty$. For the sake of clarity, we denote by p_N^+ the underlying truncated probability and by $\mathbb{P}_{p_N^+}$ the corresponding probability.

Theorem 10.2 *Assume that p^+ has a finite support. Let $0 < \beta < 1$, let $C > 0$ such that $C \geq \sqrt{-(9/2)\log(\beta)} - q_\alpha$ and for all integers N let*

$$H(C, N) = \left\{ p_N^+, \text{ such that } \Delta p_{N, y_0}^+ < -\frac{2C}{\sqrt{N(1 - p_0)}} \text{ for some } y_0 \right\}.$$

Then for all $p_N^+ \in H(C, N)$ we have

$$\lim_{N \to \infty} \mathbb{P}_{p_N^+}\left(S_n \geq q_\alpha^*\right) \leq \beta.$$

Proof of Theorem 10.2 Let $p^+ \in H(C, N)$. Thanks to (10.20), it suffices to prove that

$$\lim_{N \to \infty} \mathbb{P}_{p_N^+}\left(S_n \geq q_\alpha - \delta\right) \leq \beta \tag{10.21}$$

for an arbitrary $\delta > 0$. Consider the conditional probability given n. There exists y_0 such that on the event $\{n \geq N(1 - p_0)/2\}$ we have for $\delta > 0$ small enough (precisely, we consider $\delta \leq C(\sqrt{2} - 1)$) that

$$\mathbb{P}_{p_N^+}(S_n \geq q_\alpha - \delta | n) \leq \mathbb{P}_{p_N^+}(\Delta f_{y0}/\sqrt{n} \geq q_\alpha - \delta | n)$$
$$\leq \mathbb{P}_{p_N^+}\left(\sqrt{n}(\frac{\Delta f_{y0}}{n} - \Delta p_{y0}^+) \geq q_\alpha + C | n\right)$$
$$= \mathbb{P}_{p_N^+}\left(\frac{1}{\sqrt{n}}\sum_{i=1}^{n}\Delta\left[1_{Y_i=y_0} - \mathbb{E}(1_{Y_i=y_0})\right] \geq q_\alpha + C | n\right).$$

It is easy to show that

$$\Delta 1_{Y_i=y_0} = 1_{Y_i=y_0-1} - 21_{Y_i=y_0} + 1_{Y_i=y_0+1}$$

lies in the interval $\in [-2, 1]$. Since $q_\alpha + C$ is positive, using the Hoeffding inequality we then get

$$\mathbb{P}_{p_N^+}\left(\frac{1}{\sqrt{n}}\sum_{i=1}^{n}\Delta\left[1_{Y_i=y_0} - \mathbb{E}(1_{Y_i=y_0})\right] \geq q_\alpha + C | n\right) \leq \exp\left\{-2\frac{(q_\alpha + C)^2}{9}\right\}.$$

Because n/N converges almost surely to $1 - p_0$, the probability of the event $\{n \geq N(1 - p_0)/2\}$ tends to one as $N \to \infty$, so integrating the previous inequality yields

$$\mathbb{P}_{p_N^+}(S_n \geq q_\alpha - \delta) \leq \exp\left\{-2\frac{(q_\alpha + C)^2}{9}\right\} + o(1)$$
$$\leq \beta + o(1)$$

where the last inequality follows from the definition of C. This proves (10.21) for $\delta \leq C(\sqrt{2} - 1)$, which completes the proof of Theorem 10.2. $\qquad\square$

10.6.2 Confidence intervals and standard errors

The aim of the section is to prove Theorem 10.3 below. The theorem provides the precise description of the asymptotic distribution of $\widehat{N}$ and states that $\widehat{g} + \widehat{\Omega}$ has the appropriate conditional limiting distributions as claimed in Section 10.4. Here, we use the same notation as in Section 10.4.

In the following, we assume that p is a convex abundance distribution and that either p^+ has a finite support, or there exists $y \geq 2$ such that both $\Delta p_y^+ > 0$ and $\Delta p_{y+1}^+ > 0$ with Δ being the discrete Laplacian defined in (10.4). We consider $v \in \mathbb{R}$ such that $n^{-1/2} \ll v \ll 1$. More formally, we assume that $v \to 0$ and $N^{1/2}v \to \infty$ as $N \to \infty$ (recall that it follows from the strong law of large numbers that $N(1 - p_0)/n \to 1$ with probability one, as $N \to \infty$).

Let us introduce some more notation. Under our assumptions, we can find a finite $k \in \mathbb{N}$ such that either the maximal point of support τ of p^+ satisfies $\tau + 1 \leq k$, or there exists $y \leq k$ such that $\Delta p_y^+ > 0$ and $\Delta p_{y+1}^+ > 0$. In the following, k denotes the smallest such integer. Next, for all vectors $t = (t_1, t_2, \ldots, t_k) \in \mathbb{R}^k$ we define $\Psi^k(t)$ as the minimizer of

$$\sum_{y=1}^{k}(q_y - t_y)^2$$

over the set of all sequences $q \in \mathbb{R}^k$ satisfying $\Delta q_y \geq 0$ for all $y \in \{2, \ldots, k-1\}$ such that $\Delta p_y^+ = 0$, with no constraint at other points y. The criterion is minimized over a closed convex cone in $\mathbb{R}^k$, so $\Psi^k(t)$ is uniquely defined for all $t \in \mathbb{R}^k$. Finally, we set $\theta = 1/(1 - p_0)$,

$\widehat{\theta} = 2\widehat{p}_1^+ - \widehat{p}_2^+ + 1$ and we denote by W a centered Gaussian vector in $\mathbb{R}^k$ with covariance matrix Γ defined by

$$\Gamma_{yy'} = \begin{cases} p_y^+(1 - p_y^+) & \text{if } y = y' \\ -p_y^+ p_{y'}^+ & \text{if } y \neq y' \end{cases}$$

for all $1 \leq y, y' \leq k$. Note that the conditional variance of $\widehat{\Omega}$ is $\widehat{\theta}(1 - \widehat{\theta})$. We are now in a position to state the theorem.

Theorem 10.3 *Under the above assumptions, we have:*

    1. $\sqrt{n}(\widehat{\theta} - \theta)$ converges in law to $2\Psi_1^k(W) - \Psi_2^k(W)$ as $N \to \infty$,

    2. $(\widehat{N} - N)/\sqrt{n}$ converges in law to $2\Psi_1^k(W) - \Psi_2^k(W) + \Omega$ as $N \to \infty$, where Ω is a $N(0, \theta(\theta - 1))$ variable independent of $\Psi^k(W)$.

Moreover, if $v \to 0$ and $N^{1/2}v \to \infty$ as $N \to \infty$, then we have the following convergences, conditional on $(Y_1, \ldots, Y_N)$:

    3. the conditional distribution of $\widehat{g}$ converges to $2\Psi_1^k(W) - \Psi_2^k(W)$ in probability as $N \to \infty$,

    4. the conditional distribution of $\widehat{g} + \widehat{\Omega}$ converges to $2\Psi_1^k(W) - \Psi_2^k(W) + \Omega$ as $N \to \infty$.

Proof of Theorem 10.3 The first two assertions are proved in Theorem 2 of the online supporting information of Durot et al. [104]. It follows from the first assertion that $\widehat{\theta}$ converges in probability to θ. This means that every sub-sequence has a further sub-sequence along which $\widehat{\theta}$ converges to θ with probability one. Assuming for a while that Assertion 3 is true, this implies that every sub-sequence has a further sub-sequence along which the conditional distribution of $\widehat{g} + \widehat{\Omega}$ converges to $2\Psi_1^k(W) - \Psi_2^k(W) + \Omega$. This proves Assertion 4.

It remains to prove Assertion 3. Since $n \sim \mathcal{B}(N, 1/\theta)$, it follows from the law of large numbers that n/N converges to $1/\theta$ with probability one. Hence, $n \to \infty$ with probability one as $N \to \infty$. Moreover, $Y_1, \ldots, Y_n$ are i.i.d. with distribution p^+ conditional on n so it follows from the central limit theorem that with probability one, the conditional distribution of $\sqrt{n}(\Delta f_y/n - \Delta p_y^+)$ given n converges to a non-degenerate centered Gaussian law for any $y \geq 2$, as $N \to \infty$. Hence, $\sqrt{n}(\Delta f_y/n - \Delta p_y^+)$ converges (unconditionally) to a centered Gaussian law for any $y \geq 2$, as $N \to \infty$. As a consequence, $\sqrt{N}(\Delta f_y/n - \Delta p_y^+)$ converges to a centered Gaussian law as $N \to \infty$, for any $y \geq 2$. Since $v \to 0$ and $\sqrt{N}v \to \infty$, this means that for any y such that $\Delta p_y^+ > 0$ we have

$$\lim_{N \to \infty} P(\Delta f_y/n \geq v) = 1,$$

whereas for any y such that $\Delta p_y^+ = 0$ we have

$$\lim_{N \to \infty} P(\Delta f_y/n < v) = 1.$$

Hence, if p^+ has a finite double knot, then $\widehat{\kappa} = \kappa < \infty$ with probability that tends to one. If p^+ has a finite support, then $Y_{\max} + 1 = \tau + 1$ with probability that tends to one (see Lemma 10.1). Since it is assumed that p^+ has either a finite support or (at least) a finite double knot, we conclude that $\widehat{k} = k$ with probability that tends to one. Putting this together, we conclude that

$$\widehat{g} = \Psi^k(\widehat{W})$$

with probability that tends to one. Now, with probability that tends to one, $\hat{k} = k$ and $\widehat{W}$ has the same conditional distribution as $\widehat{\Gamma}^{1/2}Z$ where Z is a standard Gaussian vector in $\mathbb{R}^k$. As we consider convergence in probability of conditional distributions, this means that the asymptotic conditional distribution of $\widehat{g}$ is similar to that of $\widehat{\Gamma}^{1/2}Z$. Since $\widehat{\Gamma}$ converges to Γ and Ψ^k is a continuous function (see (ii) in the proof of Theorem 2 in the online supporting information of Durot et al. [104]), we conclude that the conditional distribution of $\widehat{g}$ converges in probability to $\Psi^k(W)$. Assertion 3 follows, which completes the proof of Theorem 10.3. $\qquad\square$

11

Non-parametric estimation of the population size using the empirical probability generating function

Pedro Puig

Universitat Autònoma de Barcelona

CONTENTS

11.1   Introduction

Several non-parametric methods have been proposed in the literature to estimate the number, or equivalently, the probability of zero for zero-truncated count distributions. Some of the most remarkable and commonly used approaches are Chao's estimator [71]–[72], Zelterman's estimator [306], the first- and second-order jackknife estimators [65] and Turing's estimator [131]. Some of them have been described in Chapter 1, Sections 1.3 and 1.4. There are also many extensions, bias corrections and modifications of these estimators, some of them also described in Chapter 1.

It is important to remark that, in general, it is impossible to know the proportion of zeros from its corresponding zero-truncated distribution, because the probability of zero is not univocally determined. In fact, given any zero-truncated distribution $p^+ = p_1^+, p_2^+, ...,$ a non-truncated distribution can be constructed with an arbitrary probability of zero p_0, by considering $p = p_0, (1 - p_0)p_1^+, (1 - p_0)p_2^+, ...$ In spite of the impossibility, if we restrict to certain families of distributions, the problem is more manageable and some information about p_0 can be obtained from the zero-truncated distribution.

Some estimators of the non-observed number of zeros are based on lower bounds of the probability of zero p_0, and they are valid for a large family of count distributions like the

Mixed-Poisson class. This is the case and also the main advantage of Chao's lower bound estimator of the number of zeros (also estimating the population size) shown in Chapter 1, Expression (1.11). In the next section we are going to introduce a very large family of distributions where it is possible to provide meaningful lower bounds of p_0 able to be used to construct lower bound estimates of the population size.

11.2   The LC-class: A large family of count distributions

Let us first refresh an important tool used to describe any count random variable. Given a count random variable X, its *probability generating function* (pgf) $g(s)$ is defined as

$$g(s) = E(s^X) = \sum_{k=0}^{\infty} p_k s^k,$$

where the coefficients of this power series are the probabilities $p_k = P(X = k)$. The pgf characterizes the distribution of X also summarizing a large amount of information. The derivatives of $g(s)$ at $s = 0$ give the probability mass function of X and the derivatives at $s = 1$ provide a way to obtain the moments if they exist. For instance, $g'(1) = E(X) = \mu$ and $g''(1) = E(X(X-1)) = \sigma^2 + \mu^2 - \mu$, where σ^2 is the variance of X. A Poisson random variable X, with $E(X) = \lambda$, has a very simple pgf: $g(s) = \exp(\lambda(s - 1))$.

If the count random variable X has finite moments of all orders, the logarithm of its pgf can be expanded in a power series at $s = 1$, in the form,

$$\log(g(s)) = \kappa_{(1)} \frac{(s-1)}{1!} + \kappa_{(2)} \frac{(s-1)^2}{2!} + \kappa_{(3)} \frac{(s-1)^3}{3!} + \dots \qquad (11.1)$$

Coefficient $\kappa_{(i)}$ is called the i-th factorial cumulant. In particular, $\kappa_{(1)} = \mu$ and $\kappa_{(2)} = \sigma^2 - \mu$. In fact, moments, cumulants, and factorial cumulants are closely related, and any $\kappa_{(i)}$ can be written in terms of the moments or the cumulants and vice versa. Function $\log(g(s))$ is called the *factorial cumulant generating function* (fcgf).

Because the pgf $g(s)$ is a power series of positive coefficients, this is always an increasing convex function in $[0, 1]$. However, in general this is not true for $\log(g(s))$. This definition will be important in the following:

Definition 1 *The set of count random variables having a log-convex pgf or a convex fcgf in $[0, 1]$ is said to be the LC-class.*

The LC-class was first introduced in Puig and Kokonendji [237], being a very large family of count distributions containing the Compound-Poisson family, Mixed-Poisson family and others. Working over the LC-class we shall be able to provide information about p_0 given the corresponding zero-truncated distribution. It is important to remark that most of count distributions used in practice belong to the LC-class, as will be detailed in the next subsections.

11.2.1   Compound-Poisson distributions belong to the LC-class

A random variable X has a discrete Compound-Poisson distribution if it can be represented as

$$X = \sum_{i=1}^{N} \xi_i, \qquad (11.2)$$

where N is a Poisson random variable and $\xi_1, \xi_2, \ldots$ are independent, identically distributed count random variables that are also independent of N. Compound-Poisson distributions are also called *Poisson stopped sum* distributions (Johnson et al. [158]).

Many count distributions used in practice are Compound-Poisson. For instance, when the distribution of ξ_i is Poisson, the distribution of X is a Neyman type A, also called the Poisson-Poisson distribution. When ξ_i follows a logarithmic distribution, X is negative binomial distributed, and this is another distribution widely used in practice. When ξ_i follows a distribution taking a finite range of values, $0, 1, \ldots, r$, with probabilities $q_0, q_1, \ldots, q_r$, X follows an rth-order Hermite distribution (Puig and Barquinero [236]). Other examples and properties of count Compound-Poisson distributions can be found in Johnson *et al.* [158].

It is important to point out that discrete Compound-Poisson distributions are really a large family of count distributions. In fact, according to Feller's characterization, they are the only discrete distributions that are infinitely divisible. It is well known that the pgf of a Compound-Poisson random variable X has the form,

$$g(s) = \exp(-\lambda(1 - h(s))), \tag{11.3}$$

where $\lambda > 0$ and $h(s)$ is the pgf of any of the random variables ξ_i in (11.2). Note that $\log(g(s)) = -\lambda + \lambda h(s)$. Therefore, because $h(s)$ is convex, any Compound-Poisson random variable belongs to the LC-class.

11.2.2 Mixed-Poisson distributions belong to the LC-class

Mixed-Poisson distributions were introduced in Chapter 1, Section 1.4. They are specially useful to describe heterogeneous capture-recapture patterns. A random variable X is Mixed-Poisson distributed if its pgf can be represented in the form,

$$g(s) = \int_0^\infty e^{\lambda(s-1)} dF(\lambda), \tag{11.4}$$

where F is a distribution function on the positive reals. Interestingly, $\log(g(s))$ is always a convex function in $[0, 1]$ (see the proof in Puig and Kokonendji [237]) and consequently any Mixed-Poisson random variable belongs to the LC-class.

Many of the most often used count distributions belong to both Compound- and Mixed-Poisson families. For instance, the Negative Binomial, the Neyman type A, the Poisson-Inverse Gaussian, and others. However, not all Mixed-Poisson are Compound-Poisson distributions. For example, the finite mixtures of Poissons (the distribution F in (11.4) takes a finite number of values) are not Compound-Poisson distributed. Similarly, not all the Compound-Poisson are Mixed-Poisson distributions. This is, for instance, the case of the Hermite distribution (see Kemp and Kemp [163]).

11.2.3 Other distributions belonging (and not belonging) to the LC-class

The LC-class is wider than the class of the union of Compound- and Mixed-Poisson distributions. In particular, the LC-class, is closed under independent addition, that is, given two independent random variables belonging to the LC-class the sum also belongs to the LC-class (Puig and Kokonendji [236]). Then, we can sum a Mixed-Poisson and an independent Compound-Poisson random variable and the result also belongs to the LC-class.

We can also construct random variables that are not Compound and not Mixed-Poisson distributed belonging to the LC-class. For instance, consider a count random variable taking only two values 0 and 2, with probabilities equal to 1/2. Direct calculations show that its pgf $g(s) = 1/2 + s^2/2$ is a log-convex function in $[0,1]$. We can extend this member of the LC-class to a random variable taking an infinite range of values by summing an independent Poisson. It is immediate to check that the result also belongs to the LC-class.

An interesting property of the members of the LC-class is directly deduced from the second derivative of $\log(g(s))$ in (11.1), evaluated at $s = 1$. Because the second derivative has to be positive, it is clear that $\kappa_{(2)} = \sigma^2 - \mu > 0$. Consequently, any member of the LC-class is always overdispersed (or *Poisson-overdispersed*). This property automatically excludes from the LC-class those distributions which are underdispersed (for instance, the Binomial distribution).

Puig and Kokonendji [237] proved that any member X of the LC-class satisfies the Poisson zero-inflation property, that is, $p_0 \geq \exp(-\mu)$, where $p_0 = P(X = 0)$ and $E(X) = \mu$. Consequently, the zero-deflated distributions do not belong to the LC-class.

In the next section we are going to see the information on p_0 that can be obtained from the zero-truncated distribution for a member of the LC-class.

11.3 Some lower bounds of p_0 for the LC-class

Let $g(s)$ be the pgf of the count random variable X and $g^+(s)$ be the pgf of X truncated at zero, say X^+. It is immediate to deduce that $g(s) = (1-p_0)g^+(s) + p_0$ (see also Chapter 9, Section 9.1). Taking logarithms we obtain,

$$\log(g(s)) = \log(1-p_0) + \log(g^+(s) + \frac{p_0}{1-p_0}) \tag{11.5}$$

Assuming that X belongs to the LC class, the convexity of $\log(g(s))$ implies that the second derivative of the right part of (11.5) has to be greater or equal to zero for all $s \in [0,1]$. It leads to the following inequality:

$$\frac{p_0}{1-p_0} \geq \frac{[g'^+(s)]^2}{g''^+(s)} - g^+(s), \forall s \in [0,1]. \tag{11.6}$$

Because inequality (11.6) holds for all $s \in [0,1]$, in particular it holds for $s = 0$ and for $s = 1$. Using the fact that $g'^+(0) = p_1/(1-p_0)$, $g''^+(0) = 2p_2/(1-p_0)$ and $g^+(0) = 0$, direct calculations show that for $s = 0$, inequality (11.6) is equivalent to the well known Chao inequality,

$$p_0 \geq p_1^2/(2p_2).$$

On the other hand, because $g'^+(1) = \mu_+$, $g''^+(1) = \sigma_+^2 + \mu_+^2 - \mu_+$ and $g^+(1) = 1$, for $s = 1$ inequality (11.6) leads to

$$p_0 \geq \frac{\mu_+ - \sigma_+^2}{\mu_+^2}. \tag{11.7}$$

Note that the right part of (11.7) can be negative when the truncated random variable X^+ is overdispersed. In this situation this inequality is not informative.

Anyway, it is evident that in general the sharpest inequality corresponds to the choice $s = s^*$, where,

$$s^* = arg \max_{s \in [0,1]} \psi(s), \tag{11.8}$$

where

$$\psi(s) = \left\{ \frac{[g'^+(s)]^2}{g''^+(s)} - g^+(s) \right\}, \tag{11.9}$$

obtaining

$$\frac{p_0}{1 - p_0} \geq \psi(s^*). \tag{11.10}$$

The value of p_0 such that $p_0/(1 - p_0) = \psi(s^*)$ is the minimum value of p_0 which allows us to reconstruct at zero the truncated distribution, obtaining an untruncated distribution with a log-convex pgf. This is the essence of the estimation method that will be developed afterward.

Next, we are going to illustrate this kind of untruncation method working with two examples related with known distributions.

11.3.1   Example: A two-component Mixed-Poisson distribution

Consider a 50:50 mixture of two Poisson distributions having means λ_1 and λ_2 respectively. The pgf of its zero-truncated distribution has the form,

$$g^+(s) = \frac{e^{\lambda_1(s-1)} + e^{\lambda_2(s-1)} - e^{-\lambda_1} - e^{-\lambda_2}}{2 - e^{-\lambda_1} - e^{-\lambda_2}}.$$

Consider for instance a particular case where $\lambda_1 = 0.5$ and $\lambda_2 = 2.5$.

Figure 11.1 (right panel) shows the plot of $\psi(s)$ described in (11.9) that in this case is a strictly decreasing function. The maximum is attained at $s^* = 0$, giving a value of $\psi(0) = 0.2966$. It corresponds to an untruncated distribution with $p_0/(1 - p_0) = 0.2966$ or $p_0 = 0.2288$.

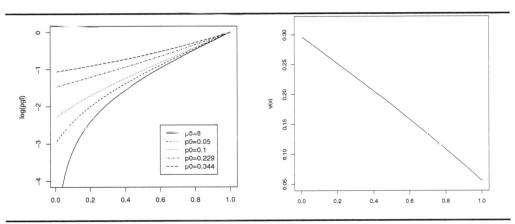

FIGURE 11.1: Log-pgf plots (left panel) for $p_0 = 0$ (zero-truncation), $p_0 = 0.05$, 0.1, 0.229 and 0.344 (untruncated Mixed-Poisson). The plot of function $\psi(s)$ (right panel) shows a maximum attained at $s^* = 0$.

Figure 11.1 (left panel) shows the log-pgf of the zero-truncated distribution ($p_0 = 0$), and the log-pgf of some untruncated extensions. Note that for $p_0 = 0.05$ and $p_0 = 0.1$ the shapes of the plots are not convex. However, for values equal to or greater than $p_0 = 0.229$ the plots become convex. The plot for $p_0 = 0.344$ corresponds to the true untruncated Mixed-Poisson distribution.

11.3.2   Example: A Hermite distribution

Consider a Hermite distribution with parameters a and b. The computation of probabilities and quantiles can be done using the *hermite* package in R (see Moriña et al. [212]–[213]).

The pgf of this distribution is $g(s) = e^{a(s-1)+b(s^2-1)}$, and the pgf of its zero-truncated counterpart has the form

$$g^+(s) = \frac{e^{a(s-1)+b(s^2-1)} - e^{-a-b}}{1 - e^{-a-b}}.$$

For illustrative purposes we are going to focus our attention on a particular case setting $a = 1$ and $b = 2$.

Figure 11.2 (right panel) shows the plot of function $\psi(s)$ described in (11.9). The maximum is attained at $s^* = 0.457$, corresponding to a value of $\psi(0.457) = 0.06479$. It leads to an untruncated distribution with $p_0/(1-p_0) = 0.06479$ or $p_0 = 0.06085$.

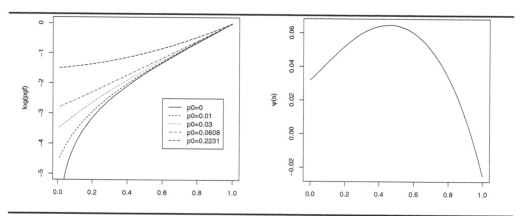

FIGURE 11.2: Hermite distribution ($a = 0.5$ and $b = 1$). Log-pgf plots (left panel) for $p_0 = 0$ (zero-truncation), $p_0 = 0.01$, 0.03, 0.0608 and 0.2231 (untruncated Hermite). The plot of function $\psi(s)$ (right panel) shows a maximum attained at $s^* = 0.457$.

Figure 11.2 (left panel) shows the log-pgf of the zero-truncated Hermite distribution ($p_0 = 0$), and the log-pgf of some untruncated versions. For $p_0 = 0.01$ and $p_0 = 0.03$ the shapes of the plots are not convex. However, for values equal to or greater than $p_0 = 0.0608$ the plots become convex and all these zero-extensions belong to the LC-class. The plot for $p_0 = 0.2231$ corresponds to the true untruncated Hermite distribution.

In the next section we are going to use the inequality (11.10) to estimate p_0 in order to obtain a lower-bound estimation of the population size.

11.4 Estimating a lower bound of the population size

Consider now a sample of zero-truncated observations $Y_1, ..., Y_n$ coming from certain distribution belonging to the LC-class. The sample can be summarized by the frequency counts $f_1, f_2, ..., f_m$, being $f_j = \sharp\{Y_i : Y_i = j\}$, $j = 1, ...m$, where m is the largest observed count $m = max_{i=1}^n\{Y_i\}$. The empirical pgf (epgf) is defined as

$$\hat{g}^+(s) = \frac{1}{n}\sum_{i=1}^m f_i s^i. \tag{11.11}$$

The epgf is an important tool in statistics; see Chapter 9, Section 9.1, for an application in the parametric framework. We propose to estimate a lower bound of the proportion of zeros p_0 of the untruncated distribution using (11.10), replacing the pgf in (11.9) by the epgf. Then, it is straightforward to see that a lower-bound estimator remains as follows:

$$\hat{p}_0 = \frac{\hat{\psi}(s^*)}{1 + \hat{\psi}(s^*)} \tag{11.12}$$

where $s^* = arg\max_{s \in [0,1]} \hat{\psi}(s)$, and

$$\hat{\psi}(s) = \frac{(\hat{g}^{+'}(s))^2}{\hat{g}^{+''}(s)} - \hat{g}^{+}(s) = \frac{(\sum_{i=1}^{m} i f_i s^{i-1})^2}{n \sum_{i=2}^{m} i(i-1) f_i s^{i-2}} - \frac{1}{n} \sum_{i=1}^{m} f_i s^i. \tag{11.13}$$

From here, the population size can be estimated using the Horvitz–Thompson estimator plugging in the corresponding estimation of p_0, and obtaining the simple expression

$$\hat{N} = \frac{n}{1 - \hat{p}_0} = n(1 + \hat{\psi}(s^*)). \tag{11.14}$$

When $s^* = 0$, $\hat{N}$ is just Chao's lower bound estimator $\hat{N}_C = n + f_1^2/(2f_2)$. Consequently, $\hat{N}$ is always greater than or equal to $\hat{N}_C$.

When $s^* = 1$, according to (11.7) we have,

$$\hat{\psi}(1) = \frac{\bar{y} - s_y^2}{s_y^2 + \bar{y}^2 - \bar{y}},$$

where $\bar{y}$ and s_y^2 are the sample mean and variance respectively.

Then, $\hat{N}$ can be expressed in terms of the sample mean and variance in the form

$$\hat{N} = \frac{n\bar{y}^2}{s_y^2 + \bar{y}^2 - \bar{y}}. \tag{11.15}$$

In general, a numerical approach is needed in order to compute $\hat{N}$ by maximizing the function $\hat{\psi}(s)$. However, for samples where only f_1 and f_2 are different from zero, a closed expression of $\hat{N}$ can be obtained. Note that in this case (11.13) remains

$$\hat{\psi}(s) = \frac{1}{f_1 + f_2} \left(\frac{(f_1 + 2f_2 s)^2}{2f_2} - f_1 s - f_2 s^2 \right) = \frac{f_1^2 + 2f_1 f_2 s + 2f_2^2}{2f_2(f_1 + f_2)}.$$

Because $\hat{\psi}(s)$ is a second-order polynomial with positive coefficients, this is an increasing function and it is clear that the maximum is attained at $s^* = 1$, resulting in

$$\hat{N} = n(1 + \hat{\psi}(s^*)) = (f_1 + f_2)(1 + \hat{\psi}(1)) - 2(f_1 + f_2) + \frac{f_1^2}{2f_2}.$$

Note that it can be expressed in terms of Chao's lower bound estimator, in the form $\hat{N} = n + \hat{N}_C$.

For the general case where function $\hat{\psi}(s)$ is not so simple, we use a script in R based in the function *optimize* in order to find the maximum.

It is not direct to find an expression of the variance of the general estimator $\hat{N}$ in (11.14). By considering n fixed, the frequencies $f_1, f_2, ..., f_m$ can be seen as a multinomial outcome, and consequently $\hat{\psi}(s^*)$ would be just a function of a multinomial. Therefore, using the delta method over $\hat{\psi}(s^*)$, a closed-form expression of the asymptotic variance of $\hat{N}$ (or more precisely, $\hat{N}|n$) can be obtained if m is not too big. The uncertainty due to n, the number of sampled individuals, can also be incorporated in the estimation of the variance of $\hat{N}$ by using a conditioning method described in Böhning [40] and Lanumteang and Böhning [173].

Anyway, for analyzing the variability of the estimates and to compute confidence intervals, we prefer to use nonparametric bootstrap from the observed truncated count data. The procedure is the following:

1. A sample of size n is drawn with replacement from the original data, n counts of $1, 2, ..., m$ with frequencies $f_1, f_2, ..., f_m$.

2. An estimate of a lower bound of the population size, $\hat{N}^b$, is calculated for the new sample using (11.14).

3. Steps (1) and (2) are repeated 5000 times obtaining $\hat{N}_1^b, \hat{N}_2^b, ..., \hat{N}_{5000}^b$.

4. Statistics of interest are computed from the estimations of the population size $\hat{N}_1^b, \hat{N}_2^b, ..., \hat{N}_{5000}^b$: mean, median, standard deviation and inter-quartile range.

5. The 95% confidence interval is calculated from the 2.5th and 97.5th percentiles of the values $\hat{N}_1^b, \hat{N}_2^b, ..., \hat{N}_{5000}^b$.

It should be noted that, sometimes, the bootstrap approach used in capture-recapture problems accounts for the original data plus the estimated frequency of zeros $\hat{f}_0$. Then, the resampled zeros in the bootstrap samples are omitted in order to calculate the estimate of the population size. This approach produces different sizes for each bootstrap sample, allowing us to account for the random variation produced by sampling n individuals. This procedure is called in Chapter 3, Section 3.4.1, the imputed bootstrap. However, because our method provides an estimator of a lower bound of f_0 (thus negatively biased), we prefer not to impute the zeros and to use only the original data for sampling. Our approach is called in Chapter 3, Section 3.4.1, the reduced bootstrap.

In the next section we are going to illustrate this procedure with some of the classical examples introduced in Chapter 1, and others related to biodosimetry.

11.5   Examples of application

We are going to analyze five examples where the maximum of $\hat{\psi}(s)$ is attained at the extremes and also inside the interval $[0, 1]$. One of these examples (Dicentrics 3.0 Gy) shows what happens when data comes from a distribution not belonging to the LC-class.

TABLE 11.1

Total number of observed subjects n, point estimates of the population size $\hat{N}$, Bootstrap Mean, Standard Deviation (SD), Median, Inter-quartile range (IQR) and 95% Bootstrap Confidence Interval. Number of bootstrap replicates=5000.

Data set	n	$\hat{N}$	Mean	SD	Median	IQR	95% CI
McKendrick's	55	92.5	99.71	13.47	97.31	15.93	$(82, 131)$
Grizzlies in 1998	33	40.4	42.27	3.89	41.21	4.07	$(37, 53)$
Grizzlies in 1999	29	39.1	44.49^1	15.73^1	39.66	14.03	$(31, 113)$
Dicentrics 0.6 Gy	158	383.9	407.53	51.63	402.01	67.24	$(329, 536)$
Dicentrics 3.0 Gy	287	654.4	658.26	40.18	657.77	52.54	$(585, 744)$

[1]73 extreme values have been excluded for computing the mean and standard deviation.

11.5.1 McKendrick's Cholera data

McKendrick's Cholera data (see Chapter 1, subsection 1.2.3) is probably one of the oldest examples related to the estimation of the number of zeros in truncated count distributions. In any case, McKendrick observed 168 households with zero cases of cholera in the Indian village under study. It is important to remark that a household with no cases of cholera could be because its members had not been exposed or because they had been exposed but they had not been infected. McKendrick wanted to estimate the number of exposed households with no cholera cases, so he ignored the 168 households with zero cases. Of course, this value is an upper bound of the number of zeros of interest (exposed households). Table 11.1 shows the estimate of the population size, which is greater than those obtained using the Turing, MLE, Chao and Chao-BC estimators (see Table 1.9 in Chapter 1). This is interesting because the Turing, MLE and Chao estimators are also non-parametric bound estimators for the LC-class (see Puig and Kokonendji [237]). On the other hand, the length of the bootstrap 95% confidence interval is similar to those in Table 1.9.

It is interesting to remark that for this example, the maximum of $\hat{\psi}(s)$ is attained at $s^* = 1$ (see Figure 11.3). According to (11.7), this means that (11.12) remains

$$\hat{p}_0 = \frac{\hat{\psi}(1)}{1 + \hat{\psi}(1)} = \frac{\bar{y} - s_y^2}{\bar{y}^2} = 0.4051,$$

where the sample mean and variance are $\bar{y} = 1.5636$ and $s_y^2 = 0.5732$. This $p_0 = 0.4051$ is the minimum value for which the log-epgf of the untruncated distribution starts to be a convex function.

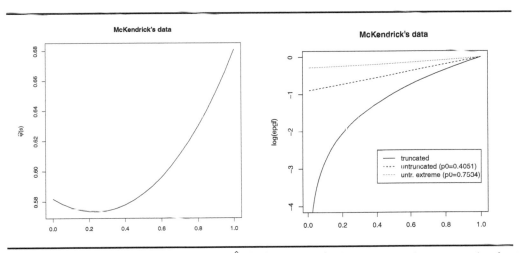

FIGURE 11.3: The plot of function $\hat{\psi}(s)$ (left panel) shows a maximum attained at $s^* = 1$. Log-epgf plots (right panel) for $p_0 = 0$ (truncated), $p_0 = 0.4051$ (untruncated) and $p_0 = 0.7534$ (the value for $f_0 = 168$ households).

Assuming $f_0 = 168$, the total number of households with zero cases, the observed proportion of zeros is $p_0 = 168/(168 + 55) = 0.7534$. Observe in Figure 11.3 that for this extreme case the log-epgf is also a convex function.

11.5.2   Abundance of grizzly bears in 1998 and 1999

Keating [162] studied the annual numbers of females with cubs-of-the-year in the Yellowstone grizzly bear population, from 1986 to 2001. Table 1.6 in Section 1.2.9 of Chapter 1 shows the complete data sets. We are going to focus our attention in two specific years, 1998 and 1999, in order to describe the behavior of $\hat{\psi}(s)$ showing where the maximum is attained. Note that for 1998 the maximum of $\hat{\psi}(s)$ is attained at $s^* = 0.504$, giving a value of $\hat{\psi}(s^*) = 0.224$ (see Figure 11.4), corresponding to an estimate of the population size $\hat{N} = 40.4$ obtained using (11.14).

For the observed sample in 1999 the maximum of function $\hat{\psi}(s)$ is attained at $s^* = 0$ and consequently $\hat{N}$ is exactly Chao's estimator. This leads to an estimate of $\hat{N} = 39.1$. This example also shows that for small samples (here $n = 29$) the bootstrap approach can cause difficulties, due to the generation of a few bootstrap samples producing extremely large values of $\hat{N}^b$. These extreme values are irrelevant for computing the median, IQR and the 95% confidence intervals because they are robust against outliers. However, it is convenient to remove the outliers to compute the mean and standard deviation. For this example, 73 of 5000 extreme values were deleted, being those where $\hat{N}^b > 300$ (see Table 11.1).

Grizzly data sets are jointly analyzed in Chapter 6, Section 6.2. The authors conclude there that the true number of bears between 1986 to 2001 varies independently and normally around 37.9 with standard deviation 14.3. Our results are in agreement with these values.

For this example, the estimated proportion of zeros obtained by (11.12) is $\hat{p}_0 = 0.1827$ in 1998 and $\hat{p}_0 = 0.2587$ in 1999. These are the minimum values for which the log-epgf of their respective untruncated distributions starts to be convex functions (see Figure 11.4).

11.5.3   Biodosimetry data

In order to check the performance of our estimator we are going to analyze two data sets where the number of zeros is known. These data sets come from biodosimetry. The main objective of biodosimetry is to quantify the dose received in individuals who have been exposed to ionizing radiation. The most widely used method is the analysis of the induced chromosome aberrations, in particular the analysis of the frequency of dicentrics observed in peripheral blood lymphocytes.

The considered data sets are the following:

1. Number of dicentric chromosomes after the exposure of a radiation dose of 0.6 Gy. Each f_i is the number of blood lymphocytes having exactly i dicentric chromosomes: $f_0 = 473$, $f_1 = 119$, $f_2 = 34$, $f_3 = 3$ and $f_4 = 2$. This data set was analyzed in Puig and Barquinero [236] using rth-order Hermite distributions (Compound-Poisson distributions) and in Oliveira et al. [224] using other distributions, all them members of the LC-class.

2. Number of dicentric chromosomes after exposure to a radiation dose of 3 Gy. The frequencies are: $f_0 = 213$, $f_1 = 192$, $f_2 = 85$, $f_3 = 9$ and $f_4 = 1$. This data set was analyzed in Pujol et al. [238]. The distribution is not a member of the LC-class because data presents a significant underdispersion.

For the first data set (0.6 Gy) the maximum of $\hat{\psi}(s)$ is attained at $s^* = 0.583$, giving a value of $\hat{\psi}(s^*) = 1.430$. It leads to an estimate of the population size $\hat{N} = 383.9$, see Table 11.1, while the observed population size (total number of blood lymphocytes) is $N = 631$, lying outside the 95% CI. The estimated proportion of zeros is $\hat{p}_0 = 0.5885$, while the observed proportion is $p_0 = 0.7496$. The performance is not very good, obtaining a relative error in the estimation of the population size about 39%. However, it has to be borne in

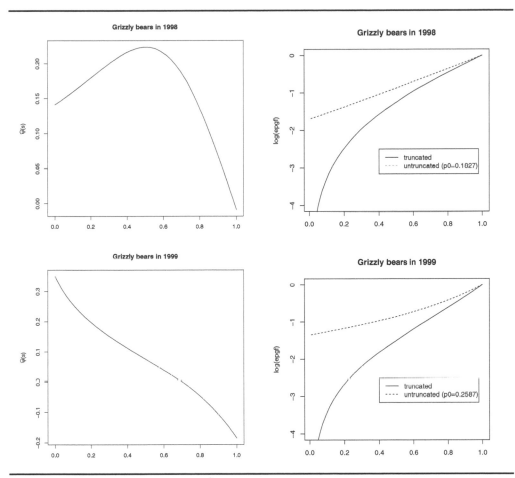

FIGURE 11.4: Plots of functions $\hat{\psi}(s)$ (left panels) showing maximums attained at $s^* = 0.504$ (1998) and $s^* = 0$ (1999). Log-epgf plots (right panels) for $p_0 = 0$ (truncated), and for $p_0 = 0.1827$ (1998) and $p_0 = 0.2587$ (1999) (untruncated) (dashed lines).

mind that our method only provides lower bound estimates, improving Chao's estimator, which can be far away from the true value. Anyway, Figure 11.5 shows that the estimated and the observed log-epgf are very similar.

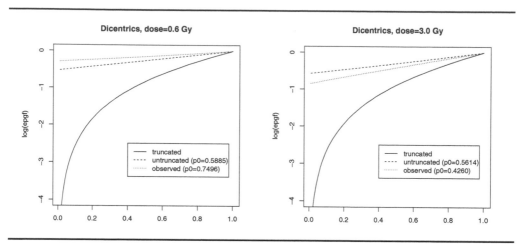

FIGURE 11.5: Log-epgf plots for Biodosimetry data sets. Log-epgfs of zero-truncated samples (solid lines), untruncated estimates (dashed lines) and observed (dotted lines).

For the second data set (3.0 Gy) the maximum of $\hat{\psi}(s)$ is attained at the right limit of the interval $[0, 1]$, that is $s^* = 1$. The sample mean and variance of the truncated distribution are $\bar{y} = 1.3693$ and $s_y^2 = 0.3166$. Therefore, according to (11.15), $\hat{N} = (n\bar{y}^2)/(s_y^2 + \bar{y}^2 - \bar{y}) = 654.4$. In this example the total number of blood lymphocytes is $N = 500$, lying to the left of the estimated 95% CI. The estimated proportion of zeros is $\hat{p}_0 = 0.5614$, while the observed proportion is $p_0 = 0.4260$. Note that $\hat{N}$ does not provide a lower bound of N. It happens because the distribution of the data does not belong to the LC class. There is a detailed discussion in Pujol et al. [238] concerning the underdispersion of these data sets. Figure 11.5 shows that the estimated log-epgf is convex while the observed log-epgf is not.

11.6   Discussion

Most capture-recapture nonparametric heterogeneity models studied in the literature concern Mixed-Poisson distributions. However, the LC-class is larger, containing Compound- and Mixed-Poisson distributions, other distributions that are not Compound- or Mixed-Poisson and their independent sums as well. The method presented here allows us to estimate a lower bound of the population size and, consequently, a lower bound of the non-observed number of zeros, for samples coming from distributions belonging to this large family. Therefore, we think that the approach presented here is very general and suitable to be applied in many circumstances. It is important to point out that $\hat{N}$ in (11.14) is actually an estimator of a lower bound of N, not an estimator of N. Because $\hat{N}$ is always greater or equal to Chao's estimator $\hat{N}_C$, it can be seen as an improvement of this estimator inside the LC-class. Despite the fact that $\hat{N}_C$ is also a lower bound estimator, it has demonstrated that it is competitive with other general estimators that not are lower bounds. The performance of $\hat{N}$ as a general estimator of the population size would be a topic of further research.

Many questions remain open and suggest further research. One of them, for instance, is to understand the log-pgf convexity from a practical point of view, or to find an equivalent but more interpretable condition. It would also be interesting to develop a goodness-of-fit test for the LC-class. All these challenges motivate further research.

12

Extending the truncated Poisson regression model to a time-at-risk model

Maarten J.L.F. Cruyff

Utrecht University

Thomas F. Husken

Utrecht University

Peter G.M. van der Heijden

Utrecht University and University of Southampton

CONTENTS

12.1 Introduction

At the beginning of this century, the Dutch government initiated a project to monitor the size of the illegal immigrant population in the Netherlands, which resulted in a series of publications with yearly estimates [111, 282, 285, 176]. The estimates are based on data extractions from police records involving illegal immigrants who had come into contact with the police during that year. These police data consist of a single record for each police contact, and also include covariates such as age and gender. By counting the number of police contacts $y_i = 1, 2, 3, \ldots$ for each individual $i = 1, \ldots, n$ in the observed data, a zero-truncated count distribution is obtained. Under the assumption that the counts follow a Poisson distribution with observed heterogeneity, i.e. heterogeneity that is completely described by observed covariates, the zero-truncated Poisson regression model can be used to estimate the total population size N.

The basic zero-truncated Poisson regression model (ZTPR) assumes a closed population. For illegal immigrant populations, however, violations of the closed population assumption are likely to occur. For one, the population is volatile in the sense that illegal immigrants may enter or leave the country at will. Secondly, illegal immigrants may be placed in detention by the police on suspicion of being illegal or of a criminal offence. If suspected of being illegal, the illegal immigrant is set free again if expulsion is impossible (e.g. if the country of origin is deemed too unsafe or does not allow the person to return), or is expelled from the country. If suspected of a criminal offence, a prison sentence may be imposed during which the person is not at risk of having new police contacts. When violations of the closed population assumption are not taken into account, the zero-truncated Poisson regression model underestimates the Poisson parameters, and consequently overestimates the population size. This mechanism is discussed in more detail in Section 12.2.1.

In the absence of data on detention times, the estimation of the illegal immigrant population was performed with a two-stage ZTPR. This model accounts for expulsion by estimating the regression coefficients for the illegal immigrants that were not expelled, and using these coefficients for population size estimation of both the expelled and non-expelled immigrants. For the year 2009, however, data on detention times became available, allowing for the specification of a time-at-risk ZTPR model that is more flexible in accommodating violations of the closed population assumption.

This chapter presents the time-at-risk ZTPR and its results for the 2009 data. Section 12.2 reviews the existing theory on the ZTPR, and shows how the two-stage and time-at-risk ZTPR models account for violations of the closed population assumption. Section 12.3 evaluates the performance of these models when the closed population assumption is violated. Section 12.4 compares the estimates of the models for the 2009 data. Section 12.5 provides a discussion.

12.2 The models

12.2.1 The ZTPR

A basic model for the analysis of count data in the presence of covariates is the Poisson regression model [68]. Let $y_i = 0, 1, 2, \ldots$ be the number of police contacts in an illegal immigrant population, for $i = 1, \ldots, N$, that is recorded during the observation period T. Under the assumption that counts follow a Poisson distribution that is conditional on the covariates, the count probabilities are described by the Poisson regression model

$$Po(y_i; \lambda_i) = \frac{e^{-\lambda_i} \lambda_i^{y_i}}{y_i!}, \tag{12.1}$$

where λ_i is the Poisson parameter of individual i. The Poisson model is a generalized linear model, with linear predictor

$$\eta_i = \mathbf{x}_i' \boldsymbol{\beta}, \tag{12.2}$$

where $\mathbf{x}_i = (1, x_{i1}, \ldots, x_{ik})'$ is the covariate vector and $\boldsymbol{\beta} = (\beta_0, \beta_1, \ldots, \beta_k)'$ the corresponding parameter vector, and the log-link function

$$\log \lambda_i = \eta_i. \tag{12.3}$$

The Poisson distribution is characterized by the equality of conditional mean and variance, so that $E(y_i | \mathbf{x}_i) = \text{var}(y_i | \mathbf{x}_i) = \lambda_i$.

For the illegal immigrant population, the individuals $i = n+1, \ldots, N$ with a count $y_i = 0$ are not observed in the data, because they did not have a police contact. The distribution of the illegal immigrants $i = 1, \ldots, n$ with at least one police contact is truncated at zero, and therefore described by the ZTPR

$$Po(y_i; \lambda_i, y_i > 0) = \frac{Po(y_i; \lambda_i)}{1 - Po(0; \lambda_i)}, \tag{12.4}$$

where the denominator ensures that the probabilities sum to unity [280, 279].

Under the ZTPR, an estimate $\hat{\lambda}_i$ can be obtained by maximization of the log-likelihood function

$$\log L(\boldsymbol{\beta}; y_1, \ldots, y_n, \mathbf{x}_1, \ldots, \mathbf{x}_n) = \sum_{i=1}^{n} \log\{Po(y_i; \lambda_i, y_i > 0)\}, \quad (12.5)$$

and given the estimate $\hat{\boldsymbol{\beta}}$, the population size estimate is obtained with the Horvitz–Thompson estimator

$$\hat{N} = \sum_{i=1}^{n} \frac{1}{1 - Po(0; \hat{\lambda}_i)}. \quad (12.6)$$

An asymptotic 95% confidence interval for $\hat{N}$ can be obtained using the method described in [280].

The closed population assumption implies that each individual in the population is at risk of having a police contact during the entire observation period T. This can be made visible by redefining the Poisson parameter λ_i as the rate or intensity, and defining a new Poisson parameter

$$\lambda_{it} = t_i \lambda_i, \quad (12.7)$$

where t_i is time at risk (also known as exposure) [68]. Under the ZTPR (12.4) t_i is assumed to equal T, for all $i = 1, \ldots, N$.

If the closed population assumption is violated, then $t_i < T$ for some individuals in the population. Because these individuals have less chance of experiencing police contacts than individuals with $t_i = T$, their count y_i is expected to be smaller. As a consequence, the ZTPR will underestimate the Poisson parameters in the population, and hence overestimate the individual probabilities of a zero count. In that case the Horvitz–Thompson estimator (12.6) will overestimate the population size.

12.2.2 The two-stage ZTPR

Under the two-stage ZTPR, the procedure for population size estimation of the expelled immigrants consists of the following two stages:

1. Estimation of the regression coefficients for the non-expelled immigrants only.

2. Population size estimation of the expelled immigrants using the regression coefficients as estimated in stage 1 for the non-expelled immigrants.

The rationale for this procedure is the assumption that expelled and non-expelled immigrants have identical regression coefficients, but that the estimates of these coefficients are biased for expelled immigrants, for whom $t_i < T$. For non-expelled immigrants, the assumption that the time at risk is $t_i = T$ is far more likely to hold (it is violated if the immigrant entered the country after the start of the observation period, or left the country before the end of the observation period). Hence, population size estimation of the expelled immigrants based on the regression coefficients of the non-expelled immigrants reduces bias due to the violation of the closed population assumption.

12.2.3 The time-at-risk ZTPR

The time-at-risk ZTPR takes into account that an illegal immigrant may not have been at risk of being apprehended by police during the entire observation period. The source of information about the times at risk is data of judicial institutions on detention times. From these data we can infer three possible time periods of not being at risk. The first one is the entry time, which applies to a detention that took place before the start of the observation time and that ended before the end of the observation time. The entry time is defined as the period that elapsed from the start of the observation period until the end of the detention. The entry time is not related to an event that occurred during the observation, because the police contact that led to the detention took place before the start of the observation period. The other two reasons for not being at risk are detention and expulsion, which both lead to a period of time of not being at risk of having a new police contact. Detention and expulsion times are event related, because they result from police contacts that took place during the observation period. Since event-related and event-unrelated times of not being at risk require different treatments, we first derive a model that accommodates entry times, and then extend the model to accommodate detention and expulsion times.

Let T denote the observation time, i.e. the time during which the police contacts are counted. For example, $T = 365$ if the observation time is one year and days are the discrete units of measurement of time. Let t_{i0} denote the first day within T that individual i is at risk of having a police contact. In general, illegal immigrants will be at risk from day 1, so that for these individuals $t_{i0} = 1$. Now consider an illegal immigrant who has been placed in detention before the start of the observation time, and who is released on day $t_{i0} > 1$. The time at risk for this individual is given by

$$t_i^* = T - t_{i0}. \tag{12.8}$$

Notice that t_{i0} is not related to the occurrence of an event within T, because the police contact that led to the detention occurred before the start of T. Since t_{i0} does not tell us anything about police contacts during T, we assume that t_{i0} is identically distributed for immigrants who are in the data and for those who are not. Under this assumption, the Poisson parameters in the population are given by

$$\lambda_{it^*} = t_i^* \lambda_i. \tag{12.9}$$

It follows that the time-at-risk ZTPR correcting for the entry times is given by

$$Po(y_i; \lambda_{it^*}, y_i > 0) = \frac{Po(y_i; \lambda_{it^*})}{1 - Po(0; \lambda_{it^*})}. \tag{12.10}$$

For the extension of the model to the detention and expulsion times, let t_{iD} be the time in days that individual i spent in detention, and let t_{iE} be the time in days from the moment that individual i had a police contact that resulted in expulsion, to the end of T. Note that detention and expulsion times are event-related, because both variables only take a non-zero value if at least one police contact occurred during T. For individuals observed in the data, the total time at risk is now given by

$$t_i = T - t_{i0} - t_{iD} - t_{iE}, \tag{12.11}$$

with Poisson parameter

$$\lambda_{it} = t_i \exp(\mathbf{x}_i' \boldsymbol{\beta}), \tag{12.12}$$

whereas for individuals not observed in the data, the total time at risk and the Poisson parameters remain as defined in (12.8) and (12.9). Since $Po(0; \lambda_{it^*})$ represents the probability of not being observed in the data, the time-at-risk ZTPR accommodating entry, detention and expulsion times is given by

$$Po(y_i; \lambda_{it}, \lambda_{it^*}, y_i > 0) = \frac{Po(y_i; \lambda_{it})}{1 - Po(0; \lambda_{it^*})}. \tag{12.13}$$

Note that model (12.13) is not a proper probability distribution, because

$$\sum_{y_i=1}^{\infty} Po(y_i; \lambda_{it}, \lambda_{it^*}, y_i > 0) \neq 1$$

if $\lambda_{it} \neq \lambda_{it^*}$. The Horvitz–Thompson estimator

$$\hat{N} = \sum_{i=1}^{n} \frac{1}{1 - Po(0; \hat{\lambda}_{it^*})}, \tag{12.14}$$

however, is unbiased if the model is specified correctly, as demonstrated by the following simulation study.

12.3 Simulation study

The simulation study evaluates the performance of the three ZTPR models by fitting these models to 500 zero-truncated samples from a population of size $N = 25,000$ and an observation time of $T = 365$ days. The populations and their zero-truncated samples are generated as follows:

- The Poisson parameters $\lambda_i = \exp(-2 + 0.75 \cdot x_i)$ are computed for 25,000 random realizations of $X_i \sim N(0, 1)$.

- The entry times $t_{0i} = d_{0i}/T$, with d_{0i} denoting the days in detention before the first event count, are drawn from the mixture $P(d_{0i} = 0) = 0.9$, and $P(0 < d_{0i} \leq 100) = 0.1$ with $d_{0i} \sim U\{1, 100\}$.

- The event counts $y_i = 0, 1, 2, \ldots$ are drawn from a Poisson distribution with Poisson parameter $\lambda_{it^*} - (1 - t_{0i})\lambda_i$,

- A random binary variable is drawn indicating whether the case can be expelled ($p = 0.2$) or not ($p = 0.8$).

- Cases with a count $y_i = 0$ are removed from the population.

- The events $j \in \{1, \ldots, y_i\}$ are randomly assigned different days of the year (to avoid multiple events on the same day), for $j \in \{0, \ldots, 364\}$.

- For those cases that can be expelled, expulsion is effectuated on the first event count. Hence y_i is set to 1, and $t_{iE} = 1 - d_{i1}/T$, with d_{i1} denoting the day of the first event count.

- For the cases that cannot be expelled, each event j is followed by a detention period of 30 days, and events that fall within the detention period of their predecessor are deleted from the data. With the remaining number of events denoted by $y_i^* \leq y_i$, the total detention time is computed as $T^{-1} \sum_{j=1}^{y_i^*} \min(30, T - d_{iy_i^*})$, where $d_{iy_i^*}$ is the day of the last event.

Table 12.1 reports the parameter estimates averaged over the 500 replications, and their corresponding RMSE. The results show that the ZTPR underestimates the intercept, which results in overestimation of the population size. As expected, the two-stage ZTPR yields a less biased estimate of the intercept, and therefore also a less biased population size estimate. The estimates of the time-at-risk ZTPR appear to be unbiased.

TABLE 12.1

Parameter estimates and RMSE of the simulation study

Model	$\hat{\beta}_0$ (RMSE)	$\hat{\beta}_1$ (RMSE)	$\hat{N}$ (RMSE)
ZTPR	-2.400 (0.409)	0.747 (0.052)	35, 958 (11, 384)
Two-stage ZTPR	-2.178 (0.197)	0.740 (0.051)	29, 184 (4, 825)
Time-at-risk ZTPR	-1.995 (0.085)	0.750 (0.051)	24, 990 (2, 055)

12.4　The application

In 2009 a total of 4, 257 illegal immigrants were observed by the police, of which 1, 854 were expelled from the country before the end of that year. Aside from the number of observations for each individual, the police data also included information on gender, age, the police region, nationality and expulsion. The descriptive statistics of the covariates are reported below in Table 12.2.

TABLE 12.2

Observed frequencies and counts of the covariates

Covariate	Category	n	Counts						
			1	2	3	4	5	6	7
Gender	Males	3709	3553	143	8	4	—	—	1
	Females	548	533	15	—	—	—	—	—
Age	> 40	662	630	131	7	4	—	—	1
	≤ 40	3595	3456	27	1	—	—	—	—
Region	A'dam	379	355	22	—	2	—	—	—
	R'dam	279	266	13	—	—	—	—	—
	Haaglanden	286	266	18	2	—	—	—	—
	Utrecht	192	189	3	—	—	—	—	—
	Rest	3121	3010	102	6	6	—	—	1
Origin	Turkey	157	155	2	—	—	—	—	—
	N-Africa	379	373	5	—	1	—	—	—
	Africa	1028	984	42	1	1	—	—	—
	Surinam	73	72	1	—	—	—	—	—
	Eastern-EU	343	325	17	—	—	—	—	—
	Asia	1021	980	39	2	—	—	—	—
	Amerika	108	104	4	—	—	—	—	—
	Unknown	1148	1093	48	5	1	—	—	1

The data were matched with data from the Office of Judicial Institutions (RJI) containing detention times. The means and standard deviations of the entry times, detention times, expulsion times and the time-at-risk (proportional to T) are reported in Table 12.3.

TABLE 12.3

Mean and standard deviations of the time variables, proportional to T

Time variables	Mean	Std
t_{i0} (entry time)	0.01	0.09
t_{iD} (detention time)	0.20	0.24
t_{iE} (expulsion time)	0.15	0.26
t_i^* (time-at-risk)	0.63	0.31

Table 12.4 reports the estimates of the ZTPR, the two-stage ZTPR and time-at-risk ZTPR. The ZTPR and two-stage ZTPR yield identical estimates for the intercept, while on the basis of the simulation study we would expect the estimate of the latter model to be smaller. However, for the two-stage model the regression coefficient for origin Surinam was not estimable, and the dummy for Surinam was therefore removed from the model. As a consequence, the intercepts of the two models have different meanings and cannot be compared. The other coefficients are also more alike for the ZTPR and the time-at-risk model, since they are estimated with the same data, while the two-stage models used only data of the 2403 non-expelled immigrants.

TABLE 12.4

Parameter and population size estimates of the three ZTPR models

Predictor	ZTPR $\hat{\beta}$ (SE)	Two-stage $\hat{\beta}$ (SE)	Time-at-risk $\hat{\beta}$ (SE)
Cons	−2.65 (0.26)	−2.65 (0.31)	−1.76 (0.25)
Gender (male)	0.42 (0.25)	0.33 (0.28)	0.41 (0.24)
Gender (female)	0 (- -)	0 (- -)	0 (- -)
Age (> 40)	0.29 (0.17)	−0.31 (0.27)	0.38 (0.16)
Age (≤ 40)	0 (- -)	0 (- -)	0 (- -)
Region (A'dam)	0.47 (0.21)	0.31 (0.33)	0.52 (0.20)
Region (R'dam)	0.15 (0.29)	0.37 (0.35)	0.26 (0.27)
Region (Haaglanden)	0.51 (0.23)	0.91 (0.28)	0.58 (0.22)
Region (Utrecht)	−1.00 (0.58)	−0.96 (0.71)	−0.95 (0.57)
Region (other)	0 (- -)	0 (- -)	0 (- -)
Origin (Turkey)	−1.66 (0.72)	−1.29 (0.73)	−0.91 (0.67)
Origin (N-Africa)	−0.89 (0.35)	−0.87 (0.53)	−0.54 (0.34)
Origin (Africa)	−0.14 (0.18)	0.16 (0.24)	−0.05 (0.17)
Origin (Surinam)*	−1.72 (1.01)	- - (- -)	−1.71 (0.99)
Origin (Eastern-EU)	−0.02 (0.25)	−0.09 (0.35)	0.07 (0.24)
Origin (Asia)	−0.17 (0.19)	0.34 (0.24)	−0.20 (0.18)
Origin (America)	−0.54 (0.51)	0.02 (0.53)	0.02 (0.46)
Origin (unknown)	0 (- -)	0 (- -)	0 (- -)
$\hat{N} \times 10^3$ (SE)	61.5 (8.8)	55.0 (9.2)	22.8 (2.6)

We globally discuss the trends that we observe in Table 12.4 (i.e. the effects need not be significant for all 3 models). The models agree in that males have a larger Poisson parameter than females, and the immigrants with police contacts in the regions A'dam, R'dam, and Haaglanden (the three biggest cities in the Netherlands) have a larger Poisson parameter than immigrants who had police contacts in the other regions in the country. Immigrants for whom the country of origin is unknown tend to have larger Poisson parameters than when the country of origin is known. The models disagree with respect to the effect of age, according to the ZTPR and time-at-risk models the younger immigrants have higher Poisson parameters, but in the two-stage model the effect is reversed. The population size estimates at the bottom of the table are in the expected order, but the effect of accounting for the times-at-risk is substantial; the estimate of the ZTPR is almost three times larger than that of the time-at-risk model.

12.5 Discussion

The analysis of the 2009 data shows the importance of taking the time at risk into consideration when estimating the size of the illegal immigrant population. However, the time-at-risk model has some restrictions that have prevented more frequent application of this to illegal immigrant data. We discuss these problems in more detail.

One restriction concerns the problematic nature of the detention time data. Given that these data are highly sensitive and are not part of the police records, special permission has to be obtained from the authorities to use these data. Obtaining permission is a lengthy process that may not always be successful. Furthermore, in the Netherlands the quality of the detention time data is questionable; the data are incomplete (about half of the individuals in the 2009 police records were not represented in the detention times data), contain some inconsistencies (e.g. police contacts during a period of detention) and have missing values (e.g. missing dates of the end of a detention) [90]. As a consequence, much time was spent on record matching and data cleaning. In view of these problems, the analysis of the 2009 data can be seen as an experiment to evaluate the feasibility of applying the time-at-risk model for the estimation of the illegal immigrant population, and it was decided that the benefits do not outweigh the costs.

Another restriction, which pertains to all regression models for count data, is the incapacity to accommodate time-varying covariates. For the illegal immigrant data, this problem is illustrated by the covariate Region, denoting the police region in which the police contact took place. Illegal immigrants with more than one police contact could have had these contacts in more than one police region, but the structure of count data with only one record for each individual does not allow for the entry of multiple police regions. An alternative that allows for time-varying covariates is the recurrent events model [85], and its zero-truncated version [152] can be used for population size estimation. This model is still in the experimental stage [90], but will be applied to the illegal immigrant data in the near future.

13

Extensions of the Chao estimator for covariate information: Poisson case

Alberto Vidal-Diez

St George's University of London

Dankmar Böhning

University of Southampton

CONTENTS

13.1 Introduction

We are interested in the setting of capturing and recapturing units in a closed population framework during a fixed time period. At the end of the period, we would have a sample of counts y_i, $i \in \{1, 2, \ldots, n, n+1, \ldots, N\}$, which represents the number of times unit i has been captured within the study period. N is the population size and our parameter of interest, whereas n is the total number of captured units. There are $N - n$ unobserved units, with $y_j = 0$, $j \in \{n+1, \ldots, N\}$ for uncaptured units. Another common notation is using the sum of frequency of frequencies $n = \sum_{y=1}^{m} f_y$, where f_y represents the number of units captured exactly y times and m is the largest number of recaptures within the period of interest. The outcome of interest is an estimate $\hat{f}_0$ of f_0, the number of unobserved units, since $\hat{N} = n + \hat{f}_0$ holds.

Chao's lower bound estimator ($\hat{N} = n + f_1^2/2f_2$) (Chao [72]) is one of the most popular estimators. See also Chapters 10, 11, and 14. It uses only units captured once or twice for the estimation of f_0. Böhning et al. [45] extended the classic Chao lower bound estimator for the Poisson case (Chao [72]) to include covariate information to explain the heterogeneity in the capture probability. Poisson truncated models with only counts of ones and twos non-truncated were applied.

In this chapter we extend the generalised Chao estimator (Böhning et al. [45]) to include more than two counts. We refer the reader to Böhning et al. [45] for the motivation and the derivation of Chao's estimator from the likelihood of a truncated Poisson distribution based on individuals captured once or twice.

Section 13.2 extends the likelihood framework to use truncated models with K counts. Thereafter, Sections 13.3 and 13.3.2 present the methodology to include covariate information to explain heterogeneity at an individual level and increase the number of counts used in the calculations. An analytical formula for the variance is presented in Section 13.3.3. Section 13.4 shows the performance of the proposed estimator under several scenarios. The chapter finishes with the application of the estimator in a case study.

13.2   Generalised Chao estimator K counts and no covariates

Analytical formulae can be obtained for the particular cases where $K = 2$ (Böhning et al. [45]) and $K = 3$. An iterative algorithm is necessary to obtain an estimate $\hat{\lambda}$ of the Poisson parameter λ for $K > 3$. The EM algorithm (Dempster et al. [100]) assumes that observed data represent only a part of the so-called complete data, where missing information should be considered as well. The EM algorithm consists of two stages, expectation and maximisation. In order to maximise the likelihood function, the posterior expectations of the complete data likelihood need to be estimated, but in order to estimate those expectations we need to obtain the estimation of the parameter of interest from the likelihood. Initial values of the parameter of interest have to be provided to start the iterative algorithm.

In our case, the complete likelihood for a Poisson distribution for K non truncated counts with m being the maximum number of captures can be written as

$$\mathcal{L}(\lambda) = \prod_{j=0}^{m} p_j^{f_j}$$

where

$$p_j = e^{-\lambda}\lambda^j/j!$$

is the probability of being captured exactly j times, while f_j is the number of units captured exactly j times. Therefore, the expected complete log-likelihood is defined as

$$\ell(\lambda) = e_0 \log(p_0) + f_1 \log(p_1) + ... + f_K \log(p_K) + e_{K+1} \log(p_{K+1}) + ... + e_m \log(p_m) \quad (13.1)$$

where $f_1, \ldots, f_K$ represent the observed frequencies of the non-truncated counts considered to obtain our estimate. Hence, the rest of the frequencies of counts are assumed to be unobserved and their expectations are used ($e_j = E(f_j|\lambda)$, $j \in \{0, K+1, \ldots, m\}$).

Replacing the probabilities in the likelihood we obtain

$$
\begin{aligned}
\ell(\lambda) &= e_0 \log(e^{-\lambda}) + f_1 \log(e^{-\lambda}\lambda) + \ldots + f_K \log(e^{-\lambda}\lambda^K/K!) \\
&\quad + e_{K+1} \log(e^{-\lambda}\lambda^{K+1}/(K+1)!) + \ldots + e_m \log(e^{-\lambda}\lambda^m/m!) \\
&= -\lambda(e_0 + f_1 + \ldots + f_K + e_{K+1} + \ldots + e_m) \\
&\quad + \log(\lambda)(f_1 + 2f_2 + \ldots + Kf_K + (K+1)e_{K+1} + \ldots + me_m) \\
&\quad - (f_2 \log(2!) + \ldots + f_K \log(K!) + e_{K+1} \log(K+1!) + \ldots + e_m \log(m!)) .
\end{aligned}
$$

M step The likelihood can be maximised by calculating the first derivative and solving the score equation $\frac{d\ell(\lambda)}{d\lambda} = 0$:

$$
\begin{aligned}
\frac{d\ell(\lambda)}{d\lambda} &= (e_0 + f_1 + \ldots + f_K + e_{K+1} + \ldots + e_m) + \\
&\quad \frac{(f_1 + 2f_2 + \ldots + Kf_K + (K+1)e_{K+1} + \ldots + me_m)}{\lambda} = 0
\end{aligned}
$$

leading to

$$
\hat{\lambda} = \frac{(f_1 + 2f_2 + \ldots + Kf_K + (K+1)e_{K+1} + \ldots + me_m)}{(e_0 + f_1 + \ldots + f_K + e_{K+1} + \ldots + e_m)} .
$$

The EM algorithm leads to an updated parameter estimate:

$$
\hat{\lambda} = \frac{(f_1 + 2f_2 + \ldots + Kf_K + (K+1)e_{K+1} + \ldots + me_m)}{(e_0 + f_1 + \ldots + f_K + e_{K+1} + \ldots + e_m)} . \tag{13.2}
$$

E step Estimates for $e_0, e_{K+1}, \ldots, e_m$ are necessary in order to calculate $\hat{\lambda}$. We write:

$$
E(f_y | f_1, \ldots, f_K; \lambda) = Po(y|\lambda)N = Po(y|\lambda)(e_0 + f_1 + \ldots + f_K + e_{K+1} + \ldots + e_m) \tag{13.3}
$$

where K is the number of non-truncated counts, m is the maximum number of captures and $Po(y|\lambda)$ is the Poisson probability of being captured exactly y times.

The next step (E step) is to find the expected values e_0 and $\sum_{j=K+1}^{m} e_j$:

$$
\begin{aligned}
e_0 + \sum_{j=K+1}^{m} e_j &= \left(1 - \sum_{i=1}^{K} Po(y|\lambda)\right)(f_1 + \ldots + f_K) \\
&\quad + \left(1 - \sum_{i=1}^{K} Po(y|\lambda)\right)(e_0 + \sum_{j=K+1}^{m} e_j).
\end{aligned}
$$

Therefore, solving for $e_0 + \sum_{j=K+1}^{m} e_j$

$$
e_0 + \sum_{j=K+1}^{m} e_j = \frac{\left(1 - \sum_{i=1}^{K} Po(y|\lambda)\right)(f_1 + \ldots + f_K)}{\sum_{i=1}^{K} Po(y|\lambda)} . \tag{13.4}
$$

We substitute (13.4) in the calculation of (13.3) to obtain:

$$
\begin{aligned}
E(f_y | f_1, \ldots, f_K; \lambda) &= Po(y|\lambda)(e_0 + f_1 + \ldots + f_K + e_{K+1} + \ldots + e_m) \\
&= Po(y|\lambda)(f_1 + \ldots + f_K) + Po(y|\lambda)\frac{1 - \sum_{y'=1}^{K} Po(y'|\lambda)}{\sum_{y'=1}^{K} Po(y'|\lambda)}[f_1 + \ldots + f_K] \\
&= \frac{Po(y|\lambda)}{\sum_{y'=1}^{K} Po(y'|\lambda)}[f_1 + \ldots + f_K] = \frac{\lambda^y/y!}{\sum_{y'=1}^{K} \lambda^{y'}/y'!}[f_1 + \ldots + f_K]. \tag{13.5}
\end{aligned}
$$

We are particularly interested in $e_0 = E(f_0|f_1, .., f_K; \lambda)$:

$$E(f_0|f_1, .., f_K; \lambda) = \frac{1}{\sum_{y'=1}^{K} \lambda^{y'}/y'!}[f_1 + ... + f_K]. \tag{13.6}$$

An initial $\hat{\lambda}$-estimate value λ_0 is first chosen. Then λ_0 is used in the expectation formulae, which is needed in the likelihood to obtain a new maximum likelihood estimate $\hat{\lambda} = \lambda_1$. The process is repeated recursively until the difference $|\lambda_{k+1} - \lambda_k|$ or $|\ell_{r+1} - \ell_r|$ is smaller than a chosen tolerance threshold.

13.3 Generalised Chao estimator Poisson case with covariates

13.3.1 Two counts

In this section, we start with the simplest case, summarising the methodology presented in Böhning et al. [45] to extend Chao's estimator to include auxiliary variables based on Poisson truncated models with 2 counts. Thereafter, we deduce the estimators based on K non-truncated counts including covariate information.

Suppose a sample where additional information for each captured individual unit i is available: $(Y_1, Z_1), ..., (Y_n, Z_n)$ with Z_i being a p-dimensional vector. The idea is to explain the heterogeneity in the capture probability of each individual by conditioning on the information collected from the captured individuals.

First, a Poisson regression with a log-link function is defined to introduce the covariate information in the likelihood framework:

$$\lambda_i = e^{\alpha + \beta' Z_i}, \tag{13.7}$$

where λ_i is the conditional Poisson mean with $P(Y_i = y) = Po(y|\lambda_i)$. $Po(y|\lambda_i)$ is, as previously defined, a truncated Poisson distribution with only one and two counts. We define the probabilities of the non-truncated counts:

$$P(Y_i = 1) = (1 - q_i) = \frac{\lambda_i e^{-\lambda_i}}{\lambda_i e^{-\lambda_i} + \frac{\lambda_i^2}{2} e^{-\lambda_i}} = \frac{1}{1 + \lambda_i/2}$$

and

$$P(Y_i = 2) = q_i = \frac{\frac{\lambda_i^2}{2} e^{-\lambda_i}}{\lambda_i e^{-\lambda_i} + \frac{\lambda_i^2}{2} e^{-\lambda_i}} = \frac{\lambda_i/2}{1 + \lambda_i/2}. \tag{13.8}$$

Let us assume that there are M different observed covariate combinations or strata with $n_1 + ... + n_M = f_1 + f_2$, where n_i is the frequency of stratum i, $n_i = \sum_{j=1}^{2} f_{ij}$ with f_{ij} the number of individuals from strata i captured j times. Continuous covariates could lead to the case where all n_i are equal to one. The truncated Poisson likelihood is defined by

$$\prod_{i=1}^{M} \left(\frac{1}{1 + \lambda_i/2}\right)^{f_{i1}} \times \left(\frac{\lambda_i/2}{1 + \lambda_i/2}\right)^{f_{i2}},$$

replacing λ_i from (13.7) we obtain

$$\prod_{i=1}^{M} \left(\frac{1}{1 + e^{\alpha + \beta' Z_i}/2}\right)^{f_{i1}} \times \left(\frac{e^{\alpha + \beta' Z_i}/2}{1 + e^{\alpha + \beta' Z_i}/2}\right)^{f_{i2}} \tag{13.9}$$

where f_{ij} are the frequencies of counts j in the ith covariate combination with $j = 1, 2$.

We observe that (13.9) is equal to a binomial logistic likelihood except for the intercept:

$$\prod_{i=1}^{M}(1 - q_i)^{f_{i1}} q_i^{f_{i2}} = \prod_{i=1}^{M}\left(\frac{1}{1 + e^{\alpha' + \beta' Z_i}}\right)^{f_{i1}} \times \left(\frac{e^{\alpha' + \beta' Z_i}}{1 + e^{\alpha' + \beta' Z_i}}\right)^{f_{i2}} \tag{13.10}$$

where $\alpha' = \log(1/2) + \alpha$. Hence a logistic regression model could be fitted to calculate the maximum likelihood estimates for the truncated Poisson model. We obtain $\hat{\alpha}'$ and $\hat{\beta}$ by maximising the binomial likelihood (13.10). We can estimate λ_i for each unit i as

$$\hat{\lambda}_i = 2\frac{\hat{q}_i}{1 - \hat{q}_i} = 2e^{\hat{\alpha}' + \hat{\beta}' Z_i} \qquad \text{for} \qquad i = 1, \ldots, M. \tag{13.11}$$

An estimate of f_0 can be obtained as the sum of the estimates for each stratum since $f_0 = \sum_{i=1}^{M} f_{i0}$. Böhning et al. [45] applied the reasoning of the E step in Section 13.2 to each stratum $\hat{f}_{0i}$. (13.6) can be obtained for each covariate combination and λ_i:

$$\hat{f}_{i0} = \frac{Po(0|\hat{\lambda}_i)}{Po(1|\hat{\lambda}_i)}(f_{i1} + f_{i2}) = \frac{e^{-\hat{\lambda}_i}}{\hat{\lambda}_i e^{-\hat{\lambda}_i} + \hat{\lambda}_i^2 e^{-\hat{\lambda}_i}/2}(f_{i1} + f_{i2}) = \frac{f_{i1} + f_{i2}}{\hat{\lambda}_i + \hat{\lambda}_i^2/2}. \tag{13.12}$$

Finally, the estimator arises summing up over all the covariate combinations. The generalised Chao estimator is asymptotically unbiased when the Poisson regression model holds ($\frac{E(\hat{N}_{GC})}{N} \xrightarrow{N \to \infty} 1$).

13.3.2 Generalised Chao estimator Poisson case: K counts and covariates

In this section, we extend the methodology presented in Section 13.2 to include covariate information working directly with a truncated Poisson likelihood rather than with the complete Poisson likelihood.

K counts are considered to be used and m is defined as the maximum number of counts in the capture distribution. Covariate information is also available and linked with λ_i as defined in the previous section. Let

$$\lambda_i = e^{\alpha + \beta' Z_i} \qquad \text{for} \quad i = 1, \ldots, M_K,$$

where M_K is the total number of covariate combinations or strata when K counts are non-truncated, and Z_i is a vector of covariates.

In this case, a Poisson likelihood truncating the counts $0, K + 1, \ldots, m$ is defined as

$$\mathcal{L}(\lambda_i | f_1, \ldots, f_K) = \prod_{i=1}^{M_K} \prod_{j=1}^{K} p_{ij}^{f_{ij}}$$

where

$$p_{iy} = \frac{\frac{e^{-\lambda_i} \lambda_i^y}{y!}}{\sum_{j=1}^{K} \frac{e^{-\lambda_i} \lambda_i^j}{j!}}, \qquad \text{for} \quad y \in \{1, \ldots, K\} \tag{13.13}$$

is the probability of being captured y times for units in the ith stratum.

Therefore, the log-likelihood becomes

$$\ell(\lambda_i | f_1, \ldots, f_K) = \sum_{i=1}^{M_K} \left[f_{i1} \times \log(p_{i1}) + \ldots + f_{iK} \times \log(p_{iK}) \right]. \tag{13.14}$$

For simplification we assign $\omega_i = \sum_{j=1}^{K} \frac{\lambda_i^j}{j!}$. Hence, the log-likelihood, after replacing the capture probabilities from (13.13) in (13.14), is

$$\ell(\lambda_i | f_{i1}, \ldots, f_{iK}) = \sum_{i=1}^{M_K} f_{i1} \log\left(\frac{\lambda_i}{\omega_i}\right) + \ldots + f_{iK} \log\left(\frac{\lambda_i^K / K!}{\omega_i}\right)$$

$$= \sum_{i=1}^{M_K} f_{iK} \log \lambda_i - f_{i1} \log(\omega_i) + \ldots + f_{iK} K \log(\lambda_i) - f_{iK} \log(\omega_i) \quad (13.15)$$

$$- f_{i2} \log(2) - \ldots - f_{iK} \log K!$$

$$= \sum_{i=1}^{M_K} \left(\sum_{j=1}^{K} j f_{ij}\right) \log(\lambda_i) - \left(\sum_{j=1}^{K} f_{ij}\right) \log(\omega_i) - \sum_{j=2}^{K} f_{ij} \log(K!). \quad (13.16)$$

Finally, the log-likelihood with respect to α and β is calculated replacing λ_i with the linear predictor. Firstly, we see that

$$\log(\omega_i) = \sum_{j=1}^{K} \log\left(\frac{e^{(\alpha + \beta' z_i)j}}{j!}\right) = \sum_{j=1}^{K} (\alpha + \beta' z_i)j - \log(j!), \quad (13.17)$$

and therefore,

$$\ell(\alpha, \beta | f_{i1}, \ldots, f_{iK}) = \sum_{i=1}^{M_K} \left[(\alpha + \beta' Z_i) \left(\sum_{j=1}^{K} j f_{ij}\right) - \sum_{j=2}^{K} f_{ij} \log(K!) \right.$$

$$\left. - \left(\sum_{j=1}^{K} f_{ij}\right) \sum_{k=1}^{K} (\alpha + \beta' Z_i)k - \log(k!) \right]. \quad (13.18)$$

At this stage, an optimisation algorithm can be used to maximise the likelihood and obtain estimators for α and β.

The calculation of $E(f_{iy}|f_{i1}, \ldots, f_{iK}; \lambda_i)$ is identical to the E-step in Section 13.2. $\hat{\alpha}$ and $\hat{\beta}$ are obtained by maximising the log-likelihood (13.18), and ultimately finding the expected value $E(f_y) = \sum_{i=1}^{M_K} E(f_{iy})$.

Indeed, we find

$$e_y = E(f_y | f_1, \ldots, f_K; \lambda_i) = \sum_{i=1}^{M_K} \frac{Po(y|\lambda_i)}{\sum_{y'=1}^{K} Po(y'|\lambda_i)} [f_{i1} + \ldots + f_{iK}]$$

$$= \sum_{i=1}^{M_K} \frac{\lambda_i^y / y!}{\omega_i} [f_{i1} + \ldots + f_{iK}]$$

$$= \sum_{i=1}^{M_K} \frac{\left(\frac{e^{(\hat{\alpha} + \hat{\beta}' Z_i)y}}{y!}\right)}{\sum_{j=1}^{K} \left(\frac{e^{(\hat{\alpha} + \hat{\beta}' Z_i)j}}{j!}\right)} [f_{i1} + \ldots + f_{iK}]. \quad (13.19)$$

Therefore,

$$e_0 = E(f_0 | f_1, \ldots, f_K; \lambda_i) = \sum_{i=1}^{M_K} \frac{1}{\sum_{y'=1}^{K} Po(y'|\lambda_i)} [f_{i1} + \ldots + f_{iK}] = \sum_{i=1}^{M_K} \frac{1}{\omega_i} [f_{i1} + \ldots + f_{iK}]$$

$$= \sum_{i=1}^{M_K} \frac{1}{\sum_{j=1}^{K} \left(\frac{e^{(\hat{\alpha} + \hat{\beta}' Z_i)j}}{j!}\right)} [f_{i1} + \ldots + f_{iK}]. \quad (13.20)$$

13.3.3 Variance estimator for N_{GC} with K non-truncated counts and covariates

For the calculation of the variance we apply the technique of conditional moments as discussed in Ross [249], Böhning [40] and Van der Heijden et al. [280]. The variance can be written as the sum of two terms:

$$Var(\hat{N}_{GC}) = Var\left[E(\hat{N}_{GC}|\Delta_i, i=1,..,N)\right] + E\left[Var(\hat{N}_{GC}|\Delta_i, i=1,..,N)\right], \quad (13.21)$$

where

$$\Delta_i = \begin{cases} 1, & y_i \in \{1,..,K\} \\ 0, & otherwise. \end{cases}$$

Our estimate $\hat{N}_{GC}$ when using K non-truncated counts and covariates can be written as

$$E(\hat{N}_{GC}|\Delta_i, i=1,\ldots,N) = E\left(n + \sum_{i=1}^{N} \frac{\Delta_i}{\hat{\lambda}_i + \hat{\lambda}_i^2/2 + ... + \hat{\lambda}_i^K/K!}\right)$$

$$= E\left(\sum_{i=1}^{N} \Delta_i + \sum_{i=1}^{N} \gamma_i + \sum_{i=1}^{N} \frac{\Delta_i}{\hat{\lambda}_i + \hat{\lambda}_i^2/2 + ... + \hat{\lambda}_i^K/K!}\right),$$

where

$$\gamma_i = \begin{cases} 1, & y_i \geq K+1 \\ 0, & otherwise \end{cases},$$

and

$$\lambda_i = e^{\alpha+\beta'Z_i}.$$

λ_i links the covariate information with the Poisson parameter.
We can also write

$$E(\hat{N}|\Delta_i, i=1,\ldots,N) \approx \sum_{i=1}^{N} \Delta_i \left(\frac{\hat{p}_i + e^{\lambda_i}}{p_i}\right) = \sum_{i=1}^{N} \Delta_i \omega_i,$$

with $\omega_i = 1 + \frac{e(\lambda_i)}{p_i}$ for simplification.

p_i is defined as the probability that $\Delta_i = 1$:

$$p_i = p(\Delta_i = 1|\lambda_i) = \lambda_i e^{-\lambda_i} + \lambda_i^2 e^{-\lambda_i}/2 + ... + \lambda_i^K e^{-\lambda_i}/K!.$$

The expected value of Δ_i is $E(\Delta_i) = p_i$ and $Var(\Delta_i) = p_i(1-p_i)$ because Δ_i follows a Bernoulli distribution with parameter p_i. Ultimately, we achieve

$$Var\left(E(\hat{N}|\Delta_i, i=1,\ldots,N)\right) \simeq \sum_{i=1}^{N} Var(\Delta_i \omega_i) \simeq \sum_{i=1}^{N} p_i(1-p_i)w_i^2.$$

The Horvitz–Thompson estimator is applied to estimate the variability:

$$\widehat{Var}(E(\hat{N}|\Delta_i, i=1,\ldots,N))) \simeq \sum_{i=1}^{N} \frac{\Delta_i}{\hat{p}_i}\hat{p}_i(1-\hat{p}_i)\hat{\omega}_i^2 = \sum_{i=1}^{f_1+f_2+...+f_K} (1-\hat{p}_i)\left[\frac{\hat{p}_i + e^{-\hat{\lambda}_i}}{\hat{p}_i}\right]^2.$$

$$(13.22)$$

The multivariate Delta method is used for calculating the second term

$$E[Var(N_{GC}|\Delta_i, i = 1, .., N)] = \nabla g(\hat{\alpha}, \hat{\beta})^T cov(\hat{\alpha}, \hat{\beta}) \nabla g(\hat{\alpha}, \hat{\beta})$$

$$(13.23)$$

where

$$\nabla g(\hat{\alpha}, \hat{\beta}) = \begin{pmatrix} \frac{\partial g}{\partial \alpha} \\ \frac{\partial g}{\partial \beta_1} \\ ... \\ \frac{\partial g}{\partial \beta_p} \end{pmatrix} = \begin{pmatrix} \sum_{i=1}^{f_1+f_2+...+f_K} \dfrac{-\sum_{j=1}^{K} \frac{\hat{\lambda}_i^j}{j-1!}}{\left(\sum_{j=1}^{K} \frac{\hat{\lambda}_i^j}{j!}\right)^2} \\ \sum_{i=1}^{f_1+f_2+...+f_K} \dfrac{-\sum_{j=1}^{K} \frac{\hat{\lambda}_i^j}{j-1!}}{\left(\sum_{j=1}^{K} \frac{\hat{\lambda}_i^j}{j!}\right)^2} z_{i1} \\ ... \\ \sum_{i=1}^{f_1+f_2+...+f_K} \dfrac{-\sum_{j=1}^{K} \frac{\hat{\lambda}_i^j}{j-1!}}{\left(\sum_{j=1}^{K} \frac{\hat{\lambda}_i^j}{j!}\right)^2} z_{ip} \end{pmatrix} .$$

$\nabla g(\alpha, \beta)$ can be also expressed in terms of $\hat{\alpha}$ and $\hat{\beta}$, which have been obtained in the maximisation of the likelihood.

The covariance matrix $cov(\hat{\alpha}, \hat{\beta})$ is calculated as the inverse of the observed Fisher information (or the inverse of the Hessian of the negative log likelihood).

$$cov(\hat{\alpha}, \hat{\beta}) = -\left(\frac{\partial}{\partial \alpha \partial \beta} \ell(\alpha, \beta)\right)^{-1}$$

$$cov(\hat{\alpha}, \hat{\beta}) = - \begin{pmatrix} \frac{\partial^2 \ell(\alpha,\beta)}{\partial \alpha^2} & \frac{\partial^2 \ell(\alpha,\beta)}{\partial \alpha \beta_1} & ... & \frac{\partial^2 \ell(\alpha,\beta)}{\partial \alpha \beta_p} \\ \frac{\partial^2 \ell(\alpha,\beta)}{\partial \alpha \beta_1} & \frac{\partial^2 \ell(\alpha,\beta)}{\partial \beta_1^2} & ... & \frac{\partial^2 \ell(\alpha,\beta)}{\partial \beta_1 \beta_p} \\ ... & ... & ... & ... \\ \frac{\partial^2 \ell(\alpha,\beta)}{\partial \alpha \beta_p} & \frac{\partial^2 \ell(\alpha,\beta)}{\partial \beta_p \beta_1} & ... & \frac{\partial^2 \ell(\alpha,\beta)}{\partial \beta_p^2} \end{pmatrix}^{-1}$$

The partial derivatives are presented here, although an approximation of the covariance matrix is commonly produced by the optimisation function of the respective statistical software. The second derivatives are

$$
\frac{\partial^2 \ell(\alpha, \beta)}{\partial \alpha^2} = -\sum_{j=1}^{K} f_j \left(\frac{\sum_{j=1}^{K-1} \frac{\hat{\lambda}_i^j j^2}{(j+1)!} \sum_{j=0}^{K-1} \frac{\hat{\lambda}_i^j}{(j+1)!} - \left(\sum_{j=1}^{K-1} \frac{\hat{\lambda}_i^j j}{(j+1)!} \right)^2}{\left(\sum_{j=0}^{K-1} \frac{\hat{\lambda}_i^j}{(j+1)!} \right)^2} \right)
$$

$$
\frac{\partial^2 \ell(\alpha, \beta)}{\partial \alpha \partial \beta_j} = -z_j \sum_{j=1}^{K} f_j \left(\frac{\sum_{j=1}^{K-1} \frac{\hat{\lambda}_i^j j^2}{(j+1)!} \sum_{j=0}^{K-1} \frac{\hat{\lambda}_i^j}{(j+1)!} - \left(\sum_{j=1}^{K-1} \frac{\hat{\lambda}_i^j j}{(j+1)!} \right)^2}{\left(\sum_{j=0}^{K-1} \frac{\hat{\lambda}_i^j}{(j+1)!} \right)^2} \right)
$$

$$
\frac{\partial^2 \ell(\alpha, \beta)}{\partial \beta_j^2} = -z_j^2 \sum_{j=1}^{J} f_j \left(\frac{\sum_{j=1}^{K-1} \frac{\hat{\lambda}_i^j j^2}{(j+1)!} \sum_{j=0}^{J-1} \frac{\hat{\lambda}_i^j}{(j+1)!} - \left(\sum_{j=1}^{J-1} \frac{\hat{\lambda}_i^j j}{(j+1)!} \right)^2}{\left(\sum_{j=0}^{J-1} \frac{\hat{\lambda}_i^j}{(j+1)!} \right)^2} \right)
$$

$$
\frac{\partial^2 \ell(\alpha, \beta)}{\partial \beta_j \partial \beta_k} = -z_j z_k \sum_{j=1}^{K} f_j \left(\frac{\sum_{j=1}^{K-1} \frac{\hat{\lambda}_i^j j^2}{(j+1)!} \sum_{j=0}^{K-1} \frac{\hat{\lambda}_i^j}{(j+1)!} - \left(\sum_{j=1}^{K-1} \frac{\hat{\lambda}_i^j j}{(j+1)!} \right)^2}{\left(\sum_{j=0}^{K-1} \frac{\hat{\lambda}_i^j}{(j+1)!} \right)^2} \right).
$$

13.4 Simulations

In this section we assess the performance of our estimator by running simulations for several scenarios. Whenever possible we compare our generalised Chao estimator (GC) with the following estimators:

- **Classic Chao lower bound estimate (Chao [72]):**

$$
\hat{N}_{Chao} = n + \frac{f_1^2}{2f_2}. \tag{13.24}
$$

- **Turing estimator.** This provides accurate estimates under homogeneity (Good [131]):

$$
\hat{N}_{Turing} = \frac{n}{1 - f_1/S} \tag{13.25}
$$

where $S = f_1 + 2f_2 + \ldots + m f_m$ and m is the maximum number of captures.

The background of the Turing estimator is based on the sample coverage estimator $(1 - f_1/S)$ (Chao et al. [74]). In the case of equal capture probability for all individuals, the sample coverage is n/N which, if equated to $(1 - f_1/S)$, leads to the $\hat{N}_{Turing}$ estimator (Darroch et al. [96]).

- **Zero-Truncated Poisson regression with covariates (ZTP).** Zero-truncated Poisson modeling has been mentioned in Van der Heijden et al. [280, 279]. The capture probability is modelled based on a zero-truncated Poisson regression model. The population size estimate is calculated using the Horvitz–Thompson estimator:

$$
\hat{N}_{ZTP} = \sum_{i=1}^{n} \frac{1}{1 - e^{-e^{\hat{\alpha} + \hat{\beta}' z_i}}}, \tag{13.26}
$$

where $\hat{\alpha} + \hat{\beta}'Z_i$ is the fitted linear predictor of a zero-truncated Poisson regression, and Z_i is a vector of covariates related to the capture-recapture probability. The estimator is asymptotically unbiased and efficient when the assumption of the Poisson distribution is true.

- **Zero-Truncated Negative binomial with covariates (ZNB) (Cruyff et al. [89]).** The heterogeneity in the probability of being captured is modelled using a zero-truncated negative binomial model with covariate information introduced in a similar way as in the ZTP model. It uses a gamma distribution for the parameter of the Poisson model.

13.4.1   Simulation 1: Including unexplained heterogeneity

We investigate a scenario where we introduce unexplained heterogeneity generating the distribution of counts of captures using two covariates, but the estimation is based only on one covariate.

The data are generated following the next steps:

1. Two vectors X_1 and X_2 of size N are generated independently following normal distributions with means 5 and 8, respectively, and variances of 64 ($X_1 \sim N(5, 64)$ and $X_2 \sim N(8, 64)$).

2. Then, the capture-recapture distribution is generated following a Poisson distribution $Y_i \sim Po(\lambda_i)$ where λ_i is calculated from a log linear model:

$$\lambda_i = e^{-0.02X_{1i} + 0.03X_{2i}} \qquad \text{with} \quad i = 1, \ldots, N.$$

3. Units which are not captured ($Y_i = 0$) are removed to obtain the sample of captured units.

4. Point and variance estimates are calculated including only X_1 into the regression models to assess the effect of unexplained heterogeneity.

5. Steps 1–4 are repeated 2000 times and the results are summarised by:
   1) The mean of the point estimations over all R repetitions ($\hat{N} = \frac{1}{R}\sum_{j=1}^{R} \hat{N}_j$).
   2) The mean of the standard deviation estimates.
   3) The mean of relative mean squared errors that are calculated as $RMSE = E(\hat{N} - N)^2/N^2$.
   4) The relative bias that is calculated as $RBias = (E(N) - N)/N$.

The bias of the generalised Chao (GC) estimator increases when more counts are used. In contrast, the standard deviation of the GC estimator decreases when more information is added to the model (Table 13.1 and Figure 13.1). The ZTP estimator has smaller standard deviation than the GC estimator as it uses all information available, but the estimator is more biased. In fact the classic Chao estimator obtained a better point estimation than the ZTP estimator. Turing's estimator underestimates severely because its assumption of homogeneity is not fulfilled.

The GC estimator becomes the best estimator for sufficiently large samples when looking at the relative mean squared error (Table 13.1). Chao's estimator also presents smaller RMSE than ZTP for the scenario where $N = 2000$. The bias of the GC estimator increase asymptotically, leading to an increase in the relative bias; in contrast to a decrease of the standard deviation, which causes the RMSE values to decrease asymptotically.

The GC estimator with 4 counts has the smallest RMSE for the scenario with the smallest population, but the best estimates for larger populations are obtained including 3 non-truncated counts.

TABLE 13.1: Point estimates (SD), RMSE and relative bias of $\hat{N}$ for the scenario with $Y_i \sim Po(e^{-0.02X_{1i}+0.03X_{2i}})$ with covariates $X_1 \sim N(5,64)$ and $X_2 \sim N(8,64)$ and the estimation process based only in X_1

Counts	N	$\hat{N}_{GC}$	$\hat{N}_{ZTP}$	$\hat{N}_{Turing}$	$\hat{N}_{Chao}$
2		497.09 (32.74)			
3		491.60 (25.09)			
4	500	487.90 (22.54)	484.62 (21.18)	481.05 (21.11)	491.21 (29.16)
5		485.88 (21.60)			
2		988.08 (45.21)			
3		979.35 (34.98)			
4	1000	973.23 (31.58)	967.23 (29.75)	960.81 (30.42)	978.91 (42.06)
5		969.85 (30.34)			
2		1973.85 (63.38)			
3		1956.69 (49.17)			
4	2000	1944.77 (44.40)	1932.27 (41.80)	1921.65 (43.13)	1958.13 (59.48)
5		1937.76 (42.65)			

Counts	N	GC	ZTP	Turing	Chao	GC	ZTP	Turing	Chao
		RMSE (x 1000)				RBias (x 100)			
2		3.93				-0.58			
3		2.80				-1.68			
4	500	2.65	2.79	3.22	3.71	-2.42	-3.08	-3.79	-1.76
5		2.72				-2.82			
2		2.10				-1.19			
3		1.68				-2.07			
4	1000	1.78	2.04	2.46	2.21	-2.68	-3.28	-3.92	-2.11
5		1.91				-3.01			
2		1.14				-1.31			
3		1.09				-2.17			
4	2000	1.29	1.63	2.00	1.32	-2.76	-3.39	-3.92	-2.09
5		1.46				-3.11			

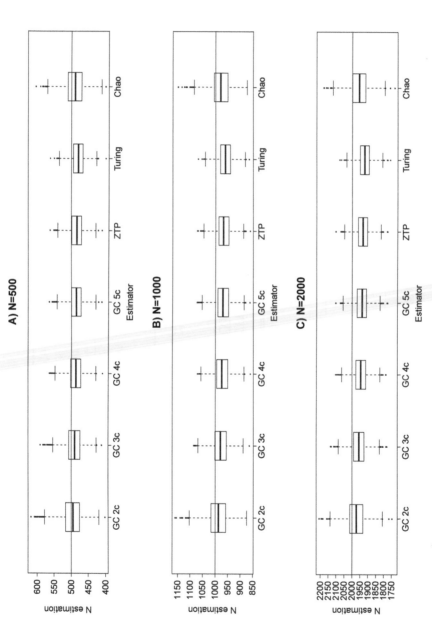

FIGURE 13.1: Boxplots of $\hat{N}$ for the scenario with $Y_i \sim Po(e^{-0.02X_{1i}+0.03X_{2i}})$ with covariates $X_1 \sim N(5, 64)$ and $X_2 \sim N(8, 64)$ and X_1 used only in the estimation process; A) $N = 500$, B) $N = 1000$, C) $N = 2000$.

TABLE 13.2: Point estimates (SD), RMSE and relative bias for a fitted model with misclassified observations $Y_i \sim Po(e^{\alpha+\beta Z_i})$ with $i = 1,..,N$. $Y_i \sim Po(0.5)$ for $Z_i = 0$ and $Y_i \sim Po(3)$ for $Z_i = 1$; the probability of $P(Z_i = 1) = 0.45$; 10% of the population misclassified

Counts	N	$\hat{N}_{GC}$	$\hat{N}_{ZTP}$	$\hat{N}_{Turing}$	$\hat{N}_{Chao}$
2	500	468.94 (33.81)			
3		440.36 (21.52)			
4		421.80 (17.25)	401.86 (10.32)	396.78 (12.27)	428.44 (21.3)
5		411.54 (15.17)			
2	1000	930.3 (47.33)			
3		876.31 (30.25)			
4		840.83 (24.44)	800.25 (14.44)	792.38 (16.91)	853.78 (29.17)
5		820.85 (21.34)			
2	2000	1855.60 (66.12)			
3		1752.62 (43.30)			
4		1680.92 (34.30)	1602.61 (20.37)	1584.2 (23.22)	1705.19 (40.85)
5		1640.62 (30.89)			

Counts	N	GC	ZTP	Turing	Chao	GC	ZTP	Turing	Chao
		RMSE (x 1000)				RBias (x 100)			
2	500	8.58				-6.21			
3		16.02				-11.93			
4		25.67	39.32	43.17	22.70	-15.64	-19.63	-20.64	-14.31
5		32.29				-17.69			
2	1000	7.15				-6.97			
3		16.37				-12.37			
4		26.11	40.28	43.93	22.85	-15.92	-19.98	-20.76	-14.62
5		32.99				-17.92			
2	2000	6.44				-7.22			
3		15.72				-12.37			
4		25.53	39.68	43.21	22.25	-15.95	-19.87	-20.79	-14.74
5		32.45				-17.97			

TABLE 13.3: Point estimates (SD), RMSE and relative bias for a fitted model with misclassified observations $Y_i \sim Po(e^{\alpha+\beta z_i})$ with $i = 1,..,N$. $Y_i \sim Po(3)$ for $Z_i = 0$ and $Y_i \sim Po(0.5)$ for $Z_i = 1$; the probability of $P(Z_i = 1) = 0.45$; 20% of the population misclassified

Counts	N	$\hat{N}_{GC}$	$\hat{N}_{ZTP}$	$\hat{N}_{Turing}$	$\hat{N}_{Chao}$
2		449.1 (27.63)			
3		422.27 (17.63)			
4	500	406.72 (14.34)	390.44 (15.07)	395.84 (11.64)	426.63 (20.02)
5		398.47 (12.92)			
2		895.02 (38.25)			
3		843.81 (24.47)			
4	1000	813.51 (20.54)	781.52 (21.03)	792.57 (17.14)	853.07 (29.4)
5		797.26 (13.69)			
2		1790.94 (54.38)			
3		1687.03 (35.33)			
4	2000	1626.97 (28.27)	1562.13 (29.77)	1585.19 (24.2)	1706.34 (41.26)
5		1594.13 (25.79)			

Counts	N	GC	ZTP	Turing	Chao	GC	ZTP	Turing	Chao
		RMSE (x 1000)				RBias (x 100)			
2		13.41				−10.18			
3		25.41				−15.55			
4	500	35.62	43.55	43.95	23.19	−18.66	−21.91	−20.83	−14.67
5		41.9				−20.31			
2		12.48				−10.5			
3		24.99				−15.62			
4	1000	35.2	43.02	43.31	22.44	−18.65	−21.85	−20.74	−14.69
5		41.45				−20.27			
2		11.67				−10.45			
3		24.8				−15.65			
4	2000	34.99	43.07	43.16	21.98	−18.65	−21.89	−20.74	−14.68
5		41.35				−20.29			

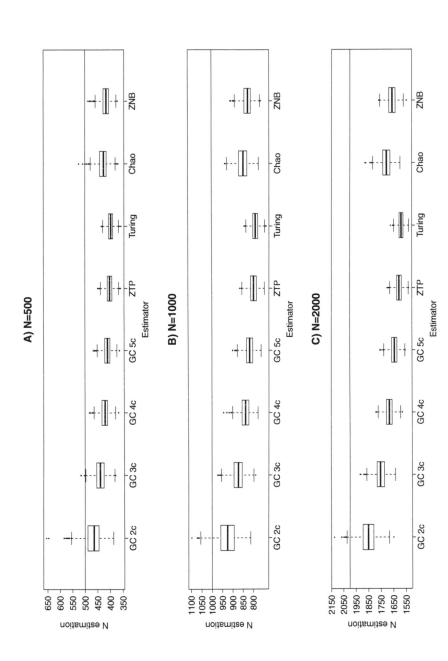

FIGURE 13.2: Scenarios with 10% misclassified individuals in the population; A) $N = 500$, B) $N = 1000$, C) $N = 2000$.

13.4.2   Simulation 2: Model with misclassification

In this section, we evaluate a situation where we misclassify individuals. We aim to assess the impact of having wrong information in our covariates. The capture distribution is simulated from a Poisson distribution $Y_i \sim Po(e^{\alpha+\beta'Z_i})$ with Z_i as a binary covariate that defines two populations. $Z_i = 0$ defines a population generated as $y = e^{\alpha} = 0.5$ and $Z_i = 1$ is a population coming from $y = e^{\alpha+\beta} = 3$. The proportions of the populations are 45% and 65% respectively. However the calculations of the estimates are made assuming that 10% and 20% of the individuals of the total population are misclassified being considered from the first component rather than the second component. Each scenario is repeated 2000 times and average estimates are calculated.

The results (Tables 13.2 and 13.3) show that all estimators underestimate the true population size, but GC was the least biased. Chao's and Turing's estimates are not affected by the misclassification because they do not use covariate information. In fact, the impact of the misclassification in the ZNB and ZTP estimators makes them inferior to Chao's estimator in this scenario. The ZNB model converges on this occasion and it seems to produce slightly better estimates than the ZTP estimator. Our estimator appears to be robust despite introducing wrong information into the logistic model (Figures 13.2 and 13.3) and there is a clear negative effect when the proportion of misclassified individuals increases. On the basis of RMSE and relative bias, GC is superior in this particular scenario and it is less sensitive to contamination in the covariate information than the other estimators. The bias of the GC estimator increases when the number of non-truncated counts increases, while the standard deviation decreases as seen in the previous simulation. In this case, the best RMSE is obtained with the model with two non-truncated counts in both scenarios of misclassification.

13.4.3   Simulation 3: Data generated from a negative binomial distribution

We evaluate the performance of our estimator under heterogeneity generated by a negative binomial distribution. The capture-recapture distribution $Y_i|Z_i \sim NB(\mu_i, \theta)$ with $\mu_i = e^{\alpha+\beta'Z_i}$, $\alpha = 0, \beta = 0.02, \theta = 3$ and $Z_i \sim N(8, 25)$. Zero counts are removed and the remaining counts represent the sampling distribution used to calculate the estimates.

The best RMSE in the GC estimator is again the estimator with 2 non-truncated counts for all simulated population sizes. The ZNB estimator presents the best relative bias because its assumptions hold for this experiment (Table 13.4). The GC with two non-truncated counts is moderately underestimating (Figure 13.4) although its standard deviation is about 0.56 times the standard deviation of the ZNB estimate. The RMSE and relative bias show that the ZNB estimator is asymptotically unbiased in comparison to the GC estimator with two non-truncated where $E(\hat{N})/N \approx 0.9$ for all population sizes included in the simulation. The classical Chao estimator performs similar to the GC estimator with 2 counts in spite of not using any covariate information. The ZTP estimator severely underestimates the true value, performing even worse than the Turing estimator. The ZNB estimator works well under the assumption that the capture distribution is a negative binomial, but in other circumstances the zero-truncated negative binomial model tends to have convergence problems like in the first experiment where the ZNB estimator could not be reported.

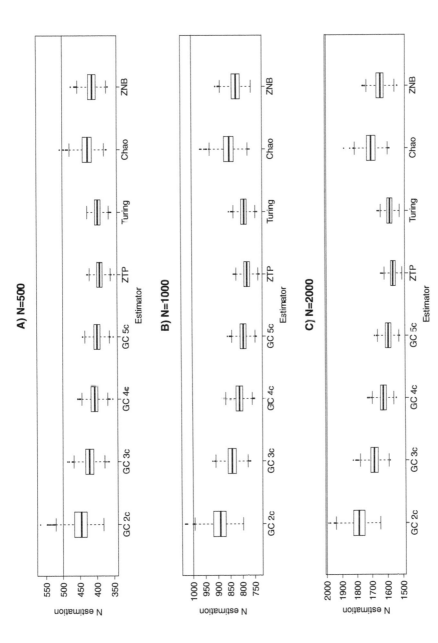

FIGURE 13.3: Scenarios with 20% misclassified individuals in the population; A) $N = 500$, B) $N = 1000$, C) $N = 2000$.

TABLE 13.4: Point estimates (SD), RMSE and relative bias for data generated using a negative binomial distribution $Y_i|Z_i \sim NB(\mu_i, \theta)$ with $\mu_i = e^{0.02Z_i}$, $Z_i \sim N(8, 25)$ and $\theta = 3$

Counts	N	$\hat{N}_{GC}$	$\hat{N}_{ZTP}$	$\hat{N}_{Turing}$	$\hat{N}_{Chao}$	$\hat{N}_{ZNB}$
2	500	458.28 (31.32)				
3		442.15 (22.81)	457.39 (18.98)	427.64 (19.58)	454.62 (29.91)	502.55 (35.88)
4		431.62 (20.31)				
5		424.94 (19.58)				
2	1000	912.59 (45.13)				
3		883.22 (34.14)	835.72 (27.41)	855.82 (29.02)	908.66 (44.26)	1002.26 (49.36)
4		862.42 (30.08)				
5		849.04 (28.48)				
2	2000	1818.14 (60.76)				
3		1762.22 (45.95)	1665.24 (37.91)	1709.87 (39.82)	1812.63 (59.88)	2002.11 (7.12)
4		1722.17 (41.05)				
5		1695.66 (39)				

Counts	N	RMSE (×1000)					RBias (×100)				
		GC	ZTP	Turing	Chao	ZNB	GC	ZTP	Turing	Chao	ZNB
2	500	1.09					−8.34				
3		1.55	2.87	2.25	1.18	0.52	−11.57	−16.52	−14.47	−9.08	0.51
4		2.04					−13.68				
5		2.41					−15.01				
2	1000	0.97					−8.74				
3		1.48	2.84	2.16	1.03	0.24	−11.68	−16.63	−14.42	−9.13	0.23
4		1.98					−13.76				
5		2.36					−15.1				
2	2000	0.92					−9.09				
3		1.47	2.84	2.14	0.97	0.11	−11.89	−16.74	−14.51	−9.37	0.11
4		1.97					−13.89				
5		2.35					−15.22				

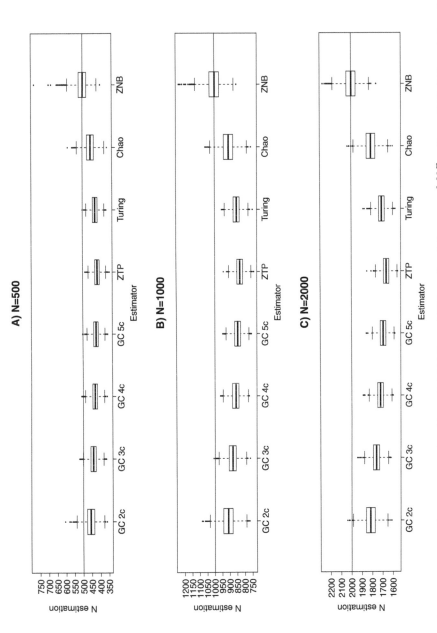

FIGURE 13.4: Simulation based on a negative binomial $Y_i | Z_i \sim NB(\mu_i, \theta)$ with $\mu_i = e^{0.02 Z_i}$, $Z_i \sim N(8, 25)$ and $\theta = 3$. Horizontal line indicates the true population size of the scenario; A) $N = 500$, B) $N = 1000$, C) $N = 2000$.

13.5 Case study: Carcass submission from animal farms in Great Britain

Private veterinary surgeons (PVS) regularly send animal submissions to the Animal and Plant Health Agency (APHA) to determine the cause of death based on a post-mortem examination, to test an animal sample to confirm a disease or to find out whether an animal needs further testing. The PVS might choose to submit or not submit a sample depending on the disease. Only notifiable diseases are compulsory to investigate and report to the authorities. The APHA could miss submissions for several reasons, for instance, a PVS might have facilities to run some diagnostic tests, or he/she might not submit a sample because there is history of a confirmed disease in the farm and the animal presents similar symptoms. The cost is also an important factor, as farmers might not even call a PVS when they believe that the disease is not going to spread to other animals. In fact, the Department of Food and Rural Affairs (DEFRA) used to subsidise some diagnostic tests but the current economic climate is leading to move all costs to farmers. Our objective is to evaluate the completeness of the farm submissions in Great Britain to understand which proportion of the general picture is being explained. In 2009, the number of farms with cattle was estimated to be 60,571 farms; 48,535 of those farms did not have any submissions that year. From the 12,036 farms that submitted, we aim to estimate the total number of farms with unknown disease that did not submit.

Three risk factors related to animal submissions were identified in previous studies carried out at APHA: holding type (beef or dairy), holding size and distance to the regional labs. Large holdings are expected to have a larger submission rate because of the potential costs involved if the disease spreads within the farm and their financial resources. The distance from the farm to the closest regional lab is also specially important for carcass samples, because farmers are obliged to cover delivery costs to the regional lab. On the positive side, a carcass sample has higher probability of identifying the disease. In this problem, the re-capture comes from the second or more submissions from the same farm, so the dependent variable is the number of submissions from each farm.

The total number of carcass submissions and the total number of submissions including other types of samples (like blood or faecal samples) are the primary endpoints. Table 13.5 contains the data in the format of frequency of frequencies. A ratio plot (Rocchetti et al. [247], Böhning et al. [44]) is initially produced to evaluate the existence of heterogeneity in the probability of submitting animal samples and to identify the right statistical distribution to model the capture-recapture probability. Table 13.6 contains the ratios and their 95% confidence limits. In our case, Figure 13.5 presents a structural heterogeneity, which questions the use of a homogeneous Poisson and suggests the use of a heterogeneous distribution, such as the negative binomial distribution. The probability of submitting any type of animal samples is found significantly related to the holding size (log-scale) and the type of the holding (dairy or beef) in all models (Table 13.7). The distance to the closest regional lab becomes significant when holdings with 4 or 5 submissions are included in the models. In contrast, the probability of submitting carcass samples was not related to any of the risk factors for models based on holdings with 2 or 3 submissions only. The distance to the regional lab and the holding size become significant once holdings submitting 4 or 5 times within the study period are included in the analysis.

There are large differences between estimates. The zero truncated Poisson model with covariates provides an estimate which is lower than the one provided by the conventional Chao estimator, which is proven to be a lower bound estimator. The Good–Turing estimator also underestimates due to the non-homogeneous captured probability. The GC estimator

TABLE 13.5: Frequency distribution of number of farms submitting any type of samples (first row) and number of farms submitting carcass samples (second row) to APHA regional laboratories in 2009

f_0	f_1	f_2	f_3	f_4	f_5	f_6	f_7	f_8	f_9	f_{10}	f_{11+}	Total
48535	6340	2520	1149	709	380	249	173	135	94	80	207	60571
58713	1532	231	51	27	6	5	2	1	3	0	0	60571

TABLE 13.6: Ratios $(r(x) = (x+1)f_{x+1}/f_x)$ and confidence bands for the ratio plot (Figure 13.5)

Ratio	Any sample $\hat{r}_x$	$\hat{r}_x$ 95% CL	Carcass $\hat{r}_x$	$\hat{r}_x$ 95% CL
r_1	0.26	(0.25-0.27)	0.05	(0.05-0.05)
r_2	1.19	(1.14-1.25)	0.45	(0.39-0.52)
r_3	1.82	(1.70-1.96)	0.88	(0.65-1.20)
r_4	3.09	(2.81-3.39)	2.65	(1.66-4.22)
r_5	3.22	(2.84-3.64)	1.33	(0.55-3.23)
r_6	4.59	(3.91-5.38)	5.83	(1.78-19.11)
r_7	5.56	(4.58-6.75)	3.20	(0.62-16.49)
r_8	7.03	(5.61-8.8)	4.50	(0.41-49.63)
r_9	6.96	(5.35-9.06)	30.00	(3.12-288.42)
r_{10}	9.36	(6.95-12.61)		
r_{11}	31.05	(23.99-40.19)		

with two non-truncated counts is significantly larger than Chao's lower bound estimator, however its confidence limits are larger as we observed in the simulated scenarios. GC estimator with three non-truncated counts is closer to Chao's estimator. The models with more than 3 counts present lower point estimates than Chao's lower bound estimator. The percentage of farms detected with all types of submissions based on the generalised Chao estimator using two counts is between $\frac{12,036\times100}{22,430}$ (53.7%) and $\frac{12,036\times100}{20,883}$ (57.6%). The completeness of carcass submissions is between 21% to 28.5%, which suggests that a further investigation should be carried out to find out the main causes of missing submissions to establish new policies for increasing submission rates.

Holling et al. [145] presented a plot based on a marginal method to assess the fitting of models with covariates for all count distributions. We use 13.19 to calculate the marginal expectations required by the method and we compare them with the observed frequencies of frequencies (Figure 13.6). We observe that $K = 3$ seems to be the optimal truncation cut-off point for the case of any submission. Although we notice the lines are close to each other, small changes in the y-axis mean large differences in frequencies because of the scale of the graph. However, it is not visually clear which model to choose for the case of carcass submissions. The conventional Chao estimator is larger than population size estimates based on models with 3 or more non-truncated counts, which suggests that those estimators are underestimating. The 95% confidence intervals for the generalised Chao estimator with 3 non-truncated counts are larger than Chao's lower bound estimator, so we could potentially choose the generalised Chao estimator with 2 and 3 non-truncated counts.

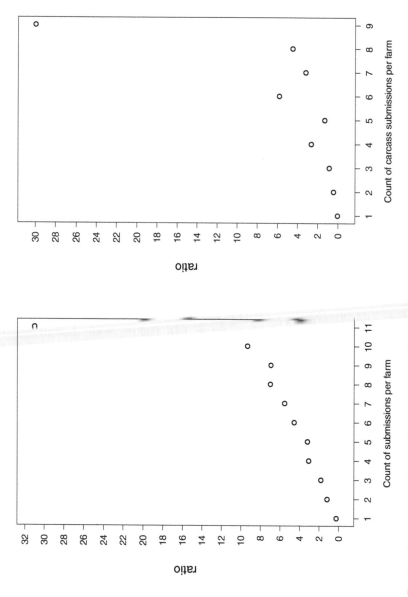

FIGURE 13.5: Ratio plot to investigate the presence of heterogeneity in the number of animal submissions and carcass submissions respectively; $r(x) = (x+1)f_{x+1}/f_x$.

TABLE 13.7: 1. Results from the logistic regressions to obtain the generalised Chao estimates 2. Point estimates and 95%CI of the number of farms with unknown disease based on any sample ard only carcass samples

1. Model summary for the generalised Chao estimator

Counts	Covariate	SUBMISSIONS coef(SE)	p-value	CARCASS SUBMISSIONS coef(SE)	p-value
2	log-size	0.333 (0.027)	< 0.0001	0.325 (0.08)	0.0891
	type(1=dairy 0=beef)	0.286 (0.052)	< 0.0001	0.06 (0.16)	0.7066
	log-distance	−0.004 (0.039)	0.9182	−0.153 (0.092)	0.1033
3	log-size	0.322 (0.018)	< 0.0001	0.34 (0.061)	0.1987
	type(1=dairy 0=beef)	0.366 (0.034)	< 0.0001	0.035 (0.123)	0.7766
	log-distance	−0.04 (0.025)	0.1055	−0.091 (0.072)	0.2170
4	log-size	0.336 (0.014)	< 0.0001	0.371 (0.053)	0.002
	type(1=dairy 0=beef)	0.395 (0.027)	< 0.0001	0.057 (0.108)	0.5975
	log-distance	−0.057 (0.019)	0.0029	−0.183 (0.059)	0.0028
5	log-size	0.352 (0.013)	< 0.0001	0.372 (0.05)	0.0108
	type(1=dairy 0=beef)	0.392 (0.024)	< 0.0001	0.136 (0.105)	0.1949
	log-distance	−0.052 (0.016)	0.0017	−0.14 (0.058)	0.01881

2. Estimated number of farms

Estimator	# SUBMISSIONS $\hat{N}$(95%CI)	# CARCASS SUBMISSIONS $\hat{N}$(95%CI)
G-Chao (2 counts)	21657 (20883,22430)	7688 (6523,8853)
G-Chao (3 counts)	20396 (19846,20947)	6851 (5999,7703)
G-Chao (4 counts)	19255 (18806,19704)	6260 (5530,6990)
G-Chao (5 counts)	18642 (18238,19047)	6110 (5406,6814)
Chao	20011 (19740,20282)	6938 (6868,7009)
Turing	15532 (15349,15716)	5279 (4645,5913)
ZTP	18346 (17932,18759)	6008 (5293,6723)

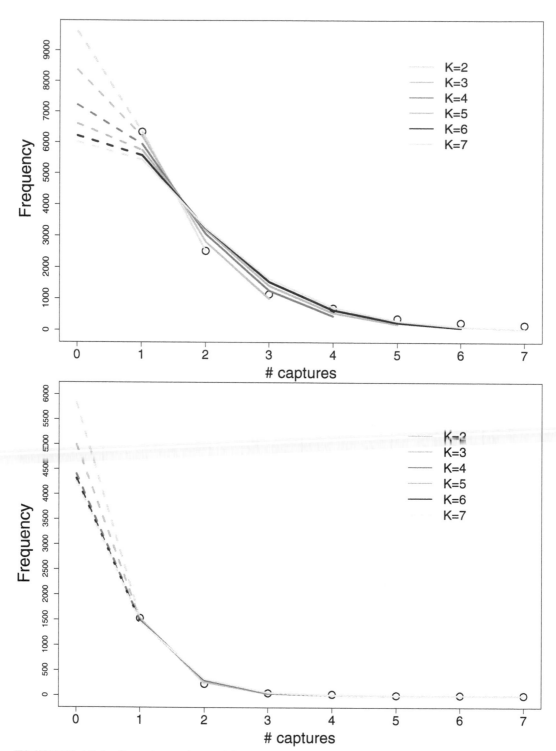

FIGURE 13.6: Covariate-adjusted frequency plot comparing observed and expected frequencies for models up to 7 counts; upper panel: all submissions; lower panel: carcass submissions.

13.6   Software

Böhning et al. [45] proved that the log-likelihood of a truncated Poisson distribution with non-truncated counts of ones and twos, could be written almost as the likelihood of a logistic regression model. Therefore, any standard software could be used to estimate the logistic regression coefficients α and β'. Thereafter, we only need to apply an easy formula to the predicted values to obtain the generalised Chao's estimator with 2 non-truncated counts.

Once we change the level of truncation in the model, there are no more simple links to known likelihoods that we could exploit to obtain estimators using standard software. Therefore, we have developed an experimental library in R, called *gchao* to obtain the estimators presented in this chapter. The library can be downloaded on http://albertovidal.org/gchao.

The syntax is simply *gchao(formula, data, subjid, dist, max_K)*, where

- **formula**: the standard way in R to indicate models. Our outcome is *number of captures*. In our case study, we would have a model like
  captures ~ log_distance + log_size + c_type.
  Estimators without covariates can be obtained using the usual R syntax captures ~ 1.

- **data**: data frame with all captured individuals, their number of times captured and their covariate information.

- **subjid**: string with the name of the column within the data frame used as the identifier.

- **dist**: Poisson, binomial or geometric distributions can be chosen. The theory for binomial and geometric distributions is in the process of being published.

- **max_K**: indicates the maximum number of non-truncated counts that we would like to produce in the output. The programme uses the maximum number of captures by default.

The variable obtained is a **gchao** object with several attributes:

- **formula**: formula used to obtain the estimates.

- **estimates**: provides all generalised Chao estimates from 2 counts to **max_K** number of counts.

- **sd**: provides the standard error of the estimates.

- **coef**: provides the coefficients for each covariate for all models with different truncation cut-off points.

- **se_coef**: provides the standard error of the coefficients.

- **llik**: provides the value of the log-likelihood for each model.

- **expected**: provides a matrix with the expected frequencies for each model.

In Section 13.5, we described a data set where we were interested in estimating the number of farms in Great Britain that have diseased cows, but the AHPA does not have any sample submitted to their regional labs. Our model included the number of cows in the farm, the distance to the closest regional lab, both in logarithmic scale, and the type of the farm (beef or dairy). Our data is saved in an R data.frame called *AHPA*. The command *gchao* expects the following format when covariates are included in the analysis.

```
> head(AHPA)
     id captures log_size log_distance c_type
[1,]  1        1 4.077537     10.83944      0
[2,]  2       12 6.369901     10.75648      1
[3,]  3        1 2.079442     10.75411      0
[4,]  4        1 3.044522     10.74261      0
[5,]  5        3 5.159055     10.99479      0
[6,]  6        3 5.899897     11.19188      0
```

The data set must have a row for each unit captured, a column that serves as unique identifier of each unit, a column with the number of times the unit has been captured in the period of interest and columns with the covariate information. The code below fits the model presented in Section (13.5) including all available covariates. The variable *results_AHPA* becomes a special **gchao** object and contains all the information described above.

Therefore, the usual R syntax can be used to access every output individually (Example: *results_AHPA$estimates*). The command *summary* can be used to obtain a summary of results as shown below.

```
##### Fit model with all covariates
> results_AHPA <-gchao(formula="captures ~ log_size + log_distance
+ c_type",data=AHPA, subjid="id", dist="POI",max_K=6)
> summary(results_AHPA)
Formula:
captures ~ log_size + log_distance + c_type
<environment: 0x000000001bf8cff8>
Generalised Chao's estimator:
GC 2 counts GC 3 counts GC 4 counts GC 5 counts GC 6 counts
  21656.55    20396.07    19254.79    18642.26    18249.79

SD GChao:
SD GC 2 counts SD GC 3 counts SD GC 4 counts SD GC 5 counts SD GC 6 counts
      394.6964       280.9132       229.2356       206.3368       194.0265

Coefficients:
         Int  log_size log_distance    c_type
2c -2.036579 0.3326375  -0.00422977 0.2864431
3c -1.506790 0.3221792  -0.04011845 0.3658361
4c -1.278085 0.3359558  -0.05713053 0.3953830
5c -1.334505 0.3524100  -0.05243374 0.3923007
6c -1.392771 0.3630323  -0.04818300 0.4234728

SE Coefficients:
         Int   log_size log_distance     c_type
2c 0.4394407 0.02660062   0.03902163 0.05160643
3c 0.2785773 0.01772500   0.02451635 0.03377262
4c 0.2167651 0.01425468   0.01895266 0.02713924
5c 0.1879633 0.01251089   0.01634677 0.02414835
6c 0.1711494 0.01143886   0.01483871 0.02257935

Expectations:
          f1       f2       f3       f4       f5       f6
2c 6340.001 2519.999       NA       NA       NA       NA
3c 6194.946 2810.068 1003.986       NA       NA       NA
4c 5941.617 3060.531 1262.925 452.9274       NA       NA
5c 5750.024 3177.953 1419.560 554.2725 196.1901       NA
6c 5577.115 3240.351 1543.923 648.8083 248.3122 88.49046

Log-likelihood:
  2 counts   3 counts    4 counts    5 counts    6 counts
 -5143.846  -8542.706 -11080.241 -12716.683 -13955.137
```

We could obtain p-values for each covariate using likelihood ratio tests between nested models with the function **lrt.gchao(model1,model2)**, where *model1* is the simplest of both models. In the following example code, we fit all models with 2 covariates and we compare them with the model with 3 covariates to obtain the p-values presented in Table 13.7. The comparison can only be done between models with the same number of non-truncated counts that share the same data. The output shows the p-values comparing the two models for every number of non-truncated counts separately.

```
##### Fit models with 2 covariates
> results1 <-gchao(formula="captures ~ c_type + log_distance",
  data=AHPA,subjid="id",dist="POI",max_K=6)
> results2 <-gchao(formula="captures ~ log_size + log_distance",
  data=AHPA,subjid="id",dist="POI",max_K=6)
> results3 <-gchao(formula="captures ~ log_size + c_type ",
  data=AHPA,subjid="id",dist="POI",max_K=6)

##### Call function lrt.gchao to obtain likelihood ratio tests
> lrt.gchao(results1,results_AHPA)
2 counts 3 counts 4 counts 5 counts 6 counts
       0        0        0        0        0
> lrt.gchao(results2,results_AHPA)
    2 counts     3 counts     4 counts     5 counts     6 counts
2.782118e-08 0.000000e+00 0.000000e+00 0.000000e+00 0.000000e+00

> lrt.gchao(results3,results_AHPA)
    2 counts     3 counts     4 counts     5 counts     6 counts
0.918217852 0.105485663 0.002891137 0.001693978 0.001366178
```

Finally, we could use the expected values to create a covariate-adjusted frequency plot. We could manually input the observed frequencies without considering covariates (Table 13.5) or we could use some basic R commands to get it from the original data (see code below). Then, we plot those observed frequencies as points and we add lines based on the expected frequencies available in the **gchao** object.

```
##### Preparing data set with the observed data ignoring covariates
> observed<- cbind(as.numeric(rownames (table(AHPA$captures))),
  table(AHPA$captures))
> colnames(observed)<-c("Number captures","Frequency")

##### First rows of the data set observed
> head(observed)
  Number captures Frequency
1               1      6340
2               2      2520
3               3      1149
4               4       709
5               5       380
6               6       249

##### Plot observed data for K=6 (number captures vs frequencies)
> plot(observed[1:6,1],observed[1:6,2],xlab="Captures", ylab="Frequency")

##### Add lines with the expected values
> lines(c(1:6),results_VLA$expected[1,],col=1)
> lines(c(1:6),results_VLA$expected[2,],col=2)
> lines(c(1:6),results_VLA$expected[3,],col=3)
> lines(c(1:6),results_VLA$expected[4,],col=4)
> lines(c(1:6),results_VLA$expected[5,],col=5)
```

This R library also includes functions to obtain the conventional Chao estimator, Turing estimator, the zero-truncated Poisson estimator and the zero-truncated negative binomial.

14

Population size estimation for one-inflated count data based upon the geometric distribution

Panicha Kaskasamkul

University of Southampton

Dankmar Böhning

University of Southampton

CONTENTS

14.1   Introduction and background

Estimation of the size of an elusive target population is of great interest in several areas such as biology, ecology, epidemiology, public health and social science. Capture-recapture methods have been applied to estimate the size of populations which are difficult to approach. They have a long history and were traditionally applied in wildlife, biology and ecology to estimate the animal abundance and the size of wildlife populations. To estimate the target population size N, capture-recapture surveys are conducted by using an identifying mechanism. The presence or absence of each individual is noted. For example, capture-recapture methods use the information available from animals captured on a number of surveys. Animals trapped are marked, released and allowed to mix with the population. After a period of time, a second survey is taken and the number of animals captured is counted and marked again. Repeated surveys are carried out and the number of animals being marked from all surveys are obtained as the capture-recapture history. This provides the observed frequency of identified individuals. Accordingly, the capture-recapture history is used to estimate the total population size or the number of cases which are never caught at any occasion. Typically, the survey is within a short period so that evolution of new cases or extinction of existing cases is unlikely to occur during the study period. This is referred to as the case of a closed population. These concepts have been applied to human populations in social science and criminology to estimate, for example, the size of an illicit drug-using population or the number of violators of a law (see e.g. van der Heijden et al. [279]; Hser [149]), in public health science for estimating the disease prevalence (see e.g. Gallay et al. [124]; Böhning et al. [35]) and to estimate the number of unreported diseases, as well as infection rate of AIDS in epidemiology (see e.g. Brookmeyer and Gail [54]), and estimating the number of unknown errors in a software in system engineering (see e.g. Lui et al. [184]). In these situations, the population size can be determined by using a number of different sources (lists) as survey occasions or identifying mechanisms such as hospital lists, treatment center registries or pharmacy records. The similarity to wildlife capture-recapture in these cases is that the role of the trap is taken by the register (cancer occurrence), the police (violations of a law) or the reviewer (software error). Typically the number of cases that do not appear in either list is unknown and needs to be estimated (see e.g. Brittain and Böhning [53]).

From the capture-recapture history, a count x as the number of individuals identified exactly x times is obtained. A counting distribution arises when a frequency table is constructed from summarizing how often a particular individual was identified. This is usually referred to as capture-recapture data in the form of *frequencies of frequencies*. However, some individuals do not appear since they have never been identified, so the zero count data are missing. The frequency count data is $\{(x, f_x) \mid x \geq 1\}$ where f_x is the frequency of individuals captured exactly x times. Consequently, the frequency distribution is a zero-truncated count distribution. Based upon a zero-truncated model, it is assumed that all individuals in the population of interest have the same parameter determining probabilities to be captured once, twice and so on. This is defined as the case of *homogeneity* and is often modeled by the Poisson or binomial distribution (see the review of Bunge and Fitzpatrick [59]). The parameter is unknown and can be estimated by various methods. If an estimate of the parameter is derived, then the probability of zero counts is obtained leading to an estimate of the hidden as well as the total size of the population. However, the homogeneous model rarely holds in practice because of the fact that the population frequently includes various subpopulations. Each subpopulation has the same distribution but different parameters. This case is the so-called *heterogeneity* case. Capture probabilities under a heterogeneous model are likely to differ for each individual. Approaches that take into account heterogeneous models are introduced by Chao [72], Zelterman [306] and Chao and Bunge [76]. The problem of heterogeneity should not be ignored as it can cause severe underestimation of the true population size (see van der Heijden et al. [280] and Böhning and Schön [37]). See also Chapter 1, Sections 1.3 and 1.4.

Throughout the years, numerous models and estimators were developed and proposed to improve inferences in capture-recapture studies which always relied on certain assumptions but, in real situations, these assumptions are often violated due to a time effect, the occurrence of heterogeneity, or behavioural response among others. In some capture-recapture studies, we can notice from data that there is some sort of *one-inflation* in the count distribution (see e.g. Farcomeni and Scacciatelli [117]). Some portion of the population is mostly captured only once. This may be a consequence of the fact that the probability of recapturing the same individual is very low, especially in large cities/areas and generally within a short survey period. Secondly, the first capture can lead to a behavioural response for some individuals to no longer be observed. For example, individuals are stressed from the first capture and learn to avoid recapture further on. Under serious law enforcement, more serious legal penalties are expected after the second time an individual is reported as a perpetrator. Individuals may lose their driver's licence, pay a fine and/or take part in treatment programs or a foreigner's entry visa may be revoked. In contrast, an individual may get, as a consequence, only a warning by the judge if they are identified the first time. It is not surprising if individuals show *trap avoidance* after the first capture. Thirdly, the frequency of count one (singleton) may not be reliably observed in some applications such as in microbial diversity. One-inflation arises, especially, in data derived from modern high-throughput DNA sequencing. A new taxa may be assigned incorrectly from the error of sequences instead of being matched to the observed taxa. This leads to an artificially inflated frequency of count one as shown in terms of one-inflation (see Bunge et al. [61]). As the result of one-inflation being present in the count data, some fitting models suffer from a boundary problem and some estimators provide extreme overestimation of the population size (see Godwin [130]), particularly for Chao's lower bound estimator which seemingly adjusted for the heterogeneity.

To illustrate the potential of large bias in a Chao estimate, we consider synthetic data of a population with size $N = 15,000$ with $10,000$ counts generated from the Poisson with parameter 2 merged with $5,000$ extra ones. The frequency distribution is $f_0 = 1,377, f_1 = 7,823, f_2 = 2,614, f_3 = 1,736, f_4 = 894, f_5 = 354, f_{6+} = 202$. In this case, the observed sample size is $n = 13,623$. We ignore the fact that f_0 is known and estimate it by the conventional Chao estimator $\hat{f}_0 = f_1^2/2f_2 = 11,706$ and finally the population size estimate is $\hat{N} = n + \hat{f}_0 = 25,329$. It can be seen clearly that Chao's estimator gives a serious overestimate of the true $f_0 = 1,377$ and $N = 15,000$, respectively. The associated ratio plot and frequency chart are presented in Figure 14.1. See also Chapter 3, Section 3.3.1. The ratio plot shows clear evidence of one-inflation since the first point is not that of a horizontal line. Here, the explanation could be that there are a lot more counts of one. Therefore, from this example, the ratio plot can be used as a rough diagnostic device of one-inflation. Additionally, we can use the ratio plot for the geometric distribution to investigate the suitability of a geometric distribution with one-inflation in a similar way. See also Chapter 4, Section 4.4.

The geometric distribution is a remarkably simple and flexible distribution. Although it has been often ignored for modeling count distributions, it is popular in survival analysis for lifetime data and also interesting for its memoryless property. Moreover, the geometric distribution provides a more flexible model than the Poisson due to the fact that it arises as a mixture of the Poisson when the Poisson parameter is mixed with an exponential distribution that allows for some heterogeneity in the count data (see Niwitpong et al. [218]). In this chapter, we will focus on models specifically designed to estimate the size of a population for one-inflated capture-recapture count data allowing for heterogeneity. These models are based upon the geometric distribution.

14.2   The geometric model with truncation

The geometric distribution has a major interesting property that turns out to be useful for the truncated process.

- Let $(1-p)^x p$ be the geometric for $x = 0, 1, ...$ Then the zero-truncated geometric is again a geometric having the form

$$\frac{(1-p)^x p}{1-p} = (1-p)^{x-1}p \tag{14.1}$$

for $x = 1, 2, ...$

There is suspicion that counts of one are inflated. Hence, it might be appropriate to exclude ones from the estimation. The density is a geometric distribution again.

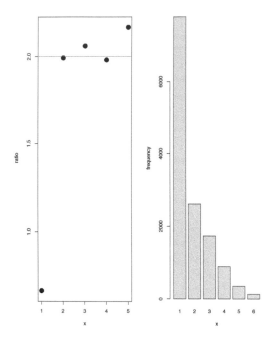

FIGURE 14.1: Ratio plot (left panel) and corresponding frequency chart (right panel) for $N = 15{,}000$ simulated Poisson counts with mean 2 and 50% one-inflation.

• *Let $(1 - p)^{x-1}p$ be the geometric for $x = 1, 2, \dots$. Then the one-truncated geometric is again a geometric of the form*

$$\frac{(1 - p)^{x-1}p}{1 - p} = (1 - p)^{x-2}p \tag{14.2}$$

for $x = 2, 3, \dots$.

This truncation process can be continued with higher counts also leading to a geometric density. The first proposed model is based on the one-truncated geometric distribution that excludes counts of ones for the estimation and uses only the other counts for estimating p. Then use the estimate $\hat{p}$ of p to find the estimate of population size:

$$\hat{N} = \frac{n}{1 - \hat{p}_0} = \frac{n}{1 - \hat{p}}, \quad \text{since} \quad p_0 = (1 - p)^0 p = p. \tag{14.3}$$

14.3   One-truncated geometric model

Under the assumption that the frequencies of count one are inflated, some estimators are developed under the one-truncated geometric model. The first proposed estimator is provided in the form of a Turing estimator and another one is developed by the maximum likelihood approach.

14.3.1 One-truncated Turing estimator

Let f_x be the frequency of individuals identified exactly x times. Also, $n = \sum_{x=1}^{m} f_x$ is the total number of observed cases in the sample, and $S = f_1 + 2f_2 + 3f_3 + \ldots + mf_m = \sum_{x=1}^{m} xf_x$ is the total number of captured cases. The estimate of p_0 and population size N can be calculated from the observed frequencies as follows:

$$\hat{p}_0 = \frac{f_1/N}{S/N} = \frac{f_1}{S}$$

$$\hat{N}_T = \frac{n}{1 - f_1/S}.$$

This is the conventional Turing estimator developed under the Poisson model, see also Chapter 1, Section 1.3. Under the geometric distribution, let $p_x = (1-p)^x p$; $x = 0, 1, 2, \ldots$. The Turing estimator of p can be derived as follows:

$$\frac{p_1}{E(X)} = \frac{(1-p)p}{(1-p)/p} = p^2,$$

or

$$\sqrt{\frac{p_1}{E(X)}} = p = p_0.$$

It follows that

$$\hat{p} = \sqrt{\frac{f_1}{S}}. \tag{14.4}$$

Consider the case of a one-truncated geometric distribution. Let us write

$$p_y = (1-p)^{y-1} p; \quad y = 1, 2, 3, \ldots$$

in the form

$$p_x = (1-p)^x p; \quad x = 0, 1, 2, \ldots$$

with $x = y - 1$. From the formula in (14.4) it follows that

$$\hat{p} = \sqrt{\frac{p_1}{E(X)}} = \sqrt{\frac{f_{x=1}}{S_x}}.$$

Transform the random variable x to y so that

$$\hat{p} = \sqrt{\frac{f_2}{0 f_{x=0} + 1 f_{x=1} + 2 f_{x=2} + \ldots + (m-1) f_{x=m-1}}}$$

$$= \sqrt{\frac{f_2}{0 f_1 + 1 f_2 + 2 f_3 + \ldots + (m-1) f_m}}.$$

Hence, the estimate of p can be calculated from the observed frequencies as

$$\hat{p}_{\text{T\_OT}} = \sqrt{\frac{f_2}{f_2 + 2f_3 + 3f_4 + \ldots + (m-1)f_m}}. \tag{14.5}$$

Thus, the one-truncated Turing estimator for estimating the population size is given by

$$\hat{N}_{\text{T\_OT}} = \frac{n}{1 - \hat{p}_{\text{T\_OT}}}. \tag{14.6}$$

The formula in (14.6) is simply derived in terms of the Horvitz–Thompson estimator in (14.3) by replacing $\hat{p}$ by $\hat{p}_{\text{T\_OT}}$ provided in (14.5), assuming there is one-inflation in the capture probability. Expanding to k-truncated geometric distribution, the k-truncated Turing estimator (T_KT) for p is of the form

$$\hat{p}_{\text{T\_KT}} = \sqrt{\frac{f_{k+1}}{\sum_{y=k+1}^{m}(y-k)f_y}}. \tag{14.7}$$

14.3.2 One-truncated maximum likelihood estimator

Let X be the number of times that a unit was identified over the study period. Count X is modelled with a geometric distribution having probability function

$$p_x = (1-p)^x p \quad ; \quad x = 0, 1, 2, \ldots$$

Since the observed sample from a capture-recapture study contains only non-zero counts, the associated probability function becomes a zero-truncated geometric. Additionally, in the sense of frequency data, the observed data are given as f_x where $x = 1, 2, 3, \ldots, m$ where m is the largest observed count. The zero-truncated geometric likelihood is of the form

$$L(p) = \prod_{x=1}^{m} \left[(1-p)^{x-1}p \right]^{f_x}.$$

The log-likelihood function is

$$\log L(p) = \log(1-p) \sum_{x=1}^{m} f_x(x-1) + \log p \sum_{x=1}^{m} f_x. \tag{14.8}$$

To find the maximum likelihood estimator (MLE) of the unknown parameter p, the derivative of (14.8) with respect to p is equated to 0:

$$\frac{dl}{dp} = -\frac{\sum_{x=1}^{m} f_x(x-1)}{1-p} + \frac{\sum_{x=1}^{m} f_x}{p} = 0.$$

This leads to

$$\hat{p}_{\text{MLE\_ZT}} = \frac{n}{S}.$$

Hence, under the assumption of a zero-truncated geometric model, the population size estimator based on the maximum likelihood estimation is

$$\hat{N}_{\text{MLE\_ZT}} = \frac{n}{1-n/S} = \frac{nS}{S-n}, \tag{14.9}$$

See also Chapter 4, Section 4.3.

Similarly, we assume that the count X is modeled as a one-truncated geometric distribution with probability function

$$p_x = (1-p)^{x-2}p \quad ; \quad x = 2, 3, 4, \ldots.$$

The log-likelihood function is

$$\log L(p) = \log(1-p) \sum_{x=2}^{m} f_x(x-2) + \log p \sum_{x=2}^{m} f_x. \tag{14.10}$$

To find the maximum likelihood estimator (MLE) of the unknown parameter p, the derivative of (14.10) with respect to p is equated to 0:

$$\frac{dl}{dp} = -\frac{\sum_{x=2}^{m} f_x(x-2)}{1-p} + \frac{n-f_1}{p} = 0$$

so that

$$\hat{p}_{\text{MLE\_OT}} = \frac{n-f_1}{S-n}$$

arises. Hence, under the assumption of a one-truncated geometric model, the population size estimator based on the maximum likelihood estimation is

$$\hat{N}_{\text{MLE\_OT}} = \frac{n}{1-(n-f_1)/(S-n)}. \tag{14.11}$$

In a similar way, the general form of the maximum likelihood estimator for the unknown parameter p under a k-truncated geometric distribution is derived as

$$\hat{p}_{\text{MLE\_KT}} = \frac{n-\sum_{x=1}^{k} f_x}{S-kn+\sum_{x=1}^{k-1}(k-x)f_x}. \tag{14.12}$$

14.4 Zero-truncated one-inflated geometric model

A one-inflated model is a statistical model based on a probability distribution which allows for frequent one observations. A one-inflated model employs two components that correspond to two one-generating processes. The first process is governed by a binary distribution that generates structural ones. The second process is generated by a probability density function of model $f_x(\theta)$ that generates counts, some of which may be one. The two components of a one-inflated model for θ are described as follows:

$$p_x = \begin{cases} \omega f_x(\theta) & , \quad \text{if} \quad x \neq 1 \\ (1-\omega) + \omega f_x(\theta) & , \quad \text{if} \quad x = 1 \end{cases}$$

where ω is an unknown weight parameter ; $0 \leq \omega \leq 1$. Assume that $x_1, x_2, ..., x_n$ are observed and drawn from a geometric distribution with mean $(1-\theta)/\theta$, where $f_x(\theta) = (1-\theta)^x\theta$; $x = 0, 1, 2, ...$. Thus, a one-inflated geometric probability density function is

$$p_x = \begin{cases} \omega(1-\theta)^x\theta & , \quad \text{if} \quad x \neq 1 \\ (1-\omega) + \omega(1-\theta)^x\theta & , \quad \text{if} \quad x = 1. \end{cases} \tag{14.13}$$

The parameter $1-\omega$ represents the proportion of extra ones present in the population which are not generated by the mechanism provided by $f_x(\theta)$ or a geometric distribution. However, due to the fact that over the study periods of capture-recapture experiments not all observed units were identified at least once, we need to incorporate zero truncation of the one-inflated geometric distribution and this results in

$$p_x^{1+} = \begin{cases} \omega(1-\theta)^x\theta/[1-\omega\theta] & , \quad \text{if} \quad x \neq 1 \\ [(1-\omega) + \omega(1-\theta)^x\theta]/[1-\omega\theta] & , \quad \text{if} \quad x = 1. \end{cases}$$

The observed, incomplete data log-likelihood for a zero-truncated one-inflated geometric distribution is

$$l_A(\omega, \theta) = \sum_{x=1}^{m} f_x \log p_x^{1+}$$

$$= f_1 \log \left\{ \frac{(1-\omega) + \omega(1-\theta)\theta)}{1-\omega\theta} \right\} + \sum_{x=2}^{m} f_x \log \left\{ \frac{\omega(1-\theta)^x \theta}{1-\omega\theta} \right\}$$

$$= f_1 \log \left\{ \frac{(1-\omega) + \omega(1-\theta)\theta)}{1-\omega\theta} \right\} + (n - f_1) \{\log \omega + \log \theta - \log(1-\omega\theta)\}$$

$$+ (S - f_1) \log(1 - \theta)$$

where $S = \sum_{x=1}^{m} x f_x$.

14.4.1 Zero-truncated one-inflated maximum likelihood estimator

The EM algorithm is a popular method for maximum likelihood estimation. McLachlan and Krishnan [206] stated that a general purpose of the EM algorithm is to cope with incomplete-data problems for maximum likelihood estimation. It consists of two steps, the Expectation (E-step) and the Maximization (M-step). In the E-step, we replace all missing data by their expected values that are calculated from the observed data and the current estimates of likelihood parameters. In the M-step, we maximize the likelihood function by using both the observed and imputed data. The EM algorithm is an iterative method, so the procedure alternates between the E-step and M-step until estimates of the likelihood parameters converge.

Here, we wish to fit the zero-truncated one-inflated geometric distribution to the frequency data gained in the capture-recapture study. The complete data log-likelihood is required. By defining the complete data as $f_x, x = 0, 1, 2, ..., m$, this situation can be viewed as a missing data problem since f_0 is unobserved. If f_0 is given, the maximum likelihood estimators are available. We can use the EM algorithm by imputing a value for f_0 and then maximize the non-zero-truncated distribution. Iterating through these two steps gives us a maximum likelihood estimate for θ and ω. The likelihood for the one-inflated distribution can be maximized by means of the EM algorithm. Embedding another EM into the M-step of the outer EM algorithm gives us a nested EM.

- **EM algorithm for the zero-truncated part (outer part)**

  The first step is to specify an initial value by letting $\hat{\omega}_{(0)} = 1/2$ and finding the initial value for $\hat{\theta}_{(0)}$ from $E(X)$ where $X \sim Geo(\theta)$. We have that

  $$E(X) = \frac{1-\theta}{\theta} = \frac{1}{\theta} - 1$$

  $$\frac{1}{\theta} = \frac{\sum_{x=0}^{m} x f_x}{n} + 1 = \frac{\sum_{x=0}^{m} x f_x + n}{n}$$

  $$\hat{\theta}_{(0)} = \frac{n}{\sum_{x=0}^{m} x f_x + n} = \frac{1}{1 + \bar{x}}.$$

Thus, the estimated probability $X = 0$ given the observed data is

$$\hat{p}_{0(0)} = \hat{\omega}_{(0)}\hat{\theta}_{(0)} = \frac{1}{2(1+\bar{x})}.$$

E-step: In order to estimate f_0, the EM algorithm is used as an instrument to solve this problem. By the E-step, the unobserved frequency f_0 is replaced by its expected value given observed frequencies, $(n = f_1 + f_2 + ... + f_m)$, and current estimates of likelihood estimators. Let $\hat{f}_0$ denote the estimate of the expected value of f_0, which can be achieved as follows:

$$\begin{aligned}
\hat{f}_0 &= E(f_0|\text{observed data}\,;\,\theta) \\
&= E(f_0|f_1, f_2, ..., f_m\,;\,\theta) \\
&= Np_0 \\
&= (n + \hat{f}_0)p_0 \\
&= np_0 + \hat{f}_0 p_0.
\end{aligned}$$

The expected frequency of zero counts is

$$\hat{f}_0 = \frac{np_0}{1 - p_0},$$

where $n = \sum_{x=1}^{m} f_x$ is the number of observed units and $\hat{N} = n + \hat{f}_0$.

M-step: The associated complete data log-likelihood is

$$l(\omega, \theta) = \sum_{x=0}^{m} f_x \log p_x$$

where p_x is a one-inflated geometric probability density function, see (14.13). We need to find $\hat{\omega}$ and $\hat{\theta}$ that maximize $l(\omega, \theta)$ to complete the M-step. Unfortunately, the M-step cannot be solved in closed form. Therefore, we use another EM algorithm to solve the M-step.

- **EM algorithm for the one-inflated part (inner part)**

  This can be accomplished by introducing a binary indicator variable z_i defined as

  $$z_i = \begin{cases} 1 & , \quad \text{if the sample value one is from the extra-ones population} \\ 0 & , \quad \text{otherwise.} \end{cases}$$

This leads to the unobserved, complete likelihood function given as

$$L(X; \omega, \theta) = \prod_{x_i=1} (1-\omega)^{z_i} [\omega(1-\theta)^{x_i}\theta]^{1-z_i} \prod_{x_i \neq 1} [\omega(1-\theta)^{x_i}\theta]. \qquad (14.14)$$

The log-likelihood is

$$\begin{aligned}
l(x; \omega, \theta) = \sum_{x_i=1} &[z_i \log(1-\omega) + (1-z_i)\log\omega + (1-z_i)x_i \log(1-\theta) \\
&+ (1-z_i)\log\theta] + \sum_{x_i \neq 1} [\log\omega + x_i \log(1-\theta) + \log\theta]
\end{aligned}$$

which can be simplified to

$$l(x; \omega, \theta) = \sum_{x_i=1} z_i[\log(1-\omega) - \log\omega] + N\log\omega + \sum_{i=1}^{N} x_i\log(1-\theta) + N\log\theta$$
$$- \sum_{x_i=1} z_i[x_i\log(1-\theta) + \log\theta]. \tag{14.15}$$

Nested E-step: The unobserved indicator z_i is treated as missing data. In the E-step, z_i is replaced by its expected value e_i conditional upon the observed data and current values of ω and θ. Moreover, e_i can be determined as the posterior probability that observation i belongs to extra-ones and can be calculated by the following version of Bayes's theorem:

$$e_i = E(z_i \mid x_i; \omega, \theta) = P(z_i = 1 \mid x_i = 1; \omega, \theta)$$

$$= \frac{P(x_i = 1 \mid z_i = 1; \omega, \theta)P(z_i = 1 \mid \omega, \theta)}{[P(x_i = 1 \mid z_i = 1)P(z_i = 1) + P(x_i = 1 \mid z_i = 0)P(z_i = 0)]}$$

$$= \frac{1-\omega}{[(1-\omega) + \omega f_1(\theta)]},$$

where $f_1(\theta)$ is the geometric probability for a one, so

$$e_i = P(z_i = 1 \mid x_i = 1; \omega, \theta) = \frac{1-\omega}{[(1-\omega) + \omega(1-\theta)\theta]}. \tag{14.16}$$

Now z_i is replaced by its expected value e_i.

Nested M-step: Let $\sum_1 = \sum_{x_i=1} e_i$. To find MLEs of ω and θ, the log-likelihood with z_i replaced by e_i in (14.15) is maximized by taking a derivative with respect to ω and setting it equal to 0,

$$\frac{\partial l}{\partial \omega} = -\frac{\sum_1}{1-\omega} - \frac{\sum_1}{\omega} + \frac{\hat{N}}{\omega} = 0$$

$$\frac{\hat{N}}{\omega} = \frac{\sum_1}{1-\omega} + \frac{\sum_1}{\omega}$$

$$1 - \omega = \frac{\sum_1}{\hat{N}}.$$

Hence,

$$\hat{\omega} = 1 - \frac{\sum_1}{\hat{N}}. \tag{14.17}$$

Then, taking a derivative with respect to θ and setting it equal to 0, we yield

$$\frac{\partial l}{\partial \theta} = -\frac{\sum_{i=1}^{\hat{N}} x_i}{1-\theta} + \frac{\hat{N}}{\theta} + \frac{\sum_1}{1-\theta} - \frac{\sum_1}{\theta} = 0,$$

or

$$\frac{\hat{N}}{\theta} - \frac{\sum_1}{\theta} = \frac{\sum_{i=1}^{\hat{N}} x_i}{1-\theta} - \frac{\sum_1}{1-\theta},$$

or

$$\frac{1-\theta}{\theta} = \frac{\sum_{i=1}^{\hat{N}} x_i - \sum_1}{\hat{N} - \sum_1},$$

and finally

$$\frac{1}{\theta} - 1 = \frac{\sum_{i=1}^{\hat{N}} x_i - \sum_1}{\hat{N} - \sum_1}.$$

Hence,

$$\hat{\theta} = \frac{\hat{N} - \sum_1}{\hat{N} + \sum_{i=1}^{\hat{N}} x_i - 2\sum_1}. \tag{14.18}$$

In summary, we have

$$\hat{\omega} = 1 - \frac{f_1}{\hat{N}}(1 - \omega)/[(1 - \omega) + \omega(1 - \theta)\theta] \tag{14.19}$$

and

$$\hat{\theta} = \frac{\hat{N} - f_1(1 - \omega)/[(1 - \omega) + \omega(1 - \theta)\theta}{\hat{N} + \sum_{i=1}^{\hat{N}} x_i - 2f_1(1 - \omega)/[(1 - \omega) + \omega(1 - \theta)\theta]}. \tag{14.20}$$

Equations (14.19) and (14.20) have to be interpreted such that ω represents the current value and $\hat{\omega}$ is the solution from the M-step for the new iteration. Note also that f_1 is the frequency of ones. Also, $\hat{N}$ refers to the current value of $\hat{f}_0$ leading to $\hat{N} = \hat{f}_0 + n$.

- **Convergence Criterion** determines when iterations are stopped. For the outer EM, iterations cease when

$$| \hat{f}_{0(k)} - \hat{f}_{0(k-1)} | < \varepsilon$$

For the inner EM, iterations cease when all parameter estimates meet the criteria

$$| \hat{\omega}_{(l)} - \hat{\omega}_{(l-1)} | < \varepsilon$$

and

$$| \hat{\theta}_{(l)} - \hat{\theta}_{(l-1)} | < \varepsilon.$$

Consequently, the population size estimator based upon a zero-truncated one-inflated geometric model through the Horvitz–Thompson approach is

$$\hat{N}_{\text{MLE\_ZTOI}} = \frac{n}{1 - \hat{p}_0} \qquad \text{where} \quad \hat{p}_0 = \hat{\omega}\hat{\theta}.$$

14.5 Simulation study

This simulation study was undertaken to investigate the performance of three proposed estimators: the Turing estimator ($\hat{N}_{\text{T\_OT}}$), the maximum likelihood estimator ($\hat{N}_{\text{MLE\_OT}}$) based on the one-truncated geometric model, and the maximum likelihood estimator based on the zero-truncated one-inflated geometric model ($\hat{N}_{\text{MLE\_ZTOI}}$). In addition, three conventional estimators, namely Chao's lower bound, the conventional Turing and maximum likelihood estimator are included to create a comprehensive comparison of all estimators affected by the one-inflation problem. The heterogeneous populations were generated from a geometric distribution (arising from the mixture of a Poisson distribution with an exponential distribution) with parameter $\theta = 0.1, 0.2, 0.3, 0.4$ and population sizes $N = 50, 100, 1000$ for two levels of one-inflation (20% and 50%). Each case is repeated 1,000 times. To evaluate the performance of estimation, the following criteria are used:

- Relative bias ($RBias(\hat{N}) = \frac{E(\hat{N}) - N}{N}$)

- Relative variance ($RVar(\hat{N}) = \frac{E(\hat{N} - E(\hat{N}))^2}{N^2}$)

- Relative mean square error ($RMSE(\hat{N}) = \frac{E(\hat{N} - N)^2}{N^2}$).

The results of the simulation study are presented in Table 14.1 to Table 14.4. Due to the fact that the results of two levels are similar, both parts are summarized. To explore preliminarily the behaviour of estimators, we consider the mean of estimates of population size. According to the results provided in Table 14.1, all of the conventional estimators (Chao, Turing and MLE) show clearly an overestimation of population size for all conditions of the study; it is particularly severe in Chao's lower bound estimator. Conventional Turing and MLE estimators are less affected by one-inflation than Chao's lower bound. All proposed estimators yield satisfying outcomes which are close to the true value of population size N with a slight tendency of overestimating except $\hat{N}_{\text{T\_OT}}$ which gives slight underestimates for small population sizes ($N = 50, 100$) in the case of 20% one-inflation. In addition, $\hat{N}_{\text{MLE\_ZTOI}}$ yields the best estimation results for almost all studied conditions. Correspondingly, $\hat{N}_{\text{MLE\_ZTOI}}$ produces the smallest RBias in all studied cases as Table 14.2 shows. We can rank the performance of proposed estimators in terms of accuracy as $\hat{N}_{\text{MLE\_ZTOI}}, \hat{N}_{\text{T\_OT}}$ and $\hat{N}_{\text{MLE\_OT}}$. This could indicate that the $\hat{N}_{\text{MLE\_ZTOI}}$ can cope with the one-inflation situation better than $\hat{N}_{\text{T\_OT}}$ and $\hat{N}_{\text{MLE\_OT}}$ in both low- and high-level, one-inflation scenarios. According to RVar (see Table 14.3), the $\hat{N}_{\text{T\_OT}}$ tends to provide the minimum RVar in the case of small population size ($N = 50, 100$) whereas $\hat{N}_{\text{MLE\_ZTOI}}$ yields the minimum RVar for the large population ($N = 1000$). However, all proposed estimators give relatively small RVar in all conditions if compared with the conventional estimators. Similar to the results of RVar, the $\hat{N}_{\text{T\_OT}}$ seems to provide the smallest RMSE for the small population size ($N = 50, 100$) whereas $\hat{N}_{\text{MLE\_ZTOI}}$ gives the smallest RMSE for the large size of population ($N = 1000$) as Table 14.4 shows. However, overall the efficiency of $\hat{N}_{\text{T\_OT}}$ seems to be reduced if the level of one-inflation is increasing, which is opposite to $\hat{N}_{\text{MLE\_ZTOI}}$. Furthermore, it can be noticed that with increase of the population size, there is a decline in the RBias, RVar and RMSE for all proposed estimators. On the other hand, with increasing geometric parameter θ there is an increase in the RBias, RVar and RMSE for all proposed estimators.

TABLE 14.1
Population size estimates under 20% and 50% one-inflation (best values give in bold)

Extra ones	N	p	Chao	Turing	MLE	T_OT	MLE_OT	MLE_ZTOI
20%	50	0.1	127.70	57.04	52.73	49.23	51.25	**50.14**
		0.2	112.46	61.63	56.61	48.98	53.22	**50.75**
		0.3	113.43	67.04	61.52	47.45	55.54	**51.33**
		0.4	122.31	75.12	69.20	45.54	59.93	**53.31**
	100	0.1	231.04	113.93	105.26	99.30	102.37	**100.15**
		0.2	205.59	122.04	111.86	98.73	105.22	**100.29**
		0.3	210.15	132.64	121.53	97.50	109.82	**101.52**
		0.4	216.72	146.61	135.11	95.24	117.00	**104.62**
	1000	0.1	2092.53	1138.30	1050.46	1017.62	1021.92	**999.64**
		0.2	1907.28	1218.46	1115.75	1038.31	1050.11	**1000.22**
		0.3	1933.49	1316.74	1203.44	1057.25	1086.36	**1000.56**
		0.4	2027.43	1445.50	1323.87	1094.42	1135.09	**1002.55**
50%	50	0.1	494.75	72.15	59.53	50.68	53.09	**50.19**
		0.2	403.30	89.65	72.78	51.48	57.27	**50.65**
		0.3	399.49	113.63	92.37	**50.56**	62.89	51.62
		0.4	420.22	152.91	127.72	**50.70**	69.75	57.09
	100	0.1	1038.28	143.13	118.28	102.20	105.94	**100.21**
		0.2	742.85	176.13	142.85	104.76	113.43	**100.62**
		0.3	684.94	219.17	178.15	106.47	123.40	**101.50**
		0.4	742.94	282.58	233.44	109.48	138.51	**103.77**
	1000	0.1	8471.62	1419.90	1173.77	1047.30	1055.66	**999.92**
		0.2	6278.96	1744.01	1412.57	1107.76	1127.99	**1001.85**
		0.3	5872.90	2138.95	1736.09	1167.56	1215.55	**1001.32**
		0.4	6253.75	2719.71	2235.84	1246.69	1339.53	**1004.02**

TABLE 14.2
Relative bias of six population size estimators under 20% and 50% one-inflation

Extra ones	N	p	Chao	Turing	MLE	T_OT	MLE_OT	MLE_ZTOI
20%	50	0.1	1.5541	0.1407	0.0546	−0.0154	0.0250	**0.0027**
		0.2	1.2492	0.2326	0.1322	−0.0204	0.0643	**0.0150**
		0.3	1.2687	0.3409	0.2304	−0.0510	0.1108	**0.0267**
		0.4	1.4463	0.5024	0.3841	−0.0893	0.1985	**0.0662**
	100	0.1	1.3104	0.1393	0.0526	−0.0070	0.0237	**0.0015**
		0.2	1.0559	0.2204	0.1186	−0.0127	0.0522	**0.0029**
		0.3	1.1015	0.3264	0.2153	−0.0250	0.0982	**0.0152**
		0.4	1.1672	0.4661	0.3511	−0.0476	0.1700	**0.0462**
	1000	0.1	1.0925	0.1383	0.0505	0.0176	0.0219	**−0.0004**
		0.2	0.9073	0.2185	0.1157	0.0383	0.0501	**0.0002**
		0.3	0.9335	0.3167	0.2034	0.0572	0.0864	**0.0006**
		0.4	1.0274	0.4455	0.3239	0.0944	0.1351	**0.0025**
50%	50	0.1	8.8950	0.4430	0.1905	0.0137	0.0619	**0.0039**
		0.2	7.0660	0.7930	0.4555	0.0296	0.1454	**0.0130**
		0.3	6.9899	1.2726	0.8474	**0.0112**	0.2578	0.0325
		0.4	7.4045	2.0582	1.5544	**0.0140**	0.4043	0.1419
	100	0.1	9.3828	0.4313	0.1828	0.0220	0.0594	**0.0021**
		0.2	6.4285	0.7613	0.4285	0.0476	0.1343	**0.0062**
		0.3	5.8494	1.1917	0.7815	0.0647	0.2340	**0.0150**
		0.4	6.4294	1.8258	1.3344	0.0948	0.3851	**0.0377**
	1000	0.1	7.4716	0.4199	0.1738	0.0473	0.0557	**−0.0001**
		0.2	5.2790	0.7440	0.4126	0.1078	0.1280	**0.0019**
		0.3	4.8729	1.1390	0.7361	0.1676	0.2156	**0.0013**
		0.4	5.2537	1.7197	1.2358	0.2467	0.3395	**0.0040**

TABLE 14.3
Relative variance of six population size estimators under 20% and 50% one-inflation

Extra ones	N	p	Chao	Turing	MLE	T_OT	MLE_OT	MLE_ZTOI
20%	50	0.1	1.9136	0.0050	0.0030	**0.0024**	0.0025	0.0024
		0.2	1.6089	0.0144	0.0092	**0.0057**	0.0069	0.0067
		0.3	1.5709	0.0401	0.0280	**0.0101**	0.0219	0.0203
		0.4	2.1363	0.0968	0.0709	**0.0154**	0.0570	0.0521
	100	0.1	0.8588	0.0023	0.0013	0.0012	0.0011	**0.0010**
		0.2	0.4538	0.0068	0.0045	**0.0032**	0.0036	0.0034
		0.3	0.9350	0.0171	0.0119	**0.0060**	0.0086	0.0086
		0.4	0.5790	0.0391	0.0292	**0.0097**	0.0225	0.0248
	1000	0.1	0.0409	0.0002	0.0001	0.0001	0.0001	**0.0001**
		0.2	0.0252	0.0007	0.0004	0.0004	0.0003	**0.0003**
		0.3	0.0261	0.0014	0.0010	0.0009	0.0007	**0.0008**
		0.4	0.0352	0.0037	0.0027	0.0018	0.0018	**0.0018**
50%	50	0.1	30.5648	0.0179	0.0067	0.0015	0.0018	**0.0013**
		0.2	27.9088	0.0749	0.0399	**0.0045**	0.0092	0.0047
		0.3	31.9849	0.2735	0.1637	**0.0077**	0.0366	0.0202
		0.4	35.7446	1.5514	1.0792	**0.0140**	0.2580	0.2362
	100	0.1	45.7641	0.0077	0.0028	0.0009	0.0009	**0.0006**
		0.2	19.6035	0.0332	0.0168	0.0029	0.0038	**0.0020**
		0.3	13.2992	0.1108	0.0627	**0.0056**	0.0119	0.0058
		0.4	20.1043	0.3522	0.2379	**0.0115**	0.0443	0.0182
	1000	0.1	1.8840	0.0007	0.0003	0.0001	0.0001	**0.0001**
		0.2	0.7029	0.0032	0.0015	0.0005	0.0003	**0.0002**
		0.3	0.4896	0.0091	0.0053	0.0011	0.0010	**0.0005**
		0.4	0.6253	0.0259	0.0175	0.0032	0.0034	**0.0013**

TABLE 14.4
Relative mean square error of six population size estimators under 20% and 50% one-inflation

Extra ones	N	p	Chao	Turing	MLE	T_OT	MLE_OT	MLE_ZTOI
20%	50	0.1	4.3268	0.0248	0.0059	0.0026	0.0031	**0.0024**
		0.2	3.1679	0.0684	0.0267	**0.0001**	0.0111	0.0060
		0.3	3.1708	0.1563	0.0811	**0.0107**	0.0341	0.0210
		0.4	—	0.3490	0.2184	**0.0234**	0.0963	0.0564
	100	0.1	2.5751	0.0217	0.0041	0.0013	0.0017	**0.0010**
		0.2	1.5682	0.0554	0.0186	**0.0033**	0.0063	0.0035
		0.3	2.1474	0.1236	0.0582	**0.0067**	0.0182	0.0089
		0.4	1.9408	0.2563	0.1525	**0.0120**	0.0514	0.0269
	1000	0.1	1.2345	0.0193	0.0027	0.0005	0.0006	**0.0001**
		0.2	0.8483	0.0484	0.0138	0.0019	0.0028	**0.0003**
		0.3	0.8975	0.1017	0.0424	0.0041	0.0082	**0.0008**
		0.4	1.0908	0.2022	0.1076	0.0107	0.0200	**0.0018**
50%	50	0.1	109.6550	0.2141	0.0430	0.0017	0.0057	**0.0013**
		0.2	77.8097	0.7036	0.2473	0.0053	0.0303	**0.0049**
		0.3	80.8113	1.8928	0.8817	**0.0079**	0.1030	0.0212
		0.4	90.5354	5.7861	3.4944	**0.0142**	0.2645	0.2561
	100	0.1	133.7561	0.1937	0.0362	0.0014	0.0044	**0.0006**
		0.2	60.9101	0.6128	0.2004	0.0051	0.0219	**0.0021**
		0.3	47.5016	1.5308	0.6734	0.0098	0.0667	**0.0060**
		0.4	61.4216	3.6855	2.0182	0.0205	0.1925	**0.0196**
	1000	0.1	57.7072	0.1770	0.0305	0.0024	0.0032	**0.0001**
		0.2	28.5697	0.5567	0.1717	0.0121	0.0167	**0.0002**
		0.3	24.2343	1.3063	0.5471	0.0292	0.0475	**0.0005**
		0.4	28.2265	2.9833	1.5448	0.0641	0.1187	**0.0013**

14.6   Real data examples

In this section, we examine the proposed estimators in some real data examples of one-inflation and compare them with conventional estimators.

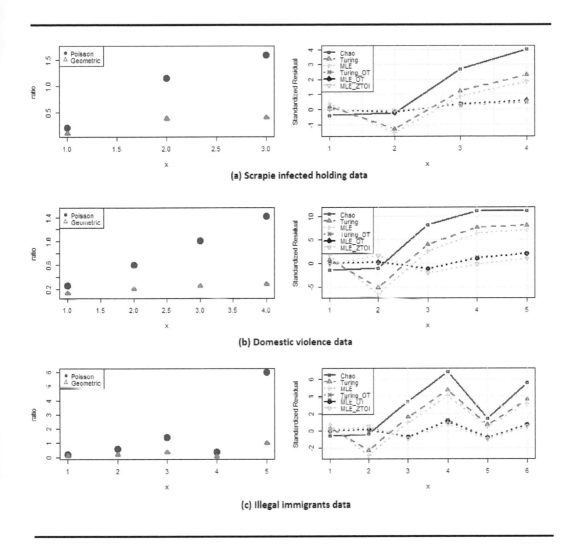

(a) Scrapie infected holding data

(b) Domestic violence data

(c) Illegal immigrants data

FIGURE 14.2: Applications: Ratio plots (left panel) and corresponding fitted value charts for all estimators (right panel).

14.6.1 Scrapie-infected holdings

In the context of animal disease surveillance, the data on scrapie-infected holdings in France are obtained from the French classical scrapie surveillance programme (Vergne et al. [289]). Here, we are interested in estimating the total number of holdings with scrapie infection in France. Table 14.5 presents the frequency distribution of detection among holdings where at least one infected animal was detected. Here f_x represents the number of detected holdings with exactly x infected sheep. The total number of detected holdings is $n = 141$. There are 121 holdings with exactly one infected sheep, 13 holdings with exactly two infected sheep and so forth.

The Figure 14.2(a) left panel shows the two ratio plots, the first one using $\hat{r}_x = (x + 1)f_{x+1}/f_x$ for the diagnosis of a Poisson and the second one using $\hat{r}'_x = f_{x+1}/f_x$ for the diagnosis of a geometric. It is clear that the ratio plot for a Poisson shows a monotone increasing pattern; in particular, we can say that it does not show a horizontal line pattern. Hence the Poisson model may not be suitable with this data whereas the ratio plot for a geometric is much closer to a horizontal line. However, it should be noticed that the first value of the geometric ratio plot $\hat{r}'_1 = f_2/f_1$ is very low if compared with the other values in the graph. This could be explained by the fact that there are a lot more holdings with one infected sheep due to one-inflation. Therefore, it is indicated to use the geometric model under one-inflation estimating the total number of holdings with scrapie infection.

Now, we can check this suspicion again by using the likelihood ratio test as follows:

H_0 : data are from a zero-truncated geometric distribution

H_A : data are from a zero-truncated one-inflated geometric distribution

Set $\alpha = 0.05$ and use the test statistic

$$LRT = -2l_0(0, \tilde{\theta}) + 2l_A(\hat{\omega}, \hat{\theta})$$
$$= -2(77.6590) + 2(75.6607)$$
$$= 4.1966,$$

with a critical value of $\chi^2_{.90,1} = 2.706$. We come to the decision to reject H_0 since $LRT > 2.706$ and $\frac{\text{p-value}}{2}(0.02025) < \alpha(0.05)$ (see the LRT details for mixture densities in Böhning et al. [33], Self and Liang [260]). We conclude that this data can be considered to arise from a zero-truncated one-inflated geometric distribution.

From the evidence provided by ratio plot and likelihood ratio test, the presence of one-inflation can be conjectured. Therefore, all proposed estimators should be appropriate for this data set. The results of estimating the total number of scrapie-infected holdings and the goodness-of-fit statistics from all estimators are shown in Table 14.6. As we expect, all suggested estimators can definitely reduce the assorted overestimation with conventional estimators by producing a distinctly smaller estimate. It clearly reveals that the MLE_ZTOI provides the smallest estimate while the estimate of T_OT is between the estimate of MLE_OT and MLE_ZTOI but slightly closer to the MLE_OT. Moreover, the

TABLE 14.5

French scrapie-infected holdings in 2006

f_1	f_2	f_3	f_4
121	13	5	2

goodness-of-fit statistics and the graph of the standardized residuals in the Figure 14.2(a) right panel show that the estimated values from MLE_ZTOI can fit the data very well and as good as the T_OT estimator.

14.6.2 Domestic violence incidents in the Netherlands

Van der Heijden et al. [284] study the prevalence of domestic violence in the Netherlands for the year 2009 by using capture-recapture methods to estimate the total population size of offenders. The study is reported with the data given in Table 14.7. The total number of observed culprits is $n = 17,662$. There are 15,169 culprits identified exactly once in a domestic violence incident, 1,957 exactly twice and so forth. From the data, it is noticed that the observed data may be in the form of one-inflation. It seems that a portion of the culprits captured for the first time changed their behaviour and will not occur again as perpetrators.

The ratio plot is shown in the Figure 14.2(b) left panel and it remains unclear if there is one-inflation, so here we investigate again by means of the likelihood ratio test and the result of testing shows the presence of one inflation; $LRT = 98.9135$ and p-value $= 0.0000$. It can be assumed that the proposed estimators are viable and suitable with this data set. The results of estimation from the classical and proposed estimators are shown in Table 14.8. The pattern of results for all proposed estimators is different from the example 1, $\hat{N}_{\text{T\_OT}} > \hat{N}_{\text{MLE\_OT}} > \hat{N}_{\text{MLE\_ZTOI}}$. Here, the estimate using MLE_ZTOI is smallest and obviously different from the estimate of T_OT and MLE_OT. This corresponds to the simulation results (large N and p induce overestimation in T_OT and MLE_OT). In terms of statistical model fitting, the Figure 14.2(b) right panel shows the fitted values for this data set with all estimators. It can be seen from the graph and the p-values of the goodness-of-fit statistics in Table 14.8 that the estimated values from all proposed estimators can fit the data reasonably well and significantly better than classical estimators.

TABLE 14.6
Results for scrapie-infected holdings in France

Estimator	$\hat{f}_0$	$\hat{N}$	Chi-square	p-value
Chao$^\$$	1126	1267	27.195	0.00000
Turing	761	902	8.487	0.01436
MLE	686	827	6.781	0.03369
T_OT	**286**	**427**	**0.283**	**0.59474**
MLE_OT	313	454	0.507	0.47644
MLE_ZTOI	121	262	0.316	0.57402

$^\$$ For GOF-test, $\hat{p}_0 = \frac{f_0}{\hat{N}}$ and $p = p_0$ for geometric model

TABLE 14.7
Frequency distribution of domestic violence in the Netherlands

f_1	f_2	f_3	f_4	f_5	f_6
15,169	1,957	393	99	28	16

14.6.3   Illegal immigrants in the Netherlands

We revisit the capture-recapture data of illegal immigrants in the Netherlands from police records (van der Heijden et al. [280]) and use this data set to compare all proposed estimators of population size with the classical estimators. The data records contain information on the number of times each illegal immigrant was apprehended by the police (see Table 14.9). It can be noticed that the number of singletons is considerably higher than the number of doubletons. This indicates that the data may experience one-inflation. Then, we look at the ratio plot as shown in the Figure 14.2(c) left panel. We find that a geometric distribution might be more suitable with this data than a Poisson distribution but we cannot see evidence of one-inflation from the ratio plot. However, the likelihood ratio test indicates that this data set undergoes one-inflation as $LRT = 20.8471$ and p-value $= 4.97 \times 10^{-6}$. Hence, all proposed estimators are applied to this data and the results of estimation from all estimators are shown in Table 14.10. Similar to the previous examples, we consider the results in two parts: estimation and model fitting. In terms of estimation, the estimates from conventional estimators are about double the size of the estimates from our proposed estimators due to the effect of one-inflation as we expected. Interestingly, the estimates of T_OT and MLE_OT are similar; $\hat{N}_{\text{T\_OT}} = 8,341$ and $\hat{N}_{\text{MLE\_OT}} = 8,191$, whereas they are about double MLE_ZTOI, namely $\hat{N}_{\text{MLE\_ZTOI}} = 4,863$. Nevertheless, the estimation of MLE_ZTOI seems to be the best in terms of model fit with $\chi^2 = 2.859$ and p-value $= 0.41388$. This corresponds to the graph in the Figure 14.2(c) right panel.

TABLE 14.8
Results for domestic violence study

Estimator	$\hat{f}_0$	$\hat{N}$	Chi-square	p-value
Chao$^\$$	117,577	135,223	317.537	0.00000
Turing	100,000	100,070	100.795	0.00000
MLE	98,788	116,434	144.797	0.00000
T_OT	65,573	83,219	7.227	0.02696
MLE_OT	64,754	82,400	6.649	0.03599
MLE_ZTOI	**35,085**	**52,731**	**8.097**	**0.01745**

TABLE 14.9
Frequency distribution of the illegal immigrants

f_1	f_2	f_3	f_4	f_5	f_6
1645	183	37	13	1	1

TABLE 14.10
Results for illegal immigrants study

Estimator	$\hat{f}_0$	$\hat{N}$	Chi-square	p-value
Chao$^\$$	14,787	16,667	92.009	0.00000
Turing	12,327	14,270	43.811	0.00000
MLE	11,588	13,468	36.326	0.00000
T_OT	6,461	8,341	3.085	0.37870
MLE_OT	6,311	8,191	2.917	0.40460
MLE_ZTOI	**2,983**	**4,863**	**2.859**	**0.41388**

14.7 Conclusion

In this chapter we have focused on the one-inflation problem that occurs when some individuals change their behaviour and will not be recaptured after the first capture. To estimate the size N of an elusive population under one-inflation, two concepts have been suggested. The first is based on a modification by truncating singletons and applying the conventional Turing and MLE approach to the one-truncated geometric data ($\hat{N}_{\text{T\_OT}}$ and $\hat{N}_{\text{MLE\_OT}}$). These are examined in Section 14.3. On the other hand, another concept, the model-based approach, focuses on developing a statistical model that describes the mechanism for extra-one generation as shown in Section 14.4. In this section, the estimator $\hat{N}_{\text{MLE\_ZTOI}}$ is developed as a maximum likelihood approach by using the nested EM algorithm based upon the zero-truncated one-inflated geometric distribution. Section 14.5 shows that all proposed estimators can solve the problem of one-inflation. $\hat{N}_{\text{MLE\_ZTOI}}$ and $\hat{N}_{\text{T\_OT}}$ perform better than $\hat{N}_{\text{MLE\_OT}}$. However, $\hat{N}_{\text{MLE\_ZTOI}}$ shows a good performance in accuracy and performs best for all conditions under study and provides the smallest variance and mean square error for the large size of population ($N = 1000$), whereas $\hat{N}_{\text{T\_OT}}$ provides the smallest variance and mean square error for the small population size ($N = 50, 100$). Overall it can be concluded that $\hat{N}_{\text{MLE\_ZTOI}}$ is better than $\hat{N}_{\text{T\_OT}}$ especially in cases of a high level of one-inflation and a large population. Furthermore, we applied the proposed estimators to three data sets. All examples show that the proposed estimators can cope with the problem of one-inflation by providing smaller estimates than conventional estimators. In terms of statistical model fitting, it is found that the fitted values of the last developed estimator can fit the data with one-inflation well and better than conventional estimators all of case studies, particularly in last example.

To sum up, it can be seen that both concepts can cope with the problem of one-inflation and each concept has a different advantage. The first concept is simpler whereas the second concept uses a model-based approach to explain the extra-ones. Although the latter approach is more complex and more computationally demanding, it produces the best estimates, especially for the large population size and high level of one-inflation. However, in case of a small population, although the first approach seems to be better than the second approach, the differences between the two are almost negligible. Hence, both approaches seem reasonable to use with a slight benefit to the first one as it is the simpler concept.

Part V

Multiple Sources

15

Dual and multiple system estimation: Fully observed and incomplete covariates

Peter G. M. van der Heijden and Maarten Cruyff

University of Southampton and Utrecht University

Joe Whittaker

University of Lancaster

Bart F.M. Bakker

Statistics Netherlands/VU University

Paul A. Smith

University of Southampton

CONTENTS

15.1   Introduction

A well-known technique[1] for estimating the size of a human population is to find two or more registers of this population, to link the individuals in the registers and estimate the number of individuals that occur in neither of the registers (Fienberg [121], Bishop, Fienberg and Holland [32], Cormack [86], International Working Group for Disease Monitoring and Forecasting, IWGDMFa [154]). For example, with two registers A and B, linkage gives a count of individuals in A but not in B, a count of individuals in B but not in A, and a count of individuals both in A and B. The counts form a contingency table denoted by $A \times B$

---

[1]This chapter is based on van der Heijden, Whittaker, Cruyff, Bakker and van der Vliet [283].

with the variable labeled A being short for 'inclusion in register A', taking the levels 'yes' and 'no', and likewise for register B. See also Chapter 1, Section 1.6.1. In this table the cell 'no,no' has a zero count by definition, and the statistical problem is to better estimate this value in the population. An improved population size estimate is obtained by adding this estimated count of missed individuals to the counts of individuals found in at least one of the registers.

With two registers, the usual assumptions under which a population size estimate is obtained are: inclusion in register A is independent of inclusion in register B; and in at least one of the two registers the inclusion probabilities are homogeneous (see Chao et al. [79], Zwane, van der Pal and van der Heijden [313] and van der Heijden et al. [283]). Interestingly it is often, but incorrectly, supposed that *both* inclusion probabilities have to be homogeneous. Other assumptions are that the population is closed, that it is possible to link the individuals in registers A and B perfectly, and there is no overcoverage due to individuals that are not part of the population (compare Chapter 17 and Chapter 18).

However it is generally agreed that these assumptions are unlikely to hold in human populations. Three approaches may be adopted to make the impact of possible violations less severe. One approach is to include covariates in the model, in particular covariates whose levels have heterogeneous inclusion probabilities for both registers (see Bishop, Fienberg and Holland [32], Baker [17], compare Pollock [235] and Chapter 16). Then log-linear models can be fitted to the higher-way contingency table of registers A and B and the covariates. The restrictive independence assumption is replaced by a less restrictive assumption of independence of A and B conditional on the covariates; and subpopulation size estimates are derived (one for every level of the covariates) that add up to a population size estimate. Another approach is to include a third register, and to analyze the three-way contingency table with log-linear models that may include one or more two-factor interactions, thus getting rid of the independence assumption. Compare also Chapter 18. Here the (less stringent) assumption made is that the three-factor interaction is absent. However, including a third register is not always possible, as it is not available, or because there is no information that makes it possible to link the individuals in the third register to both the first and to the second register. A third approach makes use of a latent variable to take heterogeneity of inclusion probabilities into account (see Fienberg, Johnson and Junker [122], Bartolucci and Forcina [24] and Chapters 19, 20, 21, and 22). Of course, these three approaches are not exclusive and may be used concurrently in one model.

When the approach is adopted to use covariates, the question is which covariates should be chosen. In the traditional approach, only covariates that are available in each of the registers can be chosen. We refer to this as fully observed covariates. Zwane and van der Heijden [311] and van der Heijden et al. [283] showed that it is also possible to use covariates that are not available in each of the registers. For example, when a covariate is available in register A but not in B, the values of the covariate missed by B are estimated under a missing-at-random assumption (Little and Rubin [183]); and the subpopulation size estimates are then derived as a by-product. We refer to this as incomplete covariates. Whether or not the covariates are available in each of the registers, the number of possible log-linear models that can be fit grows rapidly.

In this chapter we study the (in)variance of population size estimates derived from log-linear models that include covariates. Including covariates in log-linear models of population registers improves population size estimates for two reasons. Firstly, it is possible to take heterogeneity of inclusion probabilities over the levels of a covariate into account; and secondly, it allows subdivision of the estimated population by the levels of the covariates, giving insight into characteristics of individuals that are not included in any of the registers. The issue of whether or not marginalizing the full table of registers by covariates over one or more covariates leaves the estimated population size estimate invariant, is intimately related to collapsibility of contingency tables. With information from two registers it is shown that population size invariance is equivalent to the simultaneous collapsibility of each margin consisting of one register and the covariates. Covariates that are collapsible are called passive, to distinguish them from covariates that are not collapsible and are termed active. We make the case that it may be useful to include passive covariates within the estimation model, because they allow a description of the population in terms of these covariates. As an example we discuss the estimation of the population size of people born in the Middle East but residing in the Netherlands.

By focussing on population size estimates, collapsibility in log-linear models is studied in this paper from a different perspective than found in Bishop, Fienberg and Holland [32], who are interested in parametric collapsibility. Our work applies the model collapsibility of Asmussen and Edwards [15], later discussed by Whittaker [295] and Kim and Kim [165], concerning the commutativity of model fitting and marginalization. We use model collapsibility in the context of population size invariance and show that invariance requires model collapsibility of each margin consisting of one register and the covariates. A novel feature is to apply collapsibility in the context of a table containing structural zeros. We give a short path characterization of the log-linear model which describes when marginalizing over a covariate leads to different population size estimates.

The second result can be fruitfully applied in population size estimation. In a specific log-linear model, we denote covariates as passive when they are collapsible and active when they are not collapsible. In principle, many passive covariates can be included in a model. When they are available in all registers, the methodology of the first part of this chapter can be used, and when they are available in only some but not all of the registers the approach described in Zwane and van der Heijden [311] and van der Heijden [283] can be used, and this is discussed in the second part of this chapter. We make a case for including such passive covariates because they allow the description of both the observed part as well as the unobserved part of the population in terms of these covariates.

The chapter is built as follows. In Section 15.2 we discuss the data to be analyzed. These refer to the population of people with Afghan, Iranian and Iraqi nationality residing in the Netherlands. In Section 15.3 we discuss properties of the log-linear models in the context of population size estimation when all covariates are fully observed. This is first discussed in detail for the case of two registers. We illustrate the two properties of log-linear models using a number of examples, and then prove the properties using results from graphical models. For completeness we also discuss the situation when three registers are available and illustrate that the same properties apply. In Section 15.3.3 we develop the notion of active and passive covariates, and in Section 15.3.4 we present an example. In 15.4 we discuss the situation for incomplete covariates and present an example. We end with a discussion.

15.2 The population of people with Middle Eastern nationality staying in the Netherlands

In the 2011 round of the Census, several countries made use of administrative data (rather than polling) for that purpose. There were countries who were repeating this method such as Denmark, Finland and the Netherlands, and more than ten European countries that were using administrative data for the first time (Valente [286]). The administrative registers are combined by data-linking and micro-integration to clean and improve consistency. The outcome of these processes is called a statistical register or a register for short.

The most important administrative register to be used in the Netherland Census is an automated system of decentralized (municipal) population registers (in Dutch, *Gemeentelijke BasisAdminstratie*, referred to by the abbreviation *GBA*). This register is used to define the population. The GBA contains all information on people that are legally allowed to reside in the Netherlands and are registered as such. The register is accurate for that part of the population such as people with Dutch nationality and foreigners that carry documents that allow them to be in the Netherlands for work, study, asylum, and their close relatives. However, these data do not cover the total population, in particular those residing in the Netherlands but who are not allowed to stay under current Dutch law. These latter groups are sometimes referred to as undocumented foreigners or illegal immigrants.

Under Census regulations a quality report is obligatory, and one of the aspects that needs to be addressed is the undercoverage of the Census data. This asks for an estimate of the size of the population that is not included in the GBA. In this paper we approach the problem by linking the GBA to another register and then applying population size estimation methods to arrive at an estimate of the total population. Therefore, we implicitly estimate the part of the population not covered by the GBA. The second register that we employ is the central Police Recognition System or HerkenningsDienst Systeem (HKS), which is a collection of decentralized registration systems kept by 25 separate Dutch police regions. In HKS, suspects of offences are registered. Each report of an offence has a suspect identification where, if possible, information about the suspect is copied from the GBA. If a suspect does not appear in the GBA, fingerprints are taken so that he or she can be found in the HKS if apprehension at a later stage occurs.

We test the methodology described in the next sections using previously collected data of the 15- to 64-year-old age group of people with Afghan, Iranian or Iraqi nationality. For the GBA we extract the registered information of 2007. For HKS we extract information on apprehensions made during 2007. The Table 15.1 illustrates the problem. For people with Afghan, Iranian or Iraqi nationality $1,085 + 26,254 = 27,339$ are registered in the population register GBA; $1,085 + 255 = 1,340$ are registered in the police register HKS, of whom 255 are missed by the GBA. The number of people not in the GBA and not in HKS is to be estimated; this is the number of people missed by both registers. This latter estimate plus 255 should be the size of the population with Afghan, Iranian and Iraqi nationality that do not carry documents for a legal stay in the Netherlands. (We ignore the small group of persons who travel on a tourist visa, and are also not in the GBA and HKS.) This latter estimate plus $(255 + 1,085 + 26,254)$ is the size of the population with Afghan, Iranian or Iraqi nationality that stays in the Netherlands, either with or without legitimate documents.

TABLE 15.1: Linked registers *GBA* and *HKS*

| | HKS | |
GBA	Included	Not included
Included	1,085	26,254
Not included	255	-

An estimate of the number of people missed by both registers can be obtained under the assumption that inclusion in the GBA is independent of inclusion in the HKS. In other words, that the odds of being in the HKS to not being in the HKS, $(1,085 \div 26,254)$ for the people included in the GBA, also holds for the people not included in the GBA. The validity of this assumption is difficult to assess. From a rational choice perspective, people without legitimate documents do their best to stay out of the hands of the police and so make the probability of apprehension smaller for those not in the GBA. On the other hand, people without legitimate documents may be more involved in activities that lead to a higher probability of apprehension and so make the probability larger for those not in the GBA. Both perspectives have face validity but, as far as we know, there is little empirical evidence to support either.

With the data at hand we start from the independence assumption, but mitigate this by using covariates. If a covariate is related to inclusion in the GBA and in the HKS but that, conditional on the covariate, inclusion in the GBA is independent of inclusion in the HKS, then ignoring the covariate leads to dependence between inclusion in the GBA and HKS. For both registers we have gender, age (levels: 15–25, 25–35, 35–50, 50–64) and nationality (levels: Afghan, Iraqi, Iranian). For the GBA we additionally have the covariate marital status (levels: unmarried, married), and for HKS we have the covariate police region of apprehension (levels: large urban, not large urban). We will discuss the latter two covariates in the section on incomplete covariates. We start with fully observed covariates. We first study theoretical properties for the models employed and then discuss an analysis of the data.

15.3 Fully observed covariates

15.3.1 Two registers

We denote inclusion in the two registers by A and B, with levels $a, b = 1, 2$ where level 2 refers to not registered, and we assume that there are I categorical covariates denoted by X_i, where $i = 1, \ldots, I$. The contingency table classified by variables A, B and X_1 is denoted by $A \times B \times X_1$. We denote hierarchical log-linear models by their highest fitted margins using the notation of Bishop, Fienberg and Holland [32]. For example, in the absence of covariates, the independence model is denoted by $[A][B]$, and when there is one covariate X_1 the model with A and B conditionally independent given X_1 is $[AX_1][BX_1]$. In each of the models considered, the two-factor interaction between A and B is absent, as this reflects the (conditional) independence assumption discussed in the Introduction. See also Chapter 19 of this book.

Under the saturated model, the number of independent parameters is equal to the number of observed counts, and the fitted counts are equal to the observed counts. The table $A \times B$ has a single structural zero so that the saturated model is $[A][B]$. When there are I covariates, the saturated model for the table $A \times B \times X_1 \times \cdots \times X_I$ is $[AX_1 \ldots X_I][BX_1 \ldots X_I]$, where A and B are conditionally independent given the covariates.

We use the following terminology. We use the word *marginalise* to refer to the contingency table formed by considering a subset of the original variables. For example, starting with contingency table $A \times B \times X_1$, if we marginalise over X_1 we obtain the table $A \times B$. We use the word *collapse* to refer to the situation that when a table is marginalised, the population size estimate remains invariant. For example, as we see below, the table $A \times B \times X_1$ is collapsible over X_1 when the log-linear model is $[AX_1][B]$ (or is $[A][BX_1]$) as the model gives the same population size estimate as does the $[A][B]$ model for the marginal table $A \times B$.

There are two closely related properties of log-linear models that we wish to examine:

1. There exist log-linear models for which the table is collapsible over specific covariates.

2. For a given contingency table there exist different log-linear models that yield identical total population size estimates.

The properties are closely related because if Property 2 applies, for both log-linear models, the contingency table to which Property 2 refers is collapsible over the same covariates. We first illustrate the properties and then provide an explanation.

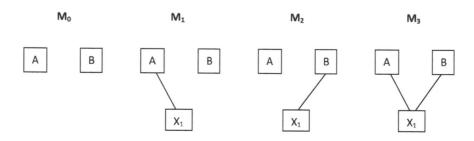

FIGURE 15.1: Interaction graphs for log-linear models with one covariate.

Example 1. Assume that there is one covariate X_1. The data are collated in a three-way contingency table $A \times B \times X_1$. The total population size estimates under log-linear models $M_1 = [AX_1][B]$ and $M_2 = [A][BX_1]$ are equal; this illustrates Property 2. Both total population size estimates are equal to the population size estimate under model $M_0 = [A][B]$ in the two-way contingency table $A \times B$. Hence the three-way table is collapsible over X_1 and this illustrates Property 1. In passing, we note that this result illustrates the second assumption of population size estimation from two registers discussed in the Introduction, namely that the inclusion probabilities only need to be homogeneous for one of the two registers. The population size estimate under log-linear model M_0 $[AX_1][BX_1]$ is different from those population size estimates. See Figure 15.1 for interaction graphs of models M_0, M_1, M_2 and M_3.

We present a numerical example in Tables 15.2 and 15.3. Here A refers to inclusion in the official register GBA, B refers to inclusion in the police register HKS and the covariate X_1 is gender. See Section 15.2 for more details. We note that, even though the total population size estimates for models M_1 and M_2 are equal, estimates of the subpopulations (i.e. males and females) for M_1 are different from those under M_2.

TABLE 15.2: Models fitted to contingency table of variables A (GBA), B (HKS) and to A, B and X_1 (gender), deviances, degrees of freedom (df) and estimated numbers missed

Model	Deviance	df	Missed
M_0: $[A][B]$	0.0	0	6,170.3
M_1: $[AX_1][B]$	548.5	1	6,170.3
M_2: $[A][BX_1]$	1.1	1	6,170.3
M_3: $[AX_1][BX_1]$	0.0	0	5,696.1

Example 2. Suppose that there are two covariates, namely X_1 and X_2. Table 15.4 presents a fairly comprehensive list of typical models including their estimated numbers missed and deviances. We note that models M_4, M_6 and M_6' have identical total population size

TABLE 15.3: Observed and fitted counts for the three-way table of A (GBA), B (HKS) and X_1 (gender); for A and B level 1 is present and for X_1 level 1 is male

A	B	X_1	obs	M_1	M_2	M_3
1	1	1	972	629.2	976.5	972.0
2	1	1	234	234.0	229.5	234.0
1	2	1	14,883	15,225.8	14,883.0	14,883.0
2	2	1	0	5,662.2	3,497.9	3,582.9
1	1	2	113	455.8	108.5	113.0
2	1	2	21	21.0	25.5	21.0
1	2	2	11,371	11,028.2	11,371.0	11,371.0
2	2	2	0	508.1	2,672.5	2,113.2

estimates. Models M_5, M_8, M_9, M_{11} and M'_{11} also have identical total population size estimates. The remaining models M_7, M_{10} and M_{12}, M'_{12} and M''_{12} have different total population size estimates.

TABLE 15.4: Models fitted in a four-way array of variables A, B, X_1 and X_2; registers A (GBA), B (HKS), covariates X_1 (gender), X_2 (age coded in four levels); deviances, degrees of freedom and estimated numbers missed

	Model	Deviance	df	Missed
M_4	$[AX_1][BX_2]$	617.6	13	6,170.3
M_5	$[AX_1][BX_1][X_2]$	228.6	15	5,696.1
M_6	$[AX_1X_2][B]$	718.2	7	6,170.3
M'_6	$[AX_1][AX_2][X_1X_2][B]$	725.6	10	6,170.3
M_7	$[AX_1][BX_2][X_1X_2]$	588.6	10	6,179.4
M_8	$[AX_1][BX_1][BX_2]$	69.1	12	5,696.1
M_9	$[AX_1][BX_1][X_1X_2]$	200.2	12	5,696.1
M_{10}	$[AX_1][BX_2][AX_2][BX_1]$	65.9	9	5,837.1
M_{11}	$[AX_1][BX_1X_2]$	4.9	6	5,696.1
M'_{11}	$[AX_1][BX_1][BX_2][[X_1X_2]$	34.4	9	5,696.1
M_{12}	$[AX_1X_2][BX_1X_2]$	0.0	0	5,910.1
M'_{12}	$[AX_1X_2][BX_1][BX_2]$	23.3	3	6,257.1
M''_{12}	$[AX_1][AX_2][BX_1][BX_2][X_1X_2]$	31.2	6	5,831.4

We discuss Properties 1 and 2 together. We use two notions from graph theory and graphical models: the path and the short path (for instance, see Whittaker [295]). The two registers A and B are connected by a *path* if there is a sequence of adjacent edges connecting the variables A and B in the graph. A *short path* from A to B is a path that does not contain a sub-path from A to B. Figures 15.1 and 15.2 illustrate these concepts.

- In models where A and B are *not* connected, so that there is no path from A to B, the contingency table can be collapsed over all of the covariates in the graph. So in Figure 15.1 the contingency table $A \times B \times X_1$ can be collapsed over X_1 in model M_1 and in model M_2. This illustrates Property 1, that under models M_1 and M_2 the population size estimate is identical to the population size estimate M_0. In this example, this also implies Property 2, that models M_1 and M_2 have identical population size estimates. The table

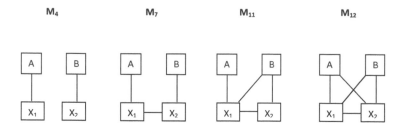

FIGURE 15.2: Interaction graphs of log-linear models with two covariates.

$A \times B \times X_1 \times X_2$ can be collapsed over both X_1 and X_2 in models M_4, M_6 and M_6' because X_1 and X_2 are not on a short path from A to B. In passing, we note that this property of model M_4 shows that the inclusion probabilities of A and of B may both be heterogeneous as long as the sources of heterogeneity, i.e. X_1 and X_2, are not related.

- In models with a short path connecting A and B, the table is not collapsible over the covariates in the path. A simple example is model M_3 of Figure 15.1, where the contingency table $A \times B \times X_1$ cannot be collapsed over X_1. Another simple example is model M_7 of Figure 15.2, where the contingency table cannot be collapsed over either X_1 or X_2.

- When the covariate X_2 is not part of any path from A to B as in models M_5 and M_8, then $A \times B \times X_1 \times X_2$ is collapsible over X_2 illustrating Property 1. Again for this example, Property 1 implies Property 2, namely that these models have identical population size estimates.

- For model M_{11} of Figure 15.2 there are two paths from A to B, $A - X_1 - B$ and $A - X_1 - X_2 - B$; however the table is collapsible over X_2 as the second path is not short, containing the unnecessary detour $X_1 - X_2 - B$.

- The other models have no covariates over which the contingency table can be collapsed. For example, in model M_{12} of Figure 15.2, and its reduced versions M_{12}' and M_{12}'', there are again two short paths, one through X_1 and one path through X_2.

15.3.2   Three registers

For completeness we give illustrative examples of the situation with three or more registers even though it is irrelevant for the data in Section 15.3.1, where there are only two. For three registers A, B and C the contingency table $A \times B \times C$ has one structural zero cell. We consider how the properties apply to the context of three registers A, B and C, and with a single covariate X. We discuss three models with their graphs displayed in Figure 15.3.

For model $M_{15} = [AX][AB][BC]$ the table $A \times B \times C \times X$ is collapsible over covariate X as it is not on any short path. This illustrates Property 1. Property 2 is illustrated by the other models where A and C are conditionally independent given B, and X is related to only one of the registers, namely models $[AB][BC][BX]$ and $[AB][BC][CX]$.

For model $M_{16} = [ABX][BCX]$ covariate X is on the short path from A to C and therefore the contingency table is not collapsible over X. For model $M_{17} = [ABX][BC][AC]$ covariate X is not on the short path from A to B, as the short path is $A - B$, and therefore the contingency table is collapsible over X.

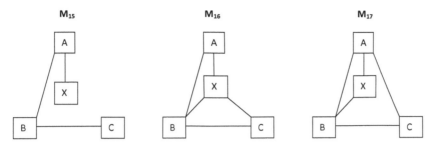

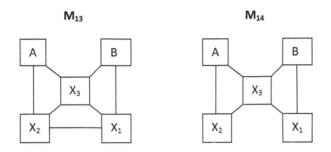

FIGURE 15.3: Interaction graphs of log-linear models with three registers and one covariate (see also next page).

FIGURE 15.4: Interaction graphs of log-linear models with partially observed covariates.

The maximal model $[ABX][BCX][ACX]$ is discussed at the end of Appendix A of van der Heijden et al. [283].

15.3.3 Active and passive covariates

In Section 15.3.1 we discussed the result that marginalising over a covariate does not necessarily lead to a change in the population size estimate. Whether the population size estimate changes or not depends on the log-linear models in the original and in the marginalised table. We term a covariate *active* if marginalising over this covariate leads to a different estimate in the reduced table, so that this covariate plays an active role in determining the population size; we call a covariate *passive* if marginalising leads to an identical estimate in the reduced table.

As an example we discuss active and passive covariates referring to Figure 15.4. Assume model M_{13}. In model M_{13} the contingency table is not collapsible over covariates X_1 and X_2, hence they are active covariates. On the other hand, in model M_{14}, by deleting the edge between X_1 and X_2, the contingency table is collapsible over X_1 and X_2, hence they are passive covariates.

While passive covariates do not affect the size estimate, which suggests that they might be ignored, a possible use is the following. A secondary objective of population size estimation is to provide estimates of the size of sub-populations, or equivalently, to break down the population size in terms of given covariates. This may well include passive covariates. Describing a population breakdown in terms of passive covariates is an elegant way to tackle this important practical problem.

We note that the introduction of many covariates may lead to sparse contingency tables and hence to numerical problems due to empty marginal cells in those margins that are fitted. Consider, for example, a saturated model such as $[AX_1X_2X_3][BX_1X_2X_3]$. In this model the conditional odds ratios between A and B are 1. However, when a zero count in one of the subtables of X_1, X_2 and X_3 occurs for the levels of A and of B, the estimate in this subtable for the missing population is infinite. One way to solve this is by setting higher-order interaction parameters equal to zero.

Another approach to tackle this numerical instability problem is as follows. We start with an analysis using only active covariates, for example, using the covariates observed in all registers in the saturated model. We may monitor the usefulness of the model by checking the size of the point estimate and its confidence interval. If the usefulness is problematic (for example, when the upper bound of the parametric bootstrap confidence interval is infinite), we may make the model more stable by choosing a more restrictive model. One way to do this is by making a covariate passive. For example, both in model $[AX_1X_2][BX_1X_2X_3]$ as well as in model $[AX_1X_2X_3][BX_1X_2]$ the covariate X_3 is passive and both models yield identical estimates and confidence intervals. When one of these two models is chosen, its size may then be increased by adding additional passive variables, such as variables that are only observed in register A or register B.

15.3.4 Example

We now discuss the analysis of the data introduced in Section 15.2. To recapitulate, A is inclusion in the municipal register GBA and B is inclusion in the police register HKS. Covariates observed in both A and B are X_1, gender, X_2, age (four levels), and X_3, nationality (1 = Iraqi; 2 = Afghan; 3 = Iranian).

A first model is model $N_1 = [AX_1X_2X_3][BX_1X_2X_3]$. This is a saturated model. For this model the estimate for the missed part of the population size is $5,504.6$, and the total population size is $33,098.6$. However, the parametric bootstrap confidence interval (Buckland and Garthwire [58]) shows that we deal with a solution that is numerically unstable, as the upper bound of the 95% percent confidence interval is infinite. The instability of the model is a consequence of too many active covariates, and a solution is to make covariate X_3 passive. Two models in which X_3 is a passive covariate are $N_2 = [AX_1X_2][BX_1X_2X_3]$ and $N_3 = [AX_1X_2X_3][BX_1X_2]$. For these models the population size estimate is $33,504.1$ (95% CI is $32,481 - 35,469$). Table 15.5 summarizes the results. Also note that model N_2 has a somewhat better fit than model N_3, but that this fit is unrelated to the estimated total population.

In the next section, Section 15.4, we discuss incomplete covariates and we will extend the analysis by including two variables that are each only observed in one of the registers.

TABLE 15.5: Models fitted to examples of variables A, B, X_1 to X_5, deviances, degrees of freedom, AICs, estimated population size and 95% confidence intervals

	Model	Dev.	df	AIC	Pop.size	CI
N_1	$[AX_1X_2X_3][BX_1X_2X_3]$	0	0	144.0	33,098.6	32,209 - ∞
N_2	$[AX_1X_2][BX_1X_2X_3]$	24.9	16	136.8	33,504.1	32,480 - 35,468
N_3	$[AX_1X_2X_3][BX_1X_2]$	28.8	16	140.7	33,504.1	32,480 - 35,468

15.4 Incomplete covariates

In Section 15.3.1 it is presumed that covariates are present in both register A as well as in register B. Recently it has been made possible to estimate the population size making use of covariates that are only observed in one of the registers (see Zwane and van der Heijden [311]; for examples, see van der Heijden, Zwane and Hessen [281], and Sutherland, Schwartz and Rivest [272]). A simple example illustrates the problem; see Panel 1 of Table 15.6, where covariate X_1 (Marital status) is only observed in register A (GBA) and covariate X_2 (Police region) is only observed in register B (HKS). As a result X_1 is missing for those observations not in A and X_2 is missing for those observations not in B. Zwane and van der Heijden [311] show that the missing observations can be estimated using the EM algorithm under a missing-at-random (MAR) assumption (Little and Rubin [183], Schafer [254] [255]) for the missing data process. After EM, in a second step, the population size estimates are obtained for each of the levels of X_1 and X_2.

TABLE 15.6: Covariate X_1 is only observed in register A and X_2 is only observed in B

Panel 1: Observed counts

		$A = 1$		$A = 2$
		$X_1 = 1$	$X_1 = 2$	X_1 missing
$B = 1$	$X_2 = 1$	259	539	13,898
	$X_2 = 2$	110	177	12,356
$B = 2$	X_2 missing	91	164	-

Panel 2: Fitted values under $[AX_2][BX_1][X_1X_2]$

		$A = 1$		$A = 2$	
		$X_1 = 1$	$X_1 = 2$	$X_1 = 1$	$X_1 = 2$
$B = 1$	$X_2 = 1$	259.0	539.0	4,510.8	9,387.2
	$X_2 = 2$	110.0	177.0	4,735.8	7,620.3
$B = 2$	$X_2 = 1$	63.9	123.5	1,112.4	2,150.2
	$X_2 = 2$	27.1	40.5	1,167.9	1,745.4

The number of observed cells is lower than in the standard situation. For example, in Panel 1 of Table 15.6 this number is 8, whereas it would have been 12 if both X_1 and X_2 were observed in both A and B. For this reason only a restricted set of log-linear models can be fit to the observed data. Zwane and van der Heijden [311] show that the most complicated model is $[AX_2][BX_1][X_1X_2]$; note that the graph is similar to the graph of M_7 in Figure 15.2, but X_1 and X_2 are interchanged. At first sight this model appears counter-intuitive as one might expect an interaction between variables A and X_1, and between B and X_2. However, the parameter for the interaction between A and X_1 (and B and X_2) cannot be identified as the levels of X_1 do not vary over individuals for which $A = 2$.

This most complicated log-linear model $[AX_2][BX_1][X_1X_2]$ is saturated, as the number of parameters is 8 (namely the general mean, four main effect parameters and three interaction parameters) and there are just 8 observed values. Consequently these 8 observed values are identical to the corresponding 8 fitted values. The fitted values under this model are presented in Panel 2 of Table 15.6. Note that, for example, the EM algorithm spreads out the observed value 13,898 over the levels of X_1 into fitted values 4,510.8 and 9,387.2; note also that the ratio 4,510.8/9,387.2 of these fitted values is identical to the ratio 259/539 of the observed values.

By comparison, when X_1 and X_2 are observed in both A and B, the saturated model is $M_{12} = [AX_1X_2][BX_1X_2]$. This is a less restrictive model than the model $[AX_2][BX_1][X_1X_2]$ and the difference is due to the MAR assumption. We now consider the more general case when there are also covariates observed in both A and B. Suppose that there is one covariate X_1 just observed in register A, one covariate X_2 just observed in register B, and one covariate X_3 observed in both registers. The most complicated model is $M_{13} = [AX_2X_3][BX_1X_3][X_1X_2X_3]$, with the graph in Figure 15.4. When X_1 and X_2 are conditionally independent given X_3, the model simplifies to $M_{14} = [AX_2X_3][BX_1X_3]$. In M_{14} there is only one short path, namely $A - X_3 - B$, and neither covariate X_1 and X_2 is part of it. Therefore we can collapse the five-way table $A \times B \times X_1 \times X_2 \times X_3$ over X_1 and X_2, which illustrates Property 1. We conclude that inclusion of covariates that are unique to specific registers only modify the total population size estimate under the model M_{13}, in which the covariates that are just in A are related to the covariates just in B.

Simplified situations exist when covariates X_1, X_2 or X_3 are not available. When X_1 is not available, M_{13} reduces to model $[AX_2X_3][BX_3]$, where the table $A \times B \times X_2 \times X_3$ is collapsible over X_2 because X_2 is not in the short path $A - X_3 - B$. Hence to improve the total population size estimate, covariates such as X_2 are not useful unless X_1 both exists and is related to X_2. Similarly, when X_2 is not available, M_{13} reduces to $[AX_3][BX_1X_3]$ where the table is collapsible over X_1. When the covariate X_3 is not available, M_{13} reduces to model $[AX_2][BX_1][X_1X_2]$, discussed earlier, where the covariates affect the population size when X_1 is related to X_2. If they are not related, the graph is similar to model M_4 and collapsing the contingency table over both X_1 and X_2 does not affect the total population size.

15.4.1 Active and passive covariates revisited

While passive covariates do not affect the size estimate, which suggests that they might be ignored, a possible use is the following. A secondary objective of population size estimation is to provide estimates of the size of sub-populations, or equivalently, to break down the population size in terms of given covariates. This may well include passive covariates. Describing a population breakdown in terms of passive covariates is an elegant way to tackle this important practical problem. This extends the approach of Zwane and van der Heijden (2007) of using register-specific covariates in the population size estimation problem.

Most registers have several covariates that are not common to other registers, because the different registers are set up with different purposes in mind. An interesting data analytic approach is therefore: first, to determine a small number of active covariates, possibly of covariates that are in both registers; and second, to set up a log-linear model structured along the lines of model M_{14}, where several passive covariates can be entered by extending X_1 or X_2, and where these covariates may or may not be register specific. Passive covariates are helpful in breaking down the population size under the assumption that the passive covariates of register A are independent of the passive covariates of register B conditional on the active covariates.

15.4.2 Example revisited

We now discuss the analysis of the data introduced in Section 15.2. To recapitulate, A is inclusion in the municipal register GBA and B is inclusion in the police register HKS. Covariates observed in both A and B are X_1, gender, X_2, age (four levels), and X_3, nationality (1 = Iraqi; 2 = Afghan; 3 = Iranian). Covariate X_4, marital status, is only observed in the municipal register GBA. Covariate X_5, police region where apprehended, with levels 1 = in one of the four largest cities of the Netherlands, and 2 = elsewhere, is only observed in the police register HKS.

TABLE 15.7: Models fitted to examples of variables A, B, X_1 to X_5, deviances, degrees of freedom, AICs, estimated population size and 95% confidence intervals

	Model	Deviance	df	AIC	Pop.size	CI
N_1	$[AX_1X_2X_3][BX_1X_2X_3]$	0	0	144.0	33,098.6	32,209 - ∞
N_2	$[AX_1X_2][BX_1X_2X_3]$	24.9	16	136.8	33,504.1	32,480 - 35,468
N_3	$[AX_1X_2X_3][BX_1X_2]$	28.8	16	140.7	33,504.1	32,480 - 35,468
N_4	$[AX_1X_2X_5][BX_1X_2X_3X_4]$	75.7	72	315.7	33,504.1	32,480 - 35,468
N_5	$[AX_1X_2X_5][BX_1X_2X_3X_4][X_4X_5]$	75.7	71	317.7	33,503.8	32,395 - 35,543
N_6	$[AX_1X_2X_3X_5][BX_1X_2X_4]$	523.8	72	763.7	33,504.1	32,480 - 35,468
N_7	$[AX_1X_2X_3X_5][BX_1X_2X_4][X_4X_5]$	289.1	71	531.4	33,510.9	32,363 - 35,432

See Table 15.7, which is an extended version of Table 15.5. Models N_2 and N_3 are both candidates to be extended by including marital status (X_4) or police region (X_5). Note that X_4 is only observed in GBA (A) and X_5 is only in HKS (B). When N_2 is extended by adding X_4 and X_5 as passive variables, we get model N_4, namely $[AX_1X_2X_5][BX_1X_2X_3X_4]$. This model yields an identical estimate for the missed part of the population illustrating that in model $[AX_1X_2X_3X_5][BX_1X_2X_3X_4]$ the covariates X_4 and X_5 are indeed passive. With 72 degrees of freedom and a deviance of 75.7 the fit is good. The AIC is 315.7. We check whether it is better to make covariates X_4 and X_5 active and we do this by adding the interaction between the covariates X_4 and X_5 to give model N_5. The deviance of this model is identical and we conclude that N_4 is a better working model than N_5. We also extend N_3 by adding X_4 and X_5 as passive variables giving N_6. Note again that the estimate for the missed part of the population is identical, however, the deviance is 523.8, so the fit is worse. Adding the interaction between X_4 and X_5 in N_7 helps as the deviance goes to 289.1, however, the deviance of N_7 is larger than the deviance of N_4, so we choose N_4 as the final model.

Our interest lies in the undocumented part of the population, i.e. in the people not registered in the GBA. Table 15.8 shows the two-way margins of the GBA with the other variables estimated under N_4. The estimates show that the undocumented population from Afghanistan, Iraq and Iran are mostly not included in the police register HKS, are more often male, between 25 and 50, from Afghanistan, unmarried and mostly not staying in the four largest cities.

TABLE 15.8: Estimates for the GBA with each of the other variables under model N_4

	In HKS	Not in HKS	Male	Female
In GBA	1,085.0	26,254.0	15,855.0	11,484.0
Not in GBA	255.0	5,910.0	3,874.7	2,290.3

	15-25	25-35	35-50	50-64
In GBA	7,234.0	8,361.0	9,185.0	2,559.0
Not in GBA	1,292.2	2,167.3	1,925.9	779.7

	Afghan	Iraqi	Iranian
In GBA	12,818.8	8,743.3	5,776.8
Not in GBA	2,950.9	1,914.5	1,299.7

	Unmarried	Married	4 Large cities	Elsewhere
In GBA	14,698.2	12.640,8	9,720.0	17,619.0
Not in GBA	3,302.3	2,862.7	2,182.6	3,982.5

15.5   Conclusion

We have demonstrated two closely related properties of log-linear models in the context of population size estimation. First, under specific log-linear models, marginalising over covariates may leave the population size estimate unchanged. Second, different log-linear models fit to the same contingency table may yield identical population size estimates. This is worked out in detail for the case of two population registers and illustrated for the three-register case.

Using the first property, we have introduced the notion of active and passive covariates. In a specific log-linear model, marginalising over an active covariate changes the population size estimate, while marginalising over a passive variable leaves the population size estimate unchanged. This idea can be particularly powerful in those situations where each of the registers has unique covariates, but a description of the full population in terms of these covariates is needed. It may then be useful to introduce these register-specific covariates as passive covariates into a model such as M_{14}. For example, if a log-linear model is proposed where the covariates unique to register A are conditionally independent of the covariates unique to register B, then the full contingency table is collapsible over these covariates and hence these covariates are passive.

Such a conditional independence assumption is strong, yet in many data sets there may not be enough power to test its correctness. It is demonstrated that a direct relation between the passive covariates of register A and those in B can only be assessed among those individuals that are in both registers A and B. If there is overlap between registers A and B, with relatively many individuals in both A and B, the relationship between the passive covariates of A and B can easily be assessed; conversely, if the overlap is small, there is little power to establish whether or not this relation should be included in the model.

This new methodology should be of use for estimating the missing population due to undercoverage in the 2011 Census of the Netherlands where the size of the total population can be estimated by application of log-linear models. It could also be applied to countries that use register information to estimate the undercoverage of their population register as well as to countries which use traditional methods. The use of passive covariates gives insight into which characteristics individuals have that are not covered by the Census and thereby illuminate the bias due to the undercoverage.

In the introduction we mentioned latent variable models that take heterogeneity of inclusion probabilities into account. For this purpose both Fienberg et al. [122] and Bartolucci and Forcina [24] proposed generalizations of the so-called Rasch model. See also Chapters 20, 21, and 22. It is beyond the scope of this chapter to study collapsibility properties for their models in the presence of covariates. However, it is interesting to note that one important specific form of the Rasch model, the so-called extended Rasch model, is mathematically equivalent to the log-linear model that includes three two-factor interactions that are identical and a three-factor interaction (see Hessen [141]; this log–linear model is also used in IWGDMFa [154], where it is referred to as a heterogeneity model; also Chapter 22). Collapsibility properties of this log–linear model can be studied using the perspective presented in this paper.

16

Population size estimation in CRC models with continuous covariates

Eugene Zwane

University of Swaziland

CONTENTS

16.1   Introduction and background

Planning in social and medical sciences requires accurate estimates of incidence and prevalence. Until recently, surveys were the primary source for prevalence data. In recent years, clinical data have been used to create disease registries. In this particular instance prevalence estimates are based on identified cases. However, for many diseases a fairly lengthy period elapses between disease onset and diagnosis. As a result, true prevalence is underestimated due to cases that are not being treated actively as well as those that have not been diagnosed. The low ascertainment is more pronounced for diseases that are harder to diagnose. Furthermore, the probability of diagnosis or even treatment (being ascertained by one of the lists or registries) is often influenced by demographics, healthcare status, and socioeconomic status.

Capture-recapture (CR) methods provide a natural way to estimate the unknown size of a *partially observed* population (see Hook and Regal [148]), through samples derived using some identification mechanism (traps, lists, registries, etc.). These methods were introduced in the wildlife setting to estimate animal abundance but have been extended to epidemiology, public health, etc., see Chao [75]. However, this technique continues to be underused, despite evidence that it can improve prevalence estimates even for diseases like diabetes that are both common and relatively well identified (Bruno et al. [57]). Use in public health has probably been discouraged by the heavy reliance in the literature on specialised programs.

In recent years though, due to the increased appreciation of the method, several papers have been published showing how this tool can be used in statistical software.

The simplest CR sampling design consists of units or individuals in some population that are captured or tagged across several sampling occasions. In these experiments, when an individual is captured for the first time it is marked or tagged so that it can be identified upon subsequent recapture. On each occasion, recaptures of individuals which have been previously marked are also noted. Thus each observed individual has a capture profile/history: a vector of 1s and 0s denoting capture/recapture and noncapture respectively. The unknown population size is then estimated using the observed capture histories and any other information collected on captured individuals.

In our analysis, we consider closed populations, where there are no births, no deaths, and no migration throughout the study period. Such an assumption is often reasonable when the overall time period is relatively short. Traditionally, CR models assume that the samples are independent, but in epidemiology, list dependence and heterogeneity (the behaviour component) are the norm and log-linear models are particularly useful in modeling these phenomena (see Schwarz and Seber [258]). The dependence may be due to capture in one list having a direct causal effect of the subject capture in another list and/or capture probabilities being influenced by a subject's characteristics. The use of covariate information (or explanatory variables) to explain heterogeneous capture probabilities in CR experiments has received considerable attention in the past 3 decades. Ignoring heterogeneous capture probabilities may lead to biased estimates of the population size (Hwang and Huggins [153]).

In this work we review work on the CR method using the multinomial (conditional) logit model (MCML). Note that Bock's multinomial logit model, which we used in our earlier analysis, is a different parametrization of the MCML. Since individual covariate information (such as age, birth/body weight, or gender) can only be collected on observed individuals, conditional likelihood models are employed (see Alho [4], Huggins [151] or Zwane and Van der Heijden [312]).

16.2   Modeling observed heterogeneity

A lot of work has been done on the topic of individual covariates in capture-recapture models (see Pollock [235]). One way suggested by Pollock et al. [234] and Darroch et al. [98] requires stratification of individuals into a finite number of discrete classes, yielding K strata with stratum population sizes $\{N_k\}_{k=1}^K$. Under this approach, the collection of N_k parameters is the object of inference. This way of analysis is suited for discrete covariates with finite support. One shortcoming is that the parameter dimension increases with the number of strata. This problem can be avoided by specifying fewer strata, but might lead to a poor approximation of the covariate effect, an issue that is compounded if more covariates are considered. Alternatively, one may model individual capture probabilities as a function of continuous covariates: Huggins [151] and Alho [4] applied a logistic model to a two-list model under independence, which was extended to multiple dependent sources by Zwane and Van der Heijden [312]. This is based on estimates of individual detectability derived from the conditional likelihood, as one conditions on the individuals seen at least once throughout the experiment, hence they allow for individual covariates to be considered in the analysis. Under this approach, N is a derived parameter, and its estimation is based on a generalized Horvitz–Thompson estimator.

In the next section we briefly describe the use of the classical log-linear model in estimating the size of an unknown population and then present the MCML model. Note that both the log-linear model and MCML model are generalized linear models (see McCullagh and Nelder [203]).

16.2.1 Notation

Suppose that L lists or samples are available for a population with unknown sample size N. Let $i = 1, \cdots, n$ index the units that appear on at least one of the lists. Let y_{ij} be an indicator that the ith population unit appears on the jth list, $j = 1, \cdots, L$. Then $y_i = (y_{i1}, \cdots, y_{iL})$ is a vector denoting the capture pattern of unit i. Let x_i be a $1 \times q$ vector of covariates associated with the ith unit. Notice that there are 2^L possible capture patterns and only the capture pattern which is a vector of all zeros (denoted by $\overrightarrow{0}$) is unobserved. In this analysis we will assume that there exists a function $r(y_i|x_i)$ that is smooth in x and satisfies $p(y_i) = r(y_i|x_i)$ (where $p(y_i)$ is the probability that unit i has capture pattern y_i). Finally, define a detection function $\psi(x) = 1 - r(\overrightarrow{0}, x)$, which is the probability that a unit with covariates x appears on at least one of the lists. The Horvitz–Thompson estimator for the population size N is

$$\hat{N} = \sum_{i=1}^{n} \frac{1}{\psi(x_i)}.$$

The estimator $\hat{N}$ relies on the inclusion probabilities only for the units that are observed (at least once).

16.2.2 Classical log-linear model

The analysis of capture-recapture data with multiple samples and observed heterogeneity can be handled in the standard framework of Poisson log-linear models (see Fienberg [121], Sanathanan [253], and Chapter 15, Section 15.3). This formulation starts from the definition of a contingency table in which subjects are grouped according to one of the following capture profiles $\{01, 10, 11\}$ in the two-sample case (where each list is treated as a binary variable and these variables are assumed to be independent), where the subjects not captured by any of the two occasions (n_{00}) are missing. The log-linear model approach allows us to model the logarithm of the expected value of the observed number of subjects in each capture profile through the following linear equation

$$\log[E(n_{y_1 y_2})] = u + u_1 I(y_1 = 1) + u_2 I(y_2 = 1),$$

where $I(A)$ is an indicator function of the event A. This model assumes that the probability of appearing on a list is constant across the whole population and that the lists operate independently. The estimate of the number of people missed by all lists (n_{00}) is given by $\hat{n}_{00} = \exp(u)$ (see Cormack [86] among others). The estimate of the unknown population size is thus $\hat{N} = n + \hat{n}_{00}$.

To take into account the effect of a categorical covariate or more specifically a *dichotomous covariate* we need to include a term for the covariate and interaction terms between the dichotomous covariates and the capture profiles in the model. The resulting model is

$$\log[E(n_{y_1 y_2|c})] = u + u_1 I(y_1 = 1) + u_2 I(y_2 = 1) + \\ u_c I(c = 1) + u_{1c} I(y_1 = 1, c = 1) + u_{2c} I(y_2 = 1, c = 1).$$

In the two levels of the covariate, the subjects never captured by any source will be given by $\hat{n}_{00|c=0} = \exp(u)$ and $\hat{n}_{00|c=1} = \exp(u + u_c)$. The estimate of the unknown population size is thus $\hat{N} = n + \hat{n}_{00|c=0} + \hat{n}_{00|c=1}$.

Generalization to more than two sources is straightforward. Also, extension to more categorical (or dichotomous) covariates is immediate. In situations where continuous covariates are available, the (Poisson) log-linear model can be used if the covariates are categorized which might not be optimal.

16.2.3   Multinomial logit model

To estimate the size of a population of interest, we use the multinomial (conditional) logit model (MCML). Unlike the log-linear model, MCML treats continuous covariates in their original measurement scale. In our context the data should represent the full capture configuration for a given individual (e.g. 01100 meaning the unit has been captured by two out of five sources, sources 2 and 3) and individual covariates which are used to account for individual observed heterogeneity (see Alho [4] and Huggins [151]). The MCML extends the logistic approach, proposed by Huggins [151] and Alho [4] for two independent sources to multiple dependent sources (see Zwane and Van der Heijden [312]).

This model allows stratification for each subject, to model different capture probabilities for each of them according to the information regarding the overlapping sources and the individual covariates. In this instance we define a function

$$\pi(y|x) = \frac{r(y|x)}{\sum_{k \neq \vec{0}} r(y|x)},$$

which is the conditional probability that a unit with covariates x has capture pattern y, given that the unit has been observed by at least one list.

In the MCML model, the vector of conditional probabilities $\boldsymbol{\pi}_i' = (\pi_{1|i}, \pi_{2|i}, \cdots, \pi_{K|i})$ of belonging to one of $K = 2^L - 1$ capture profiles (defined so that $n_{k|i} = 1$ if individual i has capture profile k and 0 otherwise) are estimated by

$$\pi_{k|i} = \frac{\exp(\sum_{h=1}^{H} \sum_{j=1}^{J} x_{ih}\lambda_{hj}y_{jk})}{\sum_{k=1}^{K} \exp(\sum_{h=1}^{H} \sum_{j=1}^{J} x_{ih}\lambda_{hj}y_{jk})}$$

where x_{ih} are elements of the covariate matrix $\mathbb{X}$, λ_{hj} are elements of the regression parameters Λ, and y_{jk} are elements of the design matrix $\mathbb{Y}$. (This in short means that the probabilities are *logit linear* in the observed auxiliary covariates.) Once these quantities have been estimated, it is possible to calculate the probability of not being captured by any source $\pi_{0|i}$ as

$$\pi_{0|i} = \frac{1}{1 + \sum_{k=1}^{K} \exp(\sum_{h=1}^{H} \sum_{j=1}^{J} x_{ih}\lambda_{hj}y_{jk})}.$$

Finally, we get the estimated population size as

$$\hat{N} = \sum_{i=1}^{n} \hat{N}_i = \sum_{i=1}^{n} \frac{1}{1 - \hat{\pi}_{0|i}},$$

where $\hat{N}_i$ represents the individual's contribution to the estimate of the unknown population size. The log-likelihood for the MCLM is $l = \sum_{i=1}^{n} \sum_{k=1}^{K} n_{k|i} \log(\pi_{k|i})$ and thus the MCML can be fitted with available software by exploiting the similarity of the likelihood with that of the stratified proportional hazards model.

16.2.4   Model selection

Selection of the best model is one of the most crucial exercises in capture-recapture modeling. It is not straightforward to evaluate all possible models as the number of models to be evaluated increases rapidly with increasing numbers of sources and/or covariates. A solution would be to consider only hierarchical models. In our analysis we only selected the best model according to the AIC for the analysis without covariates, analysis with only gender as the covariate, analysis with only birth weight as the covariate, and an analysis incorporating the two covariates (gender and birth weight).

16.2.5 Multi-model approach

Though not explored in this analysis, one way of overcoming the difficulty in selecting the best model is to use multi-model estimation (see Burnham and Anderson [66]). This approach is based on a weighted average of those models having a maximum difference ($\delta_{\max}$, which is set at 7 or 10), in terms of the AIC (or BIC) values, from the model with the minimum AIC (or BIC). Once the best model is selected according to, say, the *AIC*, the following difference $\delta_i = AIC_i - AIC_{min}$ is calculated for each ith model, and all models with $\delta_i > \delta_{\max}$ are excluded from the analysis.

The model averaged estimate of the population size is then given by

$$\hat{N} = w_i N_i$$

where

$$w_i = \frac{\exp(-\delta_i/2)}{\sum_{i=1}^{R} \exp(-\delta_i/2)}$$

where R is the total number of considered models (or the number of models with $\delta_i \leq \delta_{\max}$).

16.2.6 Bootstrap variance and confidence interval estimation

Instead of using the symmetric confidence intervals, we propose to use the parametric bootstrap confidence intervals (see Zwane and Van der Heijden [310] and Chapter 22, Section 22.6). In this instance we assume that a good estimate of the probability model exists, implying that an estimate of the unconditional variance can be computed based on the fitted inclusion probabilities. According to the fitted model, $\pi_{0|i}$ is the number of unobserved units with covariate vector x_i. However $\pi_{0|i}$ is generally not an integer, implying that $N_i = 1 + \pi_{0|i}$ is also not necessarily and integer. Following Zwane and van der Heijden [310] we use random rounding to replace N_i with a whole number. We assume that the true N_i is either $INT[\hat{N}_i]$ or $INT[\hat{N}_i + 1]$ (where $INT[\hat{N}_i]$ is the integer part of N_i). We give a higher probability to the integer close to N_i. For further information on this bootstrap approach, refer to Zwane and van der Heijden [310].

16.3 Data set

As an illustration, we estimate the number of children born with neural tube defects (NTDs) in the Netherlands, using data from three perinatal/neonatal databases (for more details on the registrations see van der Pal et al. [287]):

- *Dutch perinatal database I (LVR₁)*: This is a pregnancy and birth registry of low-risk pregnancies and births.

- *Dutch perinatal database II (LVR₂)*: This list registers anonymous data concerning the birth of a child in secondary care.

TABLE 16.1

Number of children born with neural tube defects in the
Netherlands by capture configuration

		LNR	
LVR_1	LVR_2	Not included	Included
Not included	Not included	?	16
	Included	37	17
Included	Not included	43	7
	Included	24	4

TABLE 16.2

Number of children born with neural tube defects in the Netherlands by
capture configuration

Sex	%	Mean	Variance	1st quartile	Median	3rd quartile
Female	47.6	2.973	0.870	2.675	3.235	3.565
Male	53.4	2.494	1.057	1.550	2.800	3.155

- *National neonate database (LNR)*: this list contains anonymous information about all
  admissions and readmissions of newborns to paediatric departments within the first 28
  days of life.

The delivery weight and gender of the child are recorded by each of these three lists.
Table 16.1 displays a cross-classification of the number of children born with NTDs for the
three lists for the year 2000.

The joint distribution by delivery weight and sex is summarised in Table 16.2; females
appear to be heavier and their proportion in the data is smaller.

16.4 Results

Estimation of the number of children delivered with an NTD is the main purpose of this
study. The crude AIC is used for model selection as it has been shown to select the correct
data generating model more frequently in related studies (see Stanley and Burnham [269]).
Four sets of analyses were performed on the data and the results from the two best models
in each set are shown in Table 16.3. In the first set we fitted models without covariates and
varied the dependency between lists; in the second we used gender as the only covariate (a
categorical covariate) and varied dependency between lists; in the third set of analyses we
used birth weight as the only covariate; and in the final set we used both covariates.

The log-linear model (without covariates) with the lowest AIC (AIC=522.6) assumes
that the ascertainment by LVR_1 depends on whether the child is ascertained by LVR_2 and
LNR. The estimated number of children born with NTDs for this model is 183 (95% CI:
158,224) children.

In another set of analyses we focused on log-linear models with gender as a covariate.
The model with the lowest AIC (AIC=525.8) in the set of analyses assumes that the probability of ascertainment to LVR_1 depends only on ascertainment by the LNR. The estimated
number of children for this model is 204 (95% CI: 173,248) children born with NTDs. As
could be expected, for all dependencies the incorporation of gender did not improve the
fit of the model. In the next set of analyses, birth weight was used as a covariate and a

TABLE 16.3
Estimated population size (N), bootstrapped variance,
bootstrap 95% CI

Analysis	List dependence	AIC	N	Parametric bootstrap quantile CI (95%)
No covariates	$[12, 13]$	522.6	183	$[158,224]$
	$[13, 2]$	522.9	202	$[173,242]$
Gender	$[13, 2]$	525.8	204	$[173,248]$
	$[1, 2, 3]$	526.5	221	$[185,262]$
Birth weight	$[12, 13]$	503.9	183	$[163,266]$
	$[13, 23]$	503.9	226	$[198,604]$
All covariates	$[12, 13]$	508.6	186	$[163,368]$
	$[13, 23]$	508.9	231	$[199,734]$

multinomial conditional logit model rather than a log-linear model was fitted. This set of analyses was explored in detail in Zwane and van der Heijden [312] where the issue of estimated population sizes for models with similar fits being vastly different was highlighted. In that analysis Zwane and van der Heijden [312] used the multimodel approach to circumvent that problem. The last set of analyses which includes all available covariates also highlights that gender is a redundant covariate in this analysis. See also Chapter 15, Section 15.3.3.

Table 16.3 indicates that the models with only birth weight fit the data best. As one would expect, gender does not influence the inclusion probabilities to any list. The estimates of the population size range from 183 to 231. Including dependence between list LVR_2 and the LNR (list 2 and list 3) results in a large estimate of the population size. The bootstrapped 95% confidence interval for $\hat{N}$ based on 10000 replications for the models including birth weight have a very high upper confidence limit. This is because the Horvitz–Thompson sum is unstable when the detection probability $\psi(x)$ approaches 0, even when $\psi(x)$ were known (see Alho [4]).

16.5 Conclusion

In this chapter we reviewed the conditional likelihood approach that allows for the modeling of dependence between sources for models incorporating continuous (and categorical) covariates. These models can be fitted with available software by exploiting the similarity of the likelihood with that of the stratified proportional hazards model. In the software R, Yee et al. [305] details how the VGAM package (Yee [304]) can be used for CR modeling. Log-linear models that use penalized splines to express dependence on a covariate as proposed by Zwane and van der Heijden [309] can also be fitted using the VGAM package. Bunge [62] also gives details on how to fit capture-recapture models.

In our analysis we have focused on observed heterogeneity whereby capture probabilities are allowed to vary with auxiliary variables. Other models treat heterogeneity as a latent feature, without using covariates (see Darroch et al. [98], Dorazio and Royle [102] or Pledger [232]). These models are important when the auxiliary covariates are not available or uninformative (like gender in our analysis). However, when informative covariates are available, their inclusion in the analysis adds a significant dimension to the value of a capture-recapture study as the rate of missingness can be estimated at each point in the covariate space.

17

Trimmed dual system estimation

Li-Chun Zhang

University of Southampton & Statistics Norway

John Dunne

Central Statistics Office, Ireland & University of Southampton

CONTENTS

17.1 Introduction

The production of socio-economic statistics is undergoing a paradigm shift. The traditional system has been based on a suite of repeated sample surveys interspersed by censuses that are more or less regularly spaced over time. Innovative methods that combine data from multiple sources are becoming more often the preferred approach, due to such important reasons as cost reduction and greater scope of statistics (Zhang [307]), in terms of the frequency as well as the level of detail.

A case in focus is the transformation of the population census itself. A number of European countries, including notably all the Scandinavian ones, conducted their last round of population census based entirely on administrative data sources. The population count is produced based on a Central Population Register that is lacking in most countries. Nevertheless, many other developed countries either have or are planning to replace the traditional door-to-door or mixed-mode census by administrative sources (UNECE [278]). The population statistics will instead be produced by combining population coverage surveys with statistical registers prepared on the basis of various available administrative data sources. It will be necessary to adjust for the coverage errors of such statistical registers. This is our topic here.

We use the term *list* to denote a collection of records that aims to enumerate the target population, each record for one unit, whether the source is a census, coverage survey or register. A list has under-coverage if there exist population units that do not have a corresponding record; it has over-coverage if each record does not correspond to a unique population unit. Over-coverage is the case if a record is either *duplicated*, i.e., it refers to the same unit as another record, or *erroneous*, i.e., it does not refer to a target population unit. In a census, under-coverage is often found to be the dominant of the two errors, in which case the census enumeration is a net under-count of the population. The situation can be reversed when it comes to a register enumeration, which yields then a net over-count instead. For example, the Patient Register enumeration of the population of England and Wales is over 4% higher than the Census 2011 population estimate (ONS [222]).

We briefly review the so-called dual system estimator (DSE) and the related census coverage adjustment method in Section 17.1.1. Some of the challenges of replacing the census enumeration with a register one are set out in Section 17.1.2. In Section 17.2, we propose and study a *trimmed dual system estimator (TDSE)* that can be used to explore, reduce and potentially remove the bias of the standard DSE, in the presence of erroneous enumeration, such as is the case with the Patient Register mentioned above. The relevance of both the DSE and the TDSE to the emerging census opportunity in the Irish Statistical System (Dunne [106]) will be discussed and illustrated in Section 17.3.

17.1.1 Census coverage adjustments

Capture-recapture models for population size estimation (e.g., Fienberg [121]; Cormack [86]; IWGDMFa,b [154, 155]) can be used to deal with the *under-coverage* errors that exist in multiple lists. A notable application is census under-enumeration adjustment using an independent *U-sample* coverage survey (UCS) to generate recapture data. See e.g., Wolter [301], Hogan [143], Brown *et al.* [55], Renaud [243], and Nirel and Clickman [217].

In its simplest form, the DSE of population size ($\hat{N}$) based on the census and coverage survey enumerations can be given as

$$\hat{N} = N_{census} N_{UCS} / N_{match}$$

where N_{census} is the number of population units enumerated in the census, N_{UCS} are units that are in the UCS, and N_{match} are units that are in *both*. A number of assumptions are needed, including e.g., the independence between the two captures; see Wolter [301]. In particular, neither list can contain erroneous enumeration.

For census over-coverage adjustment, therefore, the standard approach is to deploy a *separate O-sample* coverage survey (OCS), selected directly from the census reports. Non-parametric survey sampling theory can be used to yield an estimate of the number of population units in the census enumeration. Fieldwork for the O-sample can be limited or totally absent; see e.g., Renaud [243] for an account of the Swiss census. On the one hand, this helps to bring down the cost; on the other hand, spurious errors such as duplicate records and misreports of census residence area can, to a large extent, be assessed based on record matching and clerical checks without any fieldwork. However, the ability to detect erroneous enumeration, i.e., records of nonexistent or out-of-scope units, may be limited in this way.

Due to its limited sample size, a coverage survey cannot provide data across the whole country, at a detailed geographic level that is supported by the census itself. Census coverage adjustment involves then a small area–estimation problem, in the sense that *direct* population size estimates such as the DSE above cannot be produced in most of the local areas, because they are not at all represented in the UCS sample. It is necessary to smooth the adjustment factors under some statistical models (e.g., Hogan [143]), based on which the local area population size is then derived as a so-called synthetic estimate in the small area–estimation literature (Rao [240]).

17.1.2 Replacing census with administrative sources

The main part of a traditional census budget can be avoided if it is feasible to replace the census enumeration by some register enumeration based on administrative sources. Moreover, it may become possible to produce census-like population statistics at a greater frequency. See Chapters 15 and 18 for some other examples of issues with using administrative registers for population size estimation.

Apart from the countries that use a completely register-based approach, several countries have replaced the census with a statistical population register, and administered coverage surveys to provide the necessary coverage error adjustments. For instance, in Israel (Nirel and Clickman [217]), the population register yields a reliable overall count for the whole county, but coverage error adjustments are necessary primarily to cope with interregional mis-location. Both the UCS and OCS were deployed in census 2008. The UCS has a sampling fraction of 20%, which is larger than usual and alleviates the need for indirect population estimates. The fieldwork and cost are more limited when it comes to the OCS, because erroneous enumeration is not a major concern in the population register.

The situation, however, is quite different for the countries that do *not* have a population register to start with but are planning to replace the census enumeration with various administrative datasets. Erroneous enumeration can be much more prominent in the most relevant administrative sources in this context, such as the patient register, the tax register or the electoral register, etc., and the traditional approach of OCS with virtually no fieldwork is not necessarily a viable option for such countries.

Any new estimation approach that uses only a single sample, rather than the traditional two-sample approach, has then the potential for achieving cost reductions. Moreover, to avoid the need for indirect adjustments, it would be better if population estimates can be produced based on one large coverage survey than two smaller surveys that e.g., amount to the same total sample size. Finally, estimation methods that allow for erroneous enumeration in addition to under-coverage errors will provide an important extension to the existing theory of capture-recapture methods, and can be expected to find applications in many areas other than census-like population statistics.

Zhang [308] studies models that allow for erroneous enumeration in at least two lists, assuming a separate independent list that only suffers from under-coverage. This can be useful e.g., in the setting of combining census, UCS and an additional register enumeration, or two or more registers from administrative sources in combination with an independent UCS.

Here we propose and study a different approach based on the trimmed DSE. The two lists can consist of a statistical register and an UCS. This can be useful if an OCS would have been necessary otherwise, or if sampling from the register enumeration (i.e., for the OCS) is not feasible altogether due to data protection regulations. Moreover, it is sometimes possible that the TDSE can be based entirely on two suitably prepared lists, both of which stem from the administrative sources. An example of this will be discussed in Section 17.3. Important motivations for such a completely register-based estimation approach include further cost reduction, higher frequency of statistics and direct estimates at greater geographic detail. Finally, we notice that it is possible to combine the modeling approach (e.g., Zhang [308]) with the trimming that facilitates the TDSE approach to even greater effects.

17.2   Theory

17.2.1   Ideal DSE given erroneous enumeration

Let N be the unknown size of the target population, denoted by U. Let A be the *first* list enumeration that is of size x. Suppose list A is subjected to over-coverage, and the number of erroneous records is r, i.e., the size of set $\{i; i \in A \text{ and } i \notin U\}$. Suppose list A is subjected to under-coverage as well, so that $x - r < N$. Let B be the *second* list enumeration that is of the size n. Suppose list B is subjected to *only* under-coverage, so that $n < N$, but there are *no* erroneous records in B.

Suppose the records in lists A and B can be linked to each other in an error-free manner, which we refer to simply as the assumption of *matching*. This is a very common assumption, although it can be difficult to satisfy in practice if the two lists do not share a unique identifier. However, the linkage errors are not easy to adjust. A discussion of DSE in the presence of linkage errors is given in Section 17.2.5 below. For now, suppose that error-free matching between A and B gives rise to the matched list AB with m records.

Let $\delta_{iB} = 1$ if $i \in B \cap U$, and 0 otherwise. We assume that the probability $P(\delta_{iB} = 1) = \pi$ is a constant across $i \in U$. We shall refer to this as the assumption of *homogeneous capture* (of list B). See Chapter 1 for a discussion of homogeneity vs. heterogeneity in count data. Our usage here shares the same spirit. It serves as a common and useful starting point of development. An extension of heterogeneous capture is often accomplished by regression modelling of the capture probability based on the available covariates at the individual and local area level. For instance, the use of post-stratification amounts to the saturated model given the post-stratification variables, as will be illustrated in Section 17.3.

Provided the assumption of homogeneous capture, we have

$$E(n) = N\pi$$

Moreover, let $\delta_{iA} = 1$ if $i \in A \cap U$, and 0 otherwise. For any $i \in U$, we have

$$P(\delta_{iB} = 1) = P(\delta_{iB} = 1 | \delta_{iA} = 1) = P(\delta_{iB} = 1 | \delta_{iA} = 0) = \pi.$$

Notice that here we consider $\boldsymbol{\delta}_A = (\delta_{1A}, ..., \delta_{xA})$ as fixed constants, where $\sum_{i \in A} \delta_{iA} = x - r$. The above equalities are therefore merely consequences of the assumption of homogeneous capture, and formally do *not* amount to an assumption of independence between δ_{iA} and δ_{iB}.

Provided the assumptions of homogeneous capture and matching, we have

$$E(m | \boldsymbol{\delta}_A) = (x - r)\pi,$$

which is the expectation of the number of records in list AB on applying the constant capture probability π to the $x - r$ records in list A with $\delta_{iA} = 1$. Replacing $E(n)$ by n and $E(m | \boldsymbol{\delta}_A)$ by m, we obtain an *ideal* method-of-moment estimator, insofar as r is unobserved, given by

$$\tilde{N} = n(x - r)/m, \tag{17.1}$$

Meanwhile, let the naïve DSE, which ignores the erroneous enumeration in list A altogether, be given by

$$\dot{N} = nx/m.$$

It follows immediately that $\dot{N}$ can be expected to *over-estimate* N, since $n(x-r)/m < nx/m$ for any $r > 0$.

Wolter [301] lists a number of assumptions that can be used to motivate the usual DSE. In the above development of the ideal DSE $\tilde{N}$, we have retained the assumption of matching between A and B, and we have retained the assumption of homogeneous capture not of both A and B, but only of list B, and derived $\tilde{N}$ *conditional on* the realised $\boldsymbol{\delta}_A$. This is an important adaption when list A is obtained from administrative sources that may suffer from systematic under-coverage of some sub-populations. We have removed the multinomial distribution assumption of the cell counts arising from cross-clarifying U by $(\delta_{iA}, \delta_{iB})$. The estimator (17.1) is based on the method of moment instead of maximum likelihood.

Moreover, instead of the assumption that neither of the two lists contains erroneous enumeration, we allow for erroneous enumeration in list A, in order to cope with the fact that the underlying administrative sources may entail considerably higher over-coverage compared to the census, as discussed in the Introduction 17.1. Consequently we no longer need to assume that the target population is closed for both lists, as long as it is possible to correctly identify the target population units in the list B enumeration, and the matching between A and B is error-free. One only needs a particular version of the vector $\boldsymbol{\delta}_A$ that is matched to list B, even if $\boldsymbol{\delta}_A$ itself can change due to the updating of list A over time. The units with $\delta_{iA} = 1$ are, so to speak, simply the 'marks' that allow one to estimate the capture probability π of list B.

17.2.2 Trimmed DSE

The estimator (17.1) is hypothetical because one does not actually know r, i.e., the number of erroneous records in A. But one *can* (a) score some records in list A, which are most likely to be erroneous, (b) match them to list B, and then (c) calculate the DSE as if list A would have been free of erroneous enumeration once the scored records had been removed. This yields what we call the *trimmed DSE*, given by

$$\hat{N}_k = n\frac{x - k}{m - k_1} \tag{17.2}$$

where k is the number of scored records in list A, and k_1 is the number of records among them that can be matched to list B. Notice that, provided list B has only under-count, the k_1 records are indeed not erroneous, whereas the remaining $k - k_1$ records may or may not be erroneous.

The trimmed DSE can be motivated under the *same* assumptions as those for the ideal DSE (17.1), *regardless* of how systematic the scoring is in removing the records in list A, for the same reason that potential systematic under-coverage of list A does not matter to start with. For instance, had one scored all the people between 20 and 25 years old in list A, the trimmed DSE $\hat{N}_k$ would have remained a valid estimate provided all the erroneous records had been removed in this way.

As shown above, the naïve DSE, which can now be written as $\hat{N}_0$ with $k = 0$, is expected to over-estimate N. The following result is useful.

Result 1: If $k_1/m < k/x$, then $\hat{N}_k < \hat{N}_0$. If $k_1/m = k/x$, then $\hat{N}_k = \hat{N}_0$.

Now that mk/x is the expectation of k_1 under *random* scoring of k out of the x records in list A, Result 1 implies that one can expect the trimmed estimate (17.2) to be lower than the naïve estimate nx/m, as long as a relatively smaller number of scored records are confirmed to be non-erroneous, because they can be found in list B. In other words, trimming can be expected to adjust the untrimmed DSE in the right direction, as long as it is more effective at picking out the erroneous records than simple random sampling.

Meanwhile, $\tilde{N}$ would be the optimally trimmed estimate with $(k, k_1) = (r, 0)$. It seems desirable to avoid 'over-trimming' that makes the trimmed DSE (17.2) lower than the ideal DSE (17.1).

Result 2: If $k < r$, then $\tilde{N} < \hat{N}_k$.
Proof: We have $(x - r)/m < (x - k)/(m - k_1)$ if and only if $(k - r)/(x - r) < k_1/m$, which is always the case provided $k < r$ since $k_1/m \geq 0$. $\square$

In other words, Result 2 assures that over-trimming will not be case, as long as one does not score more records than the number of erroneous records that exist in list A. For instance, if it is suspected that about 10% of the records in list A may be erroneous, then over-trimming can be avoided as long as one does not score more than 10% of the records in list A.

Result 3: If all the r erroneous records are among the k scored ones, then $\hat{E}(\hat{N}_k) = \tilde{N}$.
Proof: The capture rate in list B of the $k - r$ scored non-erroneous records is π, whose estimate is $m/(x-r)$, so that $\hat{E}(k_1) = (k-r)m/(x-r)$ and $\hat{E}(\hat{N}_k) = n(x-k)/[m-\hat{E}(k_1)] = n(x - r)/m = \tilde{N}$. $\square$

To summarise, as long as one is able to score the erroneous records in list A more effectively than random scoring *and* one does not score more records than the total number of erroneous records in list A, the trimmed DSE (17.2) can be expected to reduce the bias of the naïve DSE and move it closer to the ideal DSE (17.1). Provided the scoring has succeeded in removing all the erroneous records, the expectation of the trimmed DSE would become approximately the same as the ideal DSE.

When it comes to variance estimation, consider first the ideal estimator $\tilde{N}_k = \tilde{x}n/m$ where $\tilde{x} = x - r$. As explained before, we prefer to treat the corresponding list A with $\tilde{x}$ records as fixed. To obtain the variances of n and m, we make an extra assumption of *independent capture*, such that $V(n) = N\pi(1 - \pi)$ and $V(m) = \tilde{\pi}\pi(1 - \pi)$. Moreover, let $n = m + n_{A^c}$ where n_{A^c} is the number of population units that are not in list A but are enumerated in list B. Provided independence capture, we have $Cov(n, m) = Cov(m + n_{A^c}, m) = V(m)$. Thus, by the linearisation technique, we obtain

$$V(\tilde{N}) \approx \frac{\tilde{x}^2}{E(m)^2}\left(V(n) - \frac{2E(n)}{E(m)}Cov(n, m) + \frac{E(n)^2}{E(m)^2}V(m)\right)$$
$$= N(\frac{1}{\pi} - 1)\left(\frac{N}{\tilde{x}} - 1\right).$$

Replacing N by $\tilde{x}n/m$ and π by $m/\tilde{x}$, we have

$$\tilde{v} = \hat{V}(\tilde{N}) = n(n - m)\tilde{x}(\tilde{x} - m)/m^3.$$

Notice that this is the same variance estimate as that of the standard DSE, where both lists are treated as random. Seber [301] provides adjustments in the case of $m = 0$, which is however not important in the present context.

We turn now to the trimmed DSE $\hat{N}_k = x_k n/m_k$, where $x_k = x - k$ and $m_k = m - k_1$. For variance estimation under the same assumptions as those for $\tilde{v}$ above, one needs the number of remaining erroneous records among the scored list A with x_k records, which is not possible without further work. As an approximate remedy, we propose to make an additional tacit assumption that $E(\hat{N}_k) \approx N$, i.e., all the x_k records belong to the population, so that a variance estimator of $\hat{N}_k$ can be given by

$$v_k = \hat{V}(\hat{N}_k) = n(n - m_k)x_k(x_k - m_k)/m_k^3.$$

17.2.3 Stopping rules

Notwithstanding the theoretical assurance above, one needs in practice some stopping rules for the trimming, which can tell one when to stop. Below we describe three of them, all aimed at the same stopping point.

Firstly, consider the trimmed estimate $\hat{N}_k$ itself. Starting from the naïve estimate $\hat{N}_0$, it is expected to decrease towards the ideal estimate $\tilde{N}$ as k increases, provided the scoring is more effective than random sampling. Moreover, according Result 2, we have $\hat{N}_k > \tilde{N}$ as long as $k < r$. For $k > r$, one can envisage two equilibriums.

1. According to Result 3, ideally, once all the r erroneous records have been removed, we could expect the trimmed estimate to flatten out at the level of the ideal estimate $\tilde{N}$, as k increases.

2. Or, as one gradually exhausts all the effective means, the scoring becomes more or less random at picking out the erroneous records. The trimmed estimate would then flatten out at a level higher than $\tilde{N}$, as k increases. How large the bias is depends on the proportion of the erroneous records that remain.

In practice, therefore, one could repeat the scoring to successively include more records, and to keep track of the actual $\hat{N}_k$, as k increases, to see if it flattens out at some stage.

Secondly, it is intuitive that k_1, i.e., the number of scored records that are confirmed to be non-erroneous, should be as low as possible. Denote by p the probability that a scored record is actually erroneous. Let $k_r = r/p$ be the expected number of records, in order to score the r erroneous records in list A. Then, for any $k < k_r$, the expected number of non-erroneous records is $k(1 - p)$, and homogeneous capture of list B enumeration with probability π implies that the expectation of k_1 is given by

$$E(k_1 | k, k < k_r) = k(1 - p)\pi,$$

whereas for any $k > k_r$, the expected number of non-erroneous records would be $k - r$, so that the corresponding of k_1 is given by

$$E(k_1 | k, k > k_r) = (k - r)\pi.$$

Thus, k_1 is expected to increase at a rate of $(1 - p)\pi$ as k increases towards k_r, which then changes to π after k becomes larger than k_r. On the one hand, the closer p is to one, or the more effective the scoring is at picking out the erroneous records, the bigger the change. On the other hand, in the case of random scoring or worse, we would have $p \leq r/x$ and $k_r \geq x$. Since it is not possible to score more than x records in list A, one cannot expect to detect any change in the ratio k_1/k with any such scoring method.

It should be pointed out that, in reality, it is unlikely for the probability p of scoring erroneous records to be a constant of k, i.e., the number of records scored. Still, the above consideration suggests that, in practice, one could repeat the scoring to successively include more records, and to keep track of the actual k_1/k, as k increases, for an indication of when to stop. Since it seems natural that the probability p should gradually decrease once the most probable erroneous records have been scored, k_1/k may be roughly convex, in which case the stopping point could be where the bend is most acute.

Thirdly, because the way in which k_1 changes with k is different before and after $k = k_r$, one can also expect the variance estimate v_k to behave differently before and after k_r, thus providing a third indicator.

The three stopping rules above are all aimed at the same stopping point $k_r = r/p$. Figure 17.1 provides an illustration. There are three different settings of (N, n, x, r, p), one for each

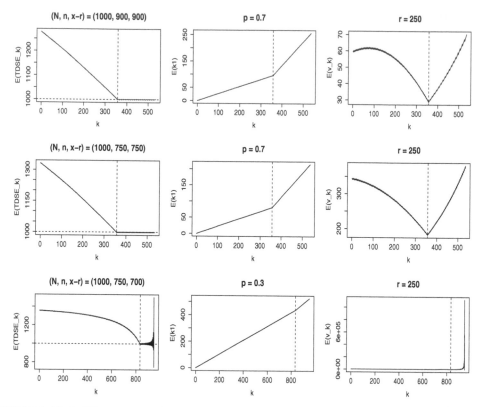

FIGURE 17.1: Illustration of three stopping rule indicators: left column: $E(\hat{N}_k)$; middle column: $E(k_1)$; right column: $E(v_k)$. Setting (N, n, x, r, p): same for each row.

row of plots. These represent, respectively, a favourable scenario with high capture probability π and reasonably high probability p of scoring erroneous records, an unfavourable scenario with both low π and p, and a scenario in between with low π but reasonably high p.

More explicitly, the target population size is $N = 1000$ in every case. The capture probability of list B is given indirectly as n/N, which is reasonably high at 0.9 in the first setting, and relatively low at 0.75 in the other two. The proportion of erroneous records in list A is given by r/x, which is relatively high at over 20% (i.e., 250/1150) in the first setting, and even higher (i.e., 250/1000 and 250/900) in the other two. The probability p of scoring erroneous records is reasonably high at 0.7 in the first two settings, but rather low at 0.3 in the last one.

Take e.g the top-left figure. The list A enumeration is $x = 900 + 250 = 1150$, and the list B enumeration is $n = 900$. The expected number of matches between list A and B is $(x - r)\pi = (1150 - 250)(900/1000) = 810$. This yields the naïve DSE $\dot{N} = 1150 \cdot 900/810 \approx 1278 = \hat{N}_0$. As k increases, the expected k_1 is calculated as described above assuming constant p, which then yields the plotted $\hat{N}_k$. The turning point is $k_r = 250/0.7 \approx 357$, after which trimming amounts to random scoring, and the TDSE remains at $N = 1000$.

It can be seen that all three stopping rules point to the same critical point $k_r = r/p$, which is 357 in the first two settings and 833 in the last one. In the first favourable setting, the trimmed DSE becomes unbiased after removing 107 $(= 357 - 250)$ extra records compared to the ideal DSE $\tilde{N}$. The standard error (SE) of $\hat{N}_{357}$, on removing all the erroneous records, is $\sqrt{v_{357}} = 5.4$, compared to that of the ideal DSE, i.e., $\sqrt{\tilde{v}} = 3.5$. Still, the loss of efficiency seems a relatively small price to pay compared to the bias of the untrimmed DSE ($\approx \hat{N}_0 - \tilde{N} = 278$).

Similarly in the second scenario with low capture probability π but reasonably high scoring probability p, the SE of the trimmed DSE is 13.6 at $k_r = 357$ compared to 10.5 of $\tilde{N}$. Again, a relatively small price to pay against the bias of the untrimmed DSE, which is approximately 332.

In the last unfavourable scenario, the bias of the naïve DSE is 357 to start with. The probability $p = 0.3$ is not much higher than random scoring (at the rate 250/950) in this case. Removing all the erroneous records at such a rate requires, on expectation, scoring 833 records out of 950 in list A, at which the SE of the trimmed DSE is 50.8 compared to 12.0 of the ideal DSE. Although this may still seem worthwhile in terms of the trade-off between bias and variance, it is unlikely that such a precision is acceptable in practice.

In summary, the performance of the trimmed DSE is, above all, determined by how effectively the scoring removes the erroneous records, provided that the trimmed DSE can yield good bias-variance trade-off compared to the naïve DSE, even when a fair number of records need to be removed from the estimator. Of course, in practice, it may be impossible to remove all the erroneous records by scoring, or one may lack very effective means of scoring. But even then the trimmed DSE can be less biased than the untrimmed one, and it can provide a useful tool for sensitivity analysis, because it is easy to compute and interpret. An example of this will be discussed in Section 17.3.

17.2.4 Discussion: Erroneous enumeration in both lists

So far we have considered the situation where erroneous enumeration is only present in list A but not B. Now, provided list B is generated by a coverage survey, it may be reasonable to assume that it is possible to keep erroneous enumeration to a negligible extent. However, to achieve further cost reduction and greater detail in the population estimates, it will be advantageous if list B can be compiled based on some suitable administrative (or other) source instead of a coverage survey. Such a case is discussed in Section 17.3. Meanwhile, it is interesting to consider in theory what happens when list B also contains over-coverage, which we do below.

For this part of the discussion, where we treat the two lists on an equal footing, let n_1 be the number of records in list A and n_2 that in B. Let n_{12} be that in the matched list AB. Let r_1, r_2 and r_{12} be the number of erroneous records in list A, B and AB, respectively. Retaining the assumption of matching and homogeneous capture (in one of the two lists), we have that the naïve DSE is given by $\dot{N} = n_1 n_2 / n_{12}$, and the ideal DSE by

$$\tilde{N} = (n_1 - r_1)(n_2 - r_2)/(n_{12} - r_{12}).$$

It is straightforward to verify that the error of $\dot{N}$ depends on the sign of

$$\lambda = \left(\frac{r_1}{n_1} + \frac{r_2}{n_2} - \frac{r_1}{n_1} \frac{r_2}{n_2} \right) - \frac{r_{12}}{n_{12}},$$

i.e., we have $\dot{N} > \tilde{N}$ if and only if $\lambda > 0$, and $\dot{N} < \tilde{N}$ if and only if $\lambda < 0$, and $\dot{N} = \tilde{N}$ if and only if $\lambda = 0$.

It is interesting to observe that the unadjusted DSE can *possibly* be equal to the ideal DSE provided the erroneous records in (A, B, AB) are related to each other in a particular way, such that $\lambda = 0$. For an interpretation of λ, imagine three separate trials: (1) select at random a record in A, the probability is $\theta_1 = r_1/n_1$ that it is erroneous; (2) select at random a record in B, the probability is $\theta_2 = r_2/n_2$ that is erroneous; (3) select at random a record in AB, the probability is $\theta_{12} = r_{12}/n_{12}$ that it is erroneous. Then, we would have $\lambda = 0$ if the chance of obtaining a positive result in the third trial is equal to that of obtaining *at least* one positive result in the first two trials, which is $\theta_1 + \theta_2 - \theta_1\theta_2$, i.e., expression in the parenthesis above for λ.

For example, let A be the electoral register and B the hospital patient register. Both may contain under- and over-counts of the target usual residents population. Let $(\theta_1, \theta_2, \theta_{12})$ be defined as above. Since $\theta_1 + \theta_2 - \theta_1\theta_2 > \max(\theta_1, \theta_2)$, we would have $\lambda > 0$, as long as the proportion of erroneous records among the people who are in *both* lists (i.e., θ_{12}) is lower than $\min(\theta_1, \theta_2)$, in which case the unadjusted DSE can be expected to be biased upwards.

Consider now the trimmed DSE defined as follows. Score k records in the list union $A \cup B$. Let k_1 be the number of records among them that are in A, and k_2 that in B, and k_{12} that in AB. The trimmed DSE is given by

$$\hat{N}_{\mathbf{k}} = (n_1 - k_1)(n_2 - k_2)/(n_{12} - k_{12})$$

where $\mathbf{k} = (k_1, k_2, k_{12})$. Again, the naïve $\dot{N}$ is the untrimmed $\hat{N}_{\mathbf{0}}$, and the ideal $\tilde{N}$ would require optimal trimming with $\mathbf{k} = (r_1, r_2, r_{12})$. Put

$$\lambda_{\mathbf{k}} = \left(\frac{k_1}{n_1} + \frac{k_2}{n_2} - \frac{k_1}{n_1}\frac{k_2}{n_2} \right) - \frac{k_{12}}{n_{12}}.$$

We have $\dot{N} > \hat{N}_{\mathbf{k}}$ if and only if $\lambda_{\mathbf{k}} > 0$, and $\dot{N} < \hat{N}_{\mathbf{k}}$ if and only if $\lambda_{\mathbf{k}} < 0$, and $\dot{N} = \hat{N}_{\mathbf{k}}$ if and only if $\lambda_{\mathbf{k}} = 0$.

It follows that, in the unlikely case of $\lambda = 0$, trimming would *introduce* bias unless $\lambda_{\mathbf{k}} = 0$. Otherwise, provided $\lambda \neq 0$ to start with, trimming *could reduce* the bias of the naïve DSE, as long as $\lambda_{\mathbf{k}}\lambda > 0$, i.e., the two have the same sign. The interpretation of λ above provides some intuition about the scoring. For instance, suppose $\dot{N}$ is an over-estimate and $\lambda > 0$, which one may be able to assert despite not knowing the true (r_1, r_2, r_{12}) as in the example of electoral and hospital patient registers above, then one should avoid any method of scoring that results in $\lambda_{\mathbf{k}} < 0$.

17.2.5   Discussion: Record linkage errors

The assumption of matching can be difficult to satisfy completely, if one lacks a unique record identifier that can be used to link the lists. Below we consider briefly the effects of linkage errors in the present context.

We resort to the situation where erroneous enumeration is only present in list A. Let (n, x, r) be defined as previously. Let m_L be the number of records in the *linked* list AB. Given the existence of linkage errors, let u be the number of *missed matches*, and let e be the number of *false links*. In other words, the true number of matches between A and B is given by

$$m = m_L - e + u.$$

Observe the distinction between the two terms match and link here. A match refers to a true pair of records in A and B that correspond to the same unit, irrespective of the actual linkage procedure and how good or bad the linkage result is. A link is an actual pair of records in the linked list AB, which may or may not be a match due to the presence of incorrect links.

We need to introduce two parameters due to the linkage errors. Put

$$\mu_L = E(m_L|d) = m - E(u|d) + E(e|d) = m - mf + q\mu_L$$

where $d = \{n, x, r, m\}$, and $f = E(u|d)/m$ is the rate of *missing (matches)*, and $q =$

$E(e|d)/E(m_L|d)$ is the rate of *false links*, and all the expectations here are with respect to the linkage errors conditional on the two lists and the true matches between them. Replacing μ_L by m_L, we obtain

$$\tilde{m} = \xi\mu_L \qquad \text{where} \quad \xi = (1-q)/(1-f)$$

as an ideal method-of-moment estimate of m, since ξ is unknown, and then

$$\tilde{N}_L = n(x-r)/(\xi m_L)$$

as the corresponding ideal *linkage DSE (LDSE)* given the linkage errors.

Meanwhile, the naïve DSE is given by $\dot{N} = nx/m_L$, such that

$$\frac{\tilde{N}_L}{\dot{N}} = \frac{x-r}{x}\frac{1-f}{1-q} = \left(1 - \frac{r}{x}\right)\left(1 + \frac{q-f}{1-q}\right).$$

Thus, provided $q > f$, the linkage errors could actually move the naïve DSE closer to the ideal LDSE, and, roughly speaking, one can even have $\dot{N} < \tilde{N}_L$ provided $r/x < (q-f)/(1-q)$.

However, in practice, the false link rate q can be more readily assessed by checking the links that can actually be made, and one is typically not willing to accept any appreciable false link rate. Assessing the missing rate f is often more difficult because it is defined against an unknown denominator, and it may be hard to reduce f because of the lack of linkage key variables or the measurement errors present in the key variables. In short, unless one purposefully accepts many false links, it may be reasonable to assume that $q < f$ is more likely the situation, in which case the linkage errors are likely to pull the naïve DSE *even* further away from the ideal LDSE.

A trimmed LDSE can possibly be given by

$$\hat{N}_k = \frac{n(x-k)}{\hat{\xi}(m_L - k_{1L})}.$$

where k is the number of records scored in A and k_{1L} that among the linked list AB, and $\hat{\xi} = (1-\hat{q})/(1-\hat{f})$ is based on the estimated linkage error parameters. But it is impossible to conclude on the properties of the trimmed LDSE *without* some strong additional assumptions involving the linkage errors.

For instance, the linkage errors may differ for the records that are scored than the rest, such that the assumption of homogeneous capture that holds for the trimmed DSE may no longer hold for the trimmed LDSE. In other words, $\hat{\xi}(m_L - k_{1L})/(x-k)$ may not be a valid estimator of π even when $(m - k_1)/(x-k)$ is. Indeed, regardless of scoring, heterogeneous linkage errors that vary across the records are likely to occur, and the determining factors of this type of heterogeneity may be different from those of the potential heterogeneity in the capture probability, making it difficult to adjust.

In summary, the violation of the assumption of matching is potentially a non-negligible source of bias and not easy to remedy. Notice that, in the traditional application of DSE for under-coverage adjustment, the violation of the assumption of matching most likely affects the DSE in the opposite direction as the violation of the assumption of independent captures, but it is in the same direction as the erroneous enumeration here.

17.3   Emerging census opportunity: Ireland

17.3.1   Background

Ireland does not have a central population register. Dunne [106] considers the possibility of conducting a census from existing data sources (administrative and survey data sources). However, in order to implement such an approach an important first step is having the capability to provide accurate estimates of the population size from administrative data sources.

Currently, the standard approach to compiling these estimates is to start with a traditional census of Population, every 5 or 10 years as the case may be, and then increment forward those estimates year by year using the demographic component method until the next census is complete. The demographic component method uses migration, births and deaths to estimate for the following year. In such a system, any errors of a given year are carried forward and if migration flows are based on using factors of the population, errors may have a compounding effect over time. For small countries with high migration flows, like Ireland, there is a practical requirement to conduct a census every 5 years at considerable cost to the taxpayer.

The access to relevant data sources and application of novel methods may provide a new solution to compiling census-like population estimates, where errors are not carried from one year to another. If this is possible, countries requiring a quinquennial census should then be able to move to a decennial census and may even be able to eliminate the need for a traditional census altogether if additional census requirements can be met from other sources.

In this section we explore census-like estimates from administrative data sources using the DSE methodology described earlier for Ireland. To start with, we provide a summary overview of the data sources and how they are used to produce the DSE, and discuss the underlying assumptions and the interpretation of the target population concept. We then apply the TDSE to obtain further insight on the DSE of the population size. This serves also to illustrate the theoretical properties elaborated earlier. In particular, three sets of estimates will be compared to each other: (1) a single register enumeration combining a set of administrative datasets, (2) a DSE constructed from the available administrative datasets, and (3) the census enumeration in 2011.

17.3.2   Overview of data sources

As outlined in Dunne [106], the basic building block is a statistical register summarising each person's activity on the main public administration systems in a given year. This Person Activity Register (PAR) is simply a rectangular dataset including the PIN (person identification number), age, gender, nationality and an activity indicator variable for each identified data source that takes a value of 1 if there is evidence of an event or transaction for that data source indicating a person was resident in the state in a given year, or 0 if insufficient or no evidence. Below is an overview of the administrative data sources included in the PAR.

Children's Benefit Universal payment made on behalf of each child generally to the mother while the child is under 18 and in full-time education. Indicators are used for both the mother and the child. *A proxy dataset based on registrations is used for 2011 as actual payment data for this year were unavailable at the time of this project.*

Early Childhood Care Each child is entitled to a year of paid childcare prior to attending primary school.

Primary Online Database Student enrolments in primary education in the state. Typically for children aged 5 to 12.

Post-Primary Pupils Database Student enrolments in secondary education. Typically for teenagers aged 12 to 18.

Higher Education Enrolments Database Student enrolments in third-level education. Typically for youth aged over 18 years.

Further Education Awards Database Student awards in further education (excluding higher education). Typically for persons aged over 16.

Employer Employee tax returns A database of paid employees (including occupational pensions) created from the employer returns to the Irish tax authorities each year.

Income Tax Returns Tax returns filed by persons for any additional taxable income other than paid employments each year.

Social Welfare Social welfare payments to recipients each year.

State Pension All those that have contributed to the state are entitled to a state pension on reaching retirement age. *A proxy based on registration has been used for 2011 as actual payment data was unavailable to the project.*

In summary, the data sources underpinning the PAR provide broad coverage of the different stages of a person's life from the cradle to the grave. The PAR, taking a 'signs of life (SoL) approach' contains records for only those people where there is evidence of that person being resident in the state in a given year. In particular, a SoL activity is admitted as evidence from the corresponding source only if the PIN can be identified.

One administrative data source not included in the PAR is the Irish drivers licence database. A significant proportion of the adult population holds a driving licence and are typically required to renew their licence every 10 years. Our second enumeration is comprised of those persons renewing their driving licence or applying for a new one in a given year, to be referred to as the driving licence dataset (DLD). Historically a person did not require their PIN to obtain or renew a driving licence, however in recent years the provision of the PIN has become mandatory. Again, a person is included in the DLD provided only the PIN is identified.

17.3.3 Underlying assumptions and population concepts

To apply the DSE and TDSE, we treat the PAR as list A and the DLD as list B. Let us consider the necessary assumptions outlined previously, including that of the matching, the homogeneous capture, the erroneous enumeration and the related assumption of closure.

Matching It is possible to make a determination without error of which individuals recorded in List A are present in List B and which are not.

In the Irish case the PIN is the official identification number used on public administration systems to identify and authenticate individual persons. A protected version of the PIN is used as the match key and the record linkage is deterministic. Linkage errors can occur where the PIN has been incorrectly recorded on an administrative source. The PIN contains a check digit which ensures that a number is authentic, so that the recorded PIN of a person is mistaken only if it inadvertently concurs with the PIN of another person. Such potential errors are further limited or eliminated where authentication is enforced. With increased validation on public administration systems in recent years, PIN-linkage errors have become increasingly rare and can safely be assumed to have only a negligible effect.

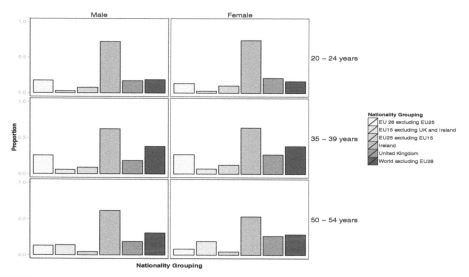

FIGURE 17.2: Proportion of identified driving licence holders on the PAR by nationality, selected age group and sex (2011).

Heterogeneous capture and post-stratification Any variable employed for post-stratification is correctly recorded for all individuals in both lists.

Figure 17.2 shows the proportion of driving licence holders identified on the PAR. Note the actual proportion is higher as only those that have renewed or applied for a licence in recent years will have been required to provide a PIN. A driving licence is typically valid for 10 years. A clear difference can be seen between nationality groupings and their propensity to hold an Irish licence. According to the rules for driving in Ireland, UK and EU licence holders may not have the strong motivation to hold an Irish licence as these driving licences are recognised in the State for a period of time, while licences originating from outside the EU do not have the same recognition as the EU licences. Post-stratification by nationality grouping, single year of age and sex is undertaken. On the one hand, post-stratification is a standard method in census population size adjustment which helps to account for the heterogeneous capture of the population. On the other hand, this provides for enhanced census-like estimates by nationality grouping, age and gender.

We notice that while the post-stratification variables can be collected in each of the public administration sources, the PAR and DLD here use the master file as the single consistent source for this information with respect to the PIN. While there may be errors in the information recorded in the master file, using this source eliminates any error associated with information being inconsistently recorded across different sources.

Spurious events and closure The target population is closed for the two lists and there are no spurious events causing erroneous enumeration.

As explained before, the assumption of spurious events and the closure are inter-related, and critical to the contrasts between the DSE and the TDSE. It is worthwhile to distinguish four relevant population concepts carefully.

Census night population (U_I) This is the *de facto* definition currently of the Irish Population Census. It includes every person that is in the State on a given date, regardless of the status or nature of the presence.

Usually resident population (U_{II}) While the exact definition of usually resident status may differ, the concept is typical and in principle feasible for register-based population counts. In the countries that have implemented completely register-based census, the usually resident population also has a specific reference date.

Hypothetical PAR population (U_A) This includes any person who have had or *in principle* could have had interactions with the relevant public administrative systems *during* a calendar year. The inclusion of the latter is necessary because the PAR is not a population register.

Hypothetical DL population (U_B) This includes any person who holds or *in principle* could hold an Irish driving licence. The latter is necessary in order to make the DLD relevant for population size estimation at all. Otherwise the actual DL population could be enumerated directly.

We notice the following. In the absence of a population register, it is in theory possible to adopt U_A as the target population concept in the Irish context, which is comparable to U_{II}, except for the usually resident persons that may have emigrated or immigrated in a given year. In reality, there is usually little cost for a person to register for a scheme and many benefits, while there can often be no incentive or requirement for a person to deregister. Using registrations as evidence of usually resident is therefore expected to lead to over-coverage. This is a potential source of error even in countries that operate with a central population register targeted at U_{II}.

As the lists A and B are based on evidence of actual transactions or SoL activities in the public administrative systems, the extent of spurious registrations is greatly reduced. Nevertheless, the compiled list A can have both over- and under-counts of U_A. For instance, discrepancy between the actual and registered dates of an event in the PAR, either due to delay or registration error, can potentially cause an over- or under-count. But it seems reasonable to assume that under-count due to the contingencies of life will dominate the net error. Some examples will be given in Section 17.3.4 below.

The net difference between the population counts of U_I and U_A depends on the balance between the non-resident population present on the census night and the dynamism of the usually resident population. Notice that U_A includes anybody that left the country that year prior to the census night and anybody that arrived in the country that year after the census night. Ireland has a relatively high degree of migration. According to the official estimates [91], emigration was 200,000 in year 2011 and immigration about 100,000. It is therefore reasonable to assume that the count of U_A is higher than that of U_I, especially in the age groups where the migration flows are strong. More details will be provided in Section 17.3.5 below.

The discussion above provides a motivation for using the DSE based on lists A and B to account for the under-coverage of U_A and U_B, which will then be compared to the Census count of U_I in year 2011. Notice that the DLD is compiled based on the relevant events in a calendar year to make U_A and U_B compatible in time. However, the extent to which U_A differs from U_B is somewhat unclear otherwise. We propose to use the TDSE to explore the matter. The idea is straightforward. Provided the set $U_A \setminus U_B$ is non-empty, then the TDSE based on trimming of lists A and AB will differ from the DSE, whether list B contains over-coverage of U_B or not. The results are given in Section 17.3.4 below. Conversely, provided the set $U_B \setminus U_A$ is non-empty, then the TDSE based on trimming of lists B and AB will differ from the DSE, whether list A contains over-coverage of U_A or not.

17.3.4 Application of TDSE

Here we apply trimming to lists A and AB. The criteria for selecting the k records to be trimmed is based on subjectively identifying those records that are most likely to contain erroneous information. In this case, the trimming method removes records for persons in list A in a number of steps where the person only has an employment record with pay less than a specified amount in EUR. So, after finding the base estimate at $\hat{N}_0$ with no trimming, step 1 requires removing records for persons with only an employment record with pay less than 1000 EUR, step 2 removes records for persons with only an employment record with pay less that 2000 EUR, and so on.

On examining the TDSE in year 2011 for different post-stratum (by age, sex, nationality group) we see that it can behave differently in different post-strata. Fig 17.3 presents 3 different cases in 3 different rows with respect to the stopping rules described in Section 17.2.3. In the first case presented in row 1, the population group relates to males aged 32 with a nationality from the most recent EU countries, referred to as EUnew, and $\hat{N}_k$ shows a distinctive fall before a general levelling off. In the second case presented in row 2, the population group relates to males aged 56 years of Irish nationality, and $\hat{N}_k$ appears to be generally level with a possible small general decline over the trimming. In the last case presented in row 3, the population group relates to females aged 28 years of Irish nationality, and $\hat{N}_k$ starts generally level before appearing to rise slowly.

More explicitly, the first stopping rule looks to see if $\hat{N}_k$ flattens out at some point indicating that the scoring method has reached an equilibrium. In considering the 3 cases, the first case looks to have a point $k_r \sim 520$ where $\hat{N}_k$ appears to flatten out at 6400. The second case has no such point while the third case appears to have a point $k_r \approx 700$ where N_k starts to rise indicating that the scoring method is removing less erroneous records than would be the case if it were removing the records at random.

The second stopping rule considers the ratio k_1/k as possibly being convex and if so the stopping point will be where the bend is most acute. The first case is the only case with a slightly convex curve with the bend being most acute at point $k_r \approx 520$, noting that k_1 is rounded. This stopping point is consistent with the first stopping rule for this case.

The third stopping rule relates to considering the behaviour of the variance estimate of $\hat{N}_k$ before and after k_r. The first case again is the only case where there is a case for stopping point at $k \approx 520$.

In terms of the estimates presented here, we see that trimming results in about 5% reduction $(1 - 6460/6780)$ in the estimate for the first case, and it appears significant with regard to the 95% confidence interval (CI) of $\hat{N}_0$. This case relates males of age 32 with a declared nationality from the most recent EU countries. Ireland has experienced significant immigration in this group in recent years, who may not have a need to apply for a driving licence immediately as their existing driving licence entitle them to drive without an Irish driving licence for a short period of time. In addition, the group may also have a relatively higher proportion of short-term workers, whether on a once off or regular basis, given the ease with which it is possible to travel between countries. These may also have no need of a driving licence but still engage with the public administration systems through paying tax. Both these groups will have a relatively high probability of being trimmed. It seems therefore plausible that the set $U_A \setminus U_B$ in this population group is non-empty, which is manifested here as "erroneous" enumeration in list A with respect to the joint set $U_A \cap U_B$.

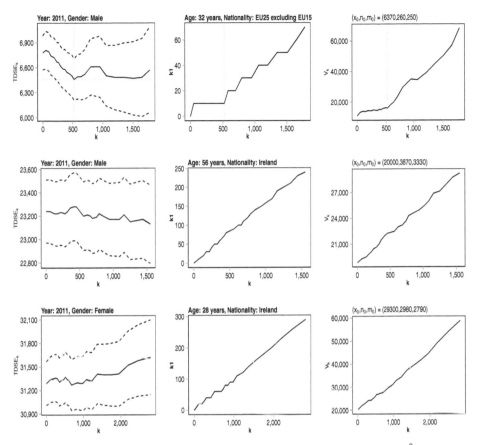

FIGURE 17.3: Illustration of TDSE in year 2011. Left column: TDSE $\hat{N}_k$ with 95% CI; middle column: k_1; right column: $V(\hat{N}_k)$. Each row presents a different population post-stratum. First row: males aged 32 years with a nationality EUnew. Second row: males aged 56 years with an Irish nationality. Third row: females aged 28 years with an Irish nationality. All figures are rounded to nearest 10.

The second case relates males of age 56 with an Irish nationality, which is relatively stable in the population. The third case refers to females of age 28 with an Irish nationality, where the resident status is more transient, due to reasons such as travel, study or work. Indeed, it may be the case that the set $U_B \setminus U_A$ is non-empty in this group. A reason for this might be that the benefit of holding a driving licence may be an incentive to a small number of these persons living abroad (intending to return home shortly) to renew their driving licence on an ongoing basis possibly using an Irish address. Nevertheless, the presence of such potential "erroneous enumeration" in U_B with respect to U_A would not by itself cause the rise in the TDSE.

A more plausible explanation for the different behaviour of the TDSE in the second and third case may lie with the different effects of trimming. For simplicity, suppose all the trimmed records are non-erroneous. Then, the TDSE will be higher than the untrimmed DSE, since $\hat{N}_k = n(x-k)/(m-k) > nx/m = \hat{N}_0$ as long as $m < x$. Of course, should it be the case that $\hat{N}_0 > \tilde{N} = n(x-r)/m$ to start with, we would also have $\hat{N}_k > \tilde{N}$. In other words, the trimming has already reached the stage where relatively more of the scored records can be found in the matched list AB among the Irish females of age 28 (before $k = 1000$) but not yet so among the Irish males of age 56 (up to $k = 1500$). Now that the TDSE is basically level to start with, there is no evidence that that $U_A \setminus U_B$ is non-empty in either of these two groups based on the chosen scoring method. Notice also that the difference between the TDSE and untrimmed DSE is not significant with regard to the 95% CIs.

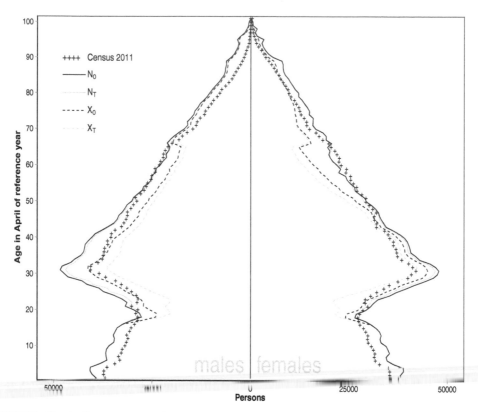

FIGURE 17.4: Comparison of various population estimates by age and sex in year 2011.

For an appreciation of the overall effects of trimming, we refer to the population tree in Figure 17.4 for the population estimates by age and sex. The DSE $\hat{N}_0$ is given by the black line, and the TDSE by the grey line, which is based on scoring all persons with an employment record and an income less than 20,000 EUR, denoted by $\hat{N}_T$. Notice that the DSE is only available for persons aged 18 and up, due to the nature of the DLD. In this instance the DSE is simply estimated as the x_0 or all those identified with activity. Where N_0 and N_T are the same, the graph shows the grey line is overwritten by the black line. Similarly for x_0 and x_T, the grey dashed line is overwritten by the black dashed line. This is observed in the over-65 age group where trimming is not expected to have an impact; nearly all aged over 65 years are retired and not in paid employment. The estimates (trimmed and untrimmed) almost do not differ from each other at all for persons aged 40 – 65, despite the actual difference between x_0 and x_T, given by the black and grey dashed lines, respectively. This suggests that the set $U_A \setminus U_B$ is essentially empty in this population group. Some difference can be detected for persons below age 40. In particular, the TDSE of the population between age 18 and 20 is close or slightly higher than the corresponding DSE. This is the age when many young people enter the work force and the number of scored records k_T is higher than the rest of population. By and large, the results suggest that the set $U_A \setminus U_B$ is nearly empty in all the relevant population groups, except for certain small groups such as in the first case presented above.

17.3.5 Comparisons with census figures

We will now compare the PAR counts (list A from U_A) with Census 2011 counts (of U_I), before looking at the DSE population estimates.

As discussed above, the count of U_A is expected to be higher than that of U_I due to the considerable migration flows in the Irish context. In year 2011 the PAR enumerates to a total 4.35 million, while the Census count is 4.5 million. This is a clear indication that the PAR overall under-counts the population U_A. Figure 17.4 shows the differences between the census count (black cross) and PAR count (black dashed line) by age and sex.

Take the youngest first. For both males and females of the ages 18–23, the PAR counts appears to fall much below the census counts, compared to those of the ages 23–40. This may be accounted for by those persons who finished second-level education and did not engage with any public system, as it is a significant transition phase in a person's life. For example, there can be expected a group of persons adjusting to adulthood who have yet to find official paid employment, are not participating in third level education, do not receive a further education award, do not avail themselves of social welfare, etc.

On examining the group aged between 23 and say 40, it looks like the PAR count matches the census count much better. But there is a difference by sex. The PAR count for males is closer to the census count, and changes from an initial higher count (up to just above age 30) to a lower one, whereas for females the PAR count is consistently higher than the census count. An immediate explanation that comes to mind is that child benefits and early childhood care are more often recorded for the mother than the father.

The profiles for persons between the ages of 40 and 65 show a progressively lower PAR count compared to the census count, which corrects at the age of 65. Moreover, the discrepancy is bigger for females. This can be explained certainly to a large extent through increasing disengagement with the administration systems as adults age between 40 and 65, possibly through early retirement or access to alternative sources of income not observed in the underlying sources. The fact that the gap is clearly larger for females may be due to a higher proportion of homemakers no longer receiving children's benefit for their offspring as they no longer qualify for the universal scheme when their children reach the age of 18 or leave full-time education.

Once a person reaches the age of 65 years, she or he qualifies for the state pension scheme. As such, this age signals another significant transition stage in life from work to retirement. From age 65, on an increasing higher PAR count of males is observed, while the PAR count seems to be more closely aligned with the census count for females, particularly so for the eldest persons. Notice also that very few persons are affected by the scoring as the scoring only relates to paid employment and not pensions. A closer analysis of the state pension and social welfare systems for the age group is needed in order to better understand the underlying factors.

In short, there are clear differences by sex when it comes to the PAR under-count with respect to U_A. Further analysis may help to gain insights of the best way to compile the PAR, and whether additional data sources can be identified to improve the direct list enumeration.

The analysis above of the TDSE suggests that the discrepancy between the conceptual populations U_A and U_B is smaller than that to the census population U_I. Provided $U_A \approx U_B$, the DSE based on both lists A and B can be considered as a means to amend the PAR under-count of U_A. The differences between the DSE and census counts are shown in Figure 17.4.

As can be expected, the DSE is greater than the census figures in the age categories 20–45 years as this age group contains a relatively large transient component. In particular, the adjustment of the direct PAR enumeration is more pronounced for males than females, and to a large extent compensates for the greater PAR under-count for males. The discrepancies with the census counts are larger than can be accounted for by the official migration flow estimates. However, these official estimates are compiled using a usually resident concept based on 12 months and as such miss the cohort of workers that may only come to work for a short period of time, who can be part of U_A by definition. In other words, one could not expect the official migration flow estimates to fully account for the discrepancies between the DSE and the census counts.

The DSE agree much better with the census count for persons aged 45–65. The gaps between the PAR and census counts observed earlier have largely been adjusted away. This seems plausible as the population in these age categories is much more stable than the younger ones.

Critical to compiling population estimates in this fashion is the underlying data sources and how they are put together to compile the PAR. The application of the DSE approach here requires that SoL eliminates the overcoverage problem. The trimming exercise has been focussed in or around ages 18 to 65 years, the age people generally seem to be engaged in employment. We have used a proxy for children's benefit (age 0 to 17 years) and for state pensions (over 65 years) and when comparing the PAR to the census for these ranges in Figure 17.4 there is evidence of overcoverage. To improve the PAR enumeration we suggest revisiting the data sources for these age categories. For instance, interactions with public health systems may, on the one hand, provide more reliable SoL evidence and, on the other hand, increase the coverage of some parts of the population that may not be able to engage with other public administration systems through long-term illness or other reasons. Such a source could be particularly valuable for older age categories. Again, the TDSE methodology can be applied to seek assurance around the robustness of population estimates in these age categories.

17.3.6 Discussion of future works

The SoL approach, based on a suitable population concept and the adoption of a common official personal identification number across the public administrative systems, provides the foundation of an alternative to the traditional census. In the Irish context, the direct PAR enumeration has overall an under-count of the targeted usually resident population, so that an appropriate estimation methodology is needed in the absence of a central population register that is of sufficient accuracy. Evidence from the TDSE further suggest that the necessary assumptions for the DSE methodology can to a large extent be met with the data sources that are already available. Overall, the DSE presented above provides a proof of concept for an alternative approach to census-like population size estimates, where the combined administrative sources are utilised to replace the traditional census. It still requires considerable development before viable census-like official population estimates can be produced. Some of the future works are briefly outlined below.

As mentioned above, additional data sources and alternative DSE-TDSE set-ups will be explored for the younger (under 15 years) and older (over 65 years) age cohorts. Moreover, the relevance of the usually resident population estimates, compiled for a given year, needs to be considered to see if it is adequate to comply with existing population concepts for register-based statistics and for general planning purposes. Among others, adjustment for migration may be required to accommodate a specific reference date, in order to bridge the difference between U_A and U_{II}. The TDSE can be explored and can potentially form part of the adjustment methodology.

Next, population estimates will need to be disaggregated by local area. However, in the case of Ireland this provides for considerable challenges, due to the lack of postcode on the public administration systems and a high degree of non-uniqueness in addresses in rural areas. The name of a person is often used in conjunction with an address to find the right house for delivering mail. Administrative data sources that hold data on the relationships between persons may provide valuable evidence for a person's correct address. Examples include medical expense reimbursement (where limits are based on household), children's benefit (where the relationship with one parent is recorded) and central tenancies register where all rental agreements are registered with details of persons signing the agreement. Nevertheless, mis-location of persons will be unavoidable, which will create erroneous enumeration at the local-area level. Both the TDSE and the modeling approach will be studied further in order to remedy the problem.

18

Estimation of non-registered usual residents in the Netherlands

Bart F. M. Bakker

Statistics Netherlands/VU University

Peter G. M. van der Heijden

Utrecht University/University of Southampton

Susanna C. Gerritse

VU University

CONTENTS

18.1   Introduction

For the 2011 Census round,[1] the population is the set of usual residents, defined as those who have lived or intend to live for a period of more than 12 months in their place of usual residence (EU [112]).[2] This definition is hard to apply if statistics are register-based. Part of the problem is that intentions are not registered at all, but also the estimation of the number of residents who lived longer than a year is difficult. The actual population differs from the population registered in the Population Register (PR, Bakker and Daas [19], Bakker, van Rooijen and Toor [20]). An important part of the difference between the registered and the actual population in the Netherlands ultimo 2010, is the group of temporary workers from Eastern Europe, in particular Poland, who do not register themselves in the population register (Gerritse, van der Heijden and Bakker [125]). Within the European Union, individuals with a European Union nationality are free to migrate and most of them are free to work

---

[1] This chapter is primarily based on previously published papers (Bakker et al. [20]; Gerritse et al. [125]; [126]). The views expressed in this paper are those of the authors and do not necessarily reflect the policies of Statistics Netherlands. We would like to thank Peter-Paul de Wolf, Eric Schulte-Nordholt, Mila van Huis and Kees Prins, all working at Statistics Netherlands, for their valuable comments on earlier versions of this paper. Please direct all correspondence to B.F.M. Bakker, bfm.bakker@cbs.nl

[2] Countries that are not able to apply the definition are allowed to restrict the population to the registered population. However, from a quality and subject matter perspective it is important to estimate the "real" number of usual residents.

without a working permit (only until January 1st 2014 Bulgarians and Romanians required a working permit). Temporary workers and other immigrants are required to register in the Dutch population register if they stay longer than four months or are planning to do so. Due to ignorance, a very small penalty, and other reasons, part of those individuals who should register do not. A second large group are refugees and asylum seekers, who provisionally stay in anticipation of the decision of the authorities for a residence permit. They are not always allowed to register in the PR. Finally, a third category are the undocumented immigrants, for a great part former asylum seekers who did not receive a residence permit but did not leave the country.

This chapter aims to show how the number of usual residents in the Netherlands ultimo September 2010 is estimated. To estimate the size of a population, capture–recapture methods are available (Fienberg [121], Bishop, Fienberg and Holland [32], International Working Group for Disease Monitoring and Forecasting [154], van der Heijden et al. [283], Baffour, Brown and Smith [16]; see also Chapter 15). Two or more registers that contain information on (parts of) the population are linked and from the overlap in the registers, the part of the population missing from all registers can be estimated. When using two registers this method makes the following assumptions:

1. the inclusion probability of being registered in the first register is independent of the inclusion probability of being registered in the second register;

2. the population should be closed, i.e. no individuals appear or disappear during the data collection period;

3. all individuals of the population should have a positive inclusion probability of being registered in each of the registers;

4. the registers do not include erroneous captures, i.e. individuals that do not belong to the population;

5. the registers are perfectly linked;

6. the registers do not include duplicates;

7. the inclusion probabilities for at least one of the registers should be homogeneous.

Violation of these assumptions could lead to biased estimates (Brown, Abott and Diamond [56], Gerritse et al. [125].

To distinguish between those who are usual residents and those who are not, we have to measure residence duration. However, in the three administrative sources used, there is no explicit measure of residence duration available. In this chapter, we present a procedure to estimate the total population of usual residents, where we tackle the problem that none of our sources measures residence duration accurately.

In this chapter we will first discuss previous research into the subject and their findings. Then we discuss the assumptions and how to meet them in our application of the method. Then we will give the results of the capture-recapture estimation, and we end this chapter with a conclusion.

For the time being, we neglect the possibility of over-coverage in the PR, because to estimate this, more extensive research is needed with the use of additional data sources which are not available yet. Therefore, we assume that those who are registered in the PR are all usual residents. Moreover, we restrict the estimation to the population 15–65 years of age because of lack of appropriate data sources for the other age groups.

18.2 Previous findings

There is previous research on the estimation of population sizes in the Netherlands, most notably Hoogteijling [147] and Bakker [18], that overlaps with the population of usual residents that we study in this paper. These estimates allow us to place the estimates found in our research in perspective. However, this previous research led to different estimated population sizes depending on the definitions of the population, different reference dates and the methods used, and therefore these studies cannot be used as a benchmark for judging the outcomes of our estimates presented in this manuscript. We have to harmonize and actualize these findings.

Hoogteijling [147] collected different estimates from earlier research in the nineties. In order to achieve an estimate of the size of the population not registered in the PR and living four months or longer in the Netherlands in 2000, she combined the available information from different sources. Neglecting some very small categories, the population can be estimated by adding undocumented immigrants, adding the balance of falsely not registered residents and falsely registered non-residents, and recently arrived asylum seekers who have non-registered because they are not allowed to do so yet. This results in an estimate of 73,000 to 149,000 missed residents, with a mean of 111,000, being less than 1% of the registered population.

Bakker [18] also used information from different sources to get an estimate of the under- and overcoverage of the PR in 2006, having the same definition of usual residence as Hoogteijling [147], so those who stay longer than four months in the Netherlands are supposed to be usual residents. He distinguishes the different categories of which it is known that they are missed or are over-counted in the PR and he estimates their numbers with different sources. He estimates the total undercoverage as 205,000 usual residents. However, there is a large uncertainty because some of the estimates are quite arbitrary. The largest contribution is from undocumented immigrants whose numbers are estimated between 74,000 and 184,000. The total number of missed persons is 236,000, where 31,000 persons are still in the population register while they have left the country or have died.

It is difficult to describe the expected value of the size of the population of usual residents in 2010, because the formerly discussed estimations are outdated and do not use the same definition. However, by harmonizing the results for the definitional differences and looking at the developments of the number of new asylum seekers and the number of immigrant workers, we can provide a range of expected outcomes. These expected outcomes could help to provide a perspective to which the current estimate of usual residents may be compared.

In the under-count of 111,000 found by Hoogteijling [147] the majority are former or present asylum seekers. Because the procedures for seeking asylum had a long duration, certainly with a mean longer than a year, we assume that most residents who were non-registered as such stayed for longer than a year in the Netherlands. Therefore we assume that 80% of the 111,000 non-registered to be usual residents, which comes down to 89,000 usual residents in 2000.

Bakker [18] estimated an overcoverage of 205,000 in 2006 and this estimate is difficult to harmonize with the definition of usual residence in this manuscript, because we do not have empirical information on the residence duration of the 10 different categories that are over-counted in the PR. However, if we assume (i) that 80% of the undocumented immigrants are usual residents because most are still former or present asylum seekers as in the period described by Hoogteijling [147], and (ii) that the same percentage is true for smaller categories like asylum seekers, diplomats and NATO military and administrative delay of newborns and immigration, and (iii) that 30% of the migrant work force and migrant students are usual residents, the same percentage as we found for the migrant work force in 2010 (see Section 18.4), then the estimated number of usual residents non-registered in the PR is 135,000 in 2006.

Two significant developments have to be mentioned to explain changes in the number of non-registered usual residents between 2000 and 2010. The first is the decline of the number of asylum seekers between 2000 and 2010 (Gerritse, Bakker and van der Heijden [126]). The numbers dropped from almost 45,000 in 2000 to 10,000 in 2004, due to changed regulations. After 2004 there is a more or less constant number of asylum requests between 7,000 and 15,000. The other one is the sharp rise of the migrant workforce from the year 2006, in particular from Eastern Europe, who did not register themselves in the population register. In 2006 this number was 121,000 and this number increased to 182,000 in 2010 (Statistics Netherlands [70]). This development was possible because under European law the civilians of these countries could enter the Netherlands without a residence permit and after 2007 for the most part they could also work without a working permit. Recently, van der Heijden, Cruyff and van Gils [285] estimated the number of undocumented immigrants on 42,000 in 2009 (95% confidence interval 21,000 to 63,000) and 36,000 in 2012–2013 (95% confidence interval 23,000 to 48,000). As the undocumented immigrants are not registered in the PR by definition, this is the absolute minimal size of the non-registered population.

We arrive at the following conclusion, knowing that we have to be cautious extrapolating earlier estimates to later periods. We expect that the number of usual residents not registered in the PR has been increased since the year 2000 to 175,000 to 200,000 in 2010. The total number is certainly much higher than the 135,000 in 2006 because of the inflow of migrant workers from Eastern Europe since then. If 30% of the 182,000 migrant employees who did not register in the PR as a usual resident, then that would increase the non-registered population by more than 60,000 to almost 200,000. On the other hand, the number of asylum seekers has been constant since 2006 and will not cause important developments.

18.3   Meeting the assumptions of the capture-recapture method

To estimate the population of usual residents, we apply the capture-recapture method. As mentioned in the introduction, the capture-recapture method has a number of assumptions (see also Chapter 15). The following section describes how these assumptions are met in our application of the method.

We will be using three registers. This relaxes the assumption 1 of independence between the registers. We now only assume that the log-linear three-factor interaction between the inclusion probabilities (to be discussed below) in the contingency table formed by the three registers, is zero. The first register is the official Netherlands' Population Register (PR). It includes the entire registered population of the Netherlands. The second one consists of the employees in registered Netherlands' companies and is called the Employment Register (ER). The third one consists of crime suspects that are registered as such by the police (CSR).

Assumption 2 that the population is closed, is easily met for the PR and ER because both registers describe a period and any common date or period can be selected. The assumption is satisfied by restricting the date to ultimo September 2010. Ultimo September is operationalized as the latest Friday in September. This date describes the status ultimo September best, because many jobs start or end in the weekend or on the first day of the month. This choice for September 2010 cannot be applied to the CSR because this register is event based: crime suspects are registered when the police make a report. The number of events on one specific day is not high enough to apply the capture-recapture method. In order to satisfy the assumption as well as possible, we restrict the period of the CSR to the second half of 2010. Note that ultimo September is in the middle of that period.

Not all elements of the population have a positive probability of being registered (assumption 3). The ER is restricted to the population of 15–65 years of age, while the CSR is restricted to persons 12 years and older. Because of these restrictions, we are not able to estimate the total population of usual residents, but only those 15–65 years of age. The youth and elderly are estimated in another way and this is not reported here.

To prevent erroneous captures (assumption 4; see also Chapter 17 for an example of erroneous captures in an application for Ireland), we removed the records from the PR, ER and CSR of persons who do not belong to the population. We have removed the following categories: (a) the few persons with the Dutch nationality not registered in the PR because we expect them to be expats working in another country and therefore not belonging to the population; (b) persons with an address in Belgium or Germany, the neighbouring countries of the Netherlands, because it is likely that they live in Belgium or Germany and are only temporarily in the Netherlands to work, to go to school, to shop or to have a short holiday; (c) persons who are reported for a crime by the border police at the airport or elsewhere because they did not enter the country at all.

Despite the removal of these categories, it is still possible that there are persons that committed a crime and are registered as such in the CSR but do not belong to the population. For example, drug runners living in France, Luxembourg, Switzerland, etc., gangs of pickpockets from Eastern Europe, or tourists arrested for drunk driving. Later on, in the estimation of the size of the population of non-registered usual residents, we will take into account that their numbers are unknown and assume that they overlap with the category that did not link to the other registers because of incomplete or unknown linkage keys.

The capture-recapture method is sensitive to linkage error. Therefore, the three registers are linked pairwise with much caution (assumption 5). In the first step, the records are linked deterministically on a personal identification number that is used in multiple administrative data sources in the Netherlands. The remaining records are linked probabilistically (Fellegi and Sunter [120], Ariel et al. [10]). To reduce the number of possible pairs, the data are blocked on variables that are assumed to be of very high quality. Data blocks are created with similar values on these high-quality variables, and only within those blocks are we searching for possible pairs. For the linkage of PR-ER and PR-CSR date of birth, sex, postal code, house number and suffix are used after blocking on postal code or date of birth. The ER and CSR are linked in a slightly different way because a large number of the

TABLE 18.1: Linkage effectiveness

		Not linked	Linked		Total
Linkage	Source		Determ.	Prob.	
		%	%	%	abs.×1000
PR ↔ ER	PR	57.6	42.4	0.0	617.3
	ER	30.1	69.9	0.0	374.8
PR ↔ CSR	PR	98.9	1.1	0.0	617.3
	CSR	43.8	54.3	1.9	12.4
ER ↔ CSR	ER	99.3	0.6	0.1	374.8
	CSR	80.2	17.8	2.0	12.4

Note: The table should be read as follows: 57.6% of all individuals in the PR are linked to the ER, 42.4% of all individuals are linked deterministically, and none are linked probabilistically. This entails a total of 617,300 individuals.

records do not contain a Dutch postal code. Therefore the linkage is done on date of birth, sex, place of residence, address and house number after blocking on date of birth, or on place of residency and month of birth or day of birth. Table 18.1 shows the results of the linkage procedures. As an example, in the linkage between the PR and the ER, 57.6% of all the PR records could not be linked to the ER, 42.4% were linked to the ER, and these were all linked deterministically. Also, 30.1% of the ER could not be linked to the PR, and 69.9% were linked to the PR. All ER individuals that could be linked to the PR were linked deterministically, none were linked probabilistically. Overall, probabilistic linkage leads to a small increase of the number of linked records, in particular CSR records are linked to the PR and the ER.

Despite the attention paid to the linkage method and the careful execution of the method, it is still possible that not all records are linked that should be linked. One of the main reasons for false negatives is that the linkage keys are incomplete or entirely missing. In Table 18.2, the number of missing values is given for different combinations of PR, ER and CSR. From the last row in the table it is clear that the records in the CSR that do not link to either the PR or the ER, contain large numbers of missing values in the linkage key. In the linkage key used for linkage to the PR this is 27.7% and in the linkage key used for linkage to the ER this is 37.7%.

The capture-recapture method is sensitive to duplicates in the registers (assumption 6). Deduplication has been applied to all three registers. However, duplication is not possible for the records in the CSR that have incomplete or entirely missing linkage keys. It is unknown how many records in the CSR are duplicates. Because duplicates have the same effect on the outcomes as erroneous captures we do take them together with that category. For reasons of readability, we restrict the name to erroneous captures.

18.1   The residence duration

Most of the assumptions of the capture-recapture method are met by making a three-register estimation, making restrictions to one day for the period-based registers and a short period for the event-based register, applying a stringent linkage method and deletion of erroneous records. However, to estimate the number of usual residents, we need to split up the estimated total number of persons into those who stay longer than a year in the Netherlands and those who do not. First, we estimate the total population and second, we estimate the number of usual residents by adding the covariate "residence duration" to the estimation model. However, residence duration is not available in all of our three registers. To solve this problem, we assume that those who are registered in the PR are usual residents. Most of them are registered for more than one year. Those who are not registered for more than one year are mostly immigrants who started a job more than one year ago and registered themselves in the PR after a while. The remaining part of the registered population consists of a varied group of diplomats, militaries, former asylum seekers, and their family members who registered themselves later than they arrived in the Netherlands.

We start the procedure with those in the ER who are not registered in the PR, because it is likely that those who have a job in the Netherlands also stay in the Netherlands during the time they have the job. Therefore, the residence duration will be derived from the jobs they had consecutively. After that, we impute the residence duration in the remaining CSR-records by using a procedure to impute missing data in a dataset called Predictive Mean

TABLE 18.2
Missing values in the linkage key in PR, ER and CSR linkage

Linked to			Sex	Birth date	House nr	Suffix	Postal code	Country	Street	Place of residence	Address1[3]	Address2[4]
PR	ER	CSR	%									
y	y	y	0.0	0.0	0.3	0.0	0.1	0.1	0.3	0.1	0.3	0.3
y	n	y	0.0	0.0	0.6	0.0	0.2	0.3	0.6	0.3	0.6	0.6
n	y	y	0.0	0.0	3.7	0.0	0.6	1.1	4.5	0.9	3.7	4.8
n	n	y	0.5	0.0	27.7	0.0	0.2	19.7	26.2	11.2	27.7	37.7

[3] Address1 is a combination of house number, postal code and suffix. A missing value in one of these variables leads to a missing value for the linkage key as a whole. This linkage key is used for the linkage with the PR.
[4] Address2 is a combination of street, house number, place of residence and country. A missing value in one of these variables leads to a missing value for the linkage key as a whole. This linkage key is used for the linkage.

Matching (PMM, van Buuren [67]; for an approach using the EM algorithm, see Chapter 15).

Crucial in this operation is to derive the residence duration from employment records. These records contain information on the starting and ending dates of sequential jobs. The residence duration of employees with only one job is estimated as the duration of that particular job. For employees with more than one job, the residence duration is defined as the period of continuous stay. However, if there are gaps between jobs, we have to decide on which duration of a gap is acceptable to assume continuous stay. In order to decide on this issue, we investigated seven scenarios that differ in the duration of the gap: 1, 8, 15, 22, 31, 62 and 93 days. For each scenario, we determined the number of employees who are assumed to have a residence duration longer than one year (Table 18.3). We distinguish seven nationality groups that are formed by shared migration motives, migration legislation and size and 7 categories: (1) EU15 (excl. Netherlands); (2) Polish; (3) Other EU; (4) Other western; (5) Turkish, Moroccan, Antillean, Surinam; (6) Iraqi, Iranian, Afghan, asylum seeker countries Africa; (7) Other Balkan, other former Soviet Union, other Asian, Latin American and not mentioned elsewhere.

There are 730,000 individuals in the union of PR and ER who do not have the Dutch nationality. Of those 730,000, 617,000 individuals are registered in both the PR and ER and we therefore assume that they are usual residents. The remaining 113,000 individuals are registered in the ER but not registered in the PR. For these 113,000 individuals we investigated the seven scenarios. See Table 18.3. The results in Table 18.3 led us to choose the scenario of 31 days, for the following reasons:

1. The biggest groups are European, in particular Eastern European. The probability that they return to their homeland if the gap is larger than 31 days becomes larger.

2. A return to their homeland will be more like a holiday than a return to live there if the period is restricted to 31 days.

3. Financially, it is relatively easy to bridge a gap of one month.

4. The differences with the scenarios of 22 and 62 days are relatively large.

5. The majority of persons with more than one job have had two jobs consecutively (not in table). If the start of the first of those two jobs is more than a year ago, they have stayed at least eleven months in the Netherlands.

According to this scenario, (29% of 113,000 is) 33,000 persons are usual residents.

After this step, we have a measure of residence duration of all persons who are registered in the PR and ER. Thus, all persons registered in the CSR who link to the PR or the ER, have been assigned a residence duration as well. However, for remaining persons registered in the CSR it is not possible to derive the residence duration because events (suspicion of committed crimes) instead of periods are registered. Therefore, we have to impute a likely value for the residence duration in the records in the CSR that do not link to the PR and ER. This will be done using PMM [67]. We will use the persons that are registered in the ER but not in the PR as donors for the residence duration. We have chosen this subpopulation because we assume that it represents individuals that are only in the CSR the most. A worse alternative would be to use the PR for donors, because those registered in the PR most of the time have settled a long time ago and for other reasons than the temporary workers or asylum seekers. In PMM we use nationality group, age and sex as the predictors. PMM assumes missingness at random (MAR, see van Buuren [67]) independence of the process of missingness conditional on the predictor variables. For age, we use four categories: (1) 15–24 (2) 25–34 (3) 35–49 and (4) 50–64 years of age.

TABLE 18.3

Seven scenarios for residence duration derived from employment records by nationality. The data concern individuals registered in the ER but not in the PR.

Nationality	Total	Residence duration >1 year in scenario						
	× 1.000	1 day	8 days	15 days	22 days	31 days	62 days	93 days
		%						
EU15 (excl. Netherlands)	18.7	40	42	42	43	47	53	58
Polish	80.7	18	19	20	21	25	32	36
Other EU	10.8	20	21	22	22	24	29	32
Other Western	0.5	42	43	43	43	44	45	48
Turkish, etc.	0.6	53	54	55	56	59	66	71
Iraqi, etc.	0.3	52	53	54	57	61	65	71
Other	1.2	39	40	41	41	43	47	51
Total	113.0	23	24	25	25	29	35	40

There are approximately 113k individuals registered in the ER but not in the PR. Of these, 23% have a residence duration when we allow one day of non-registration. 24% if we allow 8 days, and so on.

To summarize, using the approach discussed above, every person in the PR, ER and CSR now has a value on the usual residence variable:

1. Those registered in the PR are assumed to be usual residents.

2. Those registered in the ER who are not registered in the PR are usual residents if they work continuously (or with gaps of less than 32 days between jobs) for more than a year in the Netherlands.

3. Those registered in the CSR who are not registered in the PR or the ER, are usual residents if their imputed value of residence duration is longer than a year.

To estimate the variance caused by the imputation, we repeated the PMM ten times. Each of the ten datasets created in this way is input for the capture–recapture estimates.

18.5   Capture-recapture estimates

After imputing the residence duration in the remaining part of the CSR, all the necessary information is available to apply the capture–recapture method. For this we apply the standard methodology as described in IWGMDF [154] and Bishop et al. [32] that makes use of log-linear models. See also Chapter 15. The saturated log-linear model, assuming the three-factor interaction to be zero, has seven parameters and only seven observed counts:

$$\log m_{ijk} = \lambda + \lambda_i^A + \lambda_j^B + \lambda_k^C + \lambda_{ij}^{AB} + \lambda_{ik}^{AC} + \lambda_{jk}^{BC}. \tag{18.1}$$

A, B and C are the variables denoting being an element in the registers PR, ER and CSR respectively. A, B and C are indexed, respectively, by i, j and k, where a subscript is 1 (yes) if the element is in the register and 0 (no) if the element is not in the register. Expected values are denoted by m_{ijk}. Table 18.4 shows the observed values for the combination of the three registers. The count for the cell with "−" has to be estimated and split up into usual residents and non-usual residents.

According to model (18.1), i.e., a model without covariates nationality, sex and age, the size of the population missed by all registers is 946,000. Then the estimated total population that is not registered in the PR is 1,064,000 individuals: 113,000 persons who are registered in the ER but not in the PR, 5,000 persons registered in the CSR but not in the PR or ER, and estimated 946,000 who are not registered in either register. Note that this number consists of usual residents and non-usual residents, and that our research question concerns the number of usual residents missed by the PR.

TABLE 18.4

The observed values for the three registers ×1.000

PR	ER	CSR 1 yes	0 no	Total
1 yes	1 yes	2.1	259.8	261.9
1 yes	0 no	4.9	350.6	355.4
0 no	1 yes	0.4	112.5	112.9
0 no	0 no	5.1	.-	5.1
	Total	12.4	722.9	735.3

To distinguish between usual residents and non-usual residents, we add the covariate "usual residence" to model 1. To improve the plausibility of the assumption that the three-factor interaction of being observed in the registers A, B and C is zero, we add covariates age and sex.[3] To simplify estimation, we estimate models for all seven nationality groups separately. We use the variables as we defined them in Section 18.4. To prevent overfitting the models, we search for models that fit the data well and are as parsimonious as possible using the Bayesian Information Criterion (BIC). The BIC has a larger penalty for the number of parameters in the model than the Akaike Information Criterion (AIC) when the sample size is large and therefore leads to more parsimonious models when the sample size is large, as is the case here. The population size estimates for the model selected for each nationality group is shown in Table 18.5.

As can be seen from Table 18.4 there are 118,000 (0.4K + 112.5K + 5.1K) registered individuals not in the PR but in the CSR and/or the ER. Of these 118,000 individuals, we found that 35,000 are usual residents. As they are part of the known under-coverage of the PR, these individuals have to be added to the estimates from the scenarios of the missed part of the population by all registers.

After adding the covariates to the model, the estimated number of usual residents not registered in the PR is 284,000. The confidence interval is 231,000 to 347,000 (Gerritse et al. [127]). The reference point for an estimate was somewhere between 175,000 to 225,000 usual residents, and the estimate of 284,000 is larger than expected. This could be the result of a violation of some of the assumptions of CRC, in particular the assumption of no erroneous captures and perfect linkage for the records of the CSR. Because we are not able to determine to what extent these assumptions were violated, we introduce several scenarios to determine a range of possible outcomes. However, we have some knowledge that we can use to define these scenarios.

As mentioned earlier, it is possible that people committed a crime while they were not a resident and therefore are erroneous captures. Drugs runners and tourists arrested for drunk driving are possible examples of this category. Another reason for the large number of records from the CSR that cannot be linked to the other registers is that we missed links.

---

[3]An additional advantage is that this makes it possible to estimate the distribution of these variables in the total population.

TABLE 18.5
Estimated number of usual residents not registered in the PR by nationality group after PMM ×1.000

Nationality group	Missed total	>1 year	≤1 year	CI
EU15 (excl. Netherlands)	227	168	59	38-81
Polish	330	245	85	37-132
Other EU	228	164	64	31-96
Other Western	23	16	7	3-12
Turkish, etc.	4	2	2	1-4
Iraqi, etc.	11	4	7	6-9
Other	72	47	25	18-32
Total m_{000}	894	646	249	188-312
In ER, not in PR			33	
In CSR, not in ER and PR			2	
Total estimation not in PR			284	231-347

This is very likely because of the missing values in the linkage keys of part of the CSR records. However, we believe that there is a large overlap between the "criminal tourists" and the missed links because it is much harder to identify criminal tourists than usual inhabitants. In the protocol of making a police report on a crime, the police officer can check the identifying information of the suspect by looking into the Population Register. That makes it more likely that missing or incomplete linkage keys are in fact foreigners not living in the Netherlands and thus are erroneous captures (see also Zhang and Dunne Chapter 17). However, it is unknown how frequently this is done.

The experts we consulted were not able to give estimates for the number of "criminal tourists" and the other groups of individuals that may give erroneous captures. It is also very likely that the records in the CSR contain duplicates, as it is impossible to deduplicate because of the missing linkage information. This is a supportive argument to choose scenarios with a larger number of erroneous captures than linkage errors, because duplicates have the same impact on the CRC outcomes as erroneous captures. Moreover, in the CSR records that did not link to both the PR and ER with complete linkage keys, it is likely that a small percentage are erroneous captures or missed links. Therefore, we differentiate between the following scenarios.

First, we created two main scenarios. The first is a scenario where a random selection of 75% of the 37% of the CSR records that did not link to the PR and ER and had incomplete linkage key values are considered erroneous captures and the remaining 25% were considered missed links. The second main scenario considered all of the 37% of the CSR records that did not link to the PR or ER and had incomplete linkage key values to be an erroneous capture. Second, we create four subscenarios per scenario where either zero or 5% of the 63% of the CSR records that did not link to either the PR or the ER but have a known linkage key are considered an erroneous capture or a missed link (Table 18.6). For each scenario we estimated the number of usual residents.

The estimates are presented in Figure 18.1. The scenarios are reported as the assumed per cent of erroneous captures from the records with an incomplete or missing linkage key, the per cent of missed links from the records with an incomplete or missing linkage key, the per cent of erroneous captures from the records with a complete linkage key, and the per cent of missed links from the records with a complete linkage key. For example, "2. 75-25-0-5" means that in scenario 2 we assume that of the records with an incomplete or

TABLE 18.6

The 8 scenarios for the estimation of the number of usual residents

Scenario	CSR records with incomplete linkage keys (37%)		CSR records with complete linkage keys (63%)	
	Erroneous captures %	Missed links	Erroneous captures	Missed links
1	75	25	0	0
2	75	25	0	5
3	75	25	5	0
4	75	25	5	5
5	100	0	0	0
6	100	0	0	5
7	100	0	5	0
8	100	0	5	5

missing linkage key, 75% are erroneous captures and 25% are missed links and 0 and 5% of the records with a complete linkage key that do not link to the PR or ER are, respectively, erroneous captures and missed links. The outcomes are sorted according to the size of the estimated population.

The estimates are sensitive to the violation of both assumptions (see also Gerritse et al. [127]). In particular, in the scenarios in which the assumption of perfect linkage is violated, the estimates are low and implausible. The estimates vary in size from 88,000 to 185,000 usual residents, which is a large interval. Given that the reference interval is 175,000 to 225,000 individuals, and given that it is more likely that records with incomplete or missing linkage keys are erroneous captures, the higher estimates are presumably more accurate than the lower estimates.

In estimating the population of usual residents, two kinds of uncertainty can be distinguished. The first is the variance for each scenario, and the second is the uncertainty caused by the unknown size of the violation of the assumptions. If we restrict ourselves to the estimation of the variance of only one scenario, there are several sources of variance that have to be combined. One source of variance is the variance for the estimate of the missed number of usual residents. However, two more sources should be taken into account: for each scenario we simulate erroneous captures and linkage errors, for which we take samples from the CSR. Taking samples influences the imputation process: the models used in PMM depend on the observed records and the imputed values are taken from the observed records. In their turn, the set of observed records depends on the actual sample that is taken for erroneous captures and missed links. To estimate the variance due to sampling as well, we take several samples and per sample apply the PMM. Per sample we have an estimate by averaging the estimates per imputation. It has been explained in Gerritse et al. [127] how a combined variance is estimated, where the researchers extended the work of Little and Rubin [183].

However, the variance is not the only source of the uncertainty, because possible violation of the assumptions of CRC is another. To simulate with different scenarios, it is possible to have an idea what the size of uncertainty is. However, there is no formal argument to distinguish between different scenarios and only subject matter knowledge can help to decide on which scenarios should be considered and on the plausibility of the outcomes of the scenarios. As we expect that the higher estimates are the most plausible, we estimated a confidence interval only for two scenarios: the lowest estimate with the assumption of 75% erroneous captures from the CSR records with incomplete or missing linkage keys and the highest estimate (Table 18.7).

TABLE 18.7
The confidence intervals for two of the scenarios. PSE denotes the population size estimate and CI stands for confidence interval

	Scenario	m_{000}	PSE	95% CI
2.	75-25-0-5	54	88	57-151
7.	100-0-5-0	151	185	149-222

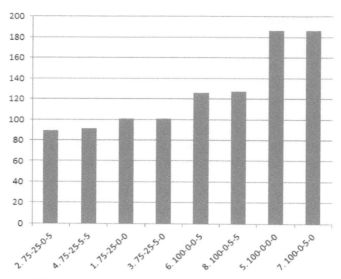

FIGURE 18.1: The estimates of the number of usual residents for each scenario.

18.6 Conclusion

In this paper, we estimated the number of non-registered usual residents in the Netherlands by using the capture-recapture method. By making a three-list estimation, making restrictions to one day for the period-based registers (PR and ER) and a short period for the event-based register (CSR), applying a stringent linkage method and deletion of erroneous records, most of the assumptions of the capture–recapture method are met. We derived the residence duration of the employed from their employment history, by choosing that usual residents had:

1. consecutive jobs for longer than a year;
2. the period between jobs is shorter than 32 days.

For those in the CSR who are not registered in the PR and ER, we imputed the residence duration based on the employment histories of the persons in the ER not registered in the PR. Finally, we estimated the number of non-registered usual residents by applying capture-recapture methodology. We divided the observed population into seven nationality groups and searched for well-fitting and parsimonious models. Because it is plausible to assume that the records in the CSR that do not link to the PR or ER are a mixture of erroneous captures, missed links and true positives, CRC has been applied under 8 scenarios. The estimates vary in size from 88,000 to 185,000 usual residents, which is a large interval.

Moreover, we estimated the confidence intervals for the scenarios with the highest and the lowest estimate. That leads to the conclusion that the population size that is non-registered in the PR is probably not higher than 222,000 thousand and will not be lower than 57,000. This is a very large interval and it is doubtful that this information is very useful for policy purposes. However, we have good reasons to believe that the true population size that is missed is closer to 222,000 than to 57,000, given that the expected outcome should lie between 175,000 and 200,000 and that it is more likely that records with incomplete or missing linkage keys are erroneous captures. Similarity with results from other methods would give more confidence in the outcomes. One of these methods is defining the categories of individuals likely to be missed by the registers used to estimate usual residents such as diplomats or children. Then, find sources to estimate the size of, and the overlap between, these categories (Bakker [18]). This is a topic of further research.

Part VI

Latent Variable Models

19

Population size estimation using a categorical latent variable

Elena Stanghellini
University of Perugia

Maria Giovanna Ranalli
University of Perugia

CONTENTS

19.1   Introduction and background

Making inference on the size of a population of interest requires assumptions on whether the population is closed or open. In the first case, the total number of individuals does not vary through births, deaths, immigration or emigration. This assumption is realistic in certain experiments where the population is not seriously exposed to demographic changes and data are collected over a short time span. When this is not reasonable, statistical models should allow for the population size to change over time.

We assume here that there are T capture occasions. Historically, studies were developed under the assumption that the population of interest is closed and $T = 2$. In this case, very simple methods, based on a proportion, can be applied to estimate the total population dimension; see International Working Group for Disease Monitoring and Forecasting [154] for a review. For $T > 2$, more refined techniques have been elaborated. A useful classification is in Otis et al. [226]. A first class of models, denoted with $\mathbf{M}_0$, assumes that the propensity of an individual to be captured does not vary over individuals and capture occasions. When this assumption turns out to be unrealistic, then a second class of models, denoted with $\mathbf{M}_t$, allows for the propensity to be captured to vary with capture occasions, but not over individuals. Empirical studies, especially on animal abundance, have shown that in some cases subjects exhibit a response to first capture and therefore the propensity depends on being captured at least once. The corresponding models are denoted by $\mathbf{M}_b$. Models can then be complicated further if the subjects' propensity varies with the pattern of captures. The corresponding class of models is then denoted by $\mathbf{M}_{tb}$.

In many studies, it is reasonable to assume that individuals (or groups of individuals) have different propensities to be captured. It then follows that the aforementioned classification can be further enlarged to take such heterogeneity into account: $\mathbf{M}_h$ contains the class of models with propensity of capture varying over individuals but not with capture occasions, while $\mathbf{M}_{hb}$ and $\mathbf{M}_{htb}$ account for propensity to vary with, in order, first capture and pattern of captures, as well as over individuals. An account can be found in Amstrup et al. [7], Ch. 1–2.

In studies on human populations, captures (or lists) are not controlled by the researchers, but are reporting systems usually set up for different purposes. It is therefore reasonable to assume that lists interact, as an individual recorded in one list may be more (or less) likely to appear on a second one than an individual that has not been recorded on that list. When lists positively (negatively) interact, models that do not include relevant parameters tend to under(over)-estimate the population size. Furthermore, it is reasonable to assume that individuals have a different behaviour with respect to appearance on one or another list. Such heterogeneity may be explained by introducing covariates in the model. However, in some studies, covariates do not fully describe the heterogeneity, and some is left unexplained. Unexplained heterogeneity induces spurious associations between lists or modifies the existing ones in a way that may be difficult to disentangle.

Agresti [2] and Darroch et al. [98] investigate the use of log-linear models with latent variables to estimate the size of a closed population when unexplained heterogeneity is present. In particular, Darroch et al. [98] exploit the close relationship between Rasch-type models (see Rasch [242]) and quasi-symmetric log-linear models and propose the use of the latter. Coull and Agresti [88] discuss the relationship between Rasch and Latent Class models further and explore the behaviour of a range of log-linear/logit models. The issue of unobserved heterogeneity has been addressed also in a Bayesian context, see e.g., King et al. [166].

In this chapter, we investigate the use of graphical log-linear models that contain one unobserved categorical variable. In particular, we assume that all covariates, i.e. observed and unobserved ones, are categorical and explore the use of concentration graphical models for contingency tables, as in Stanghellini and van der Heijden [267]. The proposed class of models permits us to disentangle the effects due to unobserved heterogeneity from the genuine associations among lists, therefore leading to a better understanding of the data. By genuine association we mean an association induced by the data generating mechanism. As a matter of fact, in this context it is likely that patients appearing in one particular reporting system are given instructions that raise/lower their probability of appearing in a second one. The proposed models can also be seen as finite mixture models as opposed to continuous ones, see Dorazio and Royle [102] and Pledger [233], and are also related to the ones proposed by Bartolucci and Forcina [24], with a different interpretation of the parameters; see also Chapter 20 of this book.

When dealing with models with unobserved variables, attention should be given to the identifiability issue. In this chapter, we present a range of models for which identifiability has been well understood (see Allman et al. [5], Stanghellini and Vantaggi [268] and Allman et al. [6]). Given the description above, our models fall into the class denoted by $\mathbf{M}_{htb}$.

The chapter is organised as follows. In Section 19.2 we introduce concepts and notation while in Section 19.3 we briefly recall the notion of graphical models for categorical data. In Section 19.4 we describe the use of log-linear models in capture-recapture problems with observed categorical covariates, while in Section 19.5 we present in detail the class of models here proposed, that can be seen as an extension of Latent Class models. In Section 19.6 we review the notion of identification while in Section 19.7 we propose a method, based on the profile log-likelihood, to construct confidence intervals for the undercounts in each stratum formed by the covariates. In Section 19.8 the proposed methods are applied to real data, in particular in Section 19.8.1 we analyse the underreporting of infants born with congenital anomaly in Massachusetts while in Section 19.8.2 we study the problem of undercount in bacterial meningitis in the Lazio region. In Section 19.9 we give some recommendations and draw conclusions.

19.2 Notation

Since we deal with log-linear models and their extensions, we need to introduce a slightly different notation. We assume to have T lists and let S_t, $t = 1, \ldots, T$, be a binary random variable such that $s_t = 1$ when an individual is enumerated on list t, and $s_t = 0$ otherwise. Let $\mathbf{s} = (s_1, \ldots, s_T)$ denote a T-dimensional string associated with a given pattern of captures. As an example, for $T = 3$, $\mathbf{s} = (1, 0, 1)$ indicates the following pattern: "captured" on list 1, "not captured" on list 2, "captured" on list 3. We further denote with $\mathbf{0}$ the string $\mathbf{0} = (0, \ldots, 0)$. Notice that each $\mathbf{s}$ identifies a cell of a 2^T contingency table obtained by cross-classifying units according to S_t. By $Y_\mathbf{s}$ we denote the random variable that counts the number of units having $\mathbf{s}$ as a pattern of captures. By definition, $Y_\mathbf{0}$ is never observed.

We assume further to have a set of observed categorical covariates that accounts for heterogeneity among individuals. Since we are not interested in modeling their joint distribution, without loss of generality we combine them into a single covariate C, with levels (or strata) indexed by $j = 1, \ldots, J$. With this in mind, we form a $J \times 2^T$ contingency table obtained from the cross-classification of units according to $(C, S_1, \ldots, S_T)$ and denote with $Y^* = \{Y_{j\mathbf{s}}\}$ the $J2^T$-dimensional vector of the random variables of the entries of the contingency table. The vector is stacked by letting the variables vary in lexicographic order, with C running slowest. By definition $Y_{j\mathbf{0}}$ is never observed. We then indicate with Y the $J(2^T - 1)$-dimensional vector of the random variables obtained by removing the first entry of each stratum of Y^*. We will refer to Y^* as the vector of cell counts of the complete table and to Y as the vector of cell counts of the uncomplete table. Finally, let $n = \{n_j\}$ be the vector of the observed counts in stratum j and $N = \{N_j\}$ be the vector of the total counts in stratum j, for $j = 1, \ldots, J$, their difference being the undercounts in each stratum.

In capture-recapture contexts, especially in human populations, it is common that available covariates do not take into account all heterogeneity among individuals in terms of their catchability. We here assume that, other than the observed covariate, there is a categorical unobserved one, denoted by U, taking on H possible levels, indexed by $h = 1, \ldots, H$. The latent variable U accounts for unobserved heterogeneity, in such a way that individuals with the same level j of the covariate C and belonging to the same class h have equal probability of appearing in each of the lists, but this probability may vary with h. Then let $X^* = \{X_{hj\mathbf{s}}\}$ be the $HJ2^T$-dimensional vector of random variables of the entries of the contingency table obtained from the cross-classification of subjects according to all variables, with U running the slowest. Let $l = HJ(2^T - 1)$, and X be the l-dimensional vector of random variables of the entries of the contingency table obtained by removing all cells with pattern of capture $\mathbf{0}$. It then follows that $Y_{j\mathbf{s}} = \sum_{h=1}^{H} X_{hj\mathbf{s}}$ are the entries of the marginal contingency table of subjects classified according to the observable variables only.

In what follows, given three random variables A, B and K, we denote with $A \perp\!\!\!\perp B \mid K$ the notion that A is independent of B given K. With $\otimes$ we denote the Kronecker product,

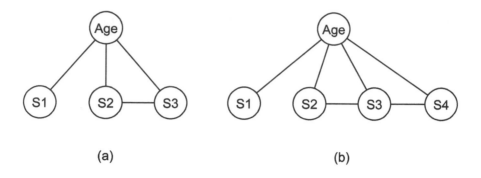

(a)   (b)

FIGURE 19.1: Concentration graphs corresponding to possible models with "Age" as a covariate and (a) $T = 3$ and (b) $T = 4$

while with $\mathbf{1}_r$ we denote the r-dimensional vector of ones and with I_r the identity matrix of order r.

19.3   Concentration graphical models

We assume that all variables, i.e. observed or unobserved covariates and lists, are categorical (continuous variables will be binned). Usually, lists interact but there is no sequential ordering among them. We therefore consider all variables on an equal footing and opt to use concentration graphical models on contingency tables obtained from the cross classification of subjects using a set of categorical random variables.

A graph $G = (V, E)$ is a mathematical object composed by the node set V and the edge set E as a subset of $(V \times V)$. Graphs can be undirected or directed. In an undirected graph, all edges are undirected, that is, if (i, j) is in E, then also (j, i) is in E, where $i, j \in V$. In a directed graph, all edges are directed, that is, if (i, j) is in the edge set E, then (j, i) is not in E. A graph can be visualised, with each vertex associated to one node. Arcs represent the edges in an undirected graph while arrows represent edges in a directed graph. We here consider undirected graphs only.

We define an (a, b)-path of length r as a sequence of $r + 1$ distinct nodes $a = a_0, a_1, \ldots, a_{r-1}, a_r = b$ such that (a_{i-1}, a_i), $i = 1, \ldots, r$, are in E. We say that K, a subset of $V \setminus \{a, b\}$, separates a and b if all (a, b)-paths intersect K. Given three disjoint subsets A, B and K of V, we say that K separates A from B if all (a, b)-paths intersect K for every node a in A and b in B. Notice that K may not be the minimum subset to separate A from B. With reference to the undirected graph in Figure 19.1(a), both Age and (Age, S_2) separate S_1 from S_3.

A subset $A \subseteq V$ of the vertex set V induces a subgraph $G_A = (A, E_A)$, where $E_A = E \cap (A \times A)$. An undirected (sub)graph is *complete* if every pair of distinct vertices is joined by an arc. A *clique* is a complete subgraph that is maximal, i.e. the inclusion of any other vertex in V destroys completeness. Notice that the list of cliques uniquely identifies a undirected graph. As an instance, the undirected graph in Figure 19.1(a) is uniquely identified by the cliques: (Age, S_1), (Age, S_2, S_3).

A concentration graph is an undirected graph such that each node represents a random variable. In Figures 19.1(a) and (b) two concentration graphs are depicted corresponding to possible models with an observed covariate C, say "Age", and, in order, 3 and 4 lists. Given a concentration graph $G = (V, E)$, a concentration graphical model is a family of joint distributions over the random variables in V that satisfy the *global Markov property* relative to G. A joint distribution satisfies the global Markov property relative to $G = (V, E)$ if, for any triple of disjoint subsets A, B and K of V such that K separates A from B in G, $A \perp\!\!\!\perp B \mid K$. In this case, we also say that the distribution *factorizes* according to G. It then follows that attached to any missing edge in the graph there is a list of conditional independence statements.

With reference to the undirected graph in Figure 19.1(a), $S_1 \perp\!\!\!\perp S_3 \mid$ Age but also $S_1 \perp\!\!\!\perp S_3 \mid (\text{Age}, S_2)$. However, often the interest in applied research is to find the minimum separating subset, i.e. the separating subset that cannot be reduced without destroying the conditional independence. With reference to the undirected graph in Figure 19.1(b), if the joint distribution factorizes accordingly, then we can say e.g., that $S_1 \perp\!\!\!\perp S_2 \mid$ Age, while $S_2 \perp\!\!\!\perp S_4 \mid (S_3, \text{Age})$. More details on concentration graphical models can be found in Lauritzen ([175], Ch. 3).

19.4 Capture-recapture estimation with graphical log-linear models with observed covariates

We here describe the log-linear approach to the analysis of incomplete contingency tables as generated by the capture-recapture context when all covariates are observed. We assume that the counts Y^* are independent Poisson random variables with expected value $E(Y^*) = \mu^*$, that is:

$$P(Y^*) = \prod_j \prod_s e^{-\mu_{js}^*} \frac{1}{y_{js}!} \mu_{js}^{* \; y_{js}}.$$

We also assume $\mu^* = \mu^*(\beta)$, i.e. the expected value is a function of a lower dimensional vector of parameters. The unknown parameters in this context are β and $N = \{N_j\}$, for $j = 1, \ldots, J$. An estimate can be derived by maximisation w.r.t. N and β of the complete data log-likelihood $\ell(y^* \mid M^*Y^* = N; \beta, N)$, with $M^* = I_J \otimes 1'_{2^T}$ where

$$P(Y^* \mid M^*Y^* = N) = \prod_j \binom{N_j}{n_j} [1 - p_j^*(\beta)]^{N_j - n_j} p_j^*(\beta)^{n_j} \; n_j! \prod_{s:s \neq 0} q_{js}(\beta)^{y_{js}} \frac{1}{y_{js}!}$$

in which $p_{js} = \frac{\mu_{js}}{\sum_s \mu_{js}}$, $p_j^* = \sum_{s:s \neq 0} p_{js}$ and $q_{js} = p_{js}/p_j^*$.

However, maximum likelihood estimates for β and N are usually derived in two steps, as follows. The factorisation above shows that the complete data log-likelihood can be written as the sum of the binary log-likelihood of a subject in stratum j to be captured and the multinomial log-likelihood, $\ell(y \mid MY = n; \beta)$, with $M = I_J \otimes 1'_{2^T - 1}$, that a subject exhibits a pattern s of captures, conditional on being captured. The maximum likelihood estimate of β is derived by maximising the conditional log-likelihood $\ell(y \mid MY = n; \beta)$; the vector of the conditional estimates of N_j is derived as the integer part of $n_j/\hat{p}_j^*$ (Bishop et al. [32], p. 237). This procedure naturally forms the basis of traditional log-linear modeling of capture-recapture experiments (see Cormack [86], Fienberg [121] and Agresti [2]); see Sanathanan [253] for the relationship between the estimates of N obtained by maximisation of the complete data log-likelihood and by the proposed procedure. In a context with no covariates, Fienberg [121] noticed that, in order for N to be identified, the log-linear model should not be saturated.

We here focus on concentration graphical models on $\ell(y \mid MY = n; \beta)$. Let $E(Y) = \mu$. Given a graph $G = (V, E)$, a concentration graphical model over the categorical random variables in V is a hierarchical linear model on $\log \mu$ with generating class the set of cliques of G. As an instance, the log-linear model with generating class: (Age, S_1),(Age, S_2, S_3) is the concentration graphical model over the counts of the contingency table obtained from the cross-classification of subjects according to the variables (Age, S_1, S_2, S_3) and such that the joint distribution factorises according to the graph in Figure 19.1(a). Given the interpretation in terms of conditional independence already presented in Section 19.3, this model describes a mechanism in which lists have different sampling effects and the sampling effect of all lists varies with Age. Moreover, lists S_2 and S_3 interact, and the interaction terms vary with the levels of Age.

Notice that the model for each clique is saturated. It then follows from standard results on maximum likelihood theory that observed and estimated counts of the corresponding marginal tables coincide. This implies that, in order for N to be identified, the graph G should not contain a clique involving all lists. This is the graphical counterpart to the condition of Fienberg [121] that the log-linear model should not be saturated.

Concentration graphical models for categorical variables are a subclass of hierarchical log-linear models, see Lauritzen [175], Ch. 4 and Edwards [108], Ch. 2. An example of a hierarchical log-linear model is the model with two-factor interaction terms only. The two-factor interaction model corresponding to Figure 19.1(a) has a generating class: (Age, S_1),(Age, S_2),(Age, S_3),(S_2, S_3). The joint distribution factorises according to G, so the interpretation in terms of conditional independence remains. However, the model describes a mechanism in which lists have different sampling effects and the sampling effect of all lists varies with Age. Moreover, lists S_2 and S_3 interact, but the interaction term is constant within the levels of Age. Hierarchical log-linear models have been used in capture-recapture data also in the Bayesian context; see Madigan and York [191]. See also Chapters 19 and 33.

19.5   Extended Latent Class models

We assume X to be a vector of independent Poisson random variables with $E(X) = m$. Let $L = \mathbf{1}'_H \otimes I_{J(2^T-1)}$. Then, the marginal entries $Y = LX$ are also Poisson random variables. Given a concentration graph $G = (V, E)$ over $V = (U, C, S_1, \ldots, S_T)$, a graphical log-linear model with the latent variable U is then defined on $\log m$. More precisely, we assume that $\log m = Z\beta$, where β is a p-dimensional vector of unknown parameters; Z is a $l \times p$ design matrix defined in a way that the joint distribution of $(U, C, S_1 \ldots, S_T)$ factorises according to G. We here adopt the corner point parametrisation; see e.g., Darroch and Speed [97]. In the following, we assume U to be a binary random variable, i.e. $H = 2$. In our experience, this class of models is large enough to explain heterogeneity in capture-recapture contexts, see also Pledger [232].

The proposed models extend the application of Latent Class techniques in capture-recapture contexts, see Coull and Agresti [88], to situations where (*i*) there are observed covariates and (*ii*) covariates and lists are not independent given U. For these reasons we refer to them as Extended Latent Class models. In Figure 19.2(a) a concentration graph representing a Latent Class model with $T = 3$ is shown. Figure 19.2(b) depicts the concentration graph corresponding to an Extended Latent Class model with four lists and no

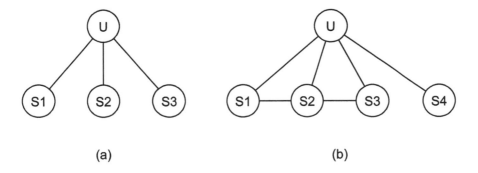

FIGURE 19.2: Concentration graphs corresponding to possible models with U as an unobserved covariate: (a) Latent Class model for $T = 3$ and (b) Extended Latent Class model for $T = 4$ with conditional associations among lists

covariates. The model allows for interaction between the lists S_1 and S_2 and between S_2 and S_3, also conditional on the latent variable U. It permits us to disentangle the effect of unobserved heterogeneity from the genuine association between lists. The first application of these models to capture-recapture data is in Biggeri et al. [31].

Maximum likelihood estimation of the Extended Latent Class model can be performed with the EM algorithm, see Dempster et al. [100]. The EM algorithm alternates the following E and M steps until convergence:

- *E-step*: the conditional expectation of the $E(X \mid LX = y)$ is computed given the current best parameter estimates of β.

- *M-step*: new parameter estimates are computed by fitting a log-linear model on X.

Due to the close similarity with the Latent Class model we refrain from describing the algorithm in detail but refer to Agresti and Lang [1].

As detailed in Section 19.6, not all log-linear models with one latent variable are locally identifiable and the identification state should then be checked. We here focus on models for which the identification state has been well understood.

19.6 Identification of Extended Latent Class models

We consider the class of concentration graphical models over the random variables: $U, C, S_1, \ldots, S_T$ and we restrict to models with U a binary random variable, i.e. $H = 2$. We denote with $\psi : \mu \to \beta$ the parametrisation map from the natural parameters, in this case the expected value of Y, to the new parameters β. Global identifiability, also known as strict identifiability, corresponds to ψ being one-to-one. Since we deal with log-linear models, the mapping ψ is polynomial. In this case, local identifiability corresponds to ψ being finite-to-one. As argued in Allman et al. [5], there may be models such that the parametrisation map is finite-to-one almost everywhere (i.e. everywhere except in a subset of null measure). In this case, we speak of generically (locally/globally) identifiable models.

When dealing with models with a categorical latent variable, attention should be paid to the fact that two models differing only in the ordering of the labels of the latent variable generate the same marginal distribution. This issue, known as "label swapping", implies that concentration graphical models with a binary unobserved variable are at most two-to-one. We consider an Extended Latent Class model identified when the mapping is two-to-one. In this case, the only source of unidentifiability is due to label swapping, a problem that is well understood and that can easily be addressed.

It is well known that for the binary Latent Class model, see Figure 19.2(a), the parametrisation map is two-to-one if and only if $T \geq 3$, see McHugh [204] and Goodman [132]. In capture-recapture contexts, however, since the information on Y_0 is missing, then, to avoid that the number of parameters is greater than the number of observed counts, it is necessary that $T \geq 4$. For Extended Latent Class models, more attention should be paid to assure that the models are identified.

By the inverse function theorem, the mapping ψ is finite-to-one if the rank of the transformation from the natural parameters μ to the new parameters β is full everywhere in the parameter space. This is equivalent to the rank of the following derivative matrix

$$D(\beta)' = \frac{\partial \mu'}{\partial \beta} = \frac{\partial (Le^{Z\beta})'}{\partial \beta} = (LRZ)' \qquad (19.1)$$

being full, where $R = \text{diag}(m)$. Note that the (i,j)-th element of $D(\beta)$ is the partial derivative of the i-th component of μ with respect to the j-th element of β.

Stanghellini and Vantaggi [268] give a graphical characterization of concentration graphical models with one binary latent variable such that the matrix $D(\beta)$ is full-rank everywhere in the parameter space, providing a sufficient condition. The condition requires several graphical notions, and it is therefore omitted. For models that violate the condition, they provide the expression of the subsets of null measure where the rank of the matrix $D(\beta)$ degenerates. The exact knowledge of the subset where identifiability breaks down is important, as standard statistical procedures may fail if the estimates of the parameters are close to the singular locus, see e.g., Drton [105].

In a different, though related, context, Allman et al. [6] focus on all possible models with a binary unobserved variable and four observed binary variables. They show that a locally identified model exists, that violates the condition of Stanghellini and Vantaggi [268] and such that the mapping is four-to-one. This implies that, further to the label-swapping issue, there are two different models, with unrelated parameters, that generate the same marginal distribution.

All these considerations lead us to consider the class of models for which the state of identification is well understood; see Stanghellini and Vantaggi [268] and Allman et al. [6]. In this case, standard model selection procedures based on likelihood ratio can be used, as the asymptotic properties are preserved. In Figure 19.2(b) the concentration graph corresponding to the most complex model with four observed lists and one binary latent variable that is identified is presented. Notice that the model allows for the identification of N, as there is no clique containing all lists.

19.7 Confidence intervals

Once a model is selected, an estimate of the undercount within each stratum provided by the covariate can be derived. Point estimates can be coupled by confidence intervals based on the profile log-likelihood method, to have an idea of all plausible values and understand how precise these estimates are.

We here use the procedure based on the unconditional multinomial profile log-likelihood presented in detail in Stanghellini and van der Heijden [267] that extends the one in Cormack [87] to the situation with observed covariates. Let $\hat{\beta}$ be the maximum likelihood estimate and $\hat{N}(\beta)$ denote the conditional estimates of N obtained as a function of $\hat{\beta}$. Let $v_r = Y_{r0}$ be the value of the first entry on stratum r and Y_r be the vector of the random variables Y augmented by Y_{r0}. With v we denote the vector obtained from n by substituting the r-th element with $n_r + v_r$. Let $\hat{\beta}_r$ be the maximum likelihood estimate of β obtained by maximisation of $\ell(y_r \mid M_r Y_r = v; \beta)$, where M_r is a matrix obtained from M after inserting a column in position $r(2^T - 1)$. The column to be inserted is the r-th column of I_J. Let $\hat{N}(\beta_r)$ be the vector with $n_r + v_r$ as the r-th element and $\hat{N}_j(\beta_r)$, as j-th element, $j \neq r$.

By an argument parallel to that of Cormack [87], Theorem 2, if $v_r = \hat{N}_r(\beta) - n_r$, then $\hat{\beta}_r = \hat{\beta}$; see Stanghellini and van der Heijden [267]. It then follows that the unconditional multinomial profile log-likelihood for v_r can be obtained as $P_M(v_r) = 2[\ell\{y^* \mid M^*Y^* = \hat{N}(\beta)\} - \ell\{y^* \mid M^*Y^* = \hat{N}(\beta_r)\}]$. A $(1-\alpha)\%$ confidence interval is given by $(n_{r1}; n_{r2})$ with $n_{r1}(n_{r2})$ being the largest (smallest) integer smaller (larger) than $\hat{n}$ such that $P_M(n_{r1}) \geq \chi^2_{1,\alpha}\{P_M(n_{r2}) \geq \chi^2_{1,\alpha}\}$ and $\chi^2_{1,\alpha}$ is the $100\alpha\%$ critical value of the χ^2 with one degree of freedom. Confidence intervals based on the profile log-likelihood are approximate, and their validity relies on asymptotic results. Corrections to improve the approximation are in Brazzale et al. [52], Ch. 2.

Details on the relationship between the unconditional Poisson profile log-likelihood and the one outlined here are in Stanghellini and van der Heijden [267]. For confidence intervals on other measures of interest in epidemiology, such as the total disease rate or the probability of ascertainment, see Farrington [118].

19.8 Example of models under unobserved heterogeneity

19.8.1 Congenital Anomaly data

We here reconsider the problem of estimating the total number of infants born with a specific congenital anomaly in Massachusetts between 1 January 1951 and 31 December 1955, and still alive on 31 December 1966. The data, first analysed by Wittes et al. [300], come from five distinct sources: obstetric records (S_1, 183 cases); other hospital records (S_2, 215 cases); a list maintained by the State Department of Public Health (S_3, 36 cases); a list maintained by the State Department of Mental Health (S_4, 263 cases) and records by special schools (S_5, 252 cases). In total, 537 infants have been ascertained. The data are in Chao et al. [79]. Preliminary analyses show a positive interaction between S_1 and S_3 and a strong negative interaction between S_4 and S_5. Fienberg [121] fits a hierarchical log-linear model with generating class: (S_1, S_2), (S_1, S_3), (S_3, S_4), (S_4, S_5), leading to an estimate of the undercount of 97.5 and, therefore, an estimated population size of 634 with $(616; 652)$ as the 95% confidence interval. All significant interaction terms in the log-linear model are positive, apart from the interaction between S_4 and S_5. Chao et al. [79] propose an estimator based on the sample coverage approach leading to an undercount of 125 and to a population total of 659 with $(606; 750)$ as the 95% confidence interval.

We here assume that there is a binary latent variable U that represents unobserved heterogeneity that may be linked to the severity of the diagnosis, and we fit an Extended Latent Class model that allows for interaction among lists also after conditioning on U. As a model selection strategy, we use a forward procedure that begins with the Latent Class model and adds at each step the most significant interaction term, paying attention that the

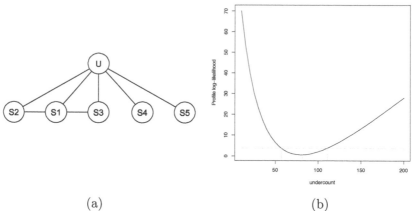

(a)   (b)

FIGURE 19.3: Infants born with Congenital Anomaly in Massachusetts: (a) concentration graph of the selected model and (b) profile log-likelihood of the undercount (point estimate 82), leading to a 95% confidence interval of $(57; 111)$

selected model satisfies the identifiability condition of Stanghellini and Vantaggi [268]. The procedure stops when there are no more significant interaction terms to add. The analyses are summarised in Table 19.1. The selected model corresponds to the concentration graph depicted in Figure 19.3(a). The estimate of the undercount is 82.3, leading to an estimated population total of 619. Notice that the unobserved heterogeneity accounts for the positive interaction between lists S_2 and S_4 and the negative one between lists S_4 and S_5. This second fact explains why in our case the estimated total is smaller than the one in Finnberg [101]. In Figure 19.3(b) the profile log-likelihood obtained for the proposed model is presented that leads to a 95% confidence interval of $(57; 111)$ for the undercounts and of $(594; 648)$ for the population total.

19.8.2   Bacterial Meningitis data

Bacterial meningitis is a serious infectious disease with a high fatality rate. Vaccine exists against some but not all aetiological agents, namely *Haemophilus influenza b*, *Streptococcus pneumonia* and *Neisseria meningitis*. Monitoring the incidence is therefore an important step to plan and evaluate preventive policies.

A surveillance system that integrates four data sources has been implemented in the Italian region of Lazio since 1999. We here report yearly data collected from the four sources between 2001 and 2005: Hospital Surveillance of Bacterial Meningitis (HSS, S_1, 355 cases), Mandatory Infectious Disease Notification (NDS, S_2, 644 cases), Laboratory Information System (LIS, S_3, 178 cases), and Hospital Information System (HIS, S_4, 826 cases). In total 944 cases were ascertained. Data are summarised in Table 19.2. Preliminary analyses show a strong positive association between S_1 and S_2, while S_3 and S_4 interact negatively. An extensive account of the data is in Giorgi Rossi et al. [133]. Many covariates are available. We merge here vaccine-preventable aetiological agents versus the other, to form a two-level covariate ("Aetiology", with levels in order "PMH" and "other"). We form a covariate C by combining the categories of "Year" and of "Aetiology".

As a preliminary analysis, we fit a hierarchical log-linear model on the observable data only. The selected model includes all two-factor interaction terms and a three-factor inter-

TABLE 19.1

Models investigated for Congenital Anomaly data; Model 12 selected

	Models	Deviance	dof	Undercount
1	$U \cdot (S_1 + S_2 + S_3 + S_4 + S_5)$	51.7	19	68
2	Model 1 $+U \cdot S_1 \cdot S_2$	39.0	17	76
3	Model 1 $+U \cdot S_1 \cdot S_3$	32.5	17	73
4	Model 1 $+U \cdot S_1 \cdot S_4$	50.7	17	55
5	Model 1 $+U \cdot S_1 \cdot S_5$	51.2	17	56
6	Model 1 $+U \cdot S_2 \cdot S_3$	47.2	17	68
7	Model 1 $+U \cdot S_2 \cdot S_4$	46.3	17	57
8	Model 1 $+U \cdot S_2 \cdot S_5$	48.7	17	58
9	Model 1 $+U \cdot S_3 \cdot S_4$	50.1	17	69
10	Model 1 $+U \cdot S_3 \cdot S_5$	51.3	17	66
11	Model 1 $+U \cdot S_4 \cdot S_5$	50.7	17	68
12	Model 3 $+U \cdot S_1 \cdot S_2$	19.4	15	82
13	Model 3 $+U \cdot S_1 \cdot S_4$	32.3	15	74
14	Model 3 $+U \cdot S_1 \cdot S_5$	32.2	15	81
15	Model 3 $+U \cdot S_2 \cdot S_3$	27.4	15	73
16	Model 3 $+U \cdot S_2 \cdot S_4$	28.1	15	66
17	Model 3 $+U \cdot S_2 \cdot S_5$	29.4	15	63
18	Model 3 $+U \cdot S_3 \cdot S_4$	32.4	15	74
19	Model 3 $|U \cdot S_3 \cdot S_5$	32.2	15	72
20	Model 3 $+U \cdot S_4 \cdot S_5$	26.7	15	75

action term between C, S_1 and S_2 (deviance 86.9 with 85 d.f.) and it leads us to an estimate of the total undercount of 534. See Table 19.3 for the estimated undercount in each level of the covariate, first row (HLLM). This model does not seem a convincing one, as from a priori knowledge, the estimated undercounts appear particularly large. Moreover, it does not allow for any conditional independence between the variables.

We believe that there is unexplained heterogeneity among individuals and we therefore fit an Extended Latent Class model, starting from a Latent Class model with C having an edge with all lists but not with U. At each step we add the most significant interaction term among the sources, paying attention that the model is identified according to the conditions discussed in Stanghellini and Vantaggi [268]. The list of models investigated is omitted for brevity. The modeling strategy leads us to select as a possible model the concentration graphical model corresponding to the graph in Figure 19.4 (deviance 91.0 with 84 d.f.). The two covariates C and U explain all associations between the lists, apart from the positive association between S_1 and S_2. The total estimated undercount is 183. The estimated undercounts in each level of the covariate are reported in Table 19.3, second row (ELCM).

In Figure 19.5 confidence intervals of the undercounts in each stratum based on the profile log-likelihood are reported. Notice that in the neighbourhood of the estimated undercount there are some terms taking negative values. This may occur when the estimated value is small. We point out that another well fitting model has been detected, which includes an edge between S_2 and S_4 instead of the one between S_1 and S_2. However, this model does not lead to acceptable confidence intervals for the undercounts and has not been considered further.

TABLE 19.2
Known cases of Bacterial Meningitis in Lazio, Italy, with covariates "Year" and "Aetiology"

HSS	S_1	0	0	0	0	0	0	0	0	1	1	1	1	1	1	1	1
NDS	S_2	0	0	0	0	1	1	1	1	0	0	0	0	1	1	1	1
LIS	S_3	0	0	1	1	0	0	1	1	0	0	1	1	0	0	1	1
HIS	S_4	0	1	0	1	0	1	0	1	0	1	0	1	0	1	0	1
2001	PMH	–	12	2	3	6	15	0	14	0	0	0	1	0	10	4	21
	Other	–	41	1	2	12	21	0	0	0	0	0	0	1	7	0	2
2002	PMH	–	11	3	0	2	19	0	4	1	3	0	3	0	20	1	18
	Other	–	37	0	1	8	19	0	3	3	0	1	0	2	13	0	3
2003	PMH	–	10	0	2	3	18	1	10	0	0	0	0	1	23	1	19
	Other	–	35	0	0	20	21	0	1	1	1	0	0	0	12	0	0
2004	PMH	–	10	2	2	0	15	0	3	0	0	0	0	0	31	0	17
	Other	–	55	0	0	5	33	0	0	0	2	0	0	2	14	0	0
2005	PMH	–	3	2	0	7	9	0	3	0	3	0	1	2	31	1	25
	Other	–	45	0	0	11	27	0	0	0	1	0	0	12	40	0	1

TABLE 19.3
Predicted undercounts of Bacterial Meningitis under the hierarchical log-linear model (HLLM) and the Extended Latent Class model (ELCM)

Year	2001		2002		2003		2004		2005	
Etiology	PMH	Other	PMH	Other	PMH	Other	PMH	Other	PMH	Other
HLLM	14	109	9	76	7	125	4	50	7	133
ELCM	6	37	4	88	3	40	1	10	3	45

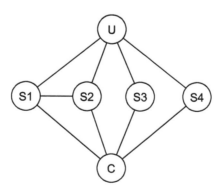

FIGURE 19.4: Concentration graph corresponding to the selected model for Bacterial Meningitis with C and U as observed and unobserved covariates

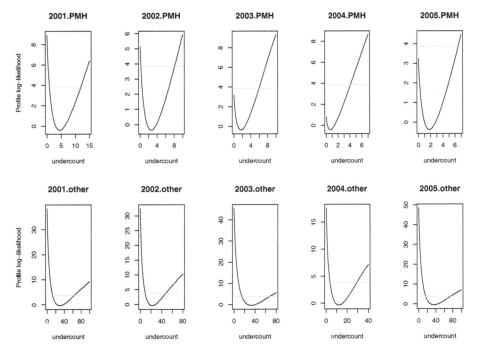

FIGURE 19.5: Profile log-likelihood and 95% confidence interval for the undercounts of Bacterial Meningitis in Lazio, Italy, for each level of the covariate

19.9 Discussion

As pointed out by many authors, see e.g., Coull and Agresti [88], the problem of estimating the size of a population is essentially a problem of forecasting. It consists in extrapolating, from the number of subjects with pattern of captures $s \neq 0$, the number of subjects never captured. It is therefore common that different models which, according to some measures of goodness of fit, well adapted to the observed data lead to rather different estimates of the undercounts. The choice therefore should be made on subject matter considerations.

In this chapter we focused on a class of models that can be applied when the observed covariates do not explain all heterogeneity of individuals in terms of their propensity to be captured, but some is left unexplained. In this case, we further assume that subjects can be grouped through a categorical latent variable and that the observed variables, i.e., the measured covariates and the lists, can also interact conditionally on the latent variable. The proposed models are concentration graphical models for categorical data with one node corresponding to an unobserved variable. They can also be seen as extensions of standard Latent Class techniques.

When dealing with models with latent variables, identifiability issues should be taken into account, as different models can generate the same marginal distribution over the observable variables. If we limit the analysis to models with only two Latent Classes, then results on identification exist and can be transferred to the capture-recapture context. The application of our methods to two data sets leads to models that adapt well to the data and give plausible estimates of the undercounts.

As a modeling strategy, we recommend to first select a good model for the observable data only, to find out which conditional associations are significant. If we believe that there is some unobserved heterogeneity that induces or confounds the associations, then the class of models here proposed may be of use. Point estimates should always be coupled with confidence intervals. The proposed strategy, based on the profile log-likelihood, provides a way to compute confidence intervals for the undercounts in each stratum. It also leads to a better understanding on how precise the estimates are.

20

Latent class: Rasch models and marginal extensions

Francesco Bartolucci

University of Perugia

Antonio Forcina

University of Perugia

CONTENTS

20.1 Introduction and background

The present chapter is focused on a class of models for the analysis of capture-recapture data that can be applied when the *full response configuration* for each unit captured at least once is available. Recall that, for unit i, with $i = 1, \ldots, N$, this configuration is denoted by $\mathbf{y}_i = (y_{i1}, \ldots, y_{iT})'$, where T is the number of capture occasions and y_{it} is a binary outcome which is equal to 1 if the unit is captured at occasion t and to 0 otherwise. As is already clear, this configuration is observed only if $y_i > 0$, where $y_i = \sum_{t=1}^{T} y_{it}$ is the number of captures or capture count.

Because y_i is a simple summary of $\mathbf{y}_i$, clearly the amount of available information is much richer when the full capture-recapture configuration $\mathbf{y}_i$ is available for each observed unit with respect to the case when only the counts y_i have been recorded. As a consequence, the models that will be discussed in this chapter are more sophisticated and, hopefully, should lead to a more precise inference on the population size N. However, in the wildlife context, in order to record the full capture history $\mathbf{y}_i$, at each occasion different markings should be used and this might be difficult to implement in practice.

When the full capture-recapture configuration is available, one can investigate whether:

(*i*) certain trapping occasions are more effective than others; (*ii*) being captured at a certain occasion may have an effect on the chance that the same unit is recaptured at a later occasion; (*iii*) individual heterogeneity is present. In addition, these features may be allowed to depend on observable covariates. Finite mixture models have also been used to account for random heterogeneity when only the overall counts are available; see Chapters 13, 14, 19, 21, or 23. However, the knowledge of capture-recapture configurations offers the possibility of modelling in a more sophisticated way the heterogeneity between units in connection with that between capture occasions and investigate in which measure it gives rise to a dependence between capture occasions.

One possible approach that can account for the aspects described above is based on adopting a Rasch model [241], combined with *latent class modelling* (Goodman [132], Lazarsfeld and Henry [174]) and a *marginal log-linear parametrisation* (Bergsma et al. [29]). This approach, on which the present chapter is focused, has been developed by the same authors in two earlier papers (Bartolucci and Forcina [24, 25]). In this chapter we provide an illustration of likelihood inference for these models and the population size on the basis of the Expectation-Maximisation (EM) algorithm (Dempster *et al.* [100]) and, in particular, we describe a specific method for constructing a confidence interval for the population size.

The chapter is organised as follows. The approach of main interest is described in detail in Section 20.2, which provides an accessible illustration of the models proposed in this chapter. Likelihood inference for such models is described in Section 20.3. The chapter ends with two applications presented in Section 20.4, which illustrate model selection and inference on the population size in two different contexts.

20.2   Latent class: Rasch models and their extensions

For a formal description of the models of interest, it is convenient to consider every outcome y_{it}, $i = 1, \ldots, N$, $t = 1, \ldots, T$, as a realisation of the binary random variable Y_t which is referred to as capture occasion t. We then introduce the corresponding random vector $\mathbf{Y} = (Y_1, \ldots, Y_T)'$.

In the following we first recall the basic latent class (LC) model as suggested in Goodman [132] and Lazarsfeld and Henry [174], and then we describe a finite mixture version of the Rasch model [241] which may be seen as a restricted version of the LC model. Both latent class and Rasch models, in their traditional formulation, rely on the assumption of conditional independence; marginal log-linear parametrisations (MLLP for short) (Bartolucci et al. [26], Bergsma et al. [29]) provide a convenient tool for allowing a limited number of violations of conditional independence and, at the same time, retaining the additive structure of the Rasch model.

20.2.1   The basic latent class model

To account for unobserved heterogeneity among units, the population of interest is conceived as the union of k unobservable sub-populations called *latent classes*. Units belonging to the same class are assumed to share the same distribution of the response variables; in addition, the assumption of *local independence* (LI) implies that the response variables, that is, the Y_t, are conditionally independent given the latent class. Within this framework, the Rasch model is simply a restricted LC model where, in addition, the difference between the logits of being captured at two different occasions does not depend on the latent trait (Lindsay et al. [179]).

Formally, we assume the existence, for each unit, of a latent variable U having a discrete distribution with k categories that, without loss of generality, may be coded with the integers from 1 to k; each class has mass probability

$$\pi_h = p(U = h), \quad h = 1, \ldots, k.$$

Next, let

$$\phi_{t|h} = p(Y_t = 1 | U = h)$$

denote the probability that a unit belonging to class h is captured at occasion t and define the conditional logits

$$\eta_{t|h} = \log \frac{\phi_{t|h}}{1 - \phi_{t|h}}.$$

Let $\phi_{\mathbf{y}|h} = p(\mathbf{Y} = \mathbf{y} \mid U = h)$ denote the conditional probability of capture configuration $\mathbf{y}$ for a unit that belongs to latent class h. The assumption of LI implies that

$$\phi_{\mathbf{y}|h} = \prod_{t=1}^{T} \phi_{t|h}^{y_t} (1 - \phi_{t|h})^{1-y_t}, \quad h = 1, \ldots, k.$$

This assumption relies on the idea that the dependence between the response variables is induced by variations of a latent variable affecting all responses simultaneously. Clearly this assumption, which will be partly relaxed later, does not imply marginal independence between the response variables.

Under the above assumptions, the *manifest distribution* for a given capture configuration has the following expression:

$$p_{\mathbf{y}} = \sum_{h=1}^{k} \phi_{\mathbf{y}|h} \pi_h; \tag{20.1}$$

this is the typical expression holding under a finite mixture model and is the key for maximum likelihood estimation.

A related distribution is the *posterior distribution*, that is, the probability that a unit with capture configuration $\mathbf{y}$ belongs to latent class h; it is based on the Bayes theorem:

$$\pi_{h|\mathbf{y}} = p(U = h | \mathbf{y}) = \frac{\phi_{\mathbf{y}|h} \pi_h}{\sum_{l=1}^{k} \phi_{\mathbf{y}|l} \pi_l}, \quad h = 1, \ldots, k.$$

This may be used to assign units to the different latent classes on the basis of their capture configuration, once a satisfactory model has been estimated. The possibility of clustering units into k homogeneous groups (latent classes) is a convenient feature of the proposed approach.

Overall, the LC model requires $k - 1$ parameters for the marginal distribution of the latent variables and $k\,T$ parameters for modelling the conditional distributions of $\mathbf{Y} \mid U$.

20.2.2 The Rasch model

This model, very popular since the sixties in psychology and education, was introduced by Rasch [241] to analyse data collected by questionnaires made of dichotomously scored items, with the objective to measure both the individual latent trait or ability and the relative difficulty of different items. As such, the model is considered one of the most important in the Item Response Theory (Hambleton and Swaminathan [138]). It has been applied to the analysis of capture-recapture data that, as is already clear, can be arranged into a table with units by row and capture occasions (items) by column. One of the first applications of the Rasch model to capture-recapture data is due to Agresti and Coull [2, 88], who used this model as a tool to account for the heterogeneity between population units in terms of their tendency to be captured, which is the latent trait of interest, and between capture occasions in terms of the effectiveness of lists. The resulting model is of type M_{th} according to the taxonomy introduced in Otis et al. [226]; see also Chapter 19 by Stanghellini and Ranalli.

A version of the Rasch model within a finite mixture framework was studied in [179]; in this formulation, the Rasch model may be seen as a restricted version of the LC model that assumes that the latent trait or ability is discrete with support points $\xi_1, \ldots, \xi_k$, while the corresponding mass probabilities remain unrestricted. The fact that the latent trait is only allowed to assume a small number k of different values may appear a restriction relative to the ordinary Rasch model where the latent ability is assumed to be continuous; however, the latter assumption is usually combined with the assumption that the latent trait distribution has a specific parametric form, usually the normal distribution.

The basic assumption of the Rasch model is that

$$\eta_{t|h} = \xi_h + \delta_t, \tag{20.2}$$

where δ_t may be interpreted as a measure of the effectiveness of trap (or list) t. This implies that the effectiveness of different lists is constant across latent types. Note that, in the original formulation of Rasch [241], the δ_t parameters are included in the model with a negative sign because they are interpreted as measures of the difficulty in answering items correctly.

20.2.3 Extensions based on marginal log-linear parametrisations

The assumption of LI, in addition to making the expression of the manifest distribution $p_{\mathbf{y}}$ much simpler, is also crucial because, otherwise, at least under the general LC approach, the unrestricted model would not be identifiable. There are, however, two typical contexts where LI may not be reasonable. One is in sociological and medical applications where, because of the way trapping occasions (lists) operate, units appearing in a given list may be more likely than the others to appear in a related list. The second is wildlife sampling where being trapped may change the behaviour of an animal that may then make efforts either to avoid being trapped again (trap shyness), or to search for a new opportunity of being trapped (trap happiness). These situations require a model denoted by M_{tbh} in the taxonomy introduced by Otis et al. [226].

MLLP, see for instance Bartolucci et al. [26] and Bergsma et al. [29], may be used to relax the assumption of LI in a flexible way and to formulate an extension of the Rasch model which combines assumption (20.2) on the univariate logits with a limited number of bivariate associations. For a practical illustration see Section 20.4, whereas for an alternative parametrisation in a similar context see Stanghellini and van der Heijden [267].

Let $\tau_{h,\mathbf{y}}$ be the probability $p(U = h, \mathbf{Y} = \mathbf{y})$ and $\boldsymbol{\tau}$ be the vector with elements equal to $\tau_{h,\mathbf{y}}$ arranged according to $h, \mathbf{y}$ in lexicographic order; this vector determines the joint distribution of the latent and the response variables. A MLLP is defined by a collection of marginal and conditional log-linear parameters which, in its most general form, may be written as

$$\boldsymbol{\eta} = \mathbf{C}\log(\mathbf{M}\boldsymbol{\tau}), \tag{20.3}$$

where $\mathbf{C}$ is a matrix of contrasts with elements summing to 0 by row and $\mathbf{M}$ is a matrix whose elements are equal to 0 or to 1, as explained below. In its unrestricted form, $\boldsymbol{\eta}$ would be a one-to-one mapping of $\boldsymbol{\tau}$ of dimension $k2^T - 1$; however, to make the model identifiable, a collection of highest-order interactions must be constrained to 0. In the following we assume, for simplicity, that the interactions defined on the full joint distribution which are constrained to 0 have been removed from $\boldsymbol{\eta}$. Additional linear restrictions may be defined by the linear model

$$\boldsymbol{\eta} = \mathbf{X}\boldsymbol{\beta}, \tag{20.4}$$

where $\mathbf{X}$ is a suitable design matrix and $\boldsymbol{\beta}$ is a vector of regression parameters. In particular, among the possible formulations of type (20.3), we consider three models of interest:

- *conditional independence* (LI): $\boldsymbol{\eta}$ contains the logits for the marginal distribution of the latent variable and, for each latent class, the conditional logits

$$\eta_{t|h} = \log p(Y_t = 1, U = h) - \log p(Y_t = 0, U = h);$$

- *selected bivariate* (SB) *associations*: in addition to the elements defined in LI above, $\boldsymbol{\eta}$ contains the log-odds ratios for a collection of specific pairs (s, t) of capture occasions; the conditional log-odds ratios are defined as

$$\eta_{s,t|h} = \log \frac{p(Y_s = 0, Y_t = 0, U = h)p(Y_s = 1, Y_t = 1, U = h)}{p(Y_s = 0, Y_t = 1, U = h)p(Y_s = 1, Y_t = 0, U = h)};$$

- *first-order Markov* (FM): for T at least equal to 2, $\boldsymbol{\eta}$ includes the conditional logits of $Y_t \mid (U - h, Y_{t-1} = y)$ as a measure of the autoregressive effect, that is,

$$\eta_{t|h,t-1}(y) = \log \frac{p(Y_t = 1|U = h, Y_{t-1} = y)}{p(Y_t = 0|U = h, Y_{t-1} = y)}, \quad t = 2, \ldots, T,$$

for $y = 0, 1$; in this way, $\boldsymbol{\eta}$ has dimension $(k - 1) + 2kT - k$.

Parametrisation SB may be suitable when capture occasions cannot be arranged in a specific order; this is typical of administrative data obtained by linking different lists which should be considered on the same footing unless there is a specific knowledge about how lists operate in practice; this may suggest, for instance, that appearing in a given list may change the probability of appearing in another list conditional on the latent variable. The FM parametrisation is appropriate when there is a temporal sequence of capture occasions based on the same system, as typical in wildlife experiments. A general algorithm for building the matrices $\mathbf{C}$ and $\mathbf{M}$ in the different contexts is described in the appendix.

Below we present a few examples to clarify the type of parametrisation proposed in this chapter. Consider, for simplicity, the case of $k = 2$ latent populations and $T = 2$ traps or lists; the probability vector in (20.3) has eight elements, that is,

$$\boldsymbol{\tau} = (\tau_{0,00} \quad \tau_{0,01} \quad \tau_{0,10} \quad \tau_{0,11} \quad \tau_{1,00} \quad \tau_{1,01} \quad \tau_{1,10} \quad \tau_{1,11})';$$

under LI, $\boldsymbol{\eta}$ has 5 elements corresponding to a matrix $\mathbf{C}$ which is block diagonal with 5 components of the form $(-1 \quad 1)$, and

$$\mathbf{M} = \begin{pmatrix} 1 & 1 & 1 & 1 & 0 & 0 & 0 & 0 \\ 0 & 0 & 0 & 0 & 1 & 1 & 1 & 1 \\ 1 & 1 & 0 & 0 & 0 & 0 & 0 & 0 \\ 0 & 0 & 1 & 1 & 0 & 0 & 0 & 0 \\ 0 & 0 & 0 & 0 & 1 & 1 & 0 & 0 \\ 0 & 0 & 0 & 0 & 0 & 0 & 1 & 1 \\ 1 & 0 & 1 & 0 & 0 & 0 & 0 & 0 \\ 0 & 1 & 0 & 1 & 0 & 0 & 0 & 0 \\ 0 & 0 & 0 & 0 & 1 & 0 & 1 & 0 \\ 0 & 0 & 0 & 0 & 0 & 1 & 0 & 1 \end{pmatrix}.$$

Since each element of $\mathbf{M}\boldsymbol{\tau}$ is the sum of the elements of $\boldsymbol{\tau}$ corresponding to the elements of $\mathbf{M}$ equal to 1, the rows of $\mathbf{M}$ required to define conditional interactions have entries equal to 1 only for the corresponding subset of cell probabilities; this is so because, in the computation of log-linear interactions, denominators cancel. In the matrix $\mathbf{M}$ above, the first two rows are used to define the logit of the latent variable, rows 3–6 define the logits for $Y_1 \mid U$, and rows 7–10 define the logits for $Y_2 \mid U$.

In the unrestricted LC model, $\mathbf{X}$ is simply an identity matrix of dimension 5; the Rasch model is obtained by setting

$$\mathbf{X} = \begin{pmatrix} 1 & 0 & 0 & 0 \\ 0 & 0 & 1 & 0 \\ 0 & 1 & 1 & 0 \\ 0 & 0 & 0 & 1 \\ 0 & 1 & 0 & 1 \end{pmatrix}.$$

In this framework, the first element of the parameter vector $\boldsymbol{\beta}$ in (20.4), denoted by β_1, corresponds to the marginal logit of the latent distribution, which would be positive when the second latent class has larger weight and negative otherwise; moreover, β_2 is positive if units in the second latent class are more likely to be captured by both lists, whereas β_3 and β_4 measure the effectiveness of lists 1 and 2, respectively. With $T = 2$ there is only one possible SB model which is equivalent to the FM model, though their parametrisations are different. The SB model may be defined by adding to $\mathbf{C}$ two more blocks of the form $\begin{pmatrix} 1 & -1 & -1 & 1 \end{pmatrix}$ for the log-odds ratio referred to $Y_1, Y_2 \mid U = h$, with $h = 1, 2$. The corresponding $\mathbf{M}$ matrix may be constructed by adding to the $\mathbf{M}$ matrix given above the rows of an identity matrix of size 8. If the resulting LC model is unrestricted, the design matrix $\mathbf{X}$ will be an identity matrix of size 7. From here, restrictions leading to the Rasch model consists of simply replacing the block corresponding to the univariate logits as above. In the LC model with bivariate associations, the assumption that the log-odds ratios of Y_1, Y_2 do not depend on the latent variable may be implemented by an $\mathbf{X}$ matrix where the two corresponding columns are replaced with their sum.

In the FM formulation we have 7 logits: one for the latent weight, two for $Y_1 \mid U = h$, and four for $Y_2 \mid U = h, Y_1 = y$. In practice, $\mathbf{C}$ has 7 blocks of the form $\begin{pmatrix} -1 & 1 \end{pmatrix}$. The $\mathbf{M}$ matrix is obtained by stacking the first six rows of the version displayed above with an identity matrix of size 8.

20.2.4   Modelling the effect of covariates

When covariates are available for the units captured at least once, these may be incorporated into the model in a suitable way. In principle, any of the parameters in the $\boldsymbol{\eta}$ vector defined above could be allowed to depend on individual covariates. For simplicity, in the following we will consider only models where the logits of the latent weights and the conditional logits of the responses given the latent variable are linear functions of the covariates.

In wildlife studies, when units are, for instance, insects captured by traps, it is unlikely that individual covariates are recorded. On the other hand, in medical or sociological applications, when an individual appears in a list, characteristics like sex, age, place of birth and others will usually be recorded and be available in the analysis.

When covariates that may affect the joint distribution of latent and response variables are available, it is convenient to collect all units with the same covariate configuration in one stratum that might also contain a single unit. Let J denote the number of such strata and write $\boldsymbol{\eta}_j$ to denote the vector of parameters whose values are now specific to the jth stratum; we may set up the linear logistic model

$$\boldsymbol{\eta}_j = \mathbf{X}\boldsymbol{\beta} + \mathbf{Z}_j\boldsymbol{\gamma}, \quad j = 1, \ldots, J,$$

where $\mathbf{X}$ is defined as in (20.4) and is used to formulate restricted models, whereas $\mathbf{Z}_j$ is a design matrix depending on the covariates. In this case, the vector $\boldsymbol{\eta}_j$ is a function of type (20.3) of the stratum-specific probability vector $\boldsymbol{\tau}_j$ having elements arranged as clarified above.

Usually $\mathbf{Z}_j$ will be block diagonal with a block for each set of logits. A covariate should be included in the block corresponding to the logits of the latent variable if we expect that the probability of belonging to one latent class rather than another depends on the value of that covariate. In traditional LC models, one would usually expect that the probability of being captured at a given occasion is uniquely determined by the latent class to which the unit belongs. In certain contexts, however, covariates may also be used to model the logits of the responses, conditional on the latent variable; for instance, being an immigrant or suffering from a certain disease may directly affect the probability of appearing in a given list.

20.3 Likelihood inference

When covariates are not available, so that units are exchangeable, let $c_{\mathbf{y}}$ denote the number of units with capture configuration $\mathbf{y}$ and $\mathbf{c}$ the vector with elements $c_{\mathbf{y}}$ arranged in lexicographic order. These data can be seen as the frequencies of a contingency table with the cell corresponding to $\mathbf{y} = \mathbf{0}$ missing; the corresponding probability distribution is described by the vector $\mathbf{p}$ of the *manifest distribution*, which is obtained by summing the elements of $\boldsymbol{\tau}$ as in (20.1). We refer to the probability of the missing cell as $r = p_{\mathbf{0}}$ and $N - n$ is the unknown frequency of units never captured. Moreover, $\dot{\mathbf{c}}$ will denote the vector $(N - n \quad \mathbf{c}')'$, that is, the vector containing the frequencies of the contingency table where the missing cell has been filled by assigning a specific value to the unknown population size N. We also write $\dot{\mathbf{p}}$ to denote the vector containing the corresponding elements of the joint distribution.

When covariates are available, let $\mathbf{c}_j$, $j = 1, \ldots J$, denote the vector containing the frequency table for the jth stratum and $\mathbf{p}_j$ denote the corresponding manifest distribution which depends on the vector $\boldsymbol{\tau}_j$ parametrised through $\boldsymbol{\eta}_j$. Though with continuous covariates each unit will have its own distribution, in the underlying latent distribution, most parameters will be common.

In the following, we illustrate maximum likelihood estimation of the model parameters and the population size N and then we describe a procedure to obtain a confidence interval for N.

20.3.1 Estimation of the model parameters

Following Sanathanan [253], we estimate the model parameters by maximising the conditional likelihood of the observed data given the overall number of captures or, when covariates are available, given the number of captures for each stratum. The log-likelihood function may be written as

$$\ell(\boldsymbol{\beta}, \boldsymbol{\gamma}) = \sum_{j=1}^{J} \mathbf{c}_j' \log \mathbf{q}_j,$$

where the vector $\mathbf{q}_j$ is obtained by dividing each element of $\mathbf{p}_j$ by $1 - r_j$, with r_j being the probability that a unit belonging to stratum j is never captured; in this way the elements of $\mathbf{q}_j$ sum to 1. Note that the above expression for $\ell(\boldsymbol{\beta}, \boldsymbol{\gamma})$ refers to the general case of stratified data; obviously, when covariates are not available, so that $J = 1$, the sum and the parameter vector γ disappear.

In order to maximise $\ell(\beta, \gamma)$, we use the EM algorithm (Dempster et al. [100]), which first reconstructs the frequencies of the latent tables and then maximises the associated complete likelihood. The frequencies of these latent tables are collected in the stratum-specific vectors d_j with elements organised as in the probability vectors τ_j. The corresponding *complete data likelihood function* has logarithm

$$\ell^*(\beta, \gamma) = \sum_j \mathbf{d}_j' \log \tau_j. \tag{20.5}$$

To maximise the target function $\ell(\beta, \gamma)$, the EM algorithm alternates the following two steps until convergence:

- *E-step*: The conditional expected value of the elements of the vectors $\mathbf{d}_j$ given the sample size n_j is computed on the basis of the current value of the parameters; the resulting vector, denoted by $\hat{\mathbf{d}}_j$, once it is substituted in (20.5) for $j = 1, \ldots, J$, gives the conditional expected value of the complete data log-likelihood, denoted by $\hat{\ell}^*(\beta, \gamma)$;

- *M-step*: Consists of maximising $\hat{\ell}^*(\beta, \gamma)$ so as to update the parameter vectors β and γ.

The two steps may be easily implemented; in particular, the model assumptions imply that $\hat{\mathbf{d}}_j = \text{diag}(\mathbf{1}_k \otimes \hat{\mathbf{f}}_j)\tau_j$, where $\hat{\mathbf{f}}_j = \text{diag}(\hat{\mathbf{p}}_j)^{-1}\hat{\mathbf{c}}_j$ with the missing cell in $\hat{\mathbf{c}}_j$ filled with $n_j/(1-\hat{r}_j)-n_j$ for the hypothetical value of the number of units in stratum j that were never captured. The M-step is based on a Fisher-scoring algorithm that may be implemented as described in Bartolucci and Forcina [25].

20.3.2   Estimation of the population size

The population size within each stratum j may be estimated as

$$\hat{N}_j = \frac{n_j}{1 - \hat{r}_j}$$

and, obviously, the estimate of the overall population size N is $\hat{N} = \sum_j \hat{N}_j$.

A confidence interval for the size N of the population may be based on the statistic $G^2(N)$ described by Bartolucci and Forcina [25] who showed that, if N_0 was the true value of N, then $G^2(N_0)$ is distributed as a $\chi^2(1)$. The procedure at issue may be implemented as follows to obtain a confidence interval of size $1 - \alpha$ for N: (i) compute $G^2(N)$ for a grid of values so that it may plotted with a reasonable accuracy; (ii) determine, by a suitable approximation, the set of values N for which the function does not exceed the quantile $\chi^2_{\alpha/2}(1)$.

Suppose that a satisfactory model has been estimated by conditional inference as described in the previous section, with $\hat{\mathbf{p}}_j$ denoting the estimated joint distribution within the jth stratum. Let $\dot{\mathbf{c}}_j$ denote the vector of frequencies where c_{j0} is set equal to $\hat{N}_j - n_j$ and

$$D^2 = \sum_j \dot{\mathbf{c}}_j' \left[\log\left(\dot{\mathbf{c}}_j/\hat{N}_j\right) - \log \hat{\mathbf{p}}_j \right].$$

Consider a hypothetical value N for the population size that we partition somehow, across strata, into $N_j \geq n_j$, $j = 1, \ldots J$, with $\sum N_j = N$. In this way, for each stratum we obtain the frequency vector $\dot{\mathbf{c}}_j(N_j)$ where the first cell is set equal to $N_j - n_j$. We may fit the same model selected before to these "artificial" data by an EM algorithm similar to the one described in Section 20.3.1, except that now we treat the first cell as if it were observed. Let $\hat{\mathbf{p}}_j(N_j)$ be the vector containing the estimated joint distribution for the jth stratum when the N_j are given and define

$$G^2(N) = D^2(N) - D^2,$$

where

$$D^2(N) = \min_{\sum N_j = N} \sum_{j=1}^{J} \dot{\mathbf{c}}_j(N_j)' \left[\log \left(\dot{\mathbf{c}}_j / N_j \right) - \log \hat{\mathbf{p}}_j(N_j) \right].$$

It can be easily shown that $G^2(N) \geq G^2(\hat{N}) = 0$ and the function increases with $\mid N - \hat{N} \mid$.

Let $r_j(N_j)$ denote the estimate of r_j when the size of the population in each stratum is assumed as known; $D^2(N)$, the deviance for the complete data when N is fixed, may be computed by an algorithm which alternates, until convergence, the following two steps:

1. with the $\hat{\mathbf{p}}_j(N_j)$ held fixed, minimise the kernel of the deviance for the complete data, that is,

$$
\begin{aligned}
d(N_1,\ldots,N_J) &= \sum_{j=1}^{J}(N_j - n_j)[\log(N_j - n_j) - \log N_j - \log r_j] \\
&\quad - \sum_{j=1}^{J} n_j \log N_j,
\end{aligned}
$$

with respect to its arguments under the constraints $\sum_j N_j = N$ and $N_j \geq n_j$, $j = 1,\ldots,J$; this task may be performed by minimising repeatedly a quadratic approximation of $d(N_1,\ldots,N_J)$ under the constraints at issue;

2. with the partition of N into $N_1,\ldots N_J$ held fixed, use the EM algorithm to update the estimate of $\hat{\mathbf{p}}_j(N_j)$ and then of $r_j(N_j)$ and go back to Step 1 until convergence.

Step 1 partitions a given population size among strata in an optimal way, thus, when there is a single stratum (covariates are not available), Step 1 is not required; in any case, even with a large number of strata, the time required to perform Step 1 is rather small relative to the time absorbed by the EM algorithm in Step 2.

20.4 Applications

20.4.1 Great Copper Butterfly

The data were collected by capturing, marking and releasing butterflies on 8 different days; see Ramsey and Severns [239] for more details about these data and the way they were collected. In this application the sample size is small relative to the number of occasions: only 45 different butterflies were captured at least once. We based model selection on the Bayes Information Criterion (BIC) (Schwarz [256]) to determine the appropriate number of latent classes, which turns out to be $k = 2$. We recall that this criterion leads us to select the model that is the best compromise between fit to the data and complexity measured in terms of the number of free parameters. The list of models considered, together with degrees of freedom, deviances, and BICs, are given in Table 20.1, where LC(k) stands for the LC model with k classes and R(k) for the Rasch model with the same number of classes.

Two of the models that fit best in comparison to their complexity are R(2)L+FM(1)c and R(2)P+FM(1)c. The first corresponds to a Rasch model with two latent classes and a linear effect of the time of capture, which is obtained by letting $\delta_t = \psi_0 + (t-1)\psi_1$ in equation (20.2); the model also includes an autoregressive effect of type FM, see Section 20.2.3, which is assumed constant over latent classes and trapping occasions. Model R(2)P+FM(1)c is a constrained version of the previous one with $\psi_0 = 0$, so that the probability of being captured in the first occasion for the units in the first latent class is equal to 0.5 and the logit of this probability is roughly proportional to the time of capture.

TABLE 20.1

Model selection for the Butterfly data: number of degrees of freedom, deviance with respect to the saturated model, BIC, and estimate of the population size

Model	df	Deviance	BIC	$\hat{N}$
LC(2)	17	51.07	370.38	108
LC(3)	26	36.74	390.31	87
LC(2)+FM(1)	31	35.55	408.15	103
R(2)+FM(1)	17	47.77	367.08	121
R(2)+FM(1)c	11	52.08	348.55	120
R(2)L+FM(1)c	5	54.14	327.76	122
R(2)P+FM(1)c	4	54.14	323.96	121
R(2)	10	56.69	349.36	90

For the selected model R(2)P+FM(1)c, the parameter estimates together with the standard errors are reported in Table 20.2. According to these estimates, butterflies belonging to the second latent class are about 97% of the total and are more than 13.5 times less likely to be captured. The estimate of the autoregressive parameter equals 1.16 with a standard error of 0.45 indicating some kind of "trap happiness".

TABLE 20.2

Parameter estimates for the Butterfly data together with the standard errors

Effect	Estimate	s.e.
Logit of second latent weight	3.492	0.753
Effect of second latent class on conditional logits	−2.608	0.389
Linear trend on capture occasions	−0.083	0.058
Autoregressive effect	1.150	0.453

The point estimate of the population size based on the final model is 121 with a confidence interval from 75 to 246, based on the plot in Figure 20.1. Note that due to the relatively small number of captured individuals, the confidence interval is quite large.

The same data used for this application were analysed by Fegatelli and Tardella [119] using a variety of models that account for serial dependence through the inclusion of different types of specifications for the autoregressive structure. The overall conclusion about the type of serial dependence is in agreement with our conclusion: there is evidence of trap happiness, though they concluded that this effect tends to decrease with the number of captures. In order to account for the estimates provided by different models, in Fegatelli and Tardella [119] an averaged estimate over different models is provided for the population size: they obtained $\hat{N} = 90$ with a confidence interval from 45 to 380 that is even wider

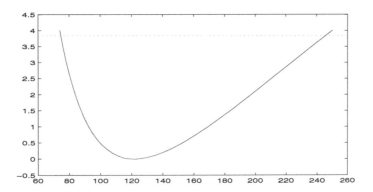

FIGURE 20.1: Plot of the $G^2(N)$ function for the Butterfly data; plausible values of the population size on the x-axis.

than the confidence interval that we obtained by our approach, despite the fact that the point estimate is smaller.

20.4.2 Bacterial meningitis

The data were provided by the research unit on infectious diseases at the local government of Lazio, a region with about 5.3 million inhabitants which includes the metropolitan area of Rome. Data were collected by four different sources: hospital surveillance of bacterial meningitis (HSS), the mandatory infectious diseases notifications (NDS), the laboratory information system (LIS), and the hospital information system (HIS). For the period 2001–2005, the records appearing in the four lists were combined into a single archive. For a detailed description of the context, see Giorgi Rossi et al. [133]; see also Section 19.8.2 of Chapter 19 by Ranalli and Stanghellini where the same data are analysed by means of models that are different but related to those used here.

The linked dataset has 944 records; in addition to the capture data, some individual covariates are available. In this application we restrict attention to the following ones which were considered to be the most relevant:

- *Age*: binary variable equal to 1 for up to 1 year old and 0 otherwise, to take into account the fact that the incidence of meningitis is much higher among very young children;

- *Aez*: binary variable for the recorded type of bacteria, which is equal to 1 for pneumococcus, meningococcus, or tuberculosis and 0 otherwise;

- *Year*: year of first appearance in a list; this is included because the functioning of certain lists has evolved during the study period.

Note that, with respect to the analysis by Stanghellini and Ranalli, we use age as an additional covariate.

We fitted several LC models with the marginal distribution of the latent variable depending on the Age and Aez covariates and the logits of the conditional probability of appearing on a list depending linearly on Year. Given the small number of available lists, we directly used $k = 2$ latent classes to avoid models that are weakly identifiable and that may lead to unreliable estimates of the populations size. With this number of classes we

TABLE 20.3

Model selection for the Meningitis data: number of degrees of freedom, deviance, BIC, and estimate of the population size

Model	df	Deviance	BIC	$\hat{N}$
LC(2)	19	426.63	3509.03	1077
LC(2)+(1,2)	20	354.51	3443.76	1041
LC(2)+(1,2)+(1,3)	21	350.87	3446.97	1042
LC(2)+(1,2)+(1,4)	21	343.90	3440.00	1069
LC(2)+(1,2)+(2,3)	21	353.32	3449.42	1043
LC(2)+(1,2)+(2,4)	21	338.76	3434.86	1386
LC(2)+(1,2)+(3,4)	21	349.09	3445.59	1046

considered different model specifications that are listed in Table 20.3, where each pair of type (s,t) refers to the SB association between lists s and t.

On the basis of prior knowledge and informal model fitting, we allowed the log-odds ratio between the first two lists to be nonzero but constant with respect to latent classes; evidence for this type of association is also provided by the data. Moreover, we included another association, between lists 2 and 4, that leads to a further reduction of BIC. In fact, the selected model, denoted by LC(2)+(1,2)+(2,4), has the smallest BIC among those considered in Table 20.3. Parameter estimates obtained under this model are reported in Table 20.4, where α_2 denotes the intercept of the logit model for the latent weights, possibly modified by the Age and/or Aez parameter; Year(g, h) is the effect of time on the g-th list conditional on latent class h; the other parameters are defined as in Section 20.2.3. These results correspond to an estimate of the population size equal to 1386 with a confidence interval from to 1142 to 1875; see Figure 20.2.

TABLE 20.4

Parameter estimates for the Meningitis data together with the standard errors

	Intercepts			Regression coefficients		
Parameter	Estimate	s.e.	Covariate	Estimate	s.e.
α_2	2.221	0.433	Age	1.194	0.428
$\eta_{1	1}$	0.746	0.157	Aez	-3.877	0.343
$\eta_{1	2}$	-2.364	0.244	Year(1,1)	0.335	0.100
$\eta_{2	1}$	1.924	0.225	Year(1,2)	0.572	0.119
$\eta_{2	2}$	-0.787	0.243	Year(2,1)	0.227	0.144
$\eta_{3	1}$	-0.107	0.140	Year(2,2)	0.285	0.091
$\eta_{3	2}$	-7.806	5.629	Year(3,1)	-0.165	0.085
$\eta_{4	1}$	2.461	0.256	Year(3,2)	-1.418	2.908
$\eta_{4	2}$	-0.085	0.309	Year(4,1)	0.148	0.174
$\eta_{1,2|1} = \eta_{1,2|2}$	2.858	0.299	Year(4,2)	0.252	0.103
$\eta_{2,4|1} = \eta_{2,4|2}$	1.571	0.354			

To better interpret the parameter estimates, it is convenient to look at how the estimates of the underlying probabilities actually change with covariates, as displayed in Table 20.5. The conditional probabilities in the lower part of the table indicate that subjects in latent class 1 are the most easily captured by all lists, though HIS and NDS are, by far, the most effective. Capture probabilities for subjects in latent class 2 are much lower; again, HIS and NDS are the most effective lists and their capability to capture individuals improves from the first to the last year of the study. The LIS list is among the least effective and its

TABLE 20.5

Latent weights depending on Age and Aez and conditional probabilities of being captured by different lists with respect to year

	Marginal weights			
	Age > 1		Age ≤ 1	
Latent class	Aez=0	Aez=1	Aez=0	Aez=1
1	0.098	0.840	0.032	0.613
2	0.902	0.160	0.968	0.387
	Conditional probabilities			
	Latent class 1		Latent class 2	
List	2001	2005	2001	2005
HSS	0.519	0.805	0.029	0.220
NDS	0.814	0.915	0.205	0.446
LIS	0.556	0.393	0.007	0.000
HIS	0.890	0.940	0.357	0.603

effectiveness even worsens across time. The upper part of the same table indicates that the kind of bacteria (Aez) is the most important factor concerning detection: the probability of belonging to latent class 1 is very high for subjects whose disease was caused by the most well-known kind of bacteria. We also note that subjects up to 1 year old are more likely to belong to latent class 1, that is, they are more often detected.

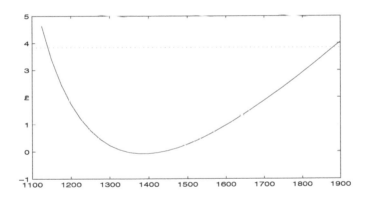

FIGURE 20.2: Plot of the $G^2(N)$ function for the meningitis data; plausible values of the population size on the x-axis.

20.5 Appendix: Matrices used in the marginal parametrisation

Below we give the procedure to construct the matrices $\mathbf{C}$ and $\mathbf{M}$ in (20.3). The first matrix is block diagonal; under LI, $\mathbf{C}$ has $(k-1) + kT$ blocks equal to $(-1 \quad 1)$. For each pair of responses associated conditionally on the latent variable, we need k additional blocks of the form $(1 \quad -1 \quad -1 \quad 1)$. Under an FM model, $\mathbf{C}$ has $(k-1) + k + 2k(T-1)$ blocks of type $(-1 \quad 1)$.

For each block in $\mathbf{C}$, there is a corresponding block of rows in $\mathbf{M}$ with each new block stacked below the others. Let $\mathbf{M}_h$, $\mathbf{M}_{h,t}$, $\mathbf{M}_{h,s,t}$ and $\mathbf{M}_{h,t|t-1,y}$ denote, respectively, the blocks of rows needed to define a marginal logit for the latent variable, the logits for $Y_t \mid U = h$, the log-odds ratios for Y_s, $Y_t \mid U = h$, and the logits for $Y_t \mid U = h, Y_{t-1} = y$. Let $\mathbf{e}_h$ also be the h-th row of an identity matrix of appropriate size and $\mathbf{E}_h$ the matrix obtained by stacking $\mathbf{e}_h$ and $\mathbf{e}_{h+1}$ one below the other; the blocks of $\mathbf{M}$ may be constructed by multiple Kronecker products as follows:

$$\mathbf{M}_h = \mathbf{E}_h \otimes \bigotimes_{j=1}^{T} (1 \quad 1), \qquad \mathbf{M}_{h,t} = \mathbf{e}_h \otimes \bigotimes_{j=1}^{T} \mathbf{A}_{j,t},$$

$$\mathbf{M}_{h,s,t} = \mathbf{e}_h \otimes \bigotimes_{j=1}^{T} \mathbf{A}_{j,s,t}, \qquad \mathbf{M}_{h,t|t-1,y} = \mathbf{e}_h \otimes \bigotimes_{j=2}^{T} \mathbf{A}_{j,t}^{(y)};$$

let $\mathbf{I}_2$ denote the identity matrix of size 2, then

$$\mathbf{A}_{j,t} = \begin{cases} \mathbf{I}_2 & \text{if } j = t, \\ (1 \quad 1) & \text{otherwise}, \end{cases} \qquad \mathbf{A}_{j,s,t} = \begin{cases} \mathbf{I}_2 & \text{if } j = s, t, \\ (1 \quad 1) & \text{otherwise}, \end{cases}$$

and $\mathbf{A}_{j,t}^{(y)}$ is equal to $(1 \quad 1)$ if $j \neq t-1, t$, otherwise to $(1 \quad 0) \otimes \mathbf{I}_2$ if $y = 0$, and to $(0 \quad 1) \otimes \mathbf{I}_2$ if $y = 1$.

21

Performance of hierarchical log-linear models for a heterogeneous population with three lists

Zhiyuan Ma

School of Public Economics and Administration, Shanghai University of Finance and Economics

Chang Xuan Mao

School of Statistics and Management, Shanghai University of Finance and Economics

Yitong Yang

School of Statistics and Management, Shanghai University of Finance and Economics

CONTENTS

21.1 Introduction

In epidemiological studies, one is constantly interested in estimating the sizes of diseased populations, such as diabetes patients and cancer patients, for the purpose of assessing the completeness of their registries. There are also disease populations whose individuals are difficult to access, such as infectious drug users and AIDS patients. For these elusive populations, their sizes are of importance by themselves. Capture-recapture techniques are frequently employed when multiple incomplete lists of individuals are available (Chao et al. [79], Hook and Regal [148],WHO [296], IWGDMFa,b [154, 155]). The most popular method for this kind of multi-list data in the literature is the family of hierarchical log-linear models introduced by Fienberg [121].

In particular, Tsay and Chao [277] studied an outbreak of hepatitis A virus (HAV) infection among students in a Taiwan college, with 271 distinct infected students identified. There were three lists from the Institute of Preventive Medicine (P-list), the National Quarantine Service (Q-list), and epidemiologists' questionnaires (E-list). The observed counts are in Table 21.1; for instance, there were 28 infected students shown in all three lists. The population size is finally found to be 545 by a screen serum test applied to all students of the college (Chao et al. [79]). From Chao et al. [79], the saturated log-linear model produces an estimate of 1313 with standard error 522 and 95% confidence interval $[683, 2904]$. The estimates from the other seven hierarchical log-linear models are smaller than 545. The large discrepancy between the estimate 1313 and the true value 545 puts some doubt on the validity of the saturated model. The dramatically different estimates from these models also motivates one to find some explanations.

Fienberg [121] applied hierarchical log-linear models to homogeneous populations. Both the vast differences among estimates and the large difference between the estimate from the saturated log-linear model and the true population size are possibly caused by the heterogeneity among individuals. For heterogeneous populations, one has to deal with the issue of singularity (Mao and Lindsay [195]) or non-identifiability (Holzmann, Munk and Zucchini [146], Link [180], Mao [194]). There is no consistent estimator available for the size of a heterogeneous population, and it is impossible to bound the population size from above (Bunge and Fitzpatrick [59], Mao and Lindsay [195]). A feasible and necessary strategy is to use some lower bound estimators and lower confidence limits (Mao [192, 193], Mao and Lindsay [195], Mao and You [199]). As a matter of fact, two lower bound estimators (Chao [71, 73]) have gained a sound reputation among practitioners (Béguinot [27], Mao, Yang and Zhong [196]).

Because there is a large percentage of epidemiological applications that involve three lists, we will investigate all eight hierarchical log-linear models one by one. To introduce the possible heterogeneity, we assume that data are generated from a nonparametric Rasch mixture model (Agresti [2], Coull and Agresti [88], Darroch et al. [98], Dobra and Fienberg [101], Fienberg, Johnson and Junker [122], Lum, Price and Banks [188], Ma, Mao and Yang [189]). See also Chapters 1, 19, 20, and 22. We will show that these hierarchical log-linear models, excluding the saturated one, produce lower bound estimators for the population size. In particular, three log-linear models with two two-way interactions produce estimators for the sharpest lower bound of the population size.

The chapter is organized as follows. The methods are presented in Section 21.2 and Section 21.3. A simulation study is reported in Section 21.4. We return to Example HAV in Section 21.5. The proofs are provided in the appendix.

21.2   Hierarchical log-linear models

A population of size s is investigated by a three-list surveillance. Let $x_{ij} = 1$ if individual i is shown in list j and $x_{ij} = 0$ otherwise. Individual i has an incidence pattern $\boldsymbol{x}_i = (x_{i1}, x_{i2}, x_{i3})^\top$. Let

$$n_{\boldsymbol{x}} = \sum_{i=1}^{s} I(\boldsymbol{x}_i = \boldsymbol{x}), \quad \boldsymbol{x} = (x_1, x_2, x_3)^\top \in \{0, 1\}^3,$$

TABLE 21.1: The observed counts in Example HAV (1: present; 0: absent)

P-list	1	0	1	0	1	0	1
Q-list	0	1	1	0	0	1	1
E-list	0	0	0	1	1	1	1
Count	69	55	21	63	17	18	28

where $I(\cdot)$ is the indicator function. The number of observed individuals is $n_+ = s - n_{0,0,0}$. The three-list saturated log-linear model can be written as

$$\log E(n_{\boldsymbol{x}}) = u + u_1 I(x_1 = 1) + u_2 I(x_2 = 1) + u_3 I(x_3 = 1)$$
$$+ u_{12} I(x_1 = x_2 = 1) + u_{13} I(x_1 = x_3 = 1) + u_{23} I(x_2 = x_3 = 1). \quad (21.1)$$

The u-terms in (21.1) are unknown parameters. Note that u is the intercept, the u_j are one-way interactions and the u_{jk} are two-way interactions. The three-way interaction u_{123} is forced to be zero, because otherwise, the parameters cannot be estimated from the observed counts (Cormack [86], Fienberg [121]). By setting some of u_{12}, u_{13} and u_{23} to be zero, one can obtain a family of log-linear models; see Table 21.2. All main effects u_1, u_2 and u_3 are included in these log-linear modes. Let M_{000} denote the model with $u_{12} = u_{13} = u_{23} = 0$, M_{100} denote the one with $u_{13} = u_{23} = 0$, and so on. Consequently, the saturated model is denoted by M_{111}.

From (21.1), $\log E(n_0) = u$ if M_{111} is the true model from which data are generated. By assuming that a log-linear model is true, we observe that u is a function of the $\log E(n_{\boldsymbol{x}})$, $\boldsymbol{x} \in \mathcal{T} = \{0,1\}^3 \backslash \{\boldsymbol{0}\}$, and we can treat $\exp(u)$ as the expected value of n_0 (Mao, Yang and You [197]). Such an idea applies to each of these eight models. By doing this, one actually defines a target of the population size s via a log-linear model. We use the same subscript to denote the target; for example, s_{111} is the target in M_{111}. In M_{ijk}, one has

$$s_{ijk} - E(n_+) + \exp\left\{\sum_{\boldsymbol{x} \in \mathcal{T}} \alpha_{ijk}(\boldsymbol{x}) \cdot \log E(n_{\boldsymbol{x}})\right\}, \quad (21.2)$$

where the coefficients $\alpha_{ijk}(\boldsymbol{x})$ are presented in Table 21.3. To be specific, we will use the saturated model M_{111} to illustrate the calculation of the coefficients. With $Y_{\boldsymbol{x}} = \log E(n_{\boldsymbol{x}})$, write (21.1) as

$$\begin{pmatrix} Y_{1,0,0} \\ Y_{0,1,0} \\ Y_{0,0,1} \\ Y_{1,1,0} \\ Y_{1,0,1} \\ Y_{0,1,1} \\ Y_{1,1,1} \end{pmatrix} = \begin{pmatrix} 1 & 1 & 0 & 0 & 0 & 0 & 0 \\ 1 & 0 & 1 & 0 & 0 & 0 & 0 \\ 1 & 0 & 0 & 1 & 0 & 0 & 0 \\ 1 & 1 & 1 & 0 & 1 & 0 & 0 \\ 1 & 1 & 0 & 1 & 0 & 1 & 0 \\ 1 & 0 & 1 & 1 & 0 & 0 & 1 \\ 1 & 1 & 1 & 1 & 1 & 1 & 1 \end{pmatrix} \begin{pmatrix} u \\ u_1 \\ u_2 \\ u_3 \\ u_{12} \\ u_{13} \\ u_{23} \end{pmatrix}.$$

Let $\boldsymbol{Y}$ denote the response vector and $\boldsymbol{X}$ denote the design matrix. Let $\boldsymbol{\alpha}_{111}^\top$ be the first row of the matrix $(\boldsymbol{X}^\top \boldsymbol{X})^{-1} \boldsymbol{X}^\top$, which contains those $\alpha_{111}(\boldsymbol{x})$. Note that $u = \boldsymbol{\alpha}_{111}^\top \boldsymbol{Y}$ and $s_{111} = E(n_+) + \exp(\boldsymbol{\alpha}_{111}^\top \boldsymbol{Y})$. In another log-linear model, one simply deletes those columns in $\boldsymbol{X}$ that correspond to those omitted two-way interactions.

TABLE 21.2: The eight hierarchical log-linear models for three lists, in which "1" means that an effect (u-term) is present and "0" means that it is absent

	M_{000}	M_{100}	M_{010}	M_{001}	M_{110}	M_{101}	M_{011}	M_{111}
u	1	1	1	1	1	1	1	1
u_1	1	1	1	1	1	1	1	1
u_2	1	1	1	1	1	1	1	1
u_3	1	1	1	1	1	1	1	1
u_{12}	0	1	0	0	1	1	0	1
u_{13}	0	0	1	0	1	0	1	1
u_{23}	0	0	0	1	0	1	1	1

Finally, the targets s_{ijk} can be estimated by the maximum likelihood method via Poisson regression given the eight design matrices. We use $\hat{s}_{ijk}$ by replacing $E(n_x)$ in (21.2) with n_x. The asymptotic variance σ_{ijk}^2 of $\hat{s}_{ijk}$ can be easily derived (Mao, Yang and You [197]) and estimated by $\hat{\sigma}_{ijk}^2$. The approximate $(1 - \alpha)$ lower confidence limit is defined to be $\hat{c}_{ijk}^{1-\alpha} = \hat{s}_{ijk} - \Phi^{-1}(1 - \alpha) \cdot \hat{\sigma}_{ijk}$, where Φ is the standard normal distribution.

21.3   Performance given Rasch mixtures

To deal with the possible heterogeneity, one may consider the Rasch model (Darroch et al. [98]), i.e., $\log\{\Pr(x_{ij} = 1)/\Pr(x_{ij} = 0)\} = \phi_i + b_j$, where ϕ_i is the individual effect and b_j is the list effect. Let ϕ_i follow a mixing distribution P (Ma, Mao and Yang [189]). With $\boldsymbol{b} = (b_1, b_2, b_3)^\top$ and $b_1 + b_2 + b_3 = 0$, the incidence patterns $\boldsymbol{x}_i$ follow a mixture

$$h_{\boldsymbol{b},P}(\boldsymbol{x}) = \int \left\{ \prod_{j=1}^{3} \frac{\exp\{x_j(b_j + \phi)\}}{1 + \exp(b_j + \phi)} \right\} dP(\phi).$$

To obtain a log-linear representation, we define a distribution G by

$$dG(\phi) \propto \left\{ \prod_{j=1}^{3} \{1 + \exp(b_j + \phi)\} \right\}^{-1} dP(\phi).$$

This is a bijective mapping between P and G. Note that G is degenerate if and only if P is degenerate. Let $r_x(G) = \log \int \exp(x\phi) \, dG(\phi)$ be the cumulant generating function of G. With $\|\boldsymbol{x}\| = x_1 + x_2 + x_3$, write

$$\log h_{\boldsymbol{b},P}(\boldsymbol{x}) = \log h_{\boldsymbol{b},P}(\boldsymbol{0}) + \boldsymbol{b}^\top \boldsymbol{x} + r_{\|\boldsymbol{x}\|}(G). \tag{21.3}$$

By treating the $r_x(G)$ as free parameters, the logarithm of a Rasch mixture admits a log-linear representation (Darroch et al. [98]), and under the constraint that $3r_1(G) - 3r_2(G) + r_3(G) = 0$, it can be fitted by Poisson regression; see Ma, Mao and Yang [189] for the standard log-linear representation.

Given the log-linear model M_{ijk}, from Table 21.3, one observes that

$$\sum_{\boldsymbol{x} \in \mathcal{T}} \alpha_{ijk}(\boldsymbol{x}) = 1, \qquad\qquad \sum_{\boldsymbol{x} \in \mathcal{T}} \alpha_{ijk}(\boldsymbol{x}) \cdot x_1 = 0,$$

$$\sum_{\boldsymbol{x} \in \mathcal{T}} \alpha_{ijk}(\boldsymbol{x}) \cdot x_2 = 0, \qquad\qquad \sum_{\boldsymbol{x} \in \mathcal{T}} \alpha_{ijk}(\boldsymbol{x}) \cdot x_3 = 0.$$

TABLE 21.3: The coefficients $\alpha_{ijk}(x_1, x_2, x_3)$ in eight log-linear models

(x_1, x_2, x_3)	M_{000}	M_{100}	M_{010}	M_{001}	M_{110}	M_{101}	M_{011}	M_{111}
$(1,0,0)$	$1/2$	$1/3$	$1/3$	1	0	1	1	1
$(0,1,0)$	$1/2$	$1/3$	1	$1/3$	1	0	1	1
$(0,0,1)$	$1/2$	1	$1/3$	$1/3$	1	1	0	1
$(1,1,0)$	0	$1/3$	$-1/3$	$-1/3$	0	0	-1	-1
$(1,0,1)$	0	$-1/3$	$1/3$	$-1/3$	0	-1	0	-1
$(0,1,1)$	0	$-1/3$	$-1/3$	$1/3$	-1	0	0	-1
$(1,1,1)$	$-1/2$	$-1/3$	$-1/3$	$-1/3$	0	0	0	1

Using $E(n_x) = s \cdot h_{b,P}(x)$, and

$$\varepsilon_{ijk}(G) = \sum_{x \in \mathcal{T}} \alpha_{ijk}(x) \cdot r_{\|x\|}(G), \tag{21.4}$$

from (21.2), (21.3) and (21.4), we can write

$$s_{ijk} = s + E(n_0) \cdot \{\exp(\varepsilon_{ijk}(G)) - 1\}. \tag{21.5}$$

By doing this, it can be much easier to tell the difference between the target s_{ijk} and the population size s. Specifically, $s_{ijk} = s$ if $\varepsilon_{ijk}(G) = 0$; $s_{ijk} < s$ if $\varepsilon_{ijk}(G) < 0$; and $s_{ijk} > s$ if $\varepsilon_{ijk}(G) > 0$.

From Table 21.3, one has $\varepsilon_{000}(G) = \nu_0$, $\varepsilon_{100}(G) = \varepsilon_{010}(G) = \varepsilon_{001}(G) = \nu_1$, $\varepsilon_{110}(G) = \varepsilon_{101}(G) = \varepsilon_{011}(G) = \nu_2$, and $\varepsilon_{111}(G) = \nu_3$, where

$$\nu_0 = \frac{3r_1(G) - r_3(G)}{2}, \qquad \nu_1 = \frac{5r_1(G) - r_2(G) - r_3(G)}{3},$$
$$\nu_2 = 2r_1(G) - r_2(G), \qquad \nu_3 = 3r_1(G) - 3r_2(G) + r_3(G).$$

It can happen that $\varepsilon_{111}(G) = 0$, $\varepsilon_{111}(G) > 0$, or $\varepsilon_{111}(G) < 0$ (Ma, Mao and Yang [189]). The saturated log-linear model can produce an estimator for the population size that can be consistent, over-estimate or under-estimate. We are interested in telling what will happen in the other seven log-linear models.

Theorem 21.1 *Given any mixing distribution G, it holds that*

$$\varepsilon_{000}(G) \leq 0, \ \varepsilon_{100}(G) = \varepsilon_{010}(G) = \varepsilon_{001}(G) \leq 0, \ \varepsilon_{110}(G) = \varepsilon_{101}(G) = \varepsilon_{011}(G) \leq 0,$$

and, if and only if G is degenerate, one has

$$\varepsilon_{000}(G) = \varepsilon_{100}(G) = \varepsilon_{010}(G) = \varepsilon_{001}(G) = \varepsilon_{110}(G) = \varepsilon_{101}(G) = \varepsilon_{011}(G) = 0.$$

Theorem 21.1 means that each of the log-linear models, except the saturated one, if used for data generated from a Rasch mixture, can produce a lower bound estimator for the population size.

To compare the estimators produced by log-linear models, we simply compare their targets; see the following theorem.

Theorem 21.2 *Given any mixing distribution G, it holds that*

$$s_{000} \leqslant s_{100} = s_{010} = s_{001} \leqslant s_{110} = s_{101} = s_{011} \leqslant s_{111}.$$

There are alternative lower bounds in the Rasch mixture model (Mao et al. [198]). In particular, the sharpest lower bound is

$$s_{\text{slb}} = E(n_+) + \frac{(E(n_{1,0,0} + n_{0,1,0} + n_{0,0,1}))^2}{E(n_{1,1,0} + n_{0,1,1} + n_{1,0,1})} \cdot \frac{\gamma_b(2)}{\gamma_b^2(1)}, \tag{21.6}$$

with $\gamma_b(0) = \gamma_b(3) = 1$, $\gamma_b(1) = \exp(b_1) + \exp(b_2) + \exp(b_3)$, and $\gamma_b(2) = \exp(b_1 + b_2) + \exp(b_1 + b_3) + \exp(b_2 + b_3)$.

Theorem 21.3 *In the Rasch mixture model, it can be shown that*

$$s_{110} = s_{101} = s_{011} = s_{\text{slb}}. \tag{21.7}$$

Given three lists, the targets s_{110}, s_{101} and s_{011} will have the best performance in terms of reducing approximation bias, although the variances of their estimators are possibly larger than those of other estimators for the sharpest lower bound. From Theorems 21.2 and 21.3, the saturated log-linear model should not be used and the other seven log-linear models are useful.

21.4   Simulation

We report simulation results from six settings with fixed population size $s = 10,000$. One setting corresponds to one pair (b, P), where $b = b_1 = (1.20, -0.52, -0.68)^\top$ or $b_2 = (0.23, 0.54, -0.77)^\top$, and $P = P_1 = \delta(0.5)$, $P_2 = 0.82\delta(0.22) + 0.18\delta(0.8)$, or $P_3 = 0.14\delta(0.84) + 0.23\delta(0.5) + 0.63\delta(0.07)$, where $\delta(\cdot)$ is the degenerated distribution. Note that P is a mixing distribution on the scale of $\pi = e^\phi/(1 + e^\phi)$ in our calculation. For each setting, 2,000 samples are generated. Table 21.4 presents the target s_{ijk}, the median and median absolute deviation of $\hat{s}_{ijk}$, and the coverage probability the approximate 95% one-sided confidence intervals $[\hat{c}_{ijk}^{95\%}, \infty)$ for each log-linear model.

For those settings with $P = P_1$ (homogeneous populations), both the target and median are close to the true value of $s = 10,000$ in each log-linear model. For all other settings (heterogeneous populations), the target of each non-saturated model is smaller than the true value, as has been shown in Theorem 21.1. The target of the saturated model can be either larger or smaller than the true value, depending on the underlying mixing distribution. The median absolute deviation of $\hat{s}_{111}$ is the largest one and differs a lot from those of other estimators. The coverage probabilities of $[\hat{c}_{ijk}^{95\%}, \infty)$ in each non-saturated model are larger than the nominal level, and the coverage probabilities of $[\hat{c}_{111}^{95\%}, \infty)$ range from zero to one.

21.5   Example

We revisit Example HAV. First we will fit the data by the Rasch mixture model. The vector b is estimated by $\hat{b} = (0.09, -0.07, -0.02)^\top$ when one maximizes the conditional likelihood of the $x_i|y_i$, where $y_i = \|x_i\|$ (Mao et al. [100]). The hypothesis $b = 0$ is not rejected at the significance level 0.05 since the p-value of the conditional likelihood ratio test is 0.58. Given $\hat{b}$, we consider estimating the mixing distribution P from the observed frequency counts $n_y = \sum_{i=1}^s I(y_i = y)$, $y = 1, 2, 3$. In Example HAV, $n_1 = 187$, $n_2 = 56$, $n_3 = 28$. Let P be on the scale $\pi = e^\phi/(1 + e^\phi)$ and define Q by

$$dQ(\pi) \propto \left\{ 1 - \frac{(1 - \pi)^3}{\sum_{j=0}^3 \gamma_b(j)\pi^j(1 - \pi)^{3-j}} \right\} dP(\pi).$$

The nonparametric maximum likelihood estimator (NPMLE) $\widehat{Q}$ for Q satisfies $\ell_{\hat{b}}(Q) \leqslant \ell_{\hat{b}}(\widehat{Q}), \forall Q$, where

$$\ell_b(Q) = n_1 \log f_{b,Q}(1) + n_2 \log f_{b,Q}(2) + n_3 \log f_{b,Q}(3),$$

$$f_{b,Q}(y) = \int \frac{\gamma_b(y)\pi^y(1 - \pi)^{3-y}}{\sum_{j=1}^3 \gamma_b(j)\pi^j(1 - \pi)^{3-j}} dQ(\pi), \; y = 1, 2, 3.$$

The fitted counts of frequencies are $\hat{n}_y = n_+ f_{\hat{b},\widehat{Q}}(y)$, and the fitted counts of incidence patterns are $\hat{n}_x = \hat{n}_{\|x\|} \exp(x^\top \hat{b})/\gamma_{\hat{b}}(\|x\|)$. The NPMLE $\widehat{Q}$ is not unique; for instance $\widehat{Q} = 0.917\delta(0.230) + 0.083\delta(1)$ and $\widehat{Q} = 0.907\delta(0.224) + 0.093\delta(0.969)$. In Example HAV, the frequency counts are fitted perfectly in the sense that $\hat{n}_y = n_y$ for $y = 1, 2, 3$. Table 21.5 presents the observed counts n_x and fitted counts $\hat{n}_x$. They are quite close with a χ^2 statistic 0.945. This indicates that there is no evidence that the Rasch mixture model is inadequate for Example HAV. The sharpest lower bound estimate is $\hat{s}_{slb} = 479$.

Next we apply eight hierarchical log-linear models to Example HAV. Table 21.6 presents the estimates $\hat{s}_{ijk}$ and approximate standard errors $\hat{\sigma}_{ijk}$. For the purpose of comparison, we also apply these log-linear models to the fitted counts $\hat{n}_x$ in Table 21.5, with corresponding estimates and standard errors denoted by $\hat{s}^\star_{ijk}$ and $\hat{\sigma}^\star_{ijk}$ respectively. It is of interest to note that

$$\hat{s}^\star_{000} < \hat{s}^\star_{100} = \hat{s}^\star_{010} = \hat{s}^\star_{001} < \hat{s}^\star_{110} = \hat{s}^\star_{101} = \hat{s}^\star_{011} < \hat{s}^\star_{111}.$$

In particular, $\hat{s}^\star_{110} = \hat{s}^\star_{101} = \hat{s}^\star_{011} = \hat{s}_{\text{slb}}$. Although neither $\{\hat{s}_{100}, \hat{s}_{010}, \hat{s}_{001}\}$ nor $\{\hat{s}_{110}, \hat{s}_{101}, \hat{s}_{011}\}$ is reduced to a single number, the estimates satisfy

$$\hat{s}_{000} < \hat{s}_{100}, \hat{s}_{010}, \hat{s}_{001} < \hat{s}_{110}, \hat{s}_{101}, \hat{s}_{011} < \hat{s}_{111}.$$

The average of $\hat{s}_{100}$, $\hat{s}_{010}$, and $\hat{s}_{001}$ is 392.5, close to $\hat{s}^\star_{100} = \hat{s}^\star_{010} = \hat{s}^\star_{001} = 392.3$; that of

TABLE 21.4: The target s_{ijk}, the median and median absolute deviation (MAD) of $\hat{s}_{ijk}$, and the coverage probability (CP) of the confidence interval $[\hat{c}^{95\%}_{ijk}, \infty)$ under various settings from each of eight log-linear models

	target	median	MAD	CP	target	median	MAD	CP
	(b_1, P_1)				(b_2, P_1)			
M_{000}	10000	9999	46	0.95	10000	10000	47	0.96
M_{100}	10000	10001	59	0.96	10000	10000	62	0.96
M_{010}	10000	10000	60	0.95	10000	9999	52	0.96
M_{001}	10000	9999	46	0.96	10000	10001	52	0.96
M_{110}	10000	9999	86	0.96	10000	10001	75	0.96
M_{101}	10000	9999	59	0.96	10000	10000	79	0.97
M_{011}	10000	9999	58	0.96	10000	10000	55	0.96
M_{111}	10000	9998	102	0.96	10000	10000	103	0.97
	(b_1, P_2)				(b_2, P_2)			
M_{000}	7804	7805	62	1	7527	7528	66	1
M_{100}	8062	8064	89	1	7825	7828	101	1
M_{010}	8062	8065	84	1	7825	7826	84	1
M_{001}	8062	8065	76	1	7825	7826	85	1
M_{110}	8797	8795	210	1	8673	8675	177	1
M_{101}	8797	8800	133	1	8673	8680	193	1
M_{011}	8797	8800	118	1	8673	8676	123	1
M_{111}	14780	14747	951	0	15543	15572	985	0
	(b_1, P_3)				(b_2, P_3)			
M_{000}	5600	5597	57	1	5335	5334	59	1
M_{100}	5718	5717	68	1	5460	5460	73	1
M_{010}	5718	5715	66	1	5460	5459	63	1
M_{001}	5718	5716	62	1	5460	5458	64	1
M_{110}	6050	6047	125	1	5807	5808	105	1
M_{101}	6050	6050	86	1	5807	5807	112	1
M_{011}	6050	6049	83	1	5807	5802	81	1
M_{111}	8644	8637	495	1	8422	8418	442	1

TABLE 21.5: The observed and fitted counts in Example HAV

x_1	1	0	0	1	1	0	1
x_2	0	1	0	1	0	1	1
x_3	0	0	1	0	1	1	1
n_x	69	55	63	21	17	18	28
$\hat{n}_x$	68	58	61	19	20	17	28

$\hat{s}_{110}$, $\hat{s}_{101}$, and $\hat{s}_{011}$ is 480.6, close to $\hat{s}_{110}^{\star} = \hat{s}_{101}^{\star} = \hat{s}_{011}^{\star} = 478.7$. The non-identifiability of $h_{b,P}(0) = E(n_0)/s$ means that the best we can do is to estimate s_{slb}, and any $s \geqslant s_{\text{slb}}$ is a possible choice. For Example HAV, we have $\hat{s}_{\text{slb}} = 479$ and $s = 545$. It is not surprising to see that the estimate $\hat{s}_{111} = 1313$ from the saturated log-linear model, is much larger than $\hat{s}_{\text{slb}} = 479$ and deviates a lot from the true population size.

21.6   Discussion

We find that when a Rasch mixture model is suitable for a real example with three lists, one may apply those log-linear models except the saturated one, because they produce lower bound estimators for the population size. While, in the real examples, the three estimates generated by the models with two two-way interactions may be different, averaging these estimates is a legitimate strategy.

Our investigation can be extended to the cases with four or more lists. The analyses are tedious because there are too many log-linear models; for example, there are 113 hierarchical log-linear models given four lists.

21.7   Appendix: Proofs of the three theorems

Given $x \geqslant 1$, by the Cauchy–Schwarz inequality, we obtain

$$r_{x-1}(G) + r_{x+1}(G) - 2r_x(G) = \log \left[\frac{\int e^{(x-1)\phi} \, dG(\phi) \int e^{(x+1)\phi} \, dG(\psi)}{\{\int e^{x\phi} \, dG(\phi)\}^2} \right] \geqslant 0.$$

By letting $x = 1$ and $x = 2$, because $r_0(G) = 0$, one has

$$r_2(G) - 2r_1(G) \geqslant 0, \quad r_1(G) + r_3(G) - 2r_2(G) \geqslant 0.$$

Note that $\nu_2 \leqslant 0$. Conclude that $\nu_0 \leqslant \nu_1 \leqslant \nu_2 \leqslant \nu_3$, $\nu_0 \leqslant 0$, $\nu_1 \leqslant 0$ because

$$-2\nu_0 = 2\{r_2(G) - 2r_1(G)\} + \{r_1(G) + r_3(G) - 2r_2(G)\},$$
$$-3\nu_1 = 3\{r_2(G) - 2r_1(G)\} + \{r_1(G) + r_3(G) - 2r_2(G)\},$$
$$6(\nu_1 - \nu_0) = 3(\nu_2 - \nu_1) = \nu_3 - \nu_2 = r_1(G) + r_3(G) - 2r_2(G).$$

These mean that Theorem 21.1 holds. Theorem 21.2 is derived from Theorem 21.1.

TABLE 21.6: The estimates $\hat{s}_{ijk}$ and standard errors $\hat{\sigma}_{ijk}$ in Example HAV. The starred estimates are obtained by applying log-linear models to the fitted counts $\hat{n}_x$ from the Rasch mixture model in Table 21.5

	M_{000}	M_{100}	M_{010}	M_{001}	M_{110}	M_{101}	M_{011}	M_{111}
$\hat{s}_{ijk}$	363	403	376	398	464	527	452	1313
$\hat{\sigma}_{ijk}$	17	29	24	28	59	78	53	518
$\hat{s}_{ijk}^{\star}$	364	392	392	392	479	479	479	1313
$\hat{\sigma}_{ijk}^{\star}$	17	27	27	27	65	61	62	518

Note that $h_{b,P}(x) = \exp(x^\top b)g(\|x\|)/\gamma_b(y)$, where

$$g(y) = \int \frac{\gamma_b(y)\exp(y\phi)}{\sum_{j=0}^{3}\gamma_b(j)\exp(j\phi)}\, dP(\phi).$$

Clearly, $(s_{\mathrm{slb}} - E(n_+))/s$ is identical to $g^2(1)\gamma_b(2)/\{g(2)\gamma_b^2(1)\}$. Write

$$\begin{aligned}
(s_{110} - E(n_+))/s &= h_{b,P}(1,0,0)h_{b,P}(0,1,0)/h_{b,P}(1,1,0) \\
&= \frac{\{\exp(b_1)g(1)/\gamma_b(1)\} \cdot \{\exp(b_2)g(1)/\gamma_b(1)\}}{\exp(b_1+b_2)g(2)/\gamma_b(2)} \\
&= \frac{g^2(1)\gamma_b(2)}{g(2)\gamma_b^2(1)}.
\end{aligned}$$

This means that $s_{110} = s_{\mathrm{slb}}$. We conclude that Theorem 21.3 holds.

22

A multidimensional Rasch model for multiple system estimation where the number of lists changes over time

Elvira Pelle

University of Trieste

David J. Hessen

University of Utrecht

Peter G. M. van der Heijden

Universities of Utrecht and Southampton

CONTENTS

22.1   Introduction

In human populations, capture-recapture methods can be used to estimate the demographic characteristics of interest using information from two or more incomplete but overlapping lists of cases from different sources. In the literature the methods are also referred to as *multiple system* and *multiple-records system* (International Working Group for Disease Monitoring and Forecasting IWGDMFa,b [154, 155]). Here, each list is viewed as a capture sample, the identification number (or name) is treated as a mark (or tag) and the statement "being captured in sample i" is replaced by "being observed in list i".

     Data are usually arranged in a 2^S contingency table (where S is the number of available lists), with one missing cell corresponding to absence in all lists; the empty cell is treated as a "structural zero", i.e. is known a priori to have a zero value and the cell must remain empty under any fitted model. Then, the contingency table is analyzed by the use of log-linear models.

Modeling dependence between lists is one of the major issues in a multiple-system framework. Dependence among lists may be due to both list dependence (inclusion in a list has a direct causal influence on inclusion to another list), or heterogeneity between individuals (differences of behaviour results in heterogeneous inclusion probabilities and may cause indirect dependence between lists) (see Chao et al. [79] for more details).

A way to model list dependence consists of adding first-order or higher-order interaction parameters in the log-linear model used (Bishop et al. [32]), while taking into account heterogeneity of inclusion probabilities; psychometric models, such as the Rasch model, can be utilised (see, among others, Darroch et al. [98], Agresti [2], Fienberg et al. [122] and Bartolucci and Forcina [24]).

The Rasch model is a model widely used by psychometricians to explain the characteristics and performances of a test; the basic idea is that the probability of a response of an individual to an item can be modelled as a function of the difficulty of the item and the latent ability of the individual. In a capture-recapture context, a correct or incorrect response to an item is replaced by the presence or absence in a list, and heterogeneity among individuals is modelled in terms of constant apparent dependence between lists (see International Working Group for Disease Monitoring and Forecasting IWGDMFa [154]), introducing into the model the first-order heterogeneity parameter H1 (all two-factor interaction terms are supposed to be equal and positive), the second-order heterogeneity parameter H2 (all three-factor interaction terms are supposed to be equal and positive), and so on.

An alternative approach is to use the log-linear multidimensional Rasch (MR) model (Pelle et al. [230]). In particular, under the assumption that lists may be viewed as indicators of the latent variables which account for correlations among lists, the probability of a generic capture profile can be easily expressed in a log-linear form, using an extension of the Dutch Identity for the multidimensional partial credit model (Hessen [141], Holland [144]). The resulting model can be used either in the case with or without a stratifying variable available.

We will apply the model in the context of a set of lists for the incidence of spina bifida, where the aim is to produce yearly estimates. A problem is that some of the lists are not observed in each year for which we want to produce estimates. This problem is solved by assuming that the lists that are not observed are missing in specific years. The EM algorithm is used to solve the missing data problem.

22.2 Data set

As an illustration of the use of the log-linear MR model in multiple system estimation, we refer to the data set on neural tube defects in the Netherlands (see van der Pal et al. [287], for details about the data). Cases of children born with neural tube defects were obtained from five different lists during the years 1988 through 1998, but lists refer to different but overlapping periods of time. In particular, before 1992, cases in only three lists were recorded (lists 1, 2 and 5), in 1992 cases in four lists were available (lists, 1, 2, 3 and 5), while in the period 1993–1998 all five lists were active. As a consequence, the resulting contingency table has 24 structural zero cells: 11 structural zero cells corresponding to cases missed by all lists for years 1988 through 1998; for each year from 1988 to 1991 there are 3 more structural zero cells, resulting from the fact that lists 3 and 4 are not operating in these years and corresponding to capture profiles (00100), (00010) and (00110), that are cases recorded only in list 3, only in list 4 and in both lists 3 and 4. Finally, there is 1 more structural zero cell in 1992 due to the fact that list 4 is not active and corresponding to capture profile (00010).

Let $\mathbf{i} = (i_1, i_2, i_3, i_4, i_5)$ be the generic capture profile, where $i_s = 0$ if the individual is not observed in list s and $i_s = 1$ if the individual is observed in list s, for $s = 1, \ldots, 5$, so that $\mathbf{i} = (1, 0, 0, 0, 0)$ denotes the capture profile of the individual observed only in list 1, $\mathbf{i} = (1, 1, 0, 0, 0)$ indicates the capture profile of observations included in list 1 and 2 but not included in lists 3, 4 and 5, and so on. In total, we have $2^5 = 32$ capture profiles, of which the capture profile $i = (0, 0, 0, 0, 0)$ is not observed.

The data are summarised in Table 22.1. Since none of the five lists record all cases of neural tube defect, the issue here is to estimate the total number of children born with a neural tube defect in the Netherlands. We will analyse the data using multidimensional Rasch models, taking into account that some of the lists were not observed for the complete period starting at 1988 and ending at 1998.

TABLE 22.1

Observed frequencies on neural tube defects in the Netherlands for each of the year from 1988 to 1998 and for each capture profile

Year	Capture profiles															
	00000	10000	01000	11000	00100	10100	01100	11100	00010	10010	01010	11010	00110	10110	01110	11110
1988	0*	4	101	24	0*				0*				0*			
1989	0*	3	114	30	0*				0*				0*			
1990	0*	3	105	43	0*				0*				0*			
1991	0*	4	95	32	0*				0*				0*			
1992	0*	9	80	27	15	0	12	7	0*				0*			
1993	0*	5	61	24	4	1	1	0	12	0	18	7	2	0	4	3
1994	0*	24	34	13	6	1	1	1	15	7	18	9	5	4	5	2
1995	0*	29	27	15	5	1	2	1	16	15	18	4	4	0	9	5
1996	0*	26	26	11	10	1	1	1	9	6	11	9	5	0	4	5
1997	0*	41	26	18	13	2	0	1	12	11	11	7	3	4	6	3
1998	0*	27	25	20	13	0	2	1	8	7	14	3	6	11	4	7

Year	Capture profiles															
	00001	10001	01001	11001	00101	10101	01101	11101	00011	10011	01011	11011	00111	10111	01111	11111
1988	9	0	5	2												
1989	3	1	8	4												
1990	7	3	5	4												
1991	3	1	7	8												
1992	10	1	3	3	0	0	2	3								
1993	3	0	0	1	0	0	0	0	1	1	7	4	0	1	0	0
1994	3	0	1	1	0	1	0	1	1	3	1	0	0	1	0	4
1995	2	1	2	0	0	1	3	0	2	2	5	1	3	0	1	0
1996	5	0	0	0	0	0	0	0	5	6	2	1	1	2	2	4
1997	4	2	0	1	2	0	0	0	1	1	1	3	1	2	0	4
1998	1	0	0	0	0	0	0	0	0	0	1	2	0	0	1	1

0* denotes structural zero cells.

22.3   Estimating population size under the log-linear multidimensional Rasch model

22.3.1   Notation and basic assumptions

Suppose that there are S lists available. Let I_s, $s = 1, \ldots, S$ be the random variables denoting the presence or absence of an individual in the corresponding list. Let $n_{i_1, \ldots, i_S}$ and $m_{i_1, \ldots, i_S}$ denote the observed and the expected frequencies, respectively, for the capture profile $\mathbf{i} = (i_1, \ldots, i_S)$.

Let $\pi_{i_1, \ldots, i_S}$ indicate the corresponding capture probability, while π_{0_s}, for $s = 1, \ldots, S$, denote the probability of not being observed in the sth list and $\pi_{1_s} = 1 - \pi_{0_s}$ is the probability of being observed in the sth list. Since $i_s = (0, 1)$ the probability of inclusion in the sth list may be written as

$$\pi_{i_s} = \left(\pi_{1_s} \right)^{i_s} \left(\pi_{0_s} \right)^{1 - i_s}. \tag{22.1}$$

Assume that there are q latent variables which explain the covariances among lists. Let $\mathbf{\Theta} = (\Theta_1, \Theta_2, \ldots, \Theta_q)$ denote the vector of latent variables and let $\boldsymbol{\theta} = (\theta_1, \theta_2, \ldots, \theta_q)$ indicate a realisation. The assumption that the covariances among lists are explained by the latent variables means that the random variables $I_s, s = 1, 2, \ldots, S$ are conditionally independent given the latent variables. As a result, the probability of a generic capture profile, given $\boldsymbol{\theta}$, can be factored in the form

$$\pi_{i_1, \ldots, i_S | \boldsymbol{\theta}} = \prod_{s=1}^{S} \pi_{i_s | \boldsymbol{\theta}}, \tag{22.2}$$

where $\pi_{i_s | \boldsymbol{\theta}} = \left(\pi_{1_s | \boldsymbol{\theta}} \right)^{i_s} \left(\pi_{0_s | \boldsymbol{\theta}} \right)^{1 - i_s}$ denotes the conditional probability of the sth list, given the vector of latent variables $\boldsymbol{\theta}$.

22.3.2   Methodology

For the sake of simplicity, consider first the situation that 3 lists are available. Data can be arranged in a 2^3 contingency table as shown in Table 22.2. Note that, since n_{000} is the frequency of individuals not observed in any list, it will be treated as a structural zero and has to be estimated in order to estimate the total unknown population size N. To get an estimate of n_{000}, the approach used has two phases: first, a log-linear model is fitted to the incomplete contingency table, that is the contingency table without the missing cell; then, the parameter estimates of the model are projected to the missing cell to predict its value.

Suppose that there are only two latent variables which explain the covariances among lists. Due to conditional independence, the probability of a generic capture profile equals

TABLE 22.2

Contingency table for three lists

		List 3			
		Observed		Not Observed	
		List 2		List 2	
		Observed	Not Observed	Observed	Not Observed
List 1	Observed	n_{111}	n_{101}	n_{110}	n_{100}
	Not Observed	n_{011}	n_{001}	n_{010}	n_{000}

$$\pi_{i_1 i_2 i_3 | \boldsymbol{\theta}} = \prod_{s=1}^{3} \pi_{i_s | \boldsymbol{\theta}} = \prod_{s=1}^{3} \left(\pi_{1_s | \boldsymbol{\theta}} \right)^{i_s} \left(\pi_{0_s | \boldsymbol{\theta}} \right)^{1 - i_s} \qquad (22.3)$$

where $\boldsymbol{\theta} = (\theta_1, \theta_2)$.

The probability of inclusion in the sth list may be expressed in a logistic form as

$$\pi_{1_s | \boldsymbol{\theta}} = \frac{e^{\mathbf{u}_s' \boldsymbol{\theta} - \delta_s}}{1 + e^{\mathbf{u}_s' \boldsymbol{\theta} - \delta_s}} \qquad (22.4)$$

where δ_s is the parameter for list s, θ_r is the parameter for the rth latent variable and $\mathbf{u}_s'$ is the row vector of the (3×2) full column rank matrix $\mathbf{U} = [u_{sr}]$ of weights for the latent variables, where

$$u_{sr} = \begin{cases} 1 & \text{if the list } S \text{ is assumed to be indicator of the } r\text{th latent variable,} \\ 0 & \text{otherwise.} \end{cases}$$

Consequently, the probability of not being observed in the sth list takes the form

$$\pi_{0_s | \boldsymbol{\theta}} = \frac{1}{1 + e^{\mathbf{u}_s' \boldsymbol{\theta} - \delta_s}} \qquad (22.5)$$

and the probability of not being observed in any of the three lists, under the assumption of conditional independence, may be written as

$$\pi_{000 | \boldsymbol{\theta}} = \prod_{s=1}^{3} \pi_{0_s | \boldsymbol{\theta}} = \prod_{s=1}^{3} \frac{1}{1 + e^{\mathbf{u}_s' \boldsymbol{\theta} - \delta_s}}. \qquad (22.6)$$

Thus, according to standard probability theory, the probability of the capture pattern $(i_1, i_2, i_3) = (0, 0, 0)$ equals

$$\pi_{000} = \int \cdots \int \pi_{000 | \boldsymbol{\theta}} f(\boldsymbol{\theta}) \, d\boldsymbol{\theta} = \int \cdots \int \prod_{s=1}^{3} \frac{1}{1 + e^{\mathbf{u}_s' \boldsymbol{\theta} - \delta_s}} f(\boldsymbol{\theta}) \, d\boldsymbol{\theta}, \qquad (22.7)$$

where $f(\boldsymbol{\theta})$ is the multivariate density of the vector of latent variables $\boldsymbol{\theta}$ in the population.

Analogously, the probability of a generic capture profile takes the form

$$\pi_{i_1 i_2 i_3} = \int \cdots \int \pi_{i_1 i_2 i_3 | \boldsymbol{\theta}} f(\boldsymbol{\theta}) \, d\boldsymbol{\theta} = \int \cdots \int \prod_{s=1}^{3} \frac{e^{i_s \left(\mathbf{u}_s' \boldsymbol{\theta} - \delta_s \right)}}{1 + e^{\mathbf{u}_s' \boldsymbol{\theta} - \delta_s}} f(\boldsymbol{\theta}) \, d\boldsymbol{\theta} \qquad (22.8)$$

and, after some algebra (see the appendix for details)

$$\pi_{i_1 i_2 i_3} = \pi_{000} e^{- \sum_s i_s \delta_s} \int \cdots \int e^{\mathbf{t} \boldsymbol{\theta}} g \left(\boldsymbol{\theta} | (i_1 i_2 i_3 = 000) \right) d\boldsymbol{\theta} \qquad (22.9)$$

where $g(\boldsymbol{\theta} | (i_1 i_2 i_3 = 000))$ is the posterior distribution of $\boldsymbol{\theta}$ given the capture pattern equals $(0, 0, 0)$ (that is the probability of not be observed in any list).

Note that

$$M_{\boldsymbol{\Theta}}(\mathbf{t}) = \int \cdots \int e^{\mathbf{t} \boldsymbol{\theta}} g \left(\boldsymbol{\theta} | (i_1 i_2 i_3 = 000) \right) d\boldsymbol{\theta}$$

is the moment generating function conditional on the capture profile $(i_1, i_2, i_3) = (0, 0, 0)$. Making an assumption about the posterior distribution of the latent variables and thus choosing a moment generating function, allows us to compute the probability in (22.8). We assume that the population of individuals not observed in any list follows a normal distribution. This is equivalent to assuming that the posterior distribution of the latent variables follows a multivariate normal distribution, for which the moment generating function takes the form

$$M_{\Theta}(\mathbf{t}) = e^{\mathbf{t}'\boldsymbol{\mu} + \frac{1}{2}\mathbf{t}'\boldsymbol{\Gamma}\mathbf{t}} \tag{22.10}$$

where $\boldsymbol{\mu}$ is the mean vector of Θ conditional on capture profile $(i_1, i_2, i_3) = (0,0,0)$ and $\boldsymbol{\Gamma}$ is the covariance matrix of Θ conditional on capture profile $(i_1, i_2, i_3) = (0,0,0)$.

Then, the probability of a generic capture profile $\pi_{i_1 i_2 i_3}$ can be expressed as:

$$
\begin{aligned}
\pi_{i_1 i_2 i_3} &= \pi_{000} \exp\left\{ \sum_{s=1}^{3} i_s \delta_s + t_1\mu_1 + t_2\mu_2 + \frac{1}{2}t_1^2\gamma_{11} + \frac{1}{2}t_2^2\gamma_{22} + t_1 t_2 \gamma_{12} \right\} \\
&= \pi_{000} \exp\left\{ \sum_{s=1}^{3} i_s \delta_s + \mathbf{t}'\boldsymbol{\mu} + \frac{1}{2}\mathbf{t}'\boldsymbol{\Gamma}\mathbf{t} \right\}
\end{aligned}
\tag{22.11}
$$

where $\mathbf{t} = (t_1, t_2)' = \mathbf{i}'\mathbf{U}$ and $\boldsymbol{\Gamma} = [\gamma_{ir}]$ is symmetric.

Let n be the total number of individuals observed in at least one list and let A indicate the set of capture profiles of individuals observed in at least one list. In the case of three lists the set A has seven elements, $A = \{(1,0,0),(0,1,0),(0,0,1),(1,1,0),(1,0,1),(0,1,1),(1,1,1)\}$. It is known that the observed frequencies $n_{100}, n_{010}, n_{001}, n_{110}, n_{101}, n_{011}, n_{111}$ have a multinomial distribution with parameters n and $\pi_{i_1 i_2 i_3}/\sum_A \pi_{i_1 i_2 i_3}$, for all $(i_1, i_2, i_3) \in A$; thus the expected frequency of the generic capture profile $n_{i_1 i_2 i_3}$ may be expressed as

$$m_{i_1 i_2 i_3} = n \frac{\pi_{i_1 i_2 i_3}}{\sum_A \pi_{i_1 i_2 i_3}}, \quad \text{for all } (i_1, i_2, i_3) \in A. \tag{22.12}$$

Substituting Equation (22.11) into Equation (22.12) and taking the logarithm yields the log linear representation of the model

$$\log m_{i_1 i_2 i_3} = \delta + \sum_{s=1}^{3} i_s \delta_s + \mathbf{t}'\boldsymbol{\mu} + \frac{1}{2}\mathbf{t}'\boldsymbol{\Gamma}\mathbf{t} \tag{22.13}$$

where
$\delta = \log(n\pi_{000}/\sum_A \pi_{i_1 i_2 i_3}) = \log\left\{ n/\sum_A \exp\left(\sum_{s=1}^{3} i_s\delta_s + \mathbf{t}'\boldsymbol{\mu} + \frac{1}{2}\mathbf{t}'\boldsymbol{\Gamma}\mathbf{t} \right) \right\}$.

Without any additional constraint, the model in Equation (22.13) cannot be identified. To overcome this problem we arbitrarily fix $\boldsymbol{\mu}$ to $\mathbf{0}$ and the model can be rewritten as

$$
\begin{aligned}
\log m_{i_1 i_2 i_3} &= \delta + \sum_{s=1}^{3} i_s \delta_s + \frac{1}{2}\mathbf{t}'\boldsymbol{\Gamma}\mathbf{t} \\
&= \delta + i_1\delta_1 + i_2\delta_2 + i_3\delta_3 + \frac{1}{2}t_1^2\gamma_{11} + \frac{1}{2}t_2^2\gamma_{22} + t_1 t_2 \gamma_{12}
\end{aligned}
\tag{22.14}
$$

where δ is a common effect parameter, δ_s is the main-effect parameter for list s, γ_{11} is the variance of the first latent variable given t_1 and t_2, γ_{22} is the variance of the second latent variable given t_1 and t_2, and γ_{12} is the covariance between the two latent variables given t_1 and t_2. Note that in the final model there are 7 parameters, $2(2+1)/2 = 3$ of which account for the two latent variables θ_1 and θ_2. In general, with q latent variables the resulting model has $q(q+1)/2$ parameters for the latent variables.

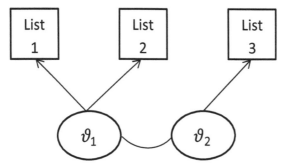

FIGURE 22.1: Model with three lists and two latent variables.

Example 1: Model with three lists and two latent variables. Consider a situation of three lists. Suppose that list 1 and list 2 are indicators of the first latent variable (named θ_1) and list 3 is an indicator of the second latent variable (θ_2). This situation may be represented as in the path diagram in Figure 22.1.

Here, the single-headed arrows from the latent variables to the lists indicate that there is a direct effect of the latent variables on these lists, while the curved line between the two latent variables indicates that there is a covariance between the two latent variables. On the other hand, since there are no double-headed arrows between pairs of lists, the lists are conditionally independent given the latent variables.

To build the model, we need the full column rank matrix $\mathbf{U}$, that in this situation is given by

$$\mathbf{U} = \begin{bmatrix} u_{11} & u_{12} \\ u_{21} & u_{22} \\ u_{31} & u_{32} \end{bmatrix} = \begin{bmatrix} 1 & 0 \\ 1 & 0 \\ 0 & 1 \end{bmatrix}.$$

In addition, we need the total scores t_1 and t_2 for each capture profile, computed by

$$\mathbf{t} = \mathbf{U}'\mathbf{i} = \begin{bmatrix} u_{11} & u_{21} & u_{31} \\ u_{12} & u_{22} & u_{32} \end{bmatrix} \begin{bmatrix} i_1 \\ i_2 \\ i_3 \end{bmatrix} = \begin{bmatrix} t_1 \\ t_2 \end{bmatrix}.$$

For example, for capture profiles $(i_1, i_2, i_3) = (1, 0, 1)$ and $(i_1, i_2, i_3) = (1, 1, 0)$ the total scores are, respectively,

$$\mathbf{t} = \mathbf{U}'\mathbf{i} = \begin{bmatrix} 1 & 1 & 0 \\ 0 & 0 & 1 \end{bmatrix} \begin{bmatrix} 1 \\ 0 \\ 1 \end{bmatrix} = \begin{bmatrix} 1 \\ 1 \end{bmatrix}$$

and

$$\mathbf{t} = \mathbf{U}'\mathbf{i} = \begin{bmatrix} 1 & 1 & 0 \\ 0 & 0 & 1 \end{bmatrix} \begin{bmatrix} 1 \\ 1 \\ 0 \end{bmatrix} = \begin{bmatrix} 2 \\ 0 \end{bmatrix}.$$

To better understand how to fit the MR model, a matrix approach may be useful. Let $\mathbf{m} = (m_{100}, m_{010}, m_{001}, m_{110}, m_{101}, m_{011}, m_{111})'$ be the vector of expected counts. In matrix terms the MR model may be written as $\log \mathbf{m} = \mathbf{X}\boldsymbol{\beta}$, where $\boldsymbol{\beta} = (\delta, \delta_1, \delta_2, \delta_3, \gamma_{11}, \gamma_{22}, \gamma_{12})'$ is the vector of parameters to be estimated and $\mathbf{X}$ is the design matrix with columns corresponding to the parameters to be estimated, that is $\mathbf{X} = (\mathbf{1}, \mathbf{i}_1, \mathbf{i}_2, \mathbf{i}_3, \mathbf{t}_1^2, \mathbf{t}_2^2, \mathbf{t}_1\mathbf{t}_2)'$.

In this example, the matrix $\mathbf{X}$ equals

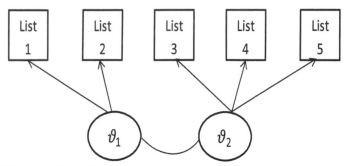

FIGURE 22.2: Model with a list in common for the two latent variables.

$$\mathbf{X} = \begin{pmatrix} 1 & 0 & 0 & 1 & 0 & 1 & 0 \\ 1 & 0 & 1 & 0 & 1 & 0 & 0 \\ 1 & 0 & 1 & 1 & 1 & 1 & 1 \\ 1 & 1 & 0 & 0 & 1 & 0 & 0 \\ 1 & 1 & 0 & 1 & 1 & 1 & 1 \\ 1 & 1 & 1 & 0 & 4 & 0 & 0 \\ 1 & 1 & 1 & 1 & 4 & 1 & 2 \end{pmatrix}.$$

The MR model can also be applied when latent variables have lists in common. For example, suppose that list 1 and list 2 are indicators of the first latent variable and that list 1 and list 3 are indicators of the second latent variable (see the path diagram in Figure 22.2).

Analogous to the previous example, in this case the full column rank matrix **U** is given by

$$\mathbf{U} = \begin{bmatrix} u_{11} & u_{12} \\ u_{21} & u_{22} \\ u_{31} & u_{32} \end{bmatrix} = \begin{bmatrix} 1 & 1 \\ 1 & 0 \\ 0 & 1 \end{bmatrix}.$$

Once the total scores are computed, it is possible to construct the design matrix **X**, which takes the form

$$\mathbf{X} = \begin{pmatrix} 1 & 0 & 0 & 1 & 0 & 1 & 0 \\ 1 & 0 & 1 & 0 & 1 & 1 & 1 \\ 1 & 0 & 1 & 1 & 1 & 4 & 2 \\ 1 & 1 & 0 & 0 & 1 & 0 & 0 \\ 1 & 1 & 0 & 1 & 1 & 1 & 1 \\ 1 & 1 & 1 & 0 & 4 & 1 & 2 \\ 1 & 1 & 1 & 1 & 4 & 4 & 4 \end{pmatrix}.$$

Example 2: Model with three lists and one latent variable (unidimensional Rasch model). The MR methodology can also be applied to model the situation of only one latent variable. Let us consider a situation of three lists which are indicators of the same latent variable, as represented in the path diagram in Figure 22.3.

Now, the full column rank matrix **U**, simplifies to

$$\mathbf{U} = \begin{bmatrix} u_{11} \\ u_{21} \\ u_{31} \end{bmatrix} = \begin{bmatrix} 1 \\ 1 \\ 1 \end{bmatrix} = \mathbf{1}$$

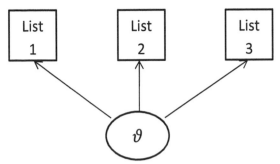

FIGURE 22.3: Model with three lists and one latent variable

and the total scores t can be computed by $t = \mathbf{1}'\mathbf{i}$.

Note that the total score t is analogous to the parameter denoted in IWGDMFa [154] as the first-order heterogeneous parameter H_1, obtained by taking all the parameters of the second order to be equal and positive. In other words, this model is a log-linear version of the unidimensional Rasch model, the simplest of the Rasch models, first introduced by Rasch [241].

22.3.3   Model with a stratifying variable

Suppose now that a stratifying variable is available. Let j be the index for the strata, so that $n_{i_1 i_2 i_3 j}$ and $\pi_{i_1 i_2 i_3 j}$ denote the observed frequency and the probability for stratum j, respectively.

For convenience, consider the situation of 3 lists recorded in two strata. The resulting contingency table has two missing cells, one corresponding to individuals not observed in any of the lists for the first stratum, and one corresponding to individuals missed by all lists in the second stratum (as shown in Table 22.3). In general, with j strata the resulting contingency table has j missing cells.

The probability of the generic capture profile (i_1, i_2, i_3) for the stratum j may be written as

$$\pi_{i_1 i_2 i_3 j} = \int \cdots \int \pi_{i_1 i_2 i_3 j \mid \boldsymbol{\theta}} f(\boldsymbol{\theta}) \, d\boldsymbol{\theta} \tag{22.15}$$

where $\pi_{i_1 i_2 i_3 j \mid \boldsymbol{\theta}}$ is the probability of the capture profile (i_1, i_2, i_3) for stratum j conditional on the vector of latent variables and $f(\boldsymbol{\theta})$ is the multivariate density of $\boldsymbol{\theta}$.

Under the assumption of a multivariate normal distribution for the posterior distribution

TABLE 22.3
Contingency table for three lists and two strata

		List 3			
		1		0	
		List 2		List 2	
Stratum	List 1	1	0	1	0
1	1	n_{1111}	n_{1011}	n_{1101}	n_{1001}
	0	n_{0111}	n_{0011}	n_{0101}	0^*
2	1	n_{1112}	n_{1012}	n_{1102}	n_{1002}
	0	n_{0112}	n_{0012}	n_{0102}	0^*

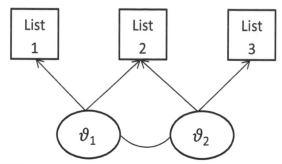

FIGURE 22.4: Structure of latent variables in stratum 1

of the vector of latent variables $\boldsymbol{\Theta}$, conditional on capture profile $(i_1, i_2, i_3) = (0, 0, 0)$, we have

$$\pi_{i_1 i_2 i_3 j} = \pi_{000j} \exp\left(\sum_{s=1}^{3} i_s \delta_{sj} + \mathbf{t}'\boldsymbol{\mu}_j + \tfrac{1}{2}\mathbf{t}'\boldsymbol{\Gamma}_j \mathbf{t}\right) \tag{22.16}$$

where $\boldsymbol{\mu}_j$ is the mean vector and $\boldsymbol{\Gamma}_j$ is the covariance matrix of $\boldsymbol{\Theta}$ in stratum j.

Let $m_{i_1 i_2 i_3 j}$ be the expected frequency of the generic capture profile (i_1, i_2, i_3) in stratum j

$$m_{i_1 i_2 i_3 j} = \frac{n \pi_{i_1 i_2 i_3 j}}{\sum_A \pi_{i_1 i_2 i_3 j}} \tag{22.17}$$

for all $(i_1, i_2, i_3) \in A$.

Substituting (22.17) in (22.16) we obtain

$$\log m_{i_1 i_2 i_3 j} = \delta_j + \sum_{s=1}^{3} i_s \delta_{sj} + \mathbf{t}'\boldsymbol{\mu}_j + \tfrac{1}{2}\mathbf{t}'\boldsymbol{\Gamma}_j \mathbf{t} \tag{22.18}$$

where $\delta_j = \log(n\pi_{000j} / \sum_A \pi_{i_1 i_2 i_3 j})$. Without any additional constraints, the model in (22.18) is not identified; setting $\boldsymbol{\mu}_j$ equal to zero for identification we have

$$\log m_{i_1 i_2 i_3 j} = \delta_j + \sum_{s=1}^{3} i_s \delta_{sj} + \tfrac{1}{2}\mathbf{t}'\boldsymbol{\Gamma}_j \mathbf{t}, \tag{22.19}$$

where δ_j is the common effect parameter in stratum j and δ_{ij} is the main-effect parameter for list i in stratum j.

With two latent variables the model for stratum j is

$$\log m_{i_1 i_2 i_3 j} = \delta_j + i_1 \delta_{1j} + i_2 \delta_{2j} + i_3 \delta_{3j} + \frac{1}{2}t_1^2 \gamma_{11j} + \frac{1}{2}t_2^2 \gamma_{22j} + t_1 t_2 \gamma_{12j}.$$

Example 3: Model with three lists, two strata and two latent variables. For the sake of simplicity, consider the situation of three lists, two strata and two latent variables. Assume that for the first stratum, list 1 and list 2 are indicators of the first latent variable and list 2 and list 3 are indicators of the second latent variable (as shown in the path diagram in Figure 22.4); furthermore, assume that for the second stratum only list 1 is an indicator of the first latent variable, while list 2 and list 3 are indicators of the second latent variable (see the path diagram in Figure 22.5).

In this situation the design matrix of the model may be written as

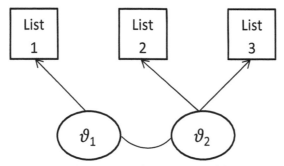

FIGURE 22.5: Structure of latent variables in stratum 2

$$\mathbf{X} = \begin{bmatrix} \mathbf{X}_1 \\ \mathbf{X}_2 \end{bmatrix}$$

where $\mathbf{X}_j, j = 1, 2$ is the design matrix for stratum j.

In particular, we have

$$\mathbf{X}_1 = \begin{pmatrix} 1 & 0 & 0 & 1 & 1 & 0 & 0 \\ 1 & 0 & 1 & 0 & 1 & 1 & 1 \\ 1 & 0 & 1 & 1 & 0 & 1 & 0 \\ 1 & 1 & 0 & 0 & 4 & 1 & 2 \\ 1 & 1 & 0 & 1 & 1 & 1 & 1 \\ 1 & 1 & 1 & 0 & 1 & 4 & 2 \\ 1 & 1 & 1 & 1 & 4 & 4 & 4 \end{pmatrix} \mathbf{X}_2 = \begin{pmatrix} 1 & 0 & 0 & 1 & 0 & 1 & 0 \\ 1 & 0 & 1 & 0 & 1 & 1 & 1 \\ 1 & 0 & 1 & 1 & 1 & 4 & 2 \\ 1 & 1 & 0 & 0 & 1 & 0 & 0 \\ 1 & 1 & 0 & 1 & 1 & 1 & 1 \\ 1 & 1 & 1 & 0 & 4 & 1 & 2 \\ 1 & 1 & 1 & 1 & 4 & 4 & 4 \end{pmatrix}.$$

22.3.4 Assumption of measurement invariance

Assume now that the MR model satisfied the assumption of measurement invariance, that means that the model measures the same construct and the latent variables have the same structure across the strata; in other words, we are assuming that the same model applies across strata. If the assumption of measurement invariance holds, then the parameters are equal across strata and we have

$$\delta_{sj} = \delta_s, \forall j$$

and the model in (22.18) becomes

$$\log m_{i_1 i_2 i_3 j} = \delta_j + \sum_{s=1}^{3} i_s \delta_s + \mathbf{t}' \boldsymbol{\mu}_j + \tfrac{1}{2} \mathbf{t}' \boldsymbol{\Gamma}_j \mathbf{t}. \tag{22.20}$$

Note that this model is not identified but, due to measurement invariance, for identification we only need to set $\boldsymbol{\mu}_j$ to $\mathbf{0}$ for one j.

Furthermore, under the assumption of measurement invariance, it can be of interest to test if the mean vector and the covariance matrix of the vector of latent variables are equal across strata. In this case, if the simultaneous hypothesis

$$\boldsymbol{\mu}_j = \boldsymbol{\mu} = \mathbf{0} \text{ and } \boldsymbol{\Gamma}_j = \boldsymbol{\Gamma}, \forall j$$

holds, then the model in (22.20) simplifies to

$$\log m_{i_1 i_2 i_3 j} = \delta_j + \sum_{s=1}^{3} i_s \delta_s + \tfrac{1}{2} \mathbf{t}' \boldsymbol{\Gamma} \mathbf{t}, \tag{22.21}$$

and the number of parameters to be estimated decreases.

22.3.5   Generalisation

The extension of the MR model to a more general situation is straightforward. Consider a situation of S lists and J strata. Let $n_{i_1...i_Sj}$ be the observed frequency of the generic capture profile $(i_1,...,i_S)$ in stratum j, $s = 1,...,S$ and $j = 1,...,J$. Let $\pi_{i_1...i_Sj}$ denote the corresponding capture probability.

Assume now that there are q latent variables which explain the covariances between the random variables $I_1,...,I_S$. Let $\mathbf{U} = [u_{sr}]$ denote the full column rank matrix of weights for the latent variables, where

$$
u_{sr} = \begin{cases} 1 & \text{if the list } S \text{ is assumed to be indicator of the } r\text{th latent variable,} \\ 0 & \text{otherwise} \end{cases}
$$

and let $\mathbf{t} = (t_1,...,t_q)$ be the vector of the total scores, where $t_r = \sum_{s=1}^{S} u_{sr}i_s$, for $r = 1,...,q$.

The probability of a generic capture profile is

$$
\pi_{i_1...i_Sj} = \int \cdots \int \pi_{i_1...i_Sj|\boldsymbol{\theta}} f(\boldsymbol{\theta}) \, d\boldsymbol{\theta} \tag{22.22}
$$

where $\pi_{i_1...i_3j|\boldsymbol{\theta}}$ is the probability of the capture profile $(i_1,...,i_S)$ for stratum j conditional on the vector of latent variables and $f(\boldsymbol{\theta})$ is the multivariate density of $\boldsymbol{\theta}$.

Analogous to the simpler situation, under the assumption of a multivariate normal posterior distribution of the latent variables (conditional on the capture pattern of individuals not observed in any list), the probability of a generic capture profile $\pi_{i_1...i_Sj}$ is equal to

$$
\pi_{i_1...i_Sj} = \pi_{0...0} \exp\left(\sum_{s=1}^{S} i_s \delta_{sj} + \mathbf{t}'\boldsymbol{\mu}_j + \tfrac{1}{2}\mathbf{t}'\boldsymbol{\Gamma}_j\mathbf{t} \right) \tag{22.23}
$$

where $\boldsymbol{\mu}_j$ is the mean vector for the jth stratum and $\boldsymbol{\Gamma}_j$ is a symmetric matrix. Let $m_{i_1...i_Sj} = n\pi_{i_1...i_Sj}$ denote the expected count of observed frequencies $n_{i_1...i_Sj}$. Then, we have the log-linear representation

$$
\log m_{i_1...i_Sj} = \delta_j + \sum_{s=1}^{S} i_s \delta_{sj} + \mathbf{t}'\boldsymbol{\mu}_j + \tfrac{1}{2}\mathbf{t}'\boldsymbol{\Gamma}_j\mathbf{t}. \tag{22.24}
$$

Without additional constraints, the model is not identified. If we set $\boldsymbol{\mu}_j$ equal to $\mathbf{0}$ for identification, then the model becomes

$$
\log m_{i_1...i_Sj} = \delta_j + \sum_{s=1}^{S} i_s \delta_{sj} + \tfrac{1}{2}\mathbf{t}'\boldsymbol{\Gamma}_j\mathbf{t}. \tag{22.25}
$$

The model in Equation (22.25) is a traditional log-linear model and, once the parameters have been estimated, an estimate of the portion of the population missed by all lists and an estimate of the total unknown population size N can be obtained.

If the assumption of measurement invariance holds, then the model in (22.25) can be written in the following way:

$$
\log m_{i_1...i_Sj} = \delta_j + \sum_{s=1}^{S} i_s \delta_s + \mathbf{t}'\boldsymbol{\mu}_j + \tfrac{1}{2}\mathbf{t}'\boldsymbol{\Gamma}_j\mathbf{t}, \tag{22.26}
$$

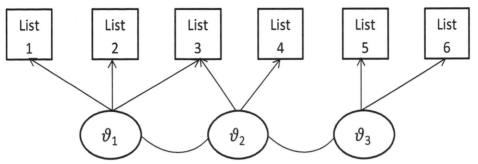

FIGURE 22.6: Model with six lists and three latent variables

where the parameters can be interpreted as before.

Example 4: Model with six lists and three latent variables. Consider the situation of six lists and three latent variables. Assume that list 1, list 2 and list 3 are indicators of the first latent variable, list 3 and list 4 are indicators of the second latent variable and list 5 and list 6 are indicators of the third latent variable (see the path diagram in Figure 22.6).

In this case the full column rank matrix **U** is given by

$$
\mathbf{U} = \begin{bmatrix} u_{11} & u_{12} & u_{13} \\ u_{21} & u_{22} & u_{23} \\ u_{31} & u_{32} & u_{33} \\ u_{41} & u_{42} & u_{43} \\ u_{51} & u_{52} & u_{53} \\ u_{61} & u_{62} & u_{63} \end{bmatrix} = \begin{bmatrix} 1 & 0 & 0 \\ 1 & 0 & 0 \\ 1 & 1 & 0 \\ 0 & 1 & 0 \\ 0 & 0 & 1 \\ 0 & 0 & 1 \end{bmatrix}.
$$

The model equals

$$
\begin{aligned}
\log m_{i_1 i_2 i_3 i_4 i_5 i_6} &= \delta + \sum_{s=1}^{6} i_s \delta_s + \frac{1}{2}\mathbf{t}'\mathbf{\Gamma}\mathbf{t} \\
&= \delta + i_1\delta_1 + i_2\delta_2 + i_3\delta_3 + i_4\delta_4 + i_5\delta_5 + i_6\delta_6 \\
&\quad + \frac{1}{2}t_1^2\gamma_{11} + \frac{1}{2}t_2^2\gamma_{22} + \frac{1}{2}t_3^2\gamma_{33} + t_1 t_2\gamma_{12} + t_1 t_3\gamma_{13} + t_2 t_3\gamma_{23}.
\end{aligned}
$$

22.4 MR model and standard log-linear model

The MR model can be reparametrised into the standard log-linear model of the second order (that is, the log-linear model that allows for the presence of an interaction parameter of each pair of lists). The re-parametrisation permits us to express the parameters of the standard log-linear model in terms of the log-linear MR model. In other words, these formulae can be helpful to better understand the MR model. In fact, in applying the log-linear unidimensional Rasch model to capture-recapture data, a standard log-linear model is assumed in which all two-factor interaction parameters are equal and positive (this model is denoted in IWGDMFa [154] as a model with a first-order heterogeneity term H_1, but the link with the log-linear unidimensional Rasch model is not explicitly made). In applying the log-linear MR model to capture-recapture data, the structure of the two-factor interaction parameters of the corresponding standard log-linear model depends on the specific assumptions about the relationships between the lists and the latent variables. Through these formulae, the parameters of the standard log-linear model can be computed, regardless the structure of the latent variables ([141]).

Let us consider the standard log-linear model in which all the two-factor interaction parameters are present

$$\log m_{i_1 \ldots i_S j} = \lambda_j + \sum_{s=1}^{S} i_s \lambda_{sj} + \sum_{s=1}^{S-1} \sum_{c=s+1}^{S} i_s i_c \lambda_{scj} \tag{22.27}$$

where λ_j denotes a main-effect parameter for stratum j, λ_{sj} denotes a main-effect parameter for the sth list in the jth stratum and λ_{scj} denotes a two-factor interaction parameter for lists c and s in stratum j.

Using that $t_r = \sum_{s=1}^{S} u_{sr} i_s$ and writing out t_r^2 and $t_r t_\nu$, after some algebra, the MR model in Equation (22.25) takes the form:

$$\log m_{i_1 \ldots i_S j} = \delta_j + \sum_{s=1}^{S} i_s \left[\delta_{sj} + \frac{1}{2} \sum_{r=1}^{q} u_{sr} \gamma_{rrj} + \sum_{r=1}^{q-1} \sum_{\nu=r+1}^{q} u_{sr} u_{s\nu} \gamma_{r\nu j} \right] \tag{22.28}$$

$$+ \sum_{s=1}^{S-1} \sum_{c=s+1}^{S} i_s i_c \left[\sum_{r=1}^{q} u_{sr} u_{cr} \gamma_{rrj} \sum_{r=1}^{q-1} \sum_{\nu=r+1}^{q} \left(u_{sr} u_{c\nu} + u_{s\nu} u_{cr} \right) \gamma_{r\nu j} \right]$$

(see the appendix for details). The parameters of the standard log-linear model can be computed starting from those of the MR model using

$$\lambda_j = \delta_j, \tag{22.29}$$

$$\lambda_{sj} = \delta_{sj} + \frac{1}{2} \sum_{r=1}^{q} u_{sr} \gamma_{rrj} + \sum_{r=1}^{q-1} \sum_{\nu=r+1}^{q} u_{sr} u_{s\nu} \gamma_{r\nu j}, \tag{22.30}$$

and

$$\lambda_{scj} = \sum_{r=1}^{q} u_{sr} u_{cr} \gamma_{rrj} + \sum_{r=1}^{q-1} \sum_{\nu=r+1}^{q} \left(u_{sr} u_{c\nu} + u_{s\nu} u_{cr} \right) \gamma_{r\nu j}. \tag{22.31}$$

If the assumption of measurement invariance holds, the standard log-linear model can be written as

$$\log m_{i_1 \ldots i_S j} = \lambda_j + \sum_{s=1}^{S} i_s \lambda_s + \sum_{s=1}^{S-1} \sum_{c=s+1}^{S} i_s i_c \lambda_{sc} \tag{22.32}$$

and the connecting formulae (22.29)–(22.31) take the form

$$\lambda_j = \delta_j, \tag{22.33}$$

$$\lambda_s = \delta_s + \frac{1}{2} \sum_{r=1}^{q} u_{sr} \gamma_{rr} + \sum_{r=1}^{q-1} \sum_{\nu=r+1}^{q} u_{sr} u_{s\nu} \gamma_{r\nu} \tag{22.34}$$

and

$$\lambda_{sc} = \sum_{r=1}^{q} u_{sr} u_{cr} \gamma_{rr} + \sum_{r=1}^{q-1} \sum_{\nu=r+1}^{q} \left(u_{sr} u_{c\nu} + u_{s\nu} u_{cr} \right) \gamma_{r\nu}. \tag{22.35}$$

Example 5: Re-parametrization of the MR model. Consider the Example 1. In the situation of Figure 22.1 the expressions (22.29)–(22.31) become

$$\lambda = \delta,$$

$$\lambda_s = \delta_s + \frac{1}{2} \sum_{r=1}^{2} u_{sr} \gamma_{rr} + u_{s1} u_{s2} \gamma_{12}$$

and

$$\lambda_{sc} = \sum_{r=1}^{2} u_{sr} u_{cr} \gamma_{rr} + (u_{s1} u_{c2} + u_{s2} u_{c1}) \gamma_{12}.$$

Applying these formulae we can compute the parameters of the traditional log-linear model starting from those of the MR model in the following way:

$$\lambda = \delta$$

$$\lambda_1 = \delta_1 + \frac{1}{2} (u_{11} \gamma_{11} + u_{12} \gamma_{22}) + u_{11} u_{12} \gamma_{12} = \delta_1 + \frac{1}{2} \gamma_{11}$$

$$\lambda_2 = \delta_2 + \frac{1}{2} (u_{21} \gamma_{11} + u_{22} \gamma_{22}) + u_{21} u_{22} \gamma_{12} = \delta_2 + \frac{1}{2} \gamma_{11}$$

$$\lambda_3 = \delta_3 + \frac{1}{2} (u_{31} \gamma_{11} + u_{32} \gamma_{22}) + u_{31} u_{32} \gamma_{12} = \delta_3 + \frac{1}{2} \gamma_{22}$$

$$\lambda_{12} = u_{11} u_{21} \gamma_{11} + u_{12} u_{22} \gamma_{22} + (u_{11} u_{22} + u_{12} u_{21}) \gamma_{12} = \gamma_{11}$$

$$\lambda_{13} = u_{11} u_{31} \gamma_{11} + u_{12} u_{32} \gamma_{22} + (u_{11} u_{32} + u_{12} u_{31}) \gamma_{12} = \gamma_{12}$$

$$\lambda_{23} = u_{21} u_{31} \gamma_{11} + u_{22} u_{32} \gamma_{22} + (u_{21} u_{32} + u_{22} u_{31}) \gamma_{12} = \gamma_{12}.$$

Thus, without lists in common between the two latent variables, the main-effect parameter (λ_s) for the list s is equal to the main-effect parameter in the MR model plus half of the variance (given the total scores) of the latent variable for which list s is an indicator. On the other hand, the two-factor interaction parameter of the standard log-linear model for those lists that are indicators of the same latent variable (λ_{12}) is equal to the variance (given the total scores); while the two-factor interaction parameters for lists which are indicators of different latent variables ($\lambda_{13}, \lambda_{23}$) are equal to the covariance (given the total scores) between the two latent variables.

However, if we construct the two latent variables differently, then we obtain a different parametrisation. Consider the situation in Figure 22.2. Applying formulae (22.29)–(22.31) we have:

$$\lambda_1 = \delta_1 + \frac{1}{2} (\gamma_{11} + \gamma_{22}) + \gamma_{12} \qquad \lambda_2 = \delta_2 + \frac{1}{2} \gamma_{11} \qquad \lambda_3 = \delta_3 + \frac{1}{2} \gamma_{22}$$

$$\lambda_{12} = \gamma_{11} + \gamma_{12} \qquad\qquad\qquad \lambda_{13} = \gamma_{22} + \gamma_{12} \qquad \lambda_{23} = \gamma_{12}.$$

22.5   EM algorithm to estimate missing entries

When using capture-recapture methods to estimate the size of a population from two or more lists of cases it is usually assumed that these lists are referring to the same population. It may happen, on the contrary, that lists relate to different populations (e.g., different periods of time or regions). In such cases, Zwane et al. [313] showed that if the fact that lists refer to different but overlapping populations is ignored, then the resulting estimates of the total population size may be biased. They presented a version of the EM algorithm to estimate the missing entries resulting from lists that are not operating in some strata.

The EM (Expectation Maximisation) algorithm is an iterative procedure proposed by Dempster et al. [100] that is useful to compute the maximum likelihood estimates when the observations can be viewed as incomplete data. Data are assumed to be "missing at random" (MAR) (Rubin [250]), that is, the missing value is conditionally independent of the actual response that would have been observed given the observed responses to other questions. Under the assumption of data MAR, it is possible to use the likelihood-based inference, as the missingness is ignorable and the maximum likelihood estimates are asymptotically unbiased if the model is true (Rice [244]). In general, the EM algorithm is composed of two steps:

- the Expectation step (E-step), in which the expected values of the log-likelihood function are calculated using the current estimates for the parameters and

- the Maximization step (M-step), in which the log-likelihood derived in the E-step is maximised to compute parameters. These parameters are then used to update data and the E-step is again computed.

The algorithm proceeds until convergence of the log-likelihood function.

In the capture-recapture context, the assumption of data MAR means that cases from strata where all lists are active and cases from strata with non-operating lists with the same characteristics do not differ systematically by strata (see Zwane et al. [313]). Let us consider a situation of S overlapping lists, such that $I = 2^S - 1$, and J strata. For convenience of notation, let $s = 1, 2, \ldots, I$ be an index denoting a cross-classification of the S lists. Application of the EM algorithm proposed by the data set is divided into two groups: one group, denoted by S_1, containing strata for which all the lists are available (completely classified cases); the other one, denoted by S_2, consists of strata for which not all the lists are available. All the partially classified cases are partitioned into g groups, so that within each group all the units have the same set of possible cells (here stratum is ignored). Let r_{gj} be the count for the partially classified cases in the jth strata which fall in the group g, and let S_{gj} be the set of cells to which the cases might belong.

In the tth iteration of the E-step, the expected frequencies of partially classified profiles are calculated according to

$$\hat{n}_{sj}^{(t)} = \frac{\sum_{p=1}^{J} \hat{\pi}_{sp}^{(t-1)} \delta(c_{sp} \in S_{gp})}{\sum_{p=1}^{J} \sum_{l=1}^{I} \hat{\pi}_{lp}^{(t-1)} \delta(c_{lp} \in S_{gp})} \times r_{gj}$$

where $\delta(c_{sj} \in S_{gj})$, for $s = (1, \ldots, I)$, and $g = (1, \ldots, G)$ is an indicator function assuming value 1 if cell c_{sj} belongs to S_{gj} and 0 otherwise.

After the tth step the data set are completed, that is all the expected frequencies corresponding to missing entries are calculated. Then, in the M-step a log-linear model is fitted to completed data, treating the cells missed by design as structurally zero, and the complete data log-likelihood

$$l^{(t)} = \sum_{c_{sj} \in S_1} n_{sj} \log \pi_{sj} + \sum_{c_{sj} \in S_{gj}} \hat{n}_{sj}^{(t)} \log \pi_{sj}$$

is maximised in order to calculate the estimated probabilities that will be used in the $(t+1)$th iteration of the E-step. Thus, the updates for the completed data are derived and the log-linear model is fitted in the M-step.

This procedure is repeated until the log-likelihood function converges. Then, the parameters estimated in the last step of the algorithm are used to estimate the expected frequencies for structural zero cells, and finally the estimation of the total population size is obtained.

22.6   Application to real data

To illustrate the application of the log-linear MR methodology to the dataset on neural tube defects in the Netherlands, the EM algorithm proposed by Zwane et al. [313] is used to analyze the data [230]. In order to better understand how to apply the EM algorithm consider, for example, the capture profile 10000 for year 1992 for which list 4 is not operating. Note that the observed frequency for this profile is 9, but it also includes cases that could have had a different capture profile if list 4 had been active. Thus, the EM algorithm has to distribute this value to capture profiles 10000 and 10010.

In this case, S_1 consists of 6 years, while S_2 consists of 5 years, that is $S_1 = (1993, 1994, 1995, 1996, 1997, 1998)$ and $S_2 = (1988, 1989, 1990, 1991, 1992)$. The (t)th E-step of the EM algorithm calculates the expectations of frequencies of capture profiles 10000 and 10010 for year 1992 in the following way:

$$\hat{n}^{(t)}_{10000|1992} = \frac{\sum_{j \in S_1} \hat{n}^{(t-1)}_{10000|j}}{\sum_{j \in S_1} \hat{n}^{(t)}_{100+0|j}} \times n_{10000|1992}$$

$$\hat{n}^{(t)}_{10010|1992} = \frac{\sum_{j \in S_1} \hat{n}^{(t-1)}_{10010|j}}{\sum_{j \in S_1} \hat{n}^{(t)}_{100+0|j}} \times n_{10000|1992}$$

where $\hat{n}^{(t)}_{100+0|j} = \hat{n}^{(t)}_{10000|j} + \hat{n}^{(t)}_{10010|j}$.

To apply the MR methodology to the data set, first of all we assume that the five lists may be divided into two sets of indicators that each measure a separate latent variable.

In order to decide which lists measure the same latent variable, in the M-step a model with an interaction parameter for each pair of lists is fitted and, after convergence of the EM algorithm, the parameter estimates of the two-factor interactions are studied. A high value of an estimate of a two-factor interaction is an indication of a positive relationship between two lists, so that they can then be viewed as indicators of the same latent variable. Table 22.4 summarises the estimates for the two factor interaction parameters among lists.

TABLE 22.4
Estimates of the two-factor interaction parameters

		1	2	c 3	4	5
	1	-				
	2	0.718424	-			
s	3	0.185740	0.024525	-		
	4	0.557406	1.055780	1.690401	-	
	5	0.633640	−0.100489	0.467334	1.725820	-

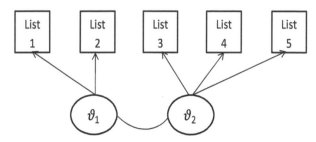

FIGURE 22.7: Path diagram of model 4

From Table 22.4, it can be assumed that lists 1 and 2 measure a first latent variable (named θ_1), and that lists 3, 4 and 5 measure a second latent variable (called θ_2). In addition, it seems also reasonable to consider a model in which list 4 measures the two latent variables. In particular, lists 1, 2 and 4 can be viewed as indicators of the same latent variable (say θ_3), and that lists 3, 4 and 5 are indicators of the another latent variable (named θ_4). For both the MR models, the assumption of measurement invariance is made. Thus, for the application several models are fitted to the data:

- **Model 1** assumes that the five lists are independent and adds another set of 10 parameters to allow the sizes of the 11 years to differ. It takes the form

$$\log m_{i_1 i_2 i_3 i_4 i_5} = \delta + \delta_j + \sum_{s=1}^{5} i_s \delta_s \qquad j = 1988, \ldots, 1997.$$

Note that it is a classical model that can be used as a baseline,

- **Model 2** adds to Model 1 an interaction parameter for each pair of lists

$$\log m_{i_1 i_2 i_3 i_4 i_5} = \delta + \delta_j + \sum_{s=1}^{5} i_s \delta_s + \sum_{s=1}^{4} \sum_{c=2}^{5} i_s i_c \delta_{sc} \qquad j = 1988, \ldots, 1997.$$

As there are five lists, 10 extra parameters are added, and the number of parameters of the model is 26.

- **Model 3** is the log-linear version of the unidimensional Rasch model, that is also found in IWGMDFa [154], and described as a log-linear model with heterogeneity of order 1 (H_1), assuming that the interaction parameters for each pair of lists are identical and positive. The model is given by

$$\log m_{i_1 i_2 i_3 i_4 i_5} = \delta + \delta_j + \sum_{s=1}^{5} i_s \delta_s + H1 \qquad j = 1988, \ldots, 1997$$

and the number of parameters included is 17.

- **Model 4**, the first of the two MR models resulting from Table 22.4, represented by the path diagram in Figure 22.7. The model takes the form

$$\log m_{i_1 i_2 i_3 i_4 i_5 j} = \delta + \delta_j + \sum_{s=1}^{5} i_s \delta_s + \frac{1}{2} \sum_{r=1}^{2} t_r^2 \gamma_{rr} + t_1 t_2 \gamma_{12}, j = 1988, \ldots, 1997,$$

where $\mathbf{t} = (t_1, t_2)' = \mathbf{i}'\mathbf{U}$ are the total scores accounting for the latent variables θ_1 and

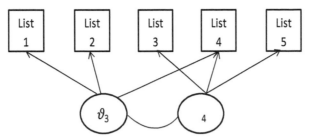

FIGURE 22.8: Path diagram of model 5

θ_2, respectively, δ is the common effect parameter and δ_j is the main-effect parameter for year j (here year 1998 was chosen as reference category). For this model, the matrix $\mathbf{U}$ of weights for the latent variables is given by

$$\mathbf{U} = \begin{bmatrix} u_{11} & u_{12} \\ u_{21} & u_{22} \\ u_{31} & u_{32} \\ u_{41} & u_{42} \\ u_{51} & u_{52} \end{bmatrix} = \begin{bmatrix} 1 & 0 \\ 1 & 0 \\ 0 & 1 \\ 0 & 1 \\ 0 & 1 \end{bmatrix}$$

and the total scores are $t_1 = i_1 + i_2$ and $t_2 = i_3 + i_4 + i_5$.

- **Model 5**, the second of the two MR models, represented by the path diagram in Figure 22.8. The model is given by

$$\log m_{i_1 i_2 i_3 i_4 i_5 j} = \delta + \delta_j + \sum_{s=1}^{5} i_s \delta_s + \frac{1}{2} \sum_{r=3}^{4} t_r^2 \gamma_{rr} + t_3 t_4 \gamma_{34}, j = 1988, \ldots, 1997.$$

In this case, the matrix $\mathbf{U}$ is given by

$$\mathbf{U} = \begin{bmatrix} u_{11} & u_{12} \\ u_{21} & u_{22} \\ u_{31} & u_{32} \\ u_{41} & u_{42} \\ u_{51} & u_{52} \end{bmatrix} = \begin{bmatrix} 1 & 0 \\ 1 & 0 \\ 0 & 1 \\ 1 & 1 \\ 0 & 1 \end{bmatrix},$$

so that the total scores are $t_3 = i_1 + i_2 + i_4$ and $t_4 = i_3 + i_4 + i_5$.

Table 22.5 summarises the results of the models fitted to the data, showing for each model the number of parameters, the degrees of freedom, the deviance, the value of AIC, the value of BIC and the estimate of the total population size $\hat{N}$ are reported. In Table 22.6, the yearly estimates $\hat{N}_j$, for $j = 1988, \ldots, 1998$, under each model are presented. Looking at Table 22.5 it is possible to observe that model 1, the log-linear model with main-effect parameters and parameters for year, does not fit the data well and has a high deviance. Model 2, the model with a different estimate for the interaction parameters between each pair of lists, has a much better fit in terms of AIC and BIC. Model 3 accomplishes a fit between Models 1 and 2: it is the unidimensional Rasch model, which uses only a single parameter H_1 instead of the 10 interaction parameters for each pair of lists (assuming that they are identical and positive). Both the MR Models 4 and 5, fit well to the data and have a smaller deviance than the unidimensional Rasch model. In particular, Model 5, where list

334 Capture-Recapture Methods for the Social and Medical Sciences

4 is an indicator for both latent variables, is the best model because it has the smallest AIC and BIC values. Therefore, this model is selected as the final model. Figure 22.9 shows the yearly estimates for each model.

Starting from the selected model, it is possible to obtain an expression for the standard log-linear model in terms of the parameters of Model 5, using the methodology discussed in Section 22.4. Here, Equations (22.30) and (22.31) simplify to

$$\lambda_s = \delta_s + \frac{1}{2}\sum_{r=3}^{4} u_{sr}\gamma_{rr} + u_{s3}u_{s4}\gamma_{34},$$

$$\lambda_{sc} = \sum_{r=3}^{4} u_{sr}u_{cr}\gamma_{rr} + (u_{s3}u_{c4} + u_{s4}u_{c3})\gamma_{34}.$$

Using these equations, the expressions for the parameters of the standard log-linear model are:

$$\lambda_1 = \delta_1 + \frac{1}{2}\gamma_{33} \qquad \lambda_2 = \delta_2 + \frac{1}{2}\gamma_{33} \qquad \lambda_3 = \delta_3 + \frac{1}{2}\gamma_{44}$$

$$\lambda_4 = \delta_4 + \frac{1}{2}(\gamma_{33} + \gamma_{44}) + \gamma_{34} \qquad \lambda_5 = \delta_5 + \frac{1}{2}\gamma_{44} \qquad \lambda_{12} = \gamma_{33}$$

$$\lambda_{13} = \gamma_{34} \qquad \lambda_{14} = \gamma_{33} + \gamma_{34} \qquad \lambda_{15} = \gamma_{34}$$

$$\lambda_{23} = \gamma_{34} \qquad \lambda_{24} = \gamma_{33} + \gamma_{34} \qquad \lambda_{25} = \gamma_{34}$$

$$\lambda_{34} = \gamma_{44} + \gamma_{34} \qquad \lambda_{35} = \gamma_{44} \qquad \lambda_{45} = \gamma_{44} + \gamma_{34}$$

Thus, the main-effect parameters are equal to the main-effect parameters for Model 5 plus half of the variance (given the total scores) of the latent variable for which the list is an indicator, except for list 4, for which it is equal to the main-effect parameter δ_4 plus half of the variance of both latent variables plus the covariance between θ_3 and θ_4, given the total scores. Note that the two-factor interaction parameters, for those lists that are

TABLE 22.5

Selected models with deviance, AIC and BIC

Model	Design matrix	Par	df*	Dev	AIC	BIC	$\hat{N}$
1	$i_1 + i_2 + i_3 + i_4 + i_5 + Y_{cat}$	16	213	400	432	487	2229
2	$1 + (i_1 i_2 + \cdots + i_4 i_5)$	26	203	298	350	439	3077
3	$1 + H_1$	17	212	349	383	441	3009
4	$1 + t_1 + t_2$	19	210	324	362	427	2793
5	$1 + t_3 + t_4$	19	210	311	349	414	3041

TABLE 22.6

Yearly estimates for the selected models

Model	$\hat{N}_{88}$	$\hat{N}_{89}$	$\hat{N}_{90}$	$\hat{N}_{91}$	$\hat{N}_{92}$	$\hat{N}_{93}$	$\hat{N}_{94}$	$\hat{N}_{95}$	$\hat{N}_{96}$	$\hat{N}_{97}$	$\hat{N}_{98}$
1	199	224	234	206	222	186	189	202	178	210	179
2	275	309	323	285	302	258	261	280	246	290	248
3	272	305	319	281	303	249	252	271	238	280	239
4	251	282	295	260	280	232	235	252	222	261	223
5	271	305	318	281	300	255	258	277	244	287	245

There are 229 observed cells
H_1 is the first-order heterogeneity term
$t_1 = i_1 + i_2$ and $t_2 = i_3 + i_4 + i_5$
$t_3 = i_1 + i_2 + i_4$ and $t_4 = i_3 + i_4 + i_5$

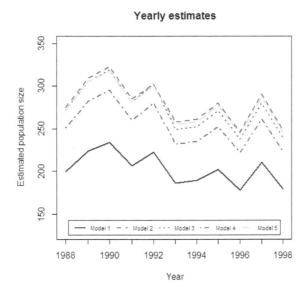

FIGURE 22.9: Yearly estimates for the five models

indicators of different latent variables (that are $\lambda_{13}, \lambda_{15}, \lambda_{23}, \lambda_{25}$), are equal to the covariance (γ_{34}) conditional on the total scores. The two-factor interaction parameters which involve lists measuring the same latent variable (except those involving list 4) are equal to the variance (given the total scores) of the corresponding latent variable, while other two-factor interaction parameters ($\lambda_{14}, \lambda_{24}, \lambda_{34}$ and λ_{45}) are equal to the covariance (given the total scores) plus the variance (given the total scores) of the latent variable for which the other list is assumed to be an indicator. Table 22.7 reports the parameter estimates for model 5 and the corresponding standard errors. In Table 22.8, the parameter estimates of the corresponding standard log-linear model are reported.

Confidence intervals are derived using the parametric bootstrap instead of asymptotic methods (compare Zwane and van der Heijden [311]). Since in this case not all the lists are active in every year, it is not easy to derive asymptotic methods. Furthermore, parametric bootstrap allows us for non-symmetric confidence intervals. To compute the confidence intervals, we first estimate the probabilities for the completed contingency table under a model, including all the cells that cannot be observed by design. These parameters are used to draw the first bootstrap sample as a multinomial sample (according to the probabilities estimates); second, the sample is then reformatted to be identical to the observed data, and the model is then fitted and the population size is estimated. This procedure is repeated 500 times, so that we obtain 500 parametric bootstrap samples, that we used with the percentile method to compute the confidence intervals for the population size estimates for each of the five models (see Table 22.9); we also computed confidence intervals for yearly estimates of the population size for Models 2 and 5. In this case, confidence intervals for the yearly estimates for Model 5 are always smaller than those of Model 2 as shown in Table 22.11. Here yearly estimates of the population size and confidence intervals for the standard log-linear model are presented. Note that the use of the EM algorithm allows to use information from other years for the lists which are not active, so that the MR approach uses the same model for each year. On the contrary, the traditional approach does not use this information and the standard log-linear models differ for each year as the number of lists differs for each year. Furthermore, estimation with a log-linear model tends to be more variable, especially for complete years.

TABLE 22.7
Parameter estimates for MR model 5

Parameter	Estimate	Std. Error
δ	4.513951	0.142557
δ_{1988}	0.101292	0.116082
δ_{1989}	0.218309	0.112754
δ_{1990}	0.260357	0.111628
δ_{1991}	0.135194	0.115088
δ_{1992}	0.201906	0.111082
δ_{1993}	0.038221	0.112887
δ_{1994}	0.050644	0.112545
δ_{1995}	0.122103	0.110637
δ_{1996}	-0.00651	0.114147
δ_{1997}	0.156004	0.109768
δ_1	-2.20858	0.14922
δ_2	-1.04768	0.142911
δ_3	-3.25652	0.124767
δ_4	-2.9981	0.176131
δ_5	-4.16525	0.145811
γ_{33}	0.618927	0.082545
γ_{44}	1.108461	0.087735
γ_{34}	0.219176	0.053513

TABLE 22.8
Calculated parameters for the standard log-linear model

Parameter	Estimate	Std. Error
λ_1	-1.89911	0.154823
λ_2	-0.73821	0.148751
λ_3	-2.70229	0.132254
λ_4	-1.91523	0.193684
λ_5	-3.61102	0.152267
λ_{12}	0.618927	0.082545
λ_{13}	0.219176	0.053513
λ_{14}	0.838102	0.098373
λ_{15}	0.219176	0.053513
λ_{23}	0.219176	0.053513
λ_{24}	0.838102	0.098373
λ_{25}	0.219176	0.053513
λ_{34}	1.327637	0.101656
λ_{35}	1.108461	0.087735
λ_{45}	1.327637	0.101656

22.7 Appendix

The probability of a generic capture profile may be written as

$$\pi_{i_1 i_2 i_3} = \pi_{000} e^{-\sum_s i_s \delta_s} \int \dots \int e^{\mathbf{t}\boldsymbol{\theta}} g\left(\boldsymbol{\theta}|(i_1 i_2 i_3 = 000)\right) d\boldsymbol{\theta}.$$

TABLE 22.9

95% Confidence intervals

Model	Design matrix	$\hat{N}$	95% C.I.
1	$i_1 + i_2 + i_3 + i_4 + i_5 + Y_{cat}$	2229	$[2164, 2297]$
2	$1 + (i_1 i_2 + \dots + i_4 i_5)$	3077	$[2724, 3571]$
3	$1 + H_1$	3009	$[2737, 3345]$
4	$1 + t_1 + t_2$	2793	$[2559, 3104]$
5	$1 + t_3 + t_4$	3041	$[2755, 3409]$

H_1 is the first-order heterogeneity term
$t_1 = i_1 + i_2$ and $t_2 = i_3 + i_4 + i_5$
$t_3 = i_1 + i_2 + i_4$ and $t_4 = i_3 + i_4 + i_5$

TABLE 22.10

95% Confidence intervals for yearly estimates of the population size for MR models

Year	Observed	Model 2		Model 5	
		$\hat{N}$	95% C.I.	$\hat{N}$	95% C.I.
1988	145	275	$[225, 333]$	271	$[221, 330]$
1989	163	309	$[256, 385]$	305	$[255, 372]$
1990	170	323	$[272, 395]$	318	$[268, 394]$
1991	150	285	$[234, 360]$	281	$[234, 344]$
1992	172	302	$[251, 367]$	300	$[254, 362]$
1993	160	258	$[211, 311]$	255	$[211, 305]$
1994	162	261	$[216, 325]$	258	$[215, 319]$
1995	174	280	$[233, 342]$	277	$[235, 329]$
1996	153	246	$[204, 308]$	244	$[203, 296]$
1997	180	290	$[243, 355]$	287	$[238, 345]$
1998	154	248	$[200, 308]$	245	$[205, 301]$

TABLE 22.11

95% Confidence intervals for yearly estimates of the population size

Year	Model	$\hat{N}$	95% C.I.
1988	$i_1 i_2 + i_5$	311	$[200, 648]$
1989	$i_1 + i_2 i_5$	174	$[161, 192]$
1990	$i_1 + i_2 i_5$	177	$[168, 189]$
1991	$i_1 i_2 + i_1 i_5$	191	$[149, 282]$
1992	$i_1 i_2 + i_2 i_3 + i_5 + H_1$	782	$[326, 2687]$
1993	$i_1 i_2 + i_1 i_5 + i_2 i_4 + i_3 i_4 + i_4 i_5$	320	$[207, 957]$
1994	$i_1 i_4 + i_1 i_5 + i_2 i_4 + i_3 i_4 + i_4 i_5$	232	$[197, 293]$
1995	$i_1 i_2 + i_1 i_3 + i_2 i_3 + i_3 i_4 + i_3 i_5 + i_4 i_5$	206	$[188, 231]$
1996	$i_1 i_2 + i_1 i_4 + i_2 i_4 + i_2 i_5 + i_3 i_4 + i_4 i_5$	317	$[220, 583]$
1997	$i_1 i_2 + i_1 i_4 + i_1 i_5 + i_2 i_4 + i_3 i_4 + i_3 i_5 + i_4 i_5$	351	$[259, 595]$
1998	$i_1 i_4 + i_2 i_3 + i_2 i_4 + i_2 i_5 + i_3 i_4 + i_4 i_5$	212	$[179, 266]$

Proof 1 *According to standard probability theory, the probability of a generic capture profile* $(i_1 i_2 i_3)$ *is*

$$
\begin{aligned}
\pi_{i_1 i_2 i_3} &= \int \ldots \int \pi_{i_1 i_2 i_3 \mid \boldsymbol{\theta}} f(\boldsymbol{\theta}) \, d\boldsymbol{\theta} \\
&= \int \ldots \int \prod_{s=1}^{3} \frac{e^{i_s \left(\mathbf{u}_s' \boldsymbol{\theta} - \delta_s \right)}}{1 + e^{\mathbf{u}_s' \boldsymbol{\theta} - \delta_s}} f(\boldsymbol{\theta}) \, d\boldsymbol{\theta}
\end{aligned}
$$

Using

$$
\pi_{000 \mid \boldsymbol{\theta}} = \prod_{s=1}^{3} \frac{1}{1 + e^{\mathbf{u}_s' \boldsymbol{\theta} - \delta_s}}
$$

we obtain

$$
\pi_{i_1 i_2 i_3} = \int \ldots \int \prod_{s=1}^{3} e^{i_s \left(\mathbf{u}_s' \boldsymbol{\theta} - \delta_s \right)} \pi_{000 \mid \boldsymbol{\theta}} f(\boldsymbol{\theta}) \, d\boldsymbol{\theta}.
$$

Multiplying and dividing by π_{000} *and using*

$$
g\left(\boldsymbol{\theta} \mid (i_1 i_2 i_3 = 000) \right) = \frac{\pi_{000 \mid \boldsymbol{\theta}} f(\boldsymbol{\theta})}{\pi_{000}}
$$

where $g\left(\boldsymbol{\theta} \mid (i_1 i_2 i_3 = 000) \right)$ *is the posterior distribution of* $\boldsymbol{\theta}$, *we have*

$$
\begin{aligned}
\pi_{i_1 i_2 i_3} &= \int \ldots \int \pi_{000} \prod_{s=1}^{3} e^{i_s \left(\mathbf{u}_s' \boldsymbol{\theta} - \delta_s \right)} g\left(\boldsymbol{\theta} \mid (i_1 i_2 i_3 = 000) \right) \, d\boldsymbol{\theta} \\
&= \pi_{000} \int \ldots \int e^{\sum_{s=1}^{3} i_s \left(\mathbf{u}_s' \boldsymbol{\theta} - \delta_s \right)} g\left(\boldsymbol{\theta} \mid (i_1 i_2 i_3 = 000) \right) \, d\boldsymbol{\theta} \\
&= \pi_{000} e^{-\sum_s i_s \delta_s} \int \ldots \int e^{\mathbf{t} \boldsymbol{\theta}} g\left(\boldsymbol{\theta} \mid (i_1 i_2 i_3 = 000) \right) \, d\boldsymbol{\theta}
\end{aligned}
$$

and this ends the proof.

$\square$

The parameters of the standard log-linear model can be computed starting from those of the MR model using

$$
\lambda_j = \delta_j,
$$

$$
\lambda_{sj} = \delta_{sj} + \tfrac{1}{2} \sum_{r=1}^{q} u_{sr} \gamma_{rrj} + \sum_{r=1}^{q-1} \sum_{\nu=r+1}^{q} u_{sr} u_{s\nu} \gamma_{r\nu j},
$$

and

$$
\lambda_{scj} = \sum_{r=1}^{q} u_{sr} u_{cr} \gamma_{rrj} + \sum_{r=1}^{q-1} \sum_{\nu=r+1}^{q} \left(u_{sr} u_{c\nu} + u_{s\nu} u_{cr} \right) \gamma_{r\nu j}.
$$

Proof 2 *Consider the MR model*

$$
\begin{aligned}
\log m_{i_1 \ldots i_s j} &= \delta_j + \sum_{s=1}^{S} i_s \delta_{sj} + \tfrac{1}{2} \mathbf{t}' \boldsymbol{\Gamma}_j \mathbf{t} \\
&= \delta_j + \sum_{s=1}^{S} i_s \delta_{sj} + \frac{1}{2} \sum_{r=1}^{q} t_r^2 \gamma_{rrj} + \sum_{r=1}^{q-1} \sum_{\nu=r+1}^{q} t_r t_\nu \gamma_{r\nu j}.
\end{aligned}
$$

Using that $t_r = \sum_{s=1}^{S} u_{sr} i_s$ *it is possible to write out* t_r^2 *and* $t_r t_\nu$ *as:*

$$t_r^2 = \sum_{s=1}^{S} u_{sr}^2 i_s^2 + 2 \sum_{s=1}^{S-1} \sum_{c=s+1}^{S} u_{sr} u_{cr} i_s i_c.$$

and

$$t_r t_\nu = \sum_{s=1}^{S} u_{sr} u_{s\nu} i_s^2 + \sum_{s=1}^{S-1} \sum_{c=s+1}^{S} (u_{sr} u_{c\nu} + u_{s\nu} u_{cr}) i_s i_c.$$

Note that $i_s^2 = i_s$ *and* $u_{sr}^2 = u_{sr}$. *We obtain*

$$\log m_{i_1 \ldots i_S j} = \delta_j + \sum_{s=1}^{S} i_s \delta_{sj} + \frac{1}{2} \sum_{r=1}^{q} \left[\sum_{s=1}^{S} u_{sr} i_s + 2 \sum_{s=1}^{S-1} \sum_{c=s+1}^{S} u_{sr} u_{cr} i_s i_c \right] \gamma_{rrj}$$

$$+ \sum_{r=1}^{q-1} \sum_{\nu=r+1}^{q} \left[\sum_{s=1}^{S} u_{sr} u_{s\nu} i_s + \sum_{s=1}^{S-1} \sum_{c=s+1}^{S} (u_{sr} u_{c\nu} + u_{s\nu} u_{cr}) i_s i_c \right] \gamma_{r\nu j}$$

so that

$$\log m_{i_1 \ldots i_S j} = \delta_j + \sum_{s=1}^{S} i_s \left[\delta_{sj} + \frac{1}{2} \sum_{r=1}^{q} u_{sr} \gamma_{rrj} + \sum_{r=1}^{q-1} \sum_{\nu=r+1}^{q} u_{sr} u_{s\nu} \gamma_{r\nu j} \right]$$

$$+ \sum_{s=1}^{S-1} \sum_{c=s+1}^{S} i_s i_c \left[\sum_{r=1}^{q} u_{sr} u_{cr} \gamma_{rrj} + \sum_{r=1}^{q-1} \sum_{\nu=r+1}^{q} (u_{sr} u_{c\nu} + u_{s\nu} u_{cr}) \gamma_{r\nu j} \right].$$

Note that the latter model is equal to the standard log-linear model, in which

$$\lambda_j = \delta_j$$

$$\lambda_{sj} = \delta_{sj} + \frac{1}{2} \sum_{r=1}^{q} u_{sr} \gamma_{rrj} + \sum_{r=1}^{q-1} \sum_{\nu=r+1}^{q} u_{sr} u_{s\nu} \gamma_{r\nu j}$$

and

$$\lambda_{scj} = \sum_{r=1}^{q} u_{sr} u_{cr} \gamma_{rrj} + \sum_{r=1}^{q-1} \sum_{\nu=r+1}^{q} (u_{sr} u_{c\nu} + u_{s\nu} u_{cr}) \gamma_{r\nu j}.$$

$\square$

23

Extending the Lincoln–Petersen estimator when both sources are counts

Rattana Lerdsuwansri

Thammasat University

Dankmar Böhning

University of Southampton

CONTENTS

23.1 Introduction

Capture-recapture methods have been widely used in enumerating a population of size N that is difficult to approach. Ordinarily, the methods are well established to estimate wildlife abundance (Borchers [49]; Seber [259]). A diversity of application areas has adopted capture-recapture methods to estimate missing units as well as the total number in the population. For instance, in social sciences, the interest is in determining the amount of illegal behavior such as driving a car without license or immigrating without permission (van der Heijden *et al.* [280]). In medical sciences/public health, there is concern about finding the number of illicit drug users as well as estimating the number of outbreaks of particular disease (Gallay *et al.* [124]; Hook and Regal [148]).

We assume that the unknown population size N remains constant during the period of the study (no birth, death or migration), which is referred to as a closed population. To formulate an estimate of population size N, a capture mechanism, e.g., trapping, register, diagnostic device, is used to identify units having a characteristic of interest. For a situation of two independent sources, the Lincoln–Petersen estimator is one of the most popular approaches used in capture-recapture studies. Each source is treated as a binary variable taking values 0 and 1 for an unidentified and identified unit, respectively and a 2×2 contingency table is formed (see Table 23.1). A population size N is partitioned into 4 groups by $n_{11}, n_{10}, n_{01}, n_{00}$, the number of units identified in both sources, in source 1 but not source 2, in source 2 but not source 1, and neither of the two sources. n_{00} is unknown because units who were never identified did not appear. Consequently, n_{00} is required to be estimated leading to an estimate of N.

If two sources were independent, the odds ratio $\dfrac{n_{11}n_{00}}{n_{10}n_{01}} \approx 1$. Under independence, we have $\widehat{n}_{00} = \dfrac{n_{10}n_{01}}{n_{11}}$. See also Chapter 1. The Lincoln–Petersen estimate is then given by

$$
\begin{aligned}
\widehat{N}_{LP} &= n_{11} + n_{10} + n_{01} + \widehat{n}_{00} \\
&= \frac{n_{11}(n_{11} + n_{10}) + n_{01}(n_{11} + n_{10})}{n_{11}} \\
&= \frac{n_{1+}n_{+1}}{n_{11}}.
\end{aligned}
\tag{23.1}
$$

See Brittain and Böhning [53] for more details.

To extend the Lincoln–Petersen approach, not only units were identified but also the same unit was identified repeatedly from both sources. This leads to bivariate counts (i, j) where counts are used to summarize how often a unit was identified by source 1 and source 2, respectively. The number of units identified exactly i times by source 1 and j times by source 2 denoted by f_{ij} is depicted in Table 23.2. It is important to note that the observed data do not include f_{00} since the units who never apprehend do not appear in both of the two sources.

Example 1. To illustrate the situation, a count distribution of heroin user contacts in Bangkok, Thailand, in 2001 is presented in Table 23.3. The data were collected by the Office of the Narcotics Control Board (ONCB), Ministry of the Prime Minister, the Royal Thai Government (see Lanumteang [172] for details). In the case here, the 1st half and 2nd half of the year 2001 are treated as source 1 and source 2, respectively. Several repeated identifications are available from both sources. There were 1401 heroin users who had no contact in the 1st half but had one contact in the 2nd half year, 1736 users had exactly one contact in the 1st half but had no contacts in the 2nd half year and so on. The total size of the observed sample is $n = 5515$. Clearly, heroin users who never contacted treatment centers did not appear in the register and hence $(0, 0)'s$ are unobserved.

TABLE 23.1

A 2×2 table of a two-source situation

		Source 2 (not identified) 0	Source 2 (identified) 1	
Source 1	(not identified) 0	n_{00}	n_{01}	n_{0+}
	(identified) 1	n_{10}	n_{11}	n_{1+}
		n_{+0}	n_{+1}	N

TABLE 23.2

Count distribution in terms of contingency table

			Source 2 Not identified 0	Source 2 Identified 1	2	—	—	m
	Not identified	0	f_{00}	f_{01}	f_{02}	—	—	f_{0m}
Source 1	Identified	1	f_{10}	f_{11}	f_{12}	—	—	f_{1m}
		2	f_{20}	f_{21}	f_{22}	—	—	f_{2m}
		—			—	—	—	
		m	f_{m0}	f_{m1}	f_{m2}	—	—	f_{mm}

Although capture-recapture contributions have experienced theoretical developments, there is not much work available for bivariate count variables. The purpose of this chapter is to estimate the missing units and the number in total of the target population in the two-source situation coping with the independence assumption and heterogeneity in the parameters of capture probabilities.

23.2 Discrete mixtures of bivariate, conditional independent Poisson distributions

Suppose that a population is closed with size N. Let $\boldsymbol{Y}_d = (Y_{d1}, Y_{d2})'$ denote the number of times that unit d is identified by source 1 and source 2 in the observational period, $d = 1, 2, \ldots, N$. $\boldsymbol{Y}_d$ is a vector of two dimensions and a bivariate count variable having values in $\{0, 1, 2, 3, \ldots\}$. If $\boldsymbol{Y}_d = \boldsymbol{0}$, then a unit is unidentified from both of two sources. Since a unit which has never been identified does not appear in the data, the observed data set is $\{\boldsymbol{Y}_d = (Y_{d1}, Y_{d2})' \mid Y_{d1} + Y_{d2} \geq 1 , d = 1, 2, \ldots, N\}$. The associated, observed distribution of counts is therefore referred to as zero-truncated distribution. Let f_{ij} denote the number of units identified i times by source 1 and j times by source 2. f_{00} is the frequency of units identified zero times by both sources. f_{00} is unknown. We have the observed size of the sampled target population $n = f_{01} + f_{10} + f_{11} + \ldots + f_{mm}$ and $N = n + f_{00}$. f_{00} requires determination in order to obtain an estimate for the population size N.

To model the data in Table 23.2, let $p_{ij} = Pr(\boldsymbol{Y}_d = (i, j))$ denote the probability of identifying a unit i times by source 1 and j times by source 2. Accordingly, p_{00} is the probability of not identifying a unit. The unobserved frequency f_{00} might be replaced by the expected value Np_{00}. If p_{00} is known then solving for $N = n + f_{00}$ leads to the well-known Horvitz–Thompson estimator (see van der Heijden et al. [280] for more details)

$$\widehat{N} = \frac{n}{1 - \hat{p}_{00}}. \tag{23.2}$$

As p_{00} is unknown, modeling for count probability p_{ij} has to be assumed. However, count data modeled by an identical parameter θ are rare in practice. An alternative model incorporating heterogeneity of the population might be more appropriate (Norris and Pollock [220], Pledger [232], Dorazio and Royle [102], Böhning and Schön [37]). Based upon repeated identifications, we postulate that $\boldsymbol{Y}$ arises from Poisson distribution having parameter $\theta = (\lambda, \mu)'$ where λ and μ are parameters of count distribution in source 1 and source 2, respectively. The independence assumption is restrictive for the case of the two-source

TABLE 23.3
Frequency distribution of heroin user contacts in 1st half year and 2nd half year of 2001 in Bangkok (Thailand)

			2nd Half year							
			Not identified	Identified						
			0	1	2	3	4	5	6	
	Not identified	0	—	1401	369	98	23	1	1	—
		1	1736	315	129	50	26	1	0	2257
1st		2	445	137	105	53	20	4	0	764
Half	Identified	3	164	89	75	49	30	1	2	410
year		4	47	25	48	34	8	0	0	162
		5	5	7	8	2	3	0	0	25
		6	1	0	1	1	0	0	0	3
		8	0	0	0	1	0	0	0	1
			—	1974	735	288	110	7	3	—

situation. To relax this assumption, we suggest conditional independence which is a weaker assumption than independence forming the Lincoln–Petersen approach. Thus, independence of Y_1 and Y_2 is assumed by conditioning on a homogeneous component. The distribution of count $\boldsymbol{Y}$ is provided as

$$f(\boldsymbol{y}; Q) = \sum_{k=1}^{s} q_k Po(y_1; \lambda_k) Po(y_2; \mu_k). \tag{23.3}$$

where Q is the distribution of an unobserved variable Z taking value $(\lambda_k, \mu_k)'$ with probability q_k. The discrete mixing distribution $Q = \begin{pmatrix} \lambda_1 & \lambda_2 & \dots & \lambda_s \\ \mu_1 & \mu_2 & \dots & \mu_s \\ q_1 & q_2 & \dots & q_s \end{pmatrix}$ gives weight q_k to parameters λ_k and μ_k for $k = 1, 2, \dots, s$ where s is the number of unobserved components. Note that $q_k \geq 0$ and $\sum_{k=1}^{s} q_k = 1$. Equation (23.3) is referred to as discrete mixtures of bivariate, conditional independent Poisson distributions. For an introduction to mixture models, see the books of McLachlan and Peel [207], Lindsay [178], and Böhning [34].

To draw the connection between the Lincoln–Petersen approach and the proposed model (23.3), let us consider $s = 1$, which is the case of independence and homogeneity. We have that $p_{ij} = Po(i; \lambda) Po(j; \mu)$. By truncating all counts larger than 1, the associated marginal probabilities for $i = 0$ and $i = 1$ are given as $e^{-\lambda}/(e^{-\lambda} + \lambda e^{-\lambda}) = 1/(1 + \lambda)$ and $\lambda e^{-\lambda}/(e^{-\lambda} + \lambda e^{-\lambda}) = \lambda/(1 + \lambda)$, respectively. Likewise, the associated marginal probabilities are $1/(1 + \mu)$ for $j = 0$ and $\mu/(1 + \mu)$ for $j = 1$.

Let $p = \lambda/(1 + \lambda)$ and $q = \mu/(1 + \mu)$. The truncated sample leads to complete log-likelihood

$$\log L(p, q) = f_{00} \log\big((1 - p)(1 - q)\big) + f_{01} \log\big((1 - p)q\big)$$
$$+ f_{10} \log\big(p(1 - q)\big) + f_{11} \log\big(pq\big). \tag{23.4}$$

which is maximized separately for $\widehat{p} = \frac{f_{1+}}{N}$ and $\widehat{q} = \frac{f_{+1}}{N}$.

Note that $N = f_{00} + f_{01} + f_{10} + f_{11}$. Unobserved f_{00} can be replaced by $E(f_{00}|p, q) = N(1 - p)(1 - q)$ leading to $N = \dfrac{f_{01} + f_{10} + f_{11}}{1 - (1 - p)(1 - q)}$ where p and q can be estimated by f_{1+}/N and f_{+1}/N, respectively. Hence,

$$N - (N - f_{1+})(1 - \frac{f_{+1}}{N}) = f_{01} + f_{10} + f_{11}$$

$$f_{+1} + f_{1+} - \frac{f_{1+}f_{+1}}{N} = f_{01} + f_{10} + f_{11}$$

$$\widehat{N} = \frac{f_{1+}f_{+1}}{f_{11}} = \widehat{N}_{LP}.$$

23.3 Maximum likelihood estimation for bivariate zero-truncated Poisson mixtures

Assume that Y_1, Y_2, ..., Y_n are observed and drawn from mixture density. The observed, incomplete data log-likelihood is of the form

$$l(Q) = \sum_{\substack{i=0 \\ }}^{m} \sum_{\substack{j=0 \\ i+j \geq 1}}^{m} f_{ij} \log \left(\frac{\sum_{k=1}^{s} q_k Po(i; \lambda_k) Po(j; \mu_k)}{1 - \sum_{k=1}^{s} q_k e^{-\lambda_k} e^{-\mu_k}} \right). \tag{23.5}$$

An estimate of Q can be achieved by maximizing the zero-truncated Poisson likelihood (23.5) leading to the nonparametric maximum likelihood estimate (NPMLE). The EM algorithm has become popular for maximum likelihood estimation particularly in connection with mixture models. To carry out the EM algorithm, the complete data log-likelihood is required. At the E-step, the unobserved frequency f_{00} is replaced by its expected value given observed frequencies and current values of Q. Let the expected value of f_{00} be denoted by $\widehat{f}_{00}$ which can be shown to be

$$\widehat{f}_{00} = E(f_{00} \,|\, \text{observed data} \,; Q) = \frac{n \sum_{k=1}^{s} q_k e^{-\lambda_k} e^{-\mu_k}}{1 - \sum_{k=1}^{s} q_k e^{-\lambda_k} e^{-\mu_k}}.$$

The associated complete data log-likelihood is given by

$$l(Q) = \widehat{f}_{00} \log \left(\sum_{k=1}^{s} q_k e^{-\lambda_k} e^{-\mu_k} \right) + \sum_{\substack{i=0 \\ }}^{m} \sum_{\substack{j=0 \\ i+j \geq 1}}^{m} f_{ij} \log \left(\sum_{k=1}^{s} q_k Po(i; \lambda_k) Po(j; \mu_k) \right). \tag{23.6}$$

To achieve the maximum likelihood estimate $\widehat{Q}$, the log-likelihood is maximized by applying the EM algorithm as well. In this case, a variable, indicating the component to which the count (i, j) belongs to, is introduced. Let z_{ijk} denote the indicator variable defined as 1 if count (i, j) was drawn from component k; 0 otherwise.

If z_{ijk} were observed, the log-likelihood for the complete data would be given by

$$l(Q) = \sum_{i=0}^{m} \sum_{j=0}^{m} f_{ij} \sum_{k=1}^{s} z_{ijk} \log(q_k) + \sum_{i=0}^{m} \sum_{j=0}^{m} f_{ij} \sum_{k=1}^{s} z_{ijk} \log \left(Po(i; \lambda_k) Po(j; \mu_k) \right). \tag{23.7}$$

At the E-step, the unobserved indicator z_{ijk} is replaced by $e_{ij,k}$ (see also Chapter 20), its expected value conditional upon the observed data and current values of Q leading to

$$e_{ij,k} = E(z_{ijk} \,|\, f_{ij}; Q) = \frac{q_k Po(i; \lambda_k) Po(j; \mu_k)}{\sum_{k=1}^{s} q_k Po(i; \lambda_k) Po(j; \mu_k)}. \tag{23.8}$$

Substituting $e_{ij,k}$ into (23.7) yields the expected log-likelihood which is of the form

$$\sum_{i=0}^{m} \sum_{j=0}^{m} f_{ij} \sum_{k=1}^{s} e_{ij,k} \log(q_k) + \sum_{i=0}^{m} \sum_{j=0}^{m} f_{ij} \sum_{k=1}^{s} e_{ij,k} \log \left(Po(i; \lambda_k) Po(j; \mu_k) \right). \tag{23.9}$$

At the M-step, the new values of $\widehat{Q}$ are updated by maximizing (23.9). The estimates of component weights q_k are achieved as

$$\widehat{q}_k = \frac{\sum\limits_{i=0}^{m}\sum\limits_{j=0}^{m} f_{ij} e_{ij,k}}{\widehat{N}} \quad, \quad \text{for } k = 1, 2, \ldots, s. \tag{23.10}$$

The estimates of component parameters are found as

$$\widehat{\lambda}_k = \frac{\sum\limits_{i=0}^{m}\sum\limits_{j=0}^{m} i f_{ij} e_{ij,k}}{\sum\limits_{i=0}^{m}\sum\limits_{j=0}^{m} f_{ij} e_{ij,k}} \quad, \quad \text{for } k = 1, 2, \ldots, s \tag{23.11}$$

$$\widehat{\mu}_k = \frac{\sum\limits_{i=0}^{m}\sum\limits_{j=0}^{m} j f_{ij} e_{ij,k}}{\sum\limits_{i=0}^{m}\sum\limits_{j=0}^{m} f_{ij} e_{ij,k}} \quad, \quad \text{for } k = 1, 2, \ldots, s. \tag{23.12}$$

Consequently, the population size estimator based upon discrete mixtures of bivariate, conditional independent Poisson models through the Horvitz–Thompson approach is

$$\widehat{N} = \frac{n}{1 - \widehat{p}_{00}} = \frac{n}{1 - \sum\limits_{k=1}^{s} q_k e^{-\widehat{\lambda}_k} e^{-\widehat{\mu}_k}}. \tag{23.13}$$

Maximum likelihood estimation discussed above is along the lines of Böhning and Schön [37]. In addition, the estimator (23.13) is derived from a likelihood based on the conditional distribution of count $(i, j)'s$ given n and referred to as a conditional approach. Typically the conditional approach faces two major problems, namely the boundary problem and a lack of identifiability in the context of mixture models. The boundary problem deals with the circumstance that the mixing distribution equates component parameters $\approx 0^+$ with positive weight (Wang and Lindsay [290], [291]). This results in a spurious estimate for the size N of a population as illustrated in Kuhnert and Böhning [170]. The boundary problem not only makes the point estimate of N not trustworthy, but a lack of identifiability also affects the inference about population size N (Link [180]; Holzmann et al. [146]). Since different models providing different estimates of N might have identical conditional distributions, identifiability of N is in question for the conditional likelihood. As a consequence, these entail inferring the unknown parameter N by using unconditional maximum likelihood.

23.4 Unconditional MLE via a profile mixture likelihood

Assume that $\boldsymbol{f} = (f_{00}, f_{01}, f_{10}, \ldots, f_{mm})$ follows a multinomial distribution with parameters N and $\boldsymbol{p}$ where $\boldsymbol{p} = (p_{00}, p_{01}, p_{10}, \ldots, p_{mm})$ and cell probabilities

$$p_{ij} = f(i, j; Q) = \sum_{k=1}^{s} q_k f(i, j; \lambda_k, \mu_k),$$

Q is an unknown mixing distribution.

The unconditional likelihood function is given by

$$L(N, Q \mid \boldsymbol{f}) = \frac{N!}{f_{00}! f_{01}! f_{10}! \cdots f_{mm}!} \prod_{i=0}^{m} \prod_{j=0}^{m} p_{ij}^{f_{ij}}.$$

As shown in Sanathanan [253], this likelihood function can be factored into two parts,

$$L(N, Q) = L_b(N, Q) \times L_c(Q) \tag{23.14}$$

where

$$L_b(N, Q) = \frac{N!}{f_{00}! (N - f_{00})!} p_{00}^{f_{00}} (1 - p_{00})^{N - f_{00}},$$

$$L_c(Q) = \frac{n!}{f_{01}! f_{10}! \cdots f_{mm}!} \prod_{\substack{i=0 \\ i+j \geq 1}}^{m} \prod_{j=0}^{m} \left(\frac{p_{ij}}{1 - p_{00}} \right)^{f_{ij}},$$

and $n = N - f_{00}$. $L_b(N, Q)$ is a binomial likelihood with size parameter N and success parameter $(1 - p_{00})$. $L_c(Q)$ is a conditional likelihood based on the conditional distribution of count $(i, j)'s$ given n. It is clearly seen that $L_c(Q)$ is independent of the unknown parameter N. Consequently, $\widehat{Q}_c$ is obtained by maximizing $L_c(Q)$, then $\widehat{N}$ is produced by maximizing $L_b(N, \widehat{Q}_c)$ with respect to N leading to

$$\widehat{N}_c = \frac{n}{1 - p_{00}(\widehat{Q}_c)}. \tag{23.15}$$

We attach to $\widehat{N}$ in (23.15) an index c because of the conditional likelihood approach, which is discussed in the preceding section to derive the estimator (23.13).

To avoid nonidentifiability of the population size N obtained from conditional likelihood, we have suggested unconditional maximum likelihood for inferring the unknown parameter N. Recall that the unconditional likelihood function is of the form

$$L(N, Q \mid \boldsymbol{f}) = \frac{N!}{f_{00}! f_{01}! f_{10}! \cdots f_{mm}!} \prod_{i=0}^{m} \prod_{j=0}^{m} p_{ij}^{f_{ij}}. \tag{23.16}$$

The likelihood (23.16) is described by full parameters (N, Q), but indeed we are interested in only N. Consequently, a nuisance parameter Q is eliminated by replacing it with its maximum likelihood estimator at each fixed value of N (Pawitan [229]). The resulting likelihood is called a profile mixture likelihood.

Since $N = n + f_{00}$, finding a profile likelihood of N is equivalent to finding a profile likelihood of f_{00}. Therefore, the full likelihood function (23.16) can be rewritten as

$$L(f_{00}, Q) = \frac{(n + f_{00})!}{f_{00}! f_{01}! f_{10}! \ldots f_{mm}!} \prod_{i=0}^{m} \prod_{j=0}^{m} p_{ij}^{f_{ij}}. \tag{23.17}$$

We simply use $l(\cdot)$ for the log of the likelihood function and omit terms that are constants. Given fixed f_{00}, the log-likelihood function of (23.17) takes the form

$$l(Q \mid f_{00}, f_{01}, f_{10}, \ldots, f_{mm}) = \sum_{i=0}^{m} \sum_{j=0}^{m} f_{ij} \log(p_{ij}). \tag{23.18}$$

To obtain the maximum likelihood estimator of Q at given fixed f_{00}, Equation (23.18) is maximized with respect to Q leading to $\widehat{Q}(f_{00})$. We elaborate the profile mixture likelihood method for Poisson model in the following way.

23.4.1   Profile likelihood of the homogeneous Poisson model

Under the homogeneous Poisson model, we have that

$$l(Q \mid f_{00}, f_{01}, f_{10}, \ldots, f_{mm}) = \sum_{i=0}^{m} \sum_{j=0}^{m} f_{ij} \log \left(Po(i; \lambda) Po(j; \mu) \right). \tag{23.19}$$

The MLEs of λ and μ are shown to be

$$\widehat{\lambda}(f_{00}) = \frac{\displaystyle\sum_{i=0}^{m} \sum_{j=0}^{m} i f_{ij}}{\displaystyle\sum_{i=0}^{m} \sum_{j=0}^{m} f_{ij}} = \frac{S_1}{n + f_{00}} \tag{23.20}$$

and

$$\widehat{\mu}(f_{00}) = \frac{\displaystyle\sum_{i=0}^{m} \sum_{j=0}^{m} j f_{ij}}{\displaystyle\sum_{i=0}^{m} \sum_{j=0}^{m} f_{ij}} = \frac{S_2}{n + f_{00}}, \tag{23.21}$$

where $S_1 = \sum_{i=0}^{m} \sum_{j=0}^{m} i f_{ij}$ and $S_2 = \sum_{i=0}^{m} \sum_{j=0}^{m} j f_{ij}$.

Hence, the profile log-likelihood for f_{00} can be achieved as

$$l(f_{00}, \widehat{Q}(f_{00})) = \log \Gamma(n + f_{00} + 1) - \sum_{i=0}^{m} \sum_{j=0}^{m} \log \Gamma(f_{ij} + 1)$$

$$+ \sum_{i=0}^{m} \sum_{j=0}^{m} f_{ij} \log Po(i; \widehat{\lambda}(f_{00})) Po(j; \widehat{\mu}(f_{00})). \tag{23.22}$$

23.4.2   Profile mixture likelihood of the heterogeneous Poisson model

The log-likelihood for the heterogeneous Poisson model is given by

$$l(Q \mid f_{00}, f_{01}, f_{10}, \ldots, f_{mm}) = \sum_{i=0}^{m} \sum_{j=0}^{m} f_{ij} \log \left(\sum_{k=1}^{s} q_k Po(i; \lambda_k) Po(j; \mu_k) \right). \tag{23.23}$$

To find maximum likelihood estimators of Q for any fixed f_{00}, we could construct an EM algorithm similar to one proposed previously. Thus, we start with the initial parameter values $\widehat{Q}^{(0)}$ and iterate the following two steps until convergence.

E-step: Compute the conditional expectations of the unobserved indicator variable Z given the observed data and current estimates $\widehat{Q}^{(r)}$:

$$e_{ij,k}^{(r)} = \frac{\widehat{q}_k^{(r)} Po(i; \widehat{\lambda}_k^{(r)}) Po(j; \widehat{\mu}_k^{(r)})}{\sum\limits_{k=1}^{s} \widehat{q}_k^{(r)} Po(i; \widehat{\lambda}_k^{(r)}) Po(j; \widehat{\mu}_k^{(r)})} \qquad \text{for } k = 1, 2, \ldots, s. \tag{23.24}$$

M-step: Update the component parameters:

$$\widehat{q}_k^{(r+1)} = \frac{\sum\limits_{i=0}^{m} \sum\limits_{j=0}^{m} f_{ij} e_{ij,k}^{(r)}}{\sum\limits_{i=0}^{m} \sum\limits_{j=0}^{m} f_{ij}} \tag{23.25}$$

$$\widehat{\lambda}_k^{(r+1)} = \frac{\sum\limits_{i=0}^{m} \sum\limits_{j=0}^{m} i f_{ij} e_{ij,k}^{(r)}}{\sum\limits_{i=0}^{m} \sum\limits_{j=0}^{m} f_{ij} e_{ij,k}^{(r)}} \tag{23.26}$$

$$\widehat{\mu}_k^{(r+1)} = \frac{\sum\limits_{i=0}^{m} \sum\limits_{j=0}^{m} j f_{ij} e_{ij,k}^{(r)}}{\sum\limits_{i=0}^{m} \sum\limits_{j=0}^{m} f_{ij} e_{ij,k}^{(r)}}. \tag{23.27}$$

The profile log-likelihood for f_{00} under the heterogeneous Poisson model is given by

$$l(f_{00}, \widehat{Q}(f_{00})) = \log \Gamma(n + f_{00} + 1) - \sum_{i=0}^{m} \sum_{j=0}^{m} \log \Gamma(f_{ij} + 1)$$

$$+ \sum_{i=0}^{m} \sum_{j=0}^{m} f_{ij} \log \left(\sum_{k=1}^{s} \widehat{q}_k Po(i; \widehat{\lambda}_k) Po(j; \widehat{\mu}_k) \right). \tag{23.28}$$

Since the profile log-likelihoods (23.22) and (23.28) are one-dimensional functions of f_{00}, plotting of $l(f_{00}, \widehat{Q}(f_{00}))$ against f_{00} can be used as a graphical device for locating the maximum point of $l(f_{00}, \widehat{Q}(f_{00}))$ as well as corresponding $\widehat{f}_{00}$. This will be explained later in detail.

Example 2. We now show how the profile likelihood method might be used for a given data set. In a simulation, counts y were simulated from a two-component mixture $0.5Po(1)Po(1) + 0.5Po(4)Po(4)$ with $N = 100$. Only zero-truncated counts are considered and $n = 91$. Determining the profile mixture likelihood is done for f_{00} ranging from 0 to 50. In order to find the maximal mode for each f_{00}, we might run the algorithm from several initial values and choose the largest likelihood. Here, the initial values for component means are sampled from the observed counts. The initial values for component weights are the proportion of marginal counts to observed counts. We also use 30 sets drawn from the combinations of component means and component weights at the beginning of the EM algorithm. As a result, the largest log-likelihood is regarded as the profile log-likelihood $l(f_{00}, \widehat{Q}(f_{00}))$. Indeed, the number of components s is unknown and has to be estimated. Given f_{00}, computing the maximum likelihood estimates of Q for each s is started from $s = 1, 2, 3, \ldots$ until the profile log-likelihood $l(f_{00}, \widehat{Q}(f_{00}))$ stops increasing. This is called the profile nonparametric maximum likelihood estimate (profile NPMLE) and the associated number of components is obtained.

Figure 23.1 shows the profile log-likelihood plot of $l(f_{00}, \widehat{Q}(f_{00}))$ against f_{00}. We can see that the highest points of $l(f_{00}, \widehat{Q}(f_{00}))$ for $s = 2$, 3, 4 correspond to an identical value of $f_{00} = 6$. Details of profile likelihood analysis for the number of components $s = 1$ to $s = 4$ are presented in Table 23.4. It is found that at $f_{00} = 6$ the profile NPML is -85.1426 with an associated two-component mixing distribution

$$\widehat{Q}(f_{00}) = \begin{pmatrix} 0.9136 & 4.0907 \\ 1.1045 & 3.6771 \\ 0.4796 & 0.5204 \end{pmatrix}.$$

TABLE 23.4
Profile maximum likelihood analysis for Example 2

s	$l(\widehat{f}_{00}, \widehat{Q})$	$\widehat{\lambda}_k$	$\widehat{\mu}_k$	$\widehat{q}_k$	$\widehat{N}_U$
1	-113.4670	2.7363	2.6044	1.0000	91
2	-85.1426	0.9136	1.1045	0.4796	97
		4.0907	3.6771	0.5204	
3	-85.1437	0.9169	1.1072	0.4811	97
		3.8758	3.6621	0.0432	
		4.1168	3.6836	0.4757	
4	-85.1583	0.9126	1.0653	0.4449	97
		0.9684	1.7032	0.0360	
		3.8909	3.6453	0.2332	
		4.2624	3.7001	0.2859	

A corresponding estimator for population size becomes $\widehat{N}_u = n + \widehat{f}_{00} = 97$. Since the value of the profile likelihood function at $N = \widehat{N}$ is equal to $L(\widehat{N}, \widehat{Q})$ in which $\widehat{N}, \widehat{Q}$ are unconditional MLEs, we attach an index u to the estimator, and call $\widehat{N}_u$ the unconditional MLE.

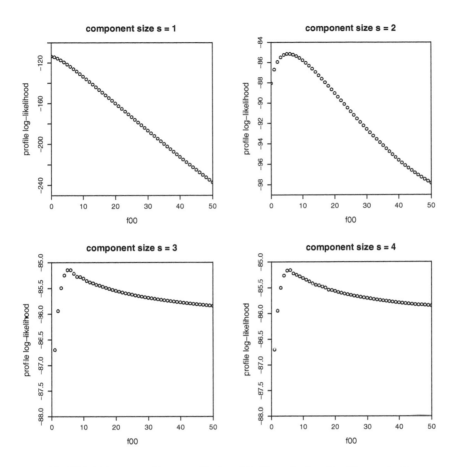

FIGURE 23.1: The profile log-likelihood plot for Example 2.

Recall that an estimator of population size based on the conditional MLE is given as

$$\widehat{N}_c = \frac{n}{1 - p_{00}(\widehat{Q}_c)} = \frac{n}{1 - \sum_{k=1}^{s} \widehat{q}_k e^{-\widehat{\lambda}_k} e^{-\widehat{\mu}_k}}.$$

To obtain the conditional maximum likelihood estimators $(\widehat{N}_c, \widehat{Q}_c)$, we start the EM algorithm suggested in the preceding section by using $\widehat{Q}(f_{00})$ as initial values. The conditional MLE of Q is

$$\widehat{Q}_c = \begin{pmatrix} 0.9036 & 4.0877 \\ 1.0924 & 3.6755 \\ 0.4807 & 0.5193 \end{pmatrix}$$

with associated population size estimate $\widehat{N}_c = 97.3827$ and $\widehat{f}_{00,c} = 6.3827$. This leads us to conclude that $\widehat{N}_u < \widehat{N}_c$, which is an illustration of the more general result by Sanathanan [253].

23.5   Confidence interval estimation for population size N based upon the profile mixture likelihood

In this section we focus our attention on inferring the unknown size N of a closed population. Several authors including Chao [73] and Cormack [87] pointed out that estimating the confidence interval of N is not an easy task for capture-recapture studies, in particular derivation of $Var(\widehat{N})$, which is important for forming the intervals. One could use a bootstrap approach such as percentile bootstrapping to construct confidence intervals. Alternatively, one might achieve a profile likelihood (see, e.g., Cormack [87]; Norris and Pollock [220],[221]). We address achieving confidence interval estimation of N based upon the profile mixture likelihood as follows.

Let $\widehat{N}$ be the profile NPMLE, then the likelihood ratio statistic is given by

$$2\left[l(\widehat{N}, \widehat{Q}) - l(N, \widehat{Q}(N))\right] \sim \chi^2(1).$$

Using the likelihood ratio statistic, all N corresponding to

$$2\left[l(\widehat{N}, \widehat{Q}) - l(N, \widehat{Q}(N))\right] \leq (z_{1-\alpha/2})^2$$

form the $100(1 - \alpha)\%$ confidence set for N. Typically, this set forms an interval for well-behaved likelihoods. Hence, a 95% confidence interval for N can be derived by considering the range of N for which

$$2\left[l(\widehat{N}, \widehat{Q}) - l(N, \widehat{Q}(N))\right] - (1.96)^2 \leq 0. \tag{23.29}$$

Example 2 (continued). We continue using results obtained from Example 2. In this situation, $n = 91$, $l(\widehat{N}, \widehat{Q}) = -85.1496$ and $l(N, \widehat{Q}(N))$ is also provided. Let

$$\delta = 2\left[l(n + \widehat{f}_{00}, \widehat{Q}) - l(n + f_{00}, \widehat{Q}(f_{00}))\right] - (1.96)^2.$$

Plotting of δ against f_{00} is shown in Figure 23.2. As can be seen, f_{00} ranging from 1 to 14 lies below the horizontal line at $\delta = 0$. Consequently, the 95% confidence interval for N is $(92, 105)$.

Confidence intervals based upon the profile mixture likelihood and normal approximation are next evaluated. We consider some traditional estimators such as $\widehat{N}_{MLE}$, $\widehat{N}_{Turing}$, $\widehat{N}_{Chao}$ and $\widehat{N}_Z$. These estimators are based upon the univariate distribution of $f_1, f_2, \ldots$ where $f_1 = f_{01} + f_{10}$, $f_2 = f_{02} + f_{20} + f_{11}$, and so forth. Assuming these estimators are normally distributed, the $100(1 - \alpha)\%$ confidence interval for N takes a form $\widehat{N} \pm z_{1-\alpha/2}\widehat{Se}(\widehat{N})$ where $z_{1-\alpha/2}$ is a $(1 - \alpha/2)^{th}$ quantile of the standard normal distribution and $\widehat{Se}(\widehat{N}) = \sqrt{\widehat{Var}(\widehat{N})}$ is an estimate of the standard error of $\widehat{N}$. Population size estimators and variance estimates of four estimators are provided in the following.

1. **Maximum Likelihood Estimator**
   Under the assumption of a homogeneous Poisson model with mean λ, the maximum likelihood estimator of the population size N

$$\widehat{N}_{MLE} = \frac{n}{1 - \exp(-\widehat{\lambda}_{MLE})} \tag{23.30}$$

where $\widehat{\lambda}_{MLE}$ is the maximum likelihood estimate for the parameter λ of the zero-truncated Poisson distribution. A simple variance estimate for $\widehat{N}_{MLE}$ can be obtained as

$$\widehat{Var}(\widehat{N}_{MLE}) = \frac{\widehat{N}_{MLE}}{(\exp(\frac{\sum_{x=1}^{m} xf_x}{\widehat{N}_{MLE}}) - \frac{\sum_{x=1}^{m} xf_x}{\widehat{N}_{MLE}} - 1)}. \quad (23.31)$$

2. **Good–Turing Estimator**
   In a case of a homogeneous Poisson distribution with mean λ, Turing's estimator is suggested and the population size can be estimated as

$$\widehat{N}_{Turing} = \frac{n}{1 - \frac{f_1}{S}} \quad (23.32)$$

where $S = \sum_{x=1}^{m} xf_x$ is the total number of observations.
As shown in Lerdsuwansri [177], the estimated variance of $\widehat{N}_{Turing}$ is given by

$$\widehat{Var}(\widehat{N}_{Turing}) = \frac{n\frac{f_1}{S}}{(1 - \frac{f_1}{S})^2} + \frac{n^2}{(1 - \frac{f_1}{S})^4} \left[\frac{f_1(1 - f_1/\widehat{N})}{S^2} + \frac{f_1^2}{S^3} \right]. \quad (23.33)$$

Derivation of $\widehat{Var}(\widehat{N}_{Turing})$ utilizes the conditioning approach which was addressed by Böhning [40] for $\widehat{Var}(\widehat{N}_{Chao})$ and $\widehat{Var}(\widehat{N}_Z)$.

3. **Chao's Lower Bound Estimator**
   Chao [72] suggested the lower bound estimator, which is

$$\widehat{N}_{Chao} = n + \frac{f_1^2}{2f_2}. \quad (23.34)$$

The estimated variance of $\widehat{N}_{Chao}$ is given by

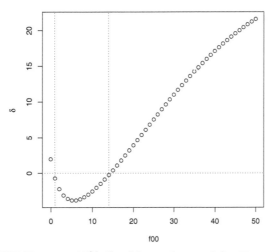

FIGURE 23.2: 95% Confidence interval for Example 2.

$$\widehat{Var}(\widehat{N}_{Chao}) = \frac{f_1^4}{4f_2^3} + \frac{f_1^3}{f_2^2} + \frac{f_1^2}{2f_2} - \frac{f_1^4}{4nf_2^2} - \frac{f_1^4}{2f_2(2nf_2 + f_1^2)}. \tag{23.35}$$

Details are given in Böhning [40].

4. **Zelterman's Estimator**
   Suggested in the case of heterogeneity is Zelterman's estimator, which is given by

$$\widehat{N}_Z = \frac{n}{1 - \exp(-\frac{2f_2}{f_1})}. \tag{23.36}$$

An estimate of the variance of the Zelterman's estimator is obtained as

$$\widehat{Var}(\widehat{N}_Z) = nG(\widehat{\lambda}) \left[1 + nG(\widehat{\lambda})(\widehat{\lambda})^2 \left(\frac{1}{f_1} + \frac{1}{f_2} \right) \right], \tag{23.37}$$

where $G(\widehat{\lambda}) = \frac{\exp(-\widehat{\lambda})}{\left(1 - \exp(-\widehat{\lambda})\right)^2}$ and $\widehat{\lambda} = \frac{2f_2}{f_1}$, see also Böhning [40].

Using the preceding simulation data from a given population size $N = 100$ with $p_{ij} = 0.5Po(i;1)Po(j;1) + 0.5Po(i;4)Po(j;4)$, comparison of confidence intervals of N based on the profile mixture likelihood and normal approximation is shown in Table 23.5. As can be seen, $\widehat{N}_{MLE}$ and $\widehat{N}_{Turing}$ are considerably underestimating the population size. Ultimately, CIs of the MLE and Turing approach do not cover the true N. The length of CI obtained from $\widehat{N}_Z$ is extremely large and becomes useless due to a large variance. Although variance estimation for the suggested method is not provided, there is evidence for success of CI based upon the profile mixture likelihood. It is interesting to notice that the suggested method produces the shortest interval among CIs which contains the true population size N. Additionally, the Profile NPMLE method provides a reasonable lower bound of the CI, which is larger than the number of observed units ($n = 91$), whereas the normal approximation method fails for this point.

23.6   A simulation study

We have carried out a limited simulation study to investigate performance of confidence intervals. Counts (i,j) were generated from the one-component Poisson model

TABLE 23.5
Estimated population size and 95% CI from various estimators (true $N = 100$)

Method	$\widehat{N}$	$\widehat{Se}(\widehat{N})$	95% CI of N	
			Approximate normal	Profile mixture likelihood
MLE	91	0.68	(90, 93)	—
Turing	93	1.67	(90, 97)	—
Chao	97	4.50	(88, 106)	—
Zelterman	105	14.05	(78, 133)	—
Profile NPMLE	97	—	—	(92, 105)

with $p_{ij} = Po(i;1)Po(j;1)$ and from the two-component Poisson model with $p_{ij} = 0.5Po(i;1)Po(j;1) + 0.5Po(i;\lambda_2)Po(j;\mu_2)$, $\lambda_2, \mu_2 \in \{2,3,4\}$ indicating weak, moderate and strong heterogeneity, respectively. Population sizes to be estimated were $N = 100$ and $N = 1000$. From 100 replications of simulated data, $\widehat{N}$ and $\widehat{Se}(\widehat{N})$ were averaged. The percentage of 100 simulated data in which the 95% CI covered the true N as well as lengths of achieved confidence intervals were recorded. The performance of the various estimators are shown in Table 23.6 and Table 23.7.

Achieving coverage probability, which is no less than the nominal confidence level and shortest length of CI, $\widehat{N}_{MLE}$ and $\widehat{N}_{Turing}$ perform the best under homogeneity. The four estimators (Chao, Zelterman, profile NPMLE, and conditional MLE) are generally comparable in the case of heterogeneity. Results from a simulation study show that the Profile NPMLE provides a value between the Chao estimator and the Conditional MLE whereas the Zelterman estimator produces the largest estimates and also the largest $\widehat{Se}(\widehat{N})$.

For a weak heterogeneous population, the profile NPMLE dominates other estimators with respect to the coverage criterion. As can be seen, only the profile NPMLE has 95% coverage probability for $N = 100$. Furthermore, coverage probabilities of the profile NPMLE and Zelterman are close to the nominal confidence level under moderate heterogeneity. However, the lengths of CI obtained from the profile NPMLE are shorter than those of $\widehat{N}_Z$. Both the profile NPMLE and $\widehat{N}_Z$ still behave well under strong heterogeneity in case of $N = 100$. In addition, the performance of the profile NPMLE remains unchanged for $N = 100$ and $N = 1000$ in that it produces the largest coverage and the shortest length of CI.

23.7 Real data example

The data introduced earlier in Example 1 relate to heroin user contacts in Bangkok, Thailand in 2001 (see Table 23.3). The list of the surveillance system is from 61 private and public treatment centers in the Bangkok metropolitan area. The information is constructed on the basis of frequencies of the treatment episodes permitted to treat drug addicts and arise from the surveillance system of the Office of the Narcotics Control Board (ONCB) of the Ministry of Public Health (Thailand). More details of the data source are provided in Lanumteang [172]. Here, the 1st half and 2nd half of the year 2001 are treated as source 1 and source 2, respectively. Based on the two-source situation, it was nnot only recorded if a drug user was identified or not, in addition, the number of times a drug user contacted treatment centers was also recorded. We have that the number of observed heroin users $n = 5515$.

Based on repeated counting of the visits per drug user to treatment centers, which occurred over a given period from January to December 2001, a total number of 5515 heroin users were observed with their contacts ranging from 1 to 11 and are summarized in Table 23.8. There were 3137 heroin users who contacted the treatment centers exactly once. There were 1129 users who visited the treatment centers twice and so forth. Clearly, f_0, the number of hidden drug users is unobserved.

As can be seen from Table 23.9, the MLE provides the smallest estimates as well as the least variation. Similarly, the Turing estimator yields small estimates and small standard deviations. In contrast, $\widehat{N}_{Chao}$ and $\widehat{N}_Z$ have the larger estimates and larger variations. $\widehat{N}_Z$ yields not only the largest variation but also the widest confidence interval. The profile NPMLE provides the estimate 11041 for the total number of heroin users with 95% CI of (10781, 11625). In addition, the corresponding MLE of the mixing distribution is a three-component distribution as follows (with a component giving large weight 0.84 to two small Poisson parameters 0.29 and 0.25)

$$\widehat{Q}(f_{00}) = \begin{pmatrix} 0.2912 & 1.5801 & 2.0340 \\ 0.2469 & 1.0712 & 1.8996 \\ 0.8419 & 0.0886 & 0.0695 \end{pmatrix}.$$

TABLE 23.6

Comparison of various estimators and 95% CI of N (true $N = 100$)

Estimator	Average $\widehat{N}$	Average $\widehat{Se}(\widehat{N})$	Coverage probability	Average length of CI
\multicolumn{5}{c}{$p_{ij} \sim Po(1)Po(1)$}				
MLE	100.97	4.95	0.95	19.40
Turing	100.28	4.86	0.93	19.06
Chao	99.79	6.64	0.94	26.04
Zelterman	99.51	8.71	0.88	34.13
Profile NPMLE	101.96	—	0.95	28.05
Conditional MLE	103.39	—	—	—
\multicolumn{5}{c}{$p_{ij} \sim 0.5Po(1)Po(1) + 0.5Po(2)Po(2)$}				
MLE	97.13	2.40	0.73	9.40
Turing	98.08	2.78	0.84	10.89
Chao	99.21	4.37	0.87	17.12
Zelterman	100.76	7.54	0.87	29.57
Profile NPMLE	100.41	—	0.95	70.40
Conditional MLE	101.38	—	—	—
\multicolumn{5}{c}{$p_{ij} \sim 0.5Po(1)Po(1) + 0.5Po(3)Po(3)$}				
MLE	94.07	1.23	**0.11**	4.82
Turing	96.00	2.04	**0.48**	7.99
Chao	99.01	4.43	0.82	17.37
Zelterman	104.24	10.36	0.94	**40.62**
Profile NPMLE	99.80	—	0.93	25.42
Conditional MLE	103.53	—	—	—
\multicolumn{5}{c}{$p_{ij} \sim 0.5Po(1)Po(1) + 0.5Po(4)Po(4)$}				
MLE	94.00	0.70	**0.02**	2.73
Turing	96.05	1.73	**0.39**	6.80
Chao	100.18	4.76	0.93	18.65
Zelterman	108.24	12.98	0.98	**50.87**
Profile NPMLE	101.54	—	0.96	25.96
Conditional MLE	104.55	—	—	—

23.8 Concluding remarks

The contribution is built upon the extension of the Lincoln–Petersen approach by modifying the binary source variable to a non-binary source variable. Rather than absence (0), presence (1), we focus on how often a unit has been identified $(0, 1, 2, 3, \ldots)$, which is practically useful and easy for practitioners to understand. Although capture-recapture contributions have experienced theoretical developments, there is not much work available for bivariate count variables. To estimate the population size in the two-source situation, we consider a bivariate count variable where counts are used to summarize how often a unit was identified

TABLE 23.7

Comparison of various estimators and 95% CI of N (true $N = 1000$)

Estimator	Average $\widehat{N}$	Average $\widehat{Se}(\widehat{N})$	Coverage probability	Average length of CI
$p_{ij} \sim Po(1)Po(1)$				
MLE	998.92	15.09	0.94	59.14
Turing	997.87	15.24	0.91	59.75
Chao	996.38	21.42	0.91	83.96
Zelterman	995.11	29.00	0.94	113.69
Profile NPMLE	1007.00	—	0.95	102.43
Conditional MLE	1021.97	—	—	—
$p_{ij} \sim 0.5Po(1)Po(1) + 0.5Po(2)Po(2)$				
MLE	968.35	7.33	**0.03**	28.71
Turing	979.91	8.74	**0.43**	34.27
Chao	992.83	14.13	0.92	55.38
Zelterman	1008.92	25.28	1.00	**99.08**
Profile NPMLE	1003.25	—	0.96	74.28
Conditional MLE	1006.56	—	—	—
$p_{ij} \sim 0.5Po(1)Po(1) + 0.5Po(3)Po(3)$				
MLE	944.58	3.86	**0.00**	15.11
Turing	964.49	6.53	**0.00**	25.58
Chao	991.81	13.78	0.88	54.02
Zelterman	1035.66	32.50	0.95	**127.41**
Profile NPMLE	1000.61	—	0.94	60.38
Conditional MLE	1002.79	—	—	—
$p_{ij} \sim 0.5Po(1)Po(1) + 0.5Po(4)Po(4)$				
MLE	936.19	2.19	**0.00**	8.59
Turing	957.16	5.56	**0.00**	21.80
Chao	991.82	13.91	0.88	54.54
Zelterman	1046.54	35.75	0.89	**140.15**
Profile NPMLE	1000.28	—	0.95	66.68
Conditional MLE	1001.61	—	—	—

TABLE 23.8

Frequency distribution of the heroin user contacts for the 1-year period in 2001 in Bangkok (Thailand)

f_0	f_1	f_2	f_3	f_4	f_5	f_6	f_7	f_8	f_9	f_{11}	n
-	3137	1129	528	314	185	127	76	12	6	1	5515

TABLE 23.9

Estimated total number of heroin users in 2001 in Bangkok (Thailand)

Estimator	$\widehat{N}$	$\widehat{Se}(\widehat{N})$	95% Confidence Interval	
			Approximate normal	Profile mixture likelihood
MLE	7115	60.35	(6997, 7234)	-
Turing	7829	79.75	(7672, 7985)	-
Chao	9873	200.08	(9481, 10265)	-
Zelterman	10747	274.01	(10210, 11284)	-
Profile NPMLE	11041	-	-	(10781, 11625)

from source/list 1 and source/list 2. Similar to the axiom of local independence in latent class analysis, independence for a homogeneous component is assumed and the mixture model is presented to model unobserved population heterogeneity. We propose discrete mixtures of bivariate, conditional independent Poisson distribution to fit the arising two-dimensional frequency table.

To estimate the size N of closed populations, two approaches, unconditional maximum likelihood and conditional maximum likelihood, can be dealt with. In this chapter, the unconditional MLE is proposed for population size estimation since identifiability of N is in question for the conditional likelihood (see, e.g., Link [180] ; Holzmann et al., [146], for identifiability of the mixture model). Profile mixture likelihood is exploited for unconditional likelihood maximization. The simulation results show that both $\widehat{N}_u$ and $\widehat{N}_c$ are positively biased and $\widehat{N}_u < \widehat{N}_c$ as proved by Sanathanan [253]. Although there is a drawback that the suggested approach is computationally intensive, the unconditional likelihood has two central properties. Firstly, we arrive at the confidence interval of population size without calculating the variance of the population size estimator. Secondly, the lower limit of the profile confidence interval is at least as large as the number of observations we have, which is not necessarily so for other methods. Additionally, CI associated with the profile mixture likelihood has satisfying coverage probabilities. With regard to long run time, bootstrap resampling techniques are not investigated and standard errors of $\widehat{N}$ are not achieved. We do not include confidence interval estimation based on the conditional likelihood approach. It might be valuable to include a stable estimator of N such as the penalized nonparametric maximum likelihood estimator suggested by Wang and Lindsay [290] for comparison. This requires further research.

In general, lists of identifying units are available from two or more sources. Identification of an anonymous person requires some criteria such as gender, date of birth and demography variables for matching. If these sources do not have matching criteria in common, it might be invalid to combine these sources to summarize the number of times that a unique unit was identified. In essence, it might be more reasonable if an informative source is considered. Based on the two-source situation, the certain source is therefore split up into time components. For instance, the 1st half and 2nd half of the year are treated as source 1 and source 2, respectively. More importantly, the crucial assumption of closed populations is definitely retained. This is the benefit of the suggested approach. With respect to subdividing time components, one might split it into quarterly or monthly intervals. Based on multiple sources, log-linear modeling is frequently used to estimate the population size N. It might be problematic since there are a great number of models to fit and select. As a result, a mixture model becomes an alternative way for estimating the population size.

Other extensions of a two-source situation with binary source variables are possible. It might be that one source identifies units in terms of absence (0) presence (1), whereas the other source provides repeated identifications $(0, 1, 2, \ldots)$. We look at the data of heroin user contacts in Bangkok, Thailand in 2001. Suppose that identifications in the 1st half year provide a binary outcome, whereas identifications in the 2nd half year provides a count outcome of how many times a unit has been identified. The associated frequency distributions are presented in Table 23.10.

TABLE 23.10

Frequencies of heroin user contacts in 2001 in Bangkok (Thailand)

			Not identified	2nd half year Identified						
			0	1	2	3	4	5	6	
1st half year	Not identified	0	—	1401	369	98	23	1	1	—
	Identified	1	2398	573	366	190	87	6	2	3622
			—	1974	735	288	110	7	3	—

Recently, Köse et al. [167] have developed a new estimator with the confidence interval of the population size under such a situation. They propose a maximum likelihood estimator of the Poisson parameter based on truncating multiple identifications larger than two. Additionally, a piece of web software called LPMultiple has been provided. For the data of heroin user contacts, the results from LPMultiple show that the new estimate is 8867 with a 95% confidence interval of 8606 9128. The diagnostic test for homogeneity of the Poisson parameter across the two groups defined by source 1 (identified versus not identified) provides invalidation for the assumption. The goodness-of-fit statistic of 305.28 is very large on 2 DF indicating a poor fit of the model. For further details, see Köse et al. [167]. Collapsing counts larger than 1 provide the Lincoln–Petersen estimate, which is 9221 with a 95% confidence interval of 8893–9548. Intuitively, the assumption of independence across two sources is rare in practice. Hence, the conclusion must be done with caution.

The mixture model could be an alternative way to deal with this situation. Let p_{ix} be the probability that a unit is unidentified/identified by source 1 and identified exactly x times by source 2. The unknown p_{00} has to be estimated. We could consider the counting distribution of the number of times X that a unit has been identified. Let the marginal distribution of X be given as

$$f(x; Q) = \sum_{k=1}^{s} q_k Po(x; \lambda_k)$$

with respect to the unobserved variable Z indicating the component membership and having distribution Q. The mixing distribution Q gives non-negative weight q_k to parameter λ_k such that $\sum_{k=1}^{s} q_k = 1$ and s is the number of unobserved components. Then, the observed, incomplete data log-likelihood function becomes

$$\log L(Q) = \sum_{x=1}^{m} f_{0x} \log \left(\frac{\sum_{k=1}^{s} q_k Po(x; \lambda_k)}{1 - \sum_{k=1}^{s} q_k e^{-\lambda_k}} \right) + \sum_{x=0}^{m} f_{1x} \log \left(\sum_{k=1}^{s} q_k Po(x; \lambda_k) \right)$$

where m is the largest occurring count. The difficulty lies in how to maximize the log-likelihood to obtain the maximum likelihood estimate of Q. The associated parameter estimates could provide $\widehat{p}_{00}$. Hence, the estimate of population size according to the Horvitz–Thompson estimator could be given as $\widehat{N} = n/(1 - \widehat{p}_{00})$. These would be left for future research.

Part VII

Bayesian Approaches

24

Objective Bayes estimation of the population size using Kemp distributions

Kathryn Barger

Tufts University

John Bunge

Cornell University

CONTENTS

24.1   Introduction and background

In this chapter we consider objective or noninformative Bayesian estimation of the population size, where the marginal model for the frequency count data is a member of the "Kemp" family of distributions. These were introduced by A. Kemp in 1968 [161] and have since received attention in a variety of settings (Dacey [93], Kemp [164]). We are interested in them here because while the simplest Kemp distributions are the Poisson and gamma-mixed Poisson or negative binomial, the other members of the family are little known, un-"named," and most importantly not necessarily mixed Poisson (Bunge and Willis [63]). Thus they represent interesting candidates for marginal count distributions that depart from the classical mixed-Poisson scenario. Furthermore, they possess an appealing property in terms of the simplicity of their ratios of probabilities $p(j+1)/p(j)$, which was exploited in Willis and Bunge [297] to produce a (frequentist) population-size estimation procedure based on nonlinear regression. We again use this property here. We wish to study objective Bayesian methods for these models due to the appealing inherent properties of such methods, and previous success with an objective Bayes approach to the population-size problem (Barger and Bunge [23]); and also because various frequentist procedures for the Kemp distributions are under development (Willis and Bunge [297], Chapter 9 (Section 9.3) of the present book). In the following, we describe the family of models, specify our objective Bayes procedure, and demonstrate the method on ten datasets from Chapter 1.

24.2   The Kemp family of distributions

The Kemp distributions are defined in terms of their probability generating functions (pgf's). These in turn are defined in terms of the generalized hypergeometric function, which is

$$_pF_q(a; b; s) = \sum_{k=0}^{\infty} \frac{(a_1)_k \cdots (a_p)_k}{(b)_k \cdots (b_q)_k} \frac{s^k}{k!},$$

where

$$(x)_n = \frac{\Gamma(x+n)}{\Gamma(x)}$$

and $p, q, a_1, \ldots, a_p$ and $b, \ldots, b_q$ are parameters. Following the notation of Dacey [93], the general Kemp pgf is then

$$g(s) = g(s; a, c, \lambda) = C_pF_q(a; b; \lambda s),$$

where $a = [a_1, \ldots, a_p], b = [b_1, \ldots, b_q]$, and

$$C^{-1} = {_pF_q}(a; b; \lambda),$$

and $\lambda > 0$ is a (further) parameter. We will write $\theta := (a, b, \lambda)$ and $\Pr(j; \theta) = g^{(j)}(0)/j! = p_\theta(j), j = 0, 1, \ldots$.

We are especially interested here in the following cases:

1. Poisson. $C_0F_0(\cdot; \cdot; \lambda s), \lambda > 0$.

2. Negative binomial. $C_1F_0(a; \cdot; \lambda s), a > 0, \lambda \in (0, 1)$.

3. (no name). $C_0F_1(\cdot; b; \lambda s), b > 0, \lambda > 0$.

In fact, it is shown in [64] that case (3), which we will call $_0F_1$ for short, is not mixed Poisson for any parameter values. This is important because most methods (even nonparametric ones) for population size estimation from frequency count data are based on mixed Poisson models. The higher-order (in p, q) Kemp distributions represent a rare departure from this scenario, and as such offer the possibility of fitting datasets that cannot be accommodated in the classical setting. Additionally, see the LC-class of distributions in Chapter 11 (Section 11.2). On the other hand, the Kemp class includes the Poisson and negative binomial, two of the most commonly used distributions in capture-recapture, and hence the class constitutes a novel direction for generalization of the classical models. In particular, these distributions admit a simple representation for the ratios $p_\theta(j+1)/p_\theta(j)$, which was exploited in Willis and Bunge [297] to produce a fitting procedure and population-size estimation method based on nonlinear regression.

24.3   The likelihood function

Denoting the likelihood function for the frequency count data $\{f_1, f_2, \ldots\}$ by L, we have (cf. Equation (1) in Barger and Bunge [23]):

$$L(N, \theta; \text{data}) = \binom{N}{n} (1 - p_\theta(0))^n (p_\theta(0))^{N-n} \frac{n!}{\prod_{j \geq 1} f_j!} \prod_{j \geq 1} \left(\frac{p_\theta(j)}{1 - p_\theta(0)} \right)^{f_j}$$

$$= \binom{N}{n} (p_\theta(0))^{N-n} \frac{n!}{\prod_{j \geq 1} f_j!} \prod_{j \geq 1} (p_\theta(j))^{f_j}. \tag{24.1}$$

Now define

$$r_\theta(j) := \frac{p_\theta(j)}{p_\theta(j-1)}, \quad j = 1, 2, \ldots,$$

and consider the telescoping product representation

$$\prod_{j \geq 1} (p_\theta(j))^{f_j} = (p_\theta(0))^n \prod_{j \geq 1} r_\theta(j)^{\sum_{i \geq j} f_j}.$$

Substituting this into L, we obtain the general likelihood expression

$$L(N, \theta; \text{data}) = \binom{N}{n} \frac{n!}{\prod_{j \geq 1} f_j!} (p_\theta(0))^N \prod_{j \geq 1} r_\theta(j)^{\sum_{i \geq j} f_j}. \tag{24.2}$$

Next we specialize to the Kemp case. From the definition of those distributions we have

$$r_\theta(j) = \frac{(a_1 + j - 1) \cdots (a_p + j - 1)\lambda}{(b_1 + j - 1) \cdots (b_q + j - 1)(j)},$$

$j = 1, 2, \ldots$, and

$$p_\theta(0) = \frac{1}{{}_pF_q[(a); (b); \lambda]} = \frac{1}{\sum_{k=0}^{\infty} \frac{(a_1)_k \cdots (a_p)_k}{(b)_k \cdots (b_q)_k} \frac{\lambda^k}{k!}}.$$

These expressions can be substituted into (24.2) to obtain a general Kemp likelihood for $\{f_1, f_2, \ldots\}$. While the result is more tractable than the original likelihood in terms of the $p_\theta(j)$, it is still not simple in general, especially with respect to numerical analysis. On the other hand, a notable advantage of the (objective) Bayesian approach is that it only requires evaluation of the likelihood, not, say differentiation. Nevertheless, we confine ourselves here to the ${}_0F_1$ case (the Poisson and negative binomial were dealt with in Barger and Bunge [23], along with some non-Kemp distributions).

For the ${}_0F_1$ case, then, we have

$$r_\theta(j) = \frac{\lambda}{(b + j - 1)j},$$

$j = 1, 2, \ldots$, and

$$p_\theta(0) = ({}_0F_1[(); (b); \lambda])^{-1} = \left(\sum_{k=0}^{\infty} \frac{1}{(b)_k} \frac{\lambda^k}{k!} \right)^{-1} = \left(\sum_{k=0}^{\infty} \frac{\Gamma(b)}{\Gamma(b+k)} \frac{\lambda^k}{k!} \right)^{-1}.$$

Finally we have

$$L(N, \theta; \text{data})$$

$$= \binom{N}{n} \frac{n!}{\prod_{j \geq 1} f_j!} \left({}_0F_1[(); (b); \lambda]\right)^{-N} \prod_{j \geq 1} \left(\frac{\lambda}{(b+j-1)j}\right)^{\sum_{i \geq j} f_j}$$

$$= \binom{N}{n} \frac{n!}{\prod_{j \geq 1} f_j!} \left({}_0F_1[(); (b); \lambda]\right)^{-N} \lambda^{\sum_{j \geq 1} j f_j} \prod_{j \geq 1} \left(\frac{1}{(b+j-1)j}\right)^{\sum_{i \geq j} f_j}.$$

24.3.1   On maximum likelihood estimation

We are interested in comparing our objective Bayes results to those produced by maximum likelihood (ML). However, finding the latter is challenging due to difficulties with numerical optimization, even for the low-order case ${}_0F_1$. But we were able to find MLEs for 6 of the 10 candidate datasets using the following method, which we outline briefly (cf. Chapter 9). First, our point estimate $\hat{N}$ will be the "conditional" MLE. In this procedure we first fit the *zero-truncated* version of the ${}_0F_1$ distribution to the frequency count data $\{f_1, f_2, \ldots\}$ to obtain a (vector) parameter estimate $\hat{\theta}$. We then calculate the empirical Horvitz-Thompson estimator

$$\hat{N} = \frac{n}{1 - p_{\hat{\theta}}(0)}.$$

This is well known to be asymptotically equivalent to the global MLE (Sanathanan [252]). Omitting the combinatorial coefficients, which now become irrelevant, the zero-truncated likelihood is proportional to

$$\left(\frac{p_\theta(0)}{1 - p_\theta(0)}\right)^n \prod_{j \geq 1} (r_\theta(j))^{\sum_{i \geq j} f_j}.$$

We next apply a further approximation to $p_\theta(0)$, based on

$${}_0F_1[(); (b); \lambda] =: {}_0F_1[b; \lambda] \approx \left(1 + \frac{\lambda}{b(b+1)}\right)^{b+1}$$

(Spanier and Oldham [264], Chapters 18 and 50), so that

$$p_\theta(0) = \left({}_0F_1[b; \lambda]\right)^{-1} \approx \left(1 + \frac{\lambda}{b(b+1)}\right)^{-b-1}.$$

The zero-truncated likelihood is thus approximately proportional to

$$\ell(\theta) := \left(\frac{\left(1 + \frac{\lambda}{b(b+1)}\right)^{-b-1}}{1 - \left(1 + \frac{\lambda}{b(b+1)}\right)^{-b-1}}\right)^n \lambda^{\sum_{j \geq 1} j f_j} \prod_{j \geq 1} \left(\frac{1}{(b+j-1)j}\right)^{\sum_{i \geq j} f_j}.$$

The problem then is to find

$$\hat{\theta} = \begin{bmatrix} \hat{\lambda} \\ \hat{b} \end{bmatrix},$$

which is the solution to

$$\begin{bmatrix} \frac{\partial}{\partial \lambda} \log \ell(\theta) &=& 0 \\ \frac{\partial}{\partial b} \log \ell(\theta) &=& 0 \end{bmatrix}.$$

To do this we use pre-implemented numerical optimization routines (Maple^{TM} version 12.0 [200]). For simplicity we will call $\hat{\theta}$, and hence $\hat{N}$, the MLEs, although $\hat{\theta}$ actually maximizes an approximation to the conditional likelihood. To compute standard errors we use the asymptotic variance approximation

$$\text{Var}(\hat{N}) = N \left(a_{00} - a_0^T A^{-1} a_0 \right)^{-1}, \tag{24.3}$$

where $a_{00} = (1 - p_\theta(0))/p_\theta(0)$, $a_0 = (1/p_\theta(0))\nabla_\theta(1 - p_\theta(0))$, and A is the Fisher information matrix of the original untruncated distribution p. The standard error is calculated by substituting $(\hat{N}, \hat{\theta})$ into (24.3) and taking the square root.

24.4 Objective Bayes procedures

We know from Barger and Bunge [23] that for a true joint reference prior, the prior for N is proportional to $N^{-1/2}$, and that the prior for θ, although difficult to derive for more complex models, is independent of that for N. We will implement a reference prior for N and a noninformative prior for θ. Here we adopt independent Cauchy priors on the positive half-line for λ and b, so we have

$$\pi(N) \propto N^{-1/2}, \quad \pi(\lambda) - \frac{2}{\pi(1 + \lambda^2)}, \quad \text{and} \quad \pi(b) = \frac{2}{\pi(1 + b^2)}.$$

The resulting joint posterior is known to be proper due the choice of a proper prior on θ and reference prior for N (Barger and Bunge [23]). Other candidate noninformative priors for λ, or for b, are $\pi(\lambda) \propto \lambda^{-1/2}$, $\pi(\lambda) \propto \lambda^{-1/2}(1 + \lambda)^{-1/2}$, and $\pi(\lambda) \propto \lambda^{-1/2}(1 + \lambda)^{-1}$. An analytic comparison of various priors on θ for the class of Kemp distributions is still needed and is a future research direction.

The joint posterior distribution is then

$$\pi(N, \lambda, b|\text{data}) \propto \pi(N, \lambda, b)L(\text{data}|N, \lambda, b)$$

$$\propto N^{-1/2}(1 + \lambda^2)^{-1}(1 + b^2)^{-1}\frac{N!}{(N - n)!} {}_0F_1[(); (b); \lambda]^{-N}$$

$$\times \left(\frac{\lambda}{b}\right)^n \prod_{j=1}^{\tau-1} \left(\frac{\lambda}{(b + j)(j + 1)}\right)^{n - \sum_{i=1}^j f_i},$$

where τ is the maximum frequency used in the model fitting procedure (cf. Section 24.5 below),

$${}_0F_1[(); (b); \lambda] = \sum_{k=0}^{\infty} \frac{\lambda^k}{(b)_k k!}$$

and the full conditional distribution for N is

$$\pi(N|\lambda, b, \text{data}) \propto N^{-1/2}\frac{N!}{(N - n)!}({}_0F_1[(); (b); \lambda])^{-N}.$$

Note that we are able to base the above development, and our objective Bayes method, on the exact likelihood, not an approximation.

Simulation from the posterior distribution of N can be achieved using Markov chain Monte Carlo methods. Samples are randomly drawn alternatively from the full conditional

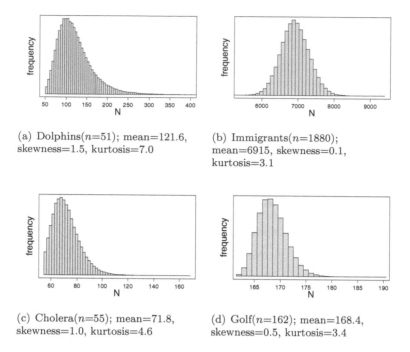

(a) Dolphins($n=51$); mean=121.6, skewness=1.5, kurtosis=7.0

(b) Immigrants($n=1880$); mean=6915, skewness=0.1, kurtosis=3.1

(c) Cholera($n=55$); mean=71.8, skewness=1.0, kurtosis=4.6

(d) Golf($n=162$); mean=168.4, skewness=0.5, kurtosis=3.4

FIGURE 24.1: Histograms of the posterior sample of N from the Kemp model $_0F_1$ showing typical shapes of the objective Bayes posterior densities. Empirical mean, empirical skewness, and empirical kurtosis from the posterior samples are shown.

posterior distributions of N and θ by Gibbs sampling, with the individual draws for N and θ implemented within a Metropolis–Hastings step. For the Bayesian estimation procedure the posterior samples are taken to have an approximate effective sample size of 2,500 (Kass et al. [159]). Acceptance rates for the Markov chain Monte Carlo are tuned to target acceptance rates between 20 and 40 percent. We use Bayesian posterior medians as point estimates and equal-tailed posterior regions based on quantiles as the corresponding interval estimates. For further details see Barger and Bunge [23].

24.5 Data analyses

When fitting parametric distributions to frequency count data, a continuing issue is the question of the upper cutoff τ. Essentially, we carry out the statistical procedure on $\{f_1, f_2, \ldots, f_\tau\}$, obtaining an estimate $\hat{N}_\tau$, and then our final estimate is $\hat{N}_\tau + \sum_{j>\tau} f_j$. There are at least two reasons for this: first, parametric distributions often do not fit complete frequency count datasets, but the fit to the first τ frequencies may be acceptable; second, heuristically speaking there may be more information about N in the lower frequency counts. See Chapter 9 (Section 9.4) for additional discussion. Here we adopt the convention of Barger and Bunge [23] (and others), and set $\tau = 10$, both for the objective Bayes and for the ML procedures.

We applied our objective Bayes procedure to ten of the datasets from Chapter 1, namely

TABLE 24.1: Objective Bayes and ML results on 10 datasets. Wald interval $= \pm 1.96 \times$ SE. Missing results $\equiv$ computation failed

Data (n)	Max. likelihood $\hat{N}$ (95% Wald interval)			Obj. Bayes $\hat{N}$ (95% credible interval)		
	Pois.	Neg. bin.	$_0F_1$	Pois.	Neg. bin.	$_0F_1$
1 Golf (162)	169 (164, 174)	275 (167, 383)	167 (162, 172)	169 (164, 175)	196 (177, 237)	168 (164, 175)
2 Homeless (222)	224 (221, 227)	252 (213, 291)	223 (220, 226)	224 (222, 227)	283 (242, 491)	224 (222, 227)
3 Cholera (55)	89 (66, 112)	1648 — —	— — —	90 (72, 120)	117 (79, 289)	71 (56, 98)
5 Scrapie (118)	170 (145, 195)	709 (−982, 2400)	— — —	171 (150, 201)	839 (349, 2879)	170 (148, 200)
6 CA drugs (20198)	26604 (26361, 26847)	149705 (115476, 183934)	— — —	26603 (26364, 26849)	— — —	— — —
7 Bangkok (3345)	15659 (14273, 17045)	463403 — —	7845 — —	15668 (14365, 17153)	— — —	15674 (14383, 17172)
10 Dolphins (51)	— — —	311 (−862, 1484)	142 (107, 177)	159 (101, 292)	252 (123, 1157)	116 (64, 230)
11 Microbial (81)	101 (88, 114)	277 (−197, 751)	90 (82, 97)	101 (90, 117)	439 (184, 3333)	101 (90, 117)
12 Immigrants (1880)	7080 (6363, 7797)	38961 (−53126, 131048)	5294 — —	7094 (6425, 7845)	— — —	6895 (6131, 7699)
13 Shakespeare (30709)	34541 (34386, 34696)	2220379 (2064217, 2376541)	32650 (32544, 32756)	34540 (34385, 34702)	— — —	34536 (34384, 34694)

golf tees (1.2.1), homeless population (1.2.2), cholera (1.2.3), scrapie (1.2.5), Los Angeles drug users (1.2.6), Bangkok methamphetamine use (1.2.7), dolphins (1.2.10), microbial diversity (1.2.11), Netherlands immigrants (1.2.12), and Shakespeare's words (1.2.13). We also calculated maximum likelihood estimates with 95% confidence intervals. Table 24.1 compares the results.

Observe first that both the objective Bayes procedure and the ML estimates may not be computable (for various numerical reasons). It is also instructive to compare the interval estimates. For the ML case we have used classical Wald intervals, $\hat{N} \pm 1.96 \times \text{SE}$, which rely on the asymptotic normality of the MLE. But the lower confidence bound for these intervals may drop below the observed n, or even below 0, indicating that the normal approximation is not accurate. An adjusted frequentist confidence interval based on a lognormal approximation has been proposed: it is guaranteed to fall above n, but its theoretical foundation is not well established (Chao [72], Bunge et al. [60]).

We can create a histogram from the simulated posterior sample of N to inspect the shape of the density. Empirical skewness and empirical kurtosis reveal posteriors in the examples of Figure 24.1 are positively skewed and leptokurtic (kurtosis greater than 3). The posterior for N is bounded below by the observed n. We have observed that when the estimate of N is close to the observed n, the posteriors become more skewed and leptokurtic. As the posterior distribution moves away from n, the distribution approaches skewness and kurtosis of the normal distribution.

The Kemp model ${}_0F_1$ is "conservative", meaning that its estimate of N is closer to n than under some other models, such as the negative binomial. In this, it is comparable to the Poisson model, at least in terms of numerical results. The latter makes an implicit assumption of homogeneity: each population unit contributes a Poisson-distributed number of observations to the data, and the mean, hence capture probability, for each unit is the same. But ${}_0F_1$ is not (mixed) Poisson, so no corresponding structural interpretation is known. Either may serve as a lower bound for N. Other studies have shown that higher-order (in p and q) Kemp models are less conservative (Willis and Bunge [297]), and implementation of objective Bayes procedures for these is a subject for future research.

Overall we find that the objective Bayes procedure yields point and interval estimates that are computable, stable, and reasonable. *Post hoc* Bayesian model selection procedures may also be applied, if the suite of candidate models is rich enough, e.g., the Kemp family for p, q up to $(1, 1)$ or higher. Outstanding areas for research include refining the choice of prior distributions for the nuisance parameters, faster computation, and expanding the family of models.

25

Bayesian population size estimation with censored counts

Danilo Alunni Fegatelli

Sapienza, University of Rome

Alessio Farcomeni

Sapienza, University of Rome

Luca Tardella

Sapienza, University of Rome

CONTENTS

25.1 Introduction

In this chapter we deal with population size estimation in a particularly interesting case. We assume that there is uncertainty regarding the fact that some observed individuals actually belong to the population of interest.

We are motivated by the Scotland Drug Injectors data set of Overstall et al. [227], where some drug users may have quit and therefore there is left-censoring for some cell counts. See also Chapter 1, Section 1.2.6.

We do so in a Bayesian framework. In the Bayesian framework (e.g., Bernardo and Smith [30]) inference is obtained via the posterior distribution of model parameters. There are clear advantages in our context: first of all, prior knowledge can be summarized by prior distributions (see also Chapter 24, Section 24.4), which also naturally provide regularization of the estimates; additionally, sampling from the posterior distribution is less cumbersome than maximizing the likelihood of a very complex model with censoring.

Left-censoring is, in our opinion, more common than one could expect especially in social science research where separate multiple lists are obtained for the investigation of the population size. In Farcomeni and Scacciatelli [117], for instance, data collection is based on the registry of subjects caught in the street carrying, buying or using cannabis. The final population size estimate is then based on the assumption that all subjects sampled actually have used cannabis at least once, while it could be possible that some of them were carrying or buying it for someone else.

The approach of Overstall et al. proceeds by modeling the counts of the target population underlying each left-censored cell via a truncated Poisson. The only assumption is that the number of subjects in a cell is only an upper bound for the actual number that should have been measured. Other approaches to the problem include Link et al. [181], where the observed counts are assumed to be affected by measurement error over a true latent multinomial count distribution, and Wright et al. [302], which is based on data augmentation.

Overstall et al. [227] focus mostly on a single choice for the prior parameters. In this chapter we revisit and extend their approach. We then compare different objective and subjective prior and model specification choices, both from a theoretical and practical point of view using the motivating data as a case study.

The rest of the chapter is organized as follows: in the next section we introduce the motivating Scotland Drug Injectors data set. We then detail log-linear models for possibly left-censored counts, and provide our first generalizations by discussing some simple forms of unobserved heterogeneity. In Section 25.4 we discuss choices for prior parameters, and their rationale; additionally we use the Deviance Information Criterion (DIC) for model choice. In Section 25.5 we briefly outline how to sample from the posterior distribution of model parameters. In Section 25.6 we illustrate several options for model specification of the Scotland Drug Injectors data set, and give concluding remarks in Section 25.7.

25.2   Scotland Drug Injectors data set

In this section we illustrate the Scotland Drug Injectors data set discussed in Overstall et al.. The aim of this study was to estimate the number of people who inject drugs (PWID) in Scotland. The data was collected in three different time points: 2003, 2006 and 2009. A total of 5670 distinct individuals were listed on four data sources: social inquiry reports (L_1^O), hospital records (L_2^O), Scottish drug misuse database (L_3^O), hepatitis C virus (HCV) database (L_1^C).

Each list is labeled with a superscript (O or C) which highlights whether that corresponds to individuals which are surely identified as PWID (data source with superscript O) or are possibly belonging to a more general population which includes drug users as a subset (data source with superscript C). In this specific example some subject might have quit shortly after data collection, but the idea is more general and applies to any situation of uncertain inclusion in the population of interest. The direct consequence is that the number of people listed in censored lists are an upper bound for the actual number of drug users identified in that specific list.

TABLE 25.1: Scotland Drug Injectors data. In bold, censored frequencies

y	L_1^O	L_2^O	L_3^O	L_1^C	X_1	X_2	X_3	y	L_1^O	L_2^O	L_3^O	L_1^C	X_1	X_2	X_3
?	0	0	0	0	G	M	Y	?	0	0	0	0	R	M	Y
97	1	0	0	0	G	M	Y	278	1	0	0	0	R	M	Y
77	0	1	0	0	G	M	Y	173	0	1	0	0	R	M	Y
3	1	1	0	0	G	M	Y	4	1	1	0	0	R	M	Y
292	0	0	1	0	G	M	Y	1379	0	0	1	0	R	M	Y
7	1	0	1	0	G	M	Y	110	1	0	1	0	R	M	Y
6	0	1	1	0	G	M	Y	39	0	1	1	0	R	M	Y
2	1	1	1	0	G	M	Y	2	1	1	1	0	R	M	Y
122	0	0	0	1	G	M	Y	**134**	0	0	0	1	R	M	Y
2	1	0	0	1	G	M	Y	6	1	0	0	1	R	M	Y
5	0	1	0	1	G	M	Y	7	0	1	0	1	R	M	Y
0	1	1	0	1	G	M	Y	1	1	1	0	1	R	M	Y
3	0	0	1	1	G	M	Y	27	0	0	1	1	R	M	Y
1	1	0	1	1	G	M	Y	5	1	0	1	1	R	M	Y
3	0	1	1	1	G	M	Y	2	0	1	1	1	R	M	Y
0	1	1	1	1	G	M	Y	0	1	1	1	1	R	M	Y
?	0	0	0	0	G	M	O	?	0	0	0	0	R	M	O
60	1	0	0	0	G	M	O	67	1	0	0	0	R	M	O
111	0	1	0	0	G	M	O	144	0	1	0	0	R	M	O
4	1	1	0	0	G	M	O	0	1	1	0	0	R	M	O
149	0	0	1	0	G	M	O	431	0	0	1	0	R	M	O
4	1	0	1	0	G	M	O	16	1	0	1	0	R	M	O
5	0	1	1	0	G	M	O	27	0	1	1	0	R	M	O
0	1	1	1	0	G	M	O	1	1	1	1	0	R	M	O
135	0	0	0	1	G	M	O	**104**	0	0	0	1	R	M	O
1	1	0	0	1	G	M	O	0	1	0	0	1	R	M	O
10	0	1	0	1	G	M	O	7	0	1	0	1	R	M	O
1	1	1	0	1	G	M	O	0	1	1	0	1	R	M	O
2	0	0	1	1	G	M	O	13	0	0	1	1	R	M	O
0	1	0	1	1	G	M	O	1	1	0	1	1	R	M	O
0	0	1	1	1	G	M	O	1	0	1	1	1	R	M	O
0	1	1	1	1	G	M	O	0	1	1	1	1	R	M	O
?	0	0	0	0	G	F	Y	?	0	0	0	0	R	F	Y
41	1	0	0	0	G	F	Y	86	1	0	0	0	R	F	Y
48	0	1	0	0	G	F	Y	108	0	1	0	0	R	F	Y
3	1	1	0	0	G	F	Y	5	1	1	0	0	R	F	Y
117	0	0	1	0	G	F	Y	584	0	0	1	0	R	F	Y
5	1	0	1	0	G	F	Y	53	1	0	1	0	R	F	Y
7	0	1	1	0	G	F	Y	24	0	1	1	0	R	F	Y
0	1	1	1	0	G	F	Y	2	1	1	1	0	R	F	Y
48	0	0	0	1	G	F	Y	**78**	0	0	0	1	R	F	Y
1	1	0	0	1	G	F	Y	3	1	0	0	1	R	F	Y
2	0	1	0	1	G	F	Y	5	0	1	0	1	R	F	Y
0	1	1	0	1	G	F	Y	0	1	1	0	1	R	F	Y
4	0	0	1	1	G	F	Y	18	0	0	1	1	R	F	Y
0	1	0	1	1	G	F	Y	0	1	0	1	1	R	F	Y
0	0	1	1	1	G	F	Y	2	0	1	1	1	R	F	Y
0	1	1	1	1	G	F	Y	0	1	1	1	1	R	F	Y
?	0	0	0	0	G	F	O	?	0	0	0	0	R	F	O
13	1	0	0	0	G	F	O	12	1	0	0	0	R	F	O
34	0	1	0	0	G	F	O	56	0	1	0	0	R	F	O
0	1	1	0	0	G	F	O	0	1	1	0	0	R	F	O
26	0	0	1	0	G	F	O	114	0	0	1	0	R	F	O
0	1	0	1	0	G	F	O	3	1	0	1	0	R	F	O
4	0	1	1	0	G	F	O	9	0	1	1	0	R	F	O
0	1	1	1	0	G	F	O	1	1	1	1	0	R	F	O
38	0	0	0	1	G	F	O	**25**	0	0	0	1	R	F	O
0	1	0	0	1	G	F	O	0	1	0	0	1	R	F	O
4	0	1	0	1	G	F	O	3	0	1	0	1	R	F	O
0	1	1	0	1	G	F	O	0	1	1	0	1	R	F	O
1	0	0	1	1	G	F	O	1	0	0	1	1	R	F	O
0	1	0	1	1	G	F	O	0	1	0	1	1	R	F	O
0	0	1	1	1	G	F	O	1	0	1	1	1	R	F	O
0	1	1	1	1	G	F	O	0	1	1	1	1	R	F	O

Three additional factors were considered: Region (X_1, categorized in two levels: Greater Glasgow and Clyde, Rest of Scotland), Gender (X_2) and Age (X_3, categorized in two levels: < 35 years, ≥35 years). In the HCV database, PWID were not actually observed. The record showed people who were newly diagnosed with the HCV and had injecting drug use history. Therefore, this data source recorded not only current PWID but also former PWID who did not belong to the target population anymore.

The Scotland Drug Injectors data set is shown in Table 25.2. The data can be represented as a 2^7 incomplete contingency table with 2^3 structurally missing cell counts (denoted with a question mark) and 2^3 censored cell counts (in bold type). The former are counts of subjects that cannot be observed by design (as being excluded from all lists). The latter are counts of subjects in the HCV database.

Whether or not a unit is observed in a specific list is treated as a categorical variable with two levels: 0 for unobserved and 1 for observed. For the additional variable Region, the levels are recorded as G (Greater Glasgow and Clyde) and R (Rest of Scotland). For the other additional factors, Gender and Age, the levels are recorded as M (Male) and F (Female), Y (Young, < 35 years) and O (Adults, ≥ 35 years) respectively.

25.3 Mathematical set-up

25.3.1 Log-linear models for possibly truncated counts

We consider a generic multiple-list study with s lists (data sources) such that

$$s = s^O + s^C$$

where s^O is the number of *observed* data sources/lists in which all the recorded units actually belong to the population of interest and s^C is the number of *censored* data sources/lists in which the recorded units may or may not belong to the population. Hence, for each of the s^C lists, the observed count represents only an upper bound of the true number of individuals belonging to the target population for that list.

We will denote with $L_1^O, \ldots, L_{s^O}^O$ and $L_1^C, \ldots, L_{s^C}^C$ the corresponding binary variable. Moreover, we consider v additional categorical predictors $X_1, \ldots, X_v$ with $l_1, \ldots, l_v$ levels respectively.

In the Scottish PWID example, there are $s = 4$ data sources/lists of which $s^O = 3$ correspond to perfectly identified *observed* PWID: social inquiry reports (L_1^O); hospital records (L_2^O); Scottish drug misuse database (L_3^O), and $s^C = 1$ list corresponding to left-*censored* counts: HCV database (L_1^C). Finally, there are $v = 3$ additional factors: age (X_1), gender (X_2) and region (X_3) and each auxiliary variable is categorized in two levels ($l_1 = l_2 = l_3 = 2$).

In this general framework, data can be expressed as an incomplete contingency table with $k = 2^s \prod_{j=1}^{v} l_j$ cells, of which $k^U = \prod_{j=1}^{v} l_j$ are unobserved, $k^C = (2^{s^C} - 1) \prod_{j=1}^{v} l_j$ are left-censored and $k^O = k - (k^U + k^C)$ are observed. In fact, for each of the possible $\prod_{j=1}^{v} l_j$ patterns defined by the auxiliary variables $X_1, \ldots, X_v$, there is one unobserved cell and $2^{s^C} - 1$ censored cells. In our example, the incomplete contingency table has $k = 2^4 \cdot 2^3 = 2^7 = 128$ cells with $k^U = k^C = 2^3 = 8$ and $k^O = 128 - (8 + 8) = 112$.

Let $\mathcal{K}$ be the set of all multidimensional indexes representing the k cross-classifications such that the generic cell index

$$\mathbf{i} = (i_1, \ldots, i_{s^O}, i_{s^O+1}, \ldots, i_{s^O+s^C}, i_{s^O+s^C+1}, \ldots, i_{s^O+s^C+v})$$

identifies the combination of levels of the $(s+v)$ available factors. The first s^O indexes refer to the *observed* data sources/lists, the next s^C ones refer to the *censored* data sources/lists and the last v ones refer to the auxiliary variables.

Let us denote with $\boldsymbol{y}$ the $k \times 1$ vector of true cell counts where the generic entry $y_{\mathbf{i}}$

represents the number of individuals belonging to the target population with the combination of factor levels identified by the index $\mathbf{i}$. Hence, the unknown population size N can be expressed as follows:

$$N = \sum_{\mathbf{i} \in \mathcal{K}} y_{\mathbf{i}}.$$

In order to distinguish between observed, unobserved and censored cells we partition the index set $\mathcal{K}$ as follows: $\mathcal{K} = \mathcal{K}^U \cup \mathcal{K}^C \cup \mathcal{K}^O$ where

$$\mathcal{K}^U = \{\mathbf{i} \in \mathcal{K} : i_1 = \cdots = i_{s^O} = i_{s^O+1} = \cdots = i_{s^O+s^C} = 0\}$$
$$\mathcal{K}^C = \{\mathbf{i} \in \mathcal{K} : i_1 = \cdots = i_{s^O} = 0\} \setminus \mathcal{K}^U$$
$$\mathcal{K}^O = \mathcal{K} \setminus (\mathcal{K}^C \cup \mathcal{K}^U).$$

The subset $\mathcal{K}^U$ is made up of all the indexes $\mathbf{i} \in \mathcal{K}$ such that the level related to each specific data source/list is 0 (unobserved) and hence, of course, the respective true counts are unobserved. In the same way, $\mathcal{K}^C$ and $\mathcal{K}^O$ represent the subsets of $\mathcal{K}$ where the corresponding true counts are censored and observed respectively. Then we can denote with $\boldsymbol{y}^U$, $\boldsymbol{y}^C$ and $\boldsymbol{y}^O$ the true counts for the unobserved, the censored and the observed cells as follows

$$\boldsymbol{y}^U = \{y_{\mathbf{i}} \in \boldsymbol{y} : \mathbf{i} \in \mathcal{K}^U\}$$
$$\boldsymbol{y}^C = \{y_{\mathbf{i}} \in \boldsymbol{y} : \mathbf{i} \in \mathcal{K}^C\}$$
$$\boldsymbol{y}^O = \{y_{\mathbf{i}} \in \boldsymbol{y} : \mathbf{i} \in \mathcal{K}^O\}.$$

Furthermore, let $\mathbf{z}^C$ be the vector representing the observed entries (upper bound) for the censored cells. Indeed, for all the indexes $\mathbf{i} \in \mathcal{K}^C$ we have that $y_{\mathbf{i}} < z_{\mathbf{i}}$. In the light of the foregoing, $\{\boldsymbol{y}^O, \mathbf{z}^C\}$ are the observed data and $\{\boldsymbol{y}^U, \boldsymbol{y}^C\}$ can be regarded as the unknown parameters to be estimated from the data.

We specify a log-linear model (Cormack [86]) where each value $y_{\mathbf{i}}$ follows an independent Poisson distribution

$$y_{\mathbf{i}} \sim \text{Poisson}(\mu_{\mathbf{i}}) \quad \forall \, \mathbf{i} \in \mathcal{K}.$$

The log of the expectation $\mu_{\mathbf{i}}$ can be written as

$$log(\mu_{\mathbf{i}}) = \phi + \mathbf{x}_{\mathbf{i}}^T \boldsymbol{\theta} \tag{25.1}$$

where ϕ is the unknown intercept parameter, $\boldsymbol{\theta}$ is the $m \times 1$ vector of log-linear parameters and $\mathbf{x}_{\mathbf{i}}$ is the $m \times 1$ design vector relating to the combination of levels identified by the index $\mathbf{i}$. In case the design matrix contains only the main effects of the additional factors, we have $m = (s + \sum_{j=1}^{v}(l_j - 1))$. We can write (25.1) in matrix form as

$$log(\mu) = \phi \mathbf{1}_k + \mathbf{X}\boldsymbol{\theta} \tag{25.2}$$

where μ is the $k \times 1$ vector with generic component $\mu_{\mathbf{i}}$, $\mathbf{1}_k$ is a $k \times 1$ vector of ones and $\mathbf{X}$ is the $k \times m$ design matrix with rows given by $\mathbf{x}_{\mathbf{i}}^T$

25.3.1.1 Unobserved heterogeneity

In the set-up above we have assumed possible *list dependence*, which is captured through interactions (specified via the design matrix $\mathbf{X}$). We have also assumed conditional equal catchability, that is, equal probability of being observed for each subject conditional on the design matrix configuration. This is sometimes a restrictive assumption, as in many cases some important covariates might not have been measured, or could even be impossible to measure. This would lead to unobserved heterogeneity. In our setting, unobserved heterogeneity can be simply detected through *overdispersion* in the cell counts (Amstrup et al.

[7]). The Poisson distribution, in fact, has the property that mean and variance coincide. Unobserved heterogeneity implies that cell counts arise from a *mixture* of Poisson distributions (conditional on unobserved covariates). This usually brings about some extra Poisson variation, leading the variance of each cell to exceed, sometimes by a large amount, the expected values.

A common solution is to model the counts through a negative binomial distribution, which can accommodate overdispersion (as its variance is in general larger than the first moment). The negative binomial therefore is able to lead to a better fit and less biased estimates. The negative binomial distribution, though, arises from the specific assumption that data are distributed according to a continuous mixture of Poisson distributions, with Gamma distributed parameters (e.g., White [294], Böhning et al. [36]). This model is less suitable for capture-recapture modeling as frequently readily assumed, as argued in Böhning [46].

More in general, we can accommodate unobserved heterogeneity by assuming there exists a vector of cell-specific intercepts distributed according to a certain distribution $F(\cdot)$, that is:

$$\phi_{\mathbf{i}} \sim F(\cdot) \quad \forall\, \mathbf{i} \in \mathcal{K},$$

for some common pre-specified mixing distribution $F(\cdot)$. There are many possible parametric choices for the mixing distribution F: continuous distributions including Gaussian, Student's T, univariate symmetric Laplace, logit-Beta and discrete distributions with finitely many support points representing latent classes. See also Chapter 20, Section 20.2.1. It is important to underline that $F(\cdot)$ may not be left unspecified due to identifiability issues. As a matter of fact, these issues arise only given that we work with the conditional likelihood (Link, Mao, Farcomeni and Tardella [180, 194, 116]).

In our Bayesian set-up the inclusion of a model component accounting for unobserved heterogeneity is straightforward, as it leads to a hierarchical model. We will compare below the classical Gaussian assumption, where

$$\phi_{\mathbf{i}} \sim N(\mu_\phi, \sigma_\phi^2),$$

and the more recommended (e.g., Pledger [233]) latent class assumption where, for some pre-specified C, we assume there exists a vector of $C \geq 2$ latent locations $\xi_1, \dots, \xi_C$, with $\xi_c \in \mathcal{R}$; and

$$\Pr(\phi = \xi_c) = \pi_C(c),$$

for some unknown probability vector π_C. The use of latent classes is more flexible, as they naturally approximate (to some extent) *any* underlying distribution $F(\cdot)$. It is well known in fact that any density can be approximated by means of discrete distributions to a certain extent and under general assumptions.

25.4   Priors and model choice

An advantage of the Bayesian approach is that prior information, if available, can be summarized and included in the analysis. In presence of large samples the results will be mostly influenced by the data, and only slightly by the prior information.

There are two important issues in our context. First of all, a desirable feature of Bayesian methods is that results are only mildly sensitive to the prior choice. In practical terms, only very strong prior information, as implied by a concentrated prior density, should drastically influence the conclusions. In our experience this is not generally the case for the prior on the parameter of interest, N. Some sensitivity to the prior choice on N is often found (e.g., Wang et al., Farcomeni and Tardella [293, 115]). In modeling left-censored counts, Overstall et al. only touch upon this issue and in fact in the companion R package `conting` it appears that the only choice $\pi(N) = 1/N$ is available. Hence we decided to pursue the present study and investigate how one can obtain other prior inputs.

A second issue with prior choices is that in many cases prior information is not available, and the user should rely on a so-called *default* prior choice. A default prior choice, in general, can be seen as a prior choice that is either convenient (e.g., conjugate) or, better, justified by the asymptotic behaviour or formal properties (e.g., invariance).

A crucial result of Bayesian inference is that the posterior distribution is proper, even in the presence of improper prior inputs. We show that this is the case in Appendix B.

25.4.1 Prior choices for the population size

Popular prior choices include:

- $\pi(N) \propto N^\lambda$. Most often one elicits a single value λ within subset $\{-2, -1, -1/2, 0\}$, possibly truncating the prior to an opportune upper bound $N_{\max}$.

- Rissanen prior, a universal prior on the integers given by $\pi(N) \propto 2^{-\log^*(N)}$, where $\log^*(x) = \log(x) + \log(\log(x)) + \log(\log(\log(x))) + \cdots$, with the sum involving only non-negative terms. See Rissanen [245] for details and more recently Berger et al. [28] for a theoretical framework for developing reference priors on integer parameters.

A more specific discussion related to capture-recapture models can be found for instance in Tardella [274] and Farcomeni and Tardella [115]. Other relevant references are Wang et al. [293] and Xu et al. [303], where extensive studies on simulated and real data are shown to compare prior choices of the kind N^λ.

25.4.1.1 Induced priors

A peculiar feature of the log-linear models considered in the previous section is that the choice of the prior for the parameter of interest N is not transparent as it does not appear explicitly in the model. The prior for N is "induced," namely, it is a consequence of the explicit prior choices for the other parameters. In particular we will point out some remarkable consequences of some specific choices of prior elicitation on ϕ. The original reasoning in Overstall et al. [227] shows that by assuming that $\pi_\phi(\phi) \propto 1$, a prior of the form N^{-1} is consequently obtained for N. In the following we generalize this reasoning.

The model specification of the observable and unobservable (structural or due to censoring) counts y_i ($i \in \mathcal{K}$), conditional on $(\phi, \boldsymbol{\theta}) = (\phi, \theta_1, ..., \theta_m)$, is specified as

$$Y_i|\phi, \theta_1, ..., \theta_m \sim Pois(\mu_i) = Pois\left(e^{\phi + x_i^T \boldsymbol{\theta}}\right).$$

It is important to note here that there is conditional independence (given the true counts Y_i) between the truncated counts and N, and (given the ϕ and $\boldsymbol{\theta}$ parameters) between σ^2 and N.

We can show that assuming a prior of the form $\pi_\phi(\phi) \propto e^{\lambda \phi}$ corresponding to $\pi_U(u) = \pi_\phi(\log u)u^{-1} \propto u^{\lambda - 1}$, where $u = \exp(\phi)$, and assuming prior independence between ϕ and $\boldsymbol{\theta}$, leads to

$$\pi(N) \propto \frac{\Gamma(N + \lambda)}{N!}, \tag{25.3}$$

which, for different choices of λ, yields some of the desired priors for N. A proof of this can be found in Appendix A.

Only a few values of λ corresponding to improper measure densities lead to handy closed-form prior choices for the integer parameter N.

More precisely, setting $\lambda = 0$ corresponds to the improper prior measure on u

$$\pi_U(u) = \pi_\phi(\log u)u^{-1} = u^{-1},$$

which is in turn equivalent to an improper constant prior on ϕ

$$\pi_\phi(\phi) = 1$$

and yields the (improper) Jeffreys prior on N

$$\pi(N) \propto \frac{1}{N}.$$

This recovers the reasoning in Overstall et al. [227].

The case $\lambda = 1$ corresponds to the improper prior measure on u

$$\pi_U(u) = \pi_\phi(\log u)u^{-1} = u \cdot u^{-1} = 1,$$

which is in turn equivalent to an improper prior on ϕ

$$\pi_\phi(\phi) = e^\phi$$

and yields an improper uniform prior on N

$$\pi(N) \propto 1.$$

The case $\lambda = 1/2$ leads to $\pi(N) \approx 1/\sqrt{N}$ for large values of N. To see this, note that

$$\frac{\Gamma(N+\lambda)}{\Gamma(N+1)} = \frac{\Gamma(N+1+(\lambda-1))}{\Gamma(N+1)} \approx N^{\lambda-1}. \tag{25.4}$$

The approximation is good only for large values of N, as in our application where N is in the order of the thousands.

Finally, a proper distribution can also be obtained with $\lambda < 0$. The resulting prior on u is still improper

$$\pi_U \propto u^{\lambda-1} \qquad u \in (0, \infty),$$

but the prior on N is proper due to (25.4). Another popular prior that can be obtained approximately is $\pi(N) \propto 1/N^2$. In fact fixing $\lambda = -1$ leads to

$$\pi(N) \propto \frac{1}{N(N-1)} \approx 1/N^2.$$

In our framework it is not possible to obtain the Rissanen prior, which takes a completely different form. On the other hand, it is well known that the Rissanen distribution is stochastically dominated (above and below, respectively) by $1/N$ and $1/N^2$ in the tails Berger et al. [28]. Consequently, we can assume that there exists a value of $\lambda \in (-1, 0)$ which might approximate the Rissanen prior. For values of N in the order of thousands we have numerically verified that $\lambda \approx -0.16$ minimizes the total variation distance between (25.3) and Rissanen prior.

The specific prior choices considered are summarized in Table 25.2.

TABLE 25.2: Induced priors as a function of λ assuming $\pi(\exp(\phi)) \propto \exp((\lambda - 1)\phi)$ and prior independence

λ	$\pi(N)$
-1	$1/N^2$
-0.16	Rissanen
0	$1/N$
$1/2$	$1/\sqrt{N}$
1	1

25.4.2   Prior choices for the other parameters

We have discussed so far prior alternative choices for the main parameter of interest, N derived as consequence of the prior choices on ϕ.

For decomposable graphical models the default prior choice for the other parameters, suggested in Madigan and York [191], leads to an almost closed-form posterior mass for N (up to a proportionality constant). However, the class of decomposable models is argued to be too restrictive in Overstall et al. [227].

In the log-linear context there are three widely used priors:

- the g-prior,

- the Sabanés-Bové and Held prior (SBH), and

- the multivariate normal prior (MVN).

Let $\mathbf{p}$ be a short-hand notation for model parameters. As in Overstall and King [228], we choose a joint prior distribution of $\mathbf{p} = (\phi, \boldsymbol{\theta})$ with independent components

$$\pi_{\mathbf{p}}(\mathbf{p}) = \pi_{\phi}(\phi)\pi_{\boldsymbol{\theta}}(\boldsymbol{\theta}) \qquad (25.5)$$

with $\pi_{\phi}(\phi) \propto e^{\lambda \phi}$ and

$$\boldsymbol{\theta} \sim N(\mathbf{0}, \mathbf{W}),$$

For the multivariate normal prior, the matrix $\mathbf{W}$ is defined as

$$\mathbf{W} = \sigma^2 \mathbf{I}$$

where $\mathbf{I}$ is the identity matrix and $\sigma^2 > 0$ is an unknown parameter with hyper-prior distribution given by

$$\sigma^2 \sim IG\left(\frac{a}{2}, \frac{b}{2}\right) \qquad (25.6)$$

where $a = b = 10^{-3}$. For the other two prior distributions (g-prior and SBH) we have

$$\mathbf{W} = \sigma^2 n \left(\mathbf{X}^T \mathbf{X}\right)^{-1}.$$

Similar to the multivariate normal case, for the SBH distribution σ^2 is still treated as an unknown parameter with hyper-prior distribution defined as in (25.6). On the other hand, for the g-prior distribution, σ^2 is a fixed hyperparameter. If unobserved heterogeneity is included in the model, we have additional parameters. In case the mixing distribution is Gaussian, the most natural priors for μ_{ϕ} and σ^2_{ϕ} are a zero-centered Gaussian and an inverse gamma. In case the mixing distribution is assumed to be discrete, we also assume $\xi_c \sim N(0, \tau^2_{\xi})$, independently, and $\pi_C \sim \text{Dirichlet}(\mathbf{1})$, where $\mathbf{1}$ denotes a column vector of ones of the appropriate length.

25.4.3 Model choice

The framework briefly sketched above is rather general and leaves us with several options regarding the choice of prior parameters, choice of the design matrix, and modeling of unobserved heterogeneity. In case a latent class model is chosen, another open issue is the choice of the number of latent classes C.

Our suggestion is to repeatedly fit the model, as in some cases estimates may depend on the combination of choices above. Our motivating example is an exception in this respect, as the sample size is very large and therefore prior parameters will not be very important.

There are several options for model choice, see for instance Ghosh [134] and Ghosh [135].

Here we use a simple device, the deviance information criterion (DIC), which summarizes data evidence and naturally adjusts for model complexity through the prior distribution. The DIC is computed as

$$DIC = \bar{D} + p_D$$

where $\bar{D} = \frac{1}{B}\sum_{i=1}^{B}(-2\log(f(\boldsymbol{y}|\mathbf{p}_i))$ is a measure of how well the model fits the data and $p_D = \bar{D} - \max_i(\log(f(\boldsymbol{y}|\mathbf{p}_i)))$ is the effective number of parameters of the model.

Alternatively, one can opt to use the marginal likelihood (ML) to select the best model. The ML is defined as

$$m(\boldsymbol{y}^U, \mathbf{z}^C, \boldsymbol{y}^O) = \int f(\boldsymbol{y}^U, \mathbf{z}^C, \boldsymbol{y}^O|\mathbf{p})d\pi_{\mathbf{p}}(\mathbf{p}), \qquad (25.7)$$

where $\pi(\cdot)$ denotes the joint prior. In practice, (25.7) corresponds to the denominator of the posterior distribution, and in general it is not available in closed form. In order to estimate the marginal likelihood, a simple device is proposed in Chib [82]. This is based on the following identity:

$$\log(m(\boldsymbol{y})) = \log(f(\boldsymbol{y}|\hat{\mathbf{p}})) - \log\left(\frac{1}{B}\sum_{i=1}^{B}\frac{f(\boldsymbol{y}|\hat{\mathbf{p}})}{f(\boldsymbol{y}|\mathbf{p}_i)}\right),$$

where $\boldsymbol{y} = (\boldsymbol{y}^U, \mathbf{z}^C, \boldsymbol{y}^O)$, $\hat{\mathbf{p}}$ is an estimate of the model parameters (e.g., the posterior expectation) and $\mathbf{p}_1, \ldots, \mathbf{p}_B$ is a sample from the posterior distribution. It should be noted that better (though more cumbersome) ways to approximate the marginal likelihood are available, see e.g., Chen and Shao [80] and references therein and thereof.

Once the DIC/log-marginal likelihood for several model specifications is obtained, the best model is the one corresponding to the smallest/largest value.

It shall be here underlined that the DIC is well defined as long as the posterior is proper (once again, see Appendix B). On the other hand, in order to have a well-defined marginal likelihood, proper prior distributions are needed.

Note that for latent class models, numerical issues might arise in computation of the marginal likelihood or of DIC. To avoid underflow we here used the summation on the log scale operator as proposed in the appendix of Farcomeni [113].

25.5 Bayesian inference

We focus on a fully Bayesian approach for carrying out inference on the main quantity of interest N as well as on the other relevant features of the proposed statistical models such as $\psi_{\mathbf{i}} = E\left[\frac{y_{\mathbf{i}}}{z_{\mathbf{i}}}\right]$ ($\mathbf{i} \in \mathcal{K}^C$), the average of the proportions of observed subjects in the censored cells which are actually members of the population of interest. Of course a prior distribution on all the unknown parameters in the model has to be specified. In our model setup where the observed quantities are

$$(\boldsymbol{y}^U, \boldsymbol{z}^C, \boldsymbol{y}^O),$$

we can represent the parameter vector as follows

$$(\phi, \boldsymbol{\theta}, \tau^C)$$

comprising the following components:

- $\phi \in \Re$ is the intercept of the log-linear model.

- $\boldsymbol{\theta} \in \Re^m$ are the remaining components of the reduced log-linear reparameterization whose dimension m depends on the possible presence of the main effects as well a higher order of interaction terms.

- τ^C is the parameter vector driving the distribution of $\boldsymbol{z}^C - \boldsymbol{y}^C$, which represents the non-negative integer number of units which are observed in the censored cells that are not part of the population of interest.

In order to set up a fully Bayesian approach, an elicitation of a prior distribution on the unknown parameters involved in the model including hyper-prior parameters as well as on the unobserved quantities such as $\boldsymbol{z}^C - \boldsymbol{y}^C$ is called for. Alternative ways of approaching prior elicitation will be detailed in the following section.

Updating prior information on all unknown/unobserved quantities using the likelihood function corresponding to the log-linear statistical model yields the posterior distribution whose complexity can be handled through standard MCMC techniques. In particular we almost completely follow Overstall et al. [227] with a Gibbs sampling algorithm which requires the derivation of the full conditionals of suitable block components of the parameter space.

In case unobserved heterogeneity is taken into account we have additional parameters linked to the mixing distribution $F(\cdot)$. These can be ignored in all full conditionals, except of course that of ϕ. All parameters are indeed independent of $F(\cdot)$, conditional on the vector ϕ. Similarly, the full conditionals of any parameter of $F(\cdot)$ involve only ϕ. If $F(\cdot)$ is assumed to be Gaussian these are the usual full conditionals for Gaussian models. If $F(\cdot)$ is a latent class model, we proceed through an augmented likelihood where we have kC (missing) indicators H_{ij}, where $H_{ij} = I(\phi_i = \xi_j)$, and obviously $\sum_j H_{ij} = 1$. In our Gibbs iterations we sample

$$\pi_C \sim \text{Dirichlet}(1 + \sum_{i \in \mathcal{K}} H_{i1}, \ldots, 1 + \sum_{i \in \mathcal{K}} H_{iC}),$$

and $H_{i1}, \ldots, H_{iC}$ from a multinomial with parameters

$$\Pr(H_{ij} = 1 | \phi_i) = \frac{F_\phi(\phi_i, \xi_j, \tau_\phi^2) \pi_C(j)}{\sum_c F_\phi(\phi_i, \xi_c, \tau_\phi^2) \pi_C(c)}.$$

Finally, ξ_c can be updated through the usual full conditional for Gaussian models, based on ϕ_i with $\{i : H_{ic} = 1\}$. As with any finite mixture model in the Bayesian setting, we finally need to take into account label switching issues. Our strategy is to simply post-process ξ_c by sorting them in increasing order within the MCMC iterations.

Indeed in Overstall et al. [227] the Bayesian approach is employed also for accounting for model uncertainty about the order of interaction among lists which can underlie the observed data. In order to do that they introduce some prior probability of selecting a range of hierarchical models and implement a reversible jump technique to explore the corresponding union of parameter spaces with different dimensions. Reversible jump for log-linear model choice goes beyond the scope of this chapter. We rather focus on modeling possible overdispersion and giving insights into prior specification when truncated counts are present. We rely on DIC to compare different model specifications.

25.6   Data analysis

In this section, we perform a Bayesian analysis on the PWID dataset focusing on the specific model

$$
\begin{aligned}
\log(\mu_i) \;=\; & \phi + \theta^{L_1^O} + \theta^{L_2^O} + \theta^{L_3^O} + \theta^{L_1^C} + \theta^{X_1} + \theta^{X_2} + \theta^{X_3} \\
& + \; \theta^{L_1^O L_3^O} + \theta^{L_1^O X_3} + \theta^{L_2^O L_1^C} + \theta^{L_2^O X_2} + \theta^{L_2^O X_3} + \theta^{L_3^O X_1} + \theta^{L_3^O X_3} \\
& + \; \theta^{X_1 X_3} + \theta^{X_2 X_3}
\end{aligned}
$$

determined using the reversible jump algorithm as described in Overstall and King [228].

We keep the model specification fixed, and vary (i) the prior for nuisance parameters, (ii) the prior for N as induced by the prior on ϕ, (iii) the occurrence and form of unobserved heterogeneity. When we take into account unobserved heterogeneity we keep prior options for N fixed as outlined above. We compare several combinations of these choices in Table 25.3, where, for nuisance parameters, we might have MVN, g-prior or SBH. For N in absence of unobserved heterogeneity we compared different choices for the hyperparameter λ, and for latent class models we specify several options for the number of latent classes C. We report the posterior expectations for the population size $E(N)$, the 95% credible interval (CI), and the deviance information criterion (DIC).

TABLE 25.3: Scotland PWID data: prior setting, unobserved heterogeneity, posterior mean, 95% highest posterior density interval and deviance information criterion

Setting	Prior		Heterogeneity	$E(N|X)$	CI	DIC
1	g-prior	$\lambda = -1$	No heterogeneity	23228	(20510 , 26336)	242671.1
2	g-prior	$\lambda = 0$	No heterogeneity	23230	(20525 , 26255)	245169.1
3	g-prior	$\lambda = 1$	No heterogeneity	23196	(20463 , 26211)	244303.3
4	SBH	$\lambda = -1$	No heterogeneity	23125	(20367 , 26176)	243831.1
5	SBH	$\lambda = 0$	No heterogeneity	23104	(20366 , 26165)	244633.3
6	SBH	$\lambda = 1$	No heterogeneity	23074	(20359 , 26132)	243088.4
7	g-prior		Gaussian	28838	(10123 , 86054)	245177.3
8	SBH		Gaussian	33643	(10717 , 108295)	267743.8
9	g-prior		2 Latent classes	23080	(19447 , 27178)	240207.7
10	g-prior		3 Latent classes	23111	(19331 , 27508)	239463.7
11	g-prior		4 Latent classes	22981	(18858 , 27772)	238102.5
12	g-prior		5 Latent classes	23412	(18465 , 31716)	237092.9
13	g-prior		10 Latent classes	23531	(18029 , 35584)	236341.5
14	g-prior		15 Latent classes	23760	(17550 , 38977)	233139.7
15	g-prior		20 Latent classes	23986	(16741 , 43262)	234621.9
16	SBH		2 Latent classes	23159	(19532 , 27305)	236594.5
17	SBH		3 Latent classes	23097	(19244 , 27557)	239391.5
18	SBH		4 Latent classes	22994	(18672 , 27919)	237971.8
10	SBH		5 Latent classes	22942	(18690 , 27903)	237085.4
20	SBH		10 Latent classes	22935	(17930 , 29650)	234892.0
21	SBH		15 Latent classes	23134	(17323 , 32203)	233611.8
22	SBH		20 Latent classes	23101	(16829 , 32769)	235743.3

It can be seen that posterior expectations of N are rather stable with respect to all choices considered. Not surprisingly the prior for N, under homogeneity, leads to a different ordering for population size estimates that one could expect. It is so, as uncertainty is dominating over prior effects. Also the length of the credible interval is almost unaffected by prior choices under homogeneity. The nice symmetry of the credible intervals confirms that asymptotic approximation of the posterior distribution with a Gaussian is very good. When unobserved heterogeneity/overdispersion is accounted for, credible interval lengths steadily increase. It can be seen that the best model is the one associated with the largest credible interval (that is, g-prior with $C = 15$ latent classes), which suggests that homogeneity might be optimistic for the data at hand. In our opinion the model with $C = 15$ latent classes provides a good uncertainty assessment, where the credible interval is reasonably large in front of a relatively low sampling fraction. The substantial increase of the DIC when adding more latent classes is an indication that a good balance between goodness-of-fit and number of parameters is achieved.

25.7   Conclusions

In this chapter we have focused on contingency tables with left-censored counts, that occur when some subjects recorded in some specific list might not actually belong to the population of interest. We have revisited the approach and motivating example of Overstall et al. [227]. We have better investigated the rationale for prior choice and provided several options. Additionally, we have extended their model set-up to account for unobserved heterogeneity in cell counts. It shall be noted that our models under unobserved heterogeneity are rather simple, but are based on the specific assumption that subjects in the same cell share the same unobserved heterogeneity parameters. A more general model would be based on the specification of subject-specific parameters to deal with unobserved heterogeneity, as in Farcomeni [114]; but this would lead to rather different modeling assumptions (e.g., to the use of negative-binomial distribution or even to model directly the subject-specific capture history, leading to a completely different modeling strategy).

25.8   Appendix A: Induced gamma-type priors on N

To see (25.3), express

$$N|\phi, \theta_1, ..., \theta_m \sim Pois\left(\sum_{i\in\mathcal{K}} \mu_i\right) = Pois\left(\sum_{i\in\mathcal{K}} e^{\phi+x_i^T\theta}\right).$$

Prior independence can be formally expressed as

$$\pi_\mathbf{p}(\mathbf{p}) = \pi_\phi(\phi)\pi_\theta(\theta).$$

Now, one can argue that

$$
\begin{aligned}
\pi(N) &= \int_{\Re^m} \int_{\Re} \pi(N|\phi,\boldsymbol{\theta})\pi_\phi(\phi)d\phi\pi(\boldsymbol{\theta})d\boldsymbol{\theta} \\
&= \int_{\Re^m} \int_{\Re} \frac{\exp\left\{-\sum_{i\in\mathcal{K}} e^{\phi+x_i^T\boldsymbol{\theta}}\right\} \left(\sum_{i\in\mathcal{K}} e^{\phi+x_i^T\boldsymbol{\theta}}\right)^N}{N!} \pi_\phi(\phi)d\phi\pi(\boldsymbol{\theta})d\boldsymbol{\theta} \\
&= \int_{\Re^m} \int_{\Re} \frac{\exp\left\{-e^\phi \sum_{i\in\mathcal{K}} e^{x_i^T\boldsymbol{\theta}}\right\} \left(e^\phi \sum_{i\in\mathcal{K}} e^{x_i^T\boldsymbol{\theta}}\right)^N}{N!} \pi_\phi(\phi)d\phi\pi(\boldsymbol{\theta})d\boldsymbol{\theta} \\
&= \int_{\Re^m} \int_{\Re} \frac{\exp\left\{-e^\phi \sum_{i\in\mathcal{K}} e^{x_i^T\boldsymbol{\theta}}\right\} \left(e^\phi\right)^N \left(\sum_{i\in\mathcal{K}} e^{x_i^T\boldsymbol{\theta}}\right)^N}{N!} \pi_\phi(\phi)d\phi\pi(\boldsymbol{\theta})d\boldsymbol{\theta}.
\end{aligned}
$$

To simplify notation, denote $H(\boldsymbol{\theta}) = \sum_{i\in\mathcal{K}} e^{x_i^T\boldsymbol{\theta}}$ and let $u = e^\phi$. Consequently, $\pi_\phi(\phi)d\phi$ becomes $\pi_U(u)du = u^{-1}\pi_\phi(\log u)du$ and the integral can be expressed as

$$
\begin{aligned}
\pi(N) &= \frac{1}{N!} \int_{\Re^m} \int_0^\infty \exp\left\{-H(\boldsymbol{\theta})u\right\} u^N H(\boldsymbol{\theta})^N u^{-1} \pi_\phi(\log u)du\pi(\boldsymbol{\theta})d\boldsymbol{\theta} \\
&= \frac{1}{N!} \int_{\Re^m} H(\boldsymbol{\theta})^N \left[\int_0^\infty \exp\left\{-H(\boldsymbol{\theta})u\right\} u^N u^{-1} \pi_\phi(\log u)du\right] \pi(\boldsymbol{\theta})d\boldsymbol{\theta}.
\end{aligned}
$$

Hence if $\pi_U(u) = u^{-1}\pi_\phi(\log u)$ is proportional to the gamma density i.e. $\pi_U(u) = \pi_\phi(\log u)u^{-1} \propto e^{-\omega u}u^{\lambda-1}$, we get a closed form expression for the inner integral as

$$
\begin{aligned}
\pi(N) &\propto \frac{1}{N!} \int_{\Re^m} H(\boldsymbol{\theta})^N \left[\int_0^\infty \exp\left\{-H(\boldsymbol{\theta})u\right\} u^N e^{-\omega u}u^{\lambda-1}du\right] \pi(\boldsymbol{\theta})d\boldsymbol{\theta} \\
&= \frac{1}{N!} \int_{\Re^m} H(\boldsymbol{\theta})^N \left[\int_0^\infty \exp\left\{-(H(\boldsymbol{\theta})+\omega)u\right\} u^{N+\lambda-1}du\right] \pi(\boldsymbol{\theta})d\boldsymbol{\theta} \\
&= \frac{1}{N!} \int_{\Re^m} H(\boldsymbol{\theta})^N \frac{\Gamma(N+\lambda)}{(H(\boldsymbol{\theta})+\omega)^{N+\lambda}} \pi(\boldsymbol{\theta})d\boldsymbol{\theta}.
\end{aligned}
$$

In particular, if we let $\omega = 0$ we get an improper $\pi_U(u) \propto u^{\lambda-1}$ and hence

$$
\begin{aligned}
\pi(N) &\propto \frac{\Gamma(N+\lambda)}{N!} \int_{\Re^m} \frac{H(\boldsymbol{\theta})^N}{(H(\boldsymbol{\theta}))^{N+\lambda}} \pi(\boldsymbol{\theta})d\boldsymbol{\theta} \\
&= \frac{\Gamma(N+\lambda)}{N!} \int_{\Re^m} \frac{1}{(H(\boldsymbol{\theta}))^\lambda} \pi(\boldsymbol{\theta})d\boldsymbol{\theta} \propto \frac{\Gamma(N+\lambda)}{N!}.
\end{aligned}
$$

Appendix B: Posterior integrability

In the presence of an improper prior on some parameter, we should verify that the use of the Bayes rule yields a proper posterior on the whole parameter space. Note that an improper prior density on ϕ such as $\pi_\phi(\phi) \propto e^{\lambda\phi}$ corresponds to an improper prior on $u = e^\phi$, namely $\pi(u) \propto u^{\lambda-1}$. We will verify the existence of suitable (possibly mild) conditions for $\lambda \geq -1$ and $\boldsymbol{y}_O$ such that the denominator of the Bayes rule is finite.

Indeed,

$$
\begin{aligned}
D &= \sum_{\{y_\mathrm{h}:h\in\mathcal{K}^U\}} \sum_{\{y_\mathrm{k}:k\in\mathcal{K}^C\}} \int_{\Re^m} \int_{\Re} \prod_{h\in\mathcal{K}^U} \pi(y_\mathrm{h}|\phi,\boldsymbol{\theta}) \prod_{k\in\mathcal{K}^C} \pi(y_\mathrm{k}|\phi,\boldsymbol{\theta}) \\
&\qquad \prod_{i\in\mathcal{K}^O} \pi(y_\mathrm{i}|\phi,\boldsymbol{\theta})\pi_\phi(\phi)d\phi\pi(\boldsymbol{\theta})d\boldsymbol{\theta} \qquad\qquad (25.8)
\end{aligned}
$$

$$
\leq \int_{\Re^m} \int_{\Re} \prod_{i\in\mathcal{K}^O} \frac{e^{-e^{\phi+x_\mathrm{i}^T\boldsymbol{\theta}}} \left(e^{\phi+x_\mathrm{i}^T\boldsymbol{\theta}}\right)^{y_\mathrm{i}}}{y_\mathrm{i}!} \pi_\phi(\phi)d\phi\pi(\boldsymbol{\theta})d\boldsymbol{\theta}
$$

$$
= \frac{1}{\prod_{i\in\mathcal{K}^O} y_\mathrm{i}!} \int_{\Re^m} \int_{\Re} \exp\left\{-e^\phi \sum_{i\in\mathcal{K}^O} e^{x_\mathrm{i}^T\boldsymbol{\theta}}\right\} \left(e^\phi\right)^{\sum_{i\in\mathcal{K}^O} y_\mathrm{i}} \qquad (25.9)
$$

$$
\left(e^{\sum_{i\in\mathcal{K}^O} y_\mathrm{i} x_\mathrm{i}^T\boldsymbol{\theta}}\right) \pi_\phi(\phi)d\phi\pi(\boldsymbol{\theta})d\boldsymbol{\theta} \qquad\qquad (25.10)
$$

where the inequality derives from the fact that the sum is

$$
\sum_{\{y_\mathrm{h}:h\in\mathcal{K}^U\}} \sum_{\{y_\mathrm{k}:k\in\mathcal{K}^C\}} \prod_{h\in\mathcal{K}^U} \pi(y_\mathrm{h}|\phi,\boldsymbol{\theta}) \prod_{k\in\mathcal{K}^C} \pi(y_\mathrm{k}|\phi,\boldsymbol{\theta}) \leq 1.
$$

Now, the latter expression in (25.8) is proportional to an integral expression where in the inner part we can change variable $u = e^\phi$ so that $\pi_\phi(\phi)d\phi = e^\phi d\phi$ changes into $\pi(u)du \propto u^{\lambda-1}du$ and we get

$$
\begin{aligned}
D &\propto \int_{\Re^m} \left(e^{\sum_{i\in\mathcal{K}^O} y_\mathrm{i} x_\mathrm{i}^T\boldsymbol{\theta}}\right) \times \\
&\qquad \times \int_{\Re} \exp\left\{-e^\phi \sum_{i\in\mathcal{K}^O} e^{x_\mathrm{i}^T\boldsymbol{\theta}}\right\} \left(e^\phi\right)^{\sum_{i\in\mathcal{K}^O} y_\mathrm{i}} \pi_\phi(\phi)d\phi\pi(\boldsymbol{\theta})d\boldsymbol{\theta} \\
&= \int_{\Re^m} e^{\sum_{i\in\mathcal{K}^O} y_\mathrm{i} x_\mathrm{i}^T\boldsymbol{\theta}} \int_0^\infty \exp\left\{-u \sum_{i\in\mathcal{K}^O} e^{x_\mathrm{i}^T\boldsymbol{\theta}}\right\} u^{\sum_{i\in\mathcal{K}^O} y_\mathrm{i}} u^{\lambda-1} \, du\pi(\boldsymbol{\theta})d\boldsymbol{\theta} \\
&= \Gamma\left(\sum_{i\in\mathcal{K}^O} y_\mathrm{i} + \lambda\right) \int_{\Re^m} e^{\sum_{i\in\mathcal{K}^O} y_\mathrm{i} x_\mathrm{i}^T\boldsymbol{\theta}} \left(\sum_{i\in\mathcal{K}^O} e^{x_\mathrm{i}^T\boldsymbol{\theta}}\right)^{-\left(\sum_{i\in\mathcal{K}^U} y_\mathrm{i}\right)-\lambda} \pi(\boldsymbol{\theta})d\boldsymbol{\theta}
\end{aligned}
$$

using the gamma integral identity $\int_0^\infty e^{-a}u^{b-1} = \Gamma(b)a^{-b}$, which holds when $b = \sum_{i\in\mathcal{K}^O} y_\mathrm{i} + \lambda > 0$ and $a = \sum_{i\in\mathcal{K}^O} e^{x_\mathrm{i}^T\boldsymbol{\theta}} > 0$.

Thus we obtain,

$$
D \propto \int_{\Re^m} \prod_{i\in\mathcal{K}^O} \left[\frac{e^{x_\mathrm{i}^T\boldsymbol{\theta}}}{\left(\sum_{i\in\mathcal{K}^O} e^{x_\mathrm{i}^T\boldsymbol{\theta}}\right)}\right]^{y_\mathrm{i}} \left(\sum_{i\in\mathcal{K}^O} e^{x_\mathrm{i}^T\boldsymbol{\theta}}\right)^{-\lambda} \pi(\boldsymbol{\theta})d\boldsymbol{\theta}. \qquad (25.11)
$$

Finally we can argue that for any $\lambda \geq 0$ we always get that the last integral is bounded by 1, hence $D < \infty$ provided that $\sum_{i\in\mathcal{K}^O} y_\mathrm{i} + \lambda > 0$. On the other hand, for $-1 \leq \lambda < 0$ one can ensure propriety as long as $\sum_{i\in\mathcal{K}^O} y_\mathrm{i} + \lambda > 0$ and $\pi(\boldsymbol{\theta})$ admits the moment generating function defined everywhere in $\boldsymbol{\theta}$ since, up to a suitable proportionality constant c

$$
D \leq c \cdot \int_{\Re^m} \left(\sum_{i\in\mathcal{K}^O} e^{x_\mathrm{i}^T\boldsymbol{\theta}}\right)^{-\lambda} \pi(\boldsymbol{\theta})d\boldsymbol{\theta} \leq \sum_{i\in\mathcal{K}^O} \int_{\Re^m} e^{x_\mathrm{i}^T\boldsymbol{\theta}} \pi(\boldsymbol{\theta})d\boldsymbol{\theta}.
$$

Part VIII

Miscellaneous Topics

26

Uncertainty assessment in capture-recapture studies and the choice of sampling effort

Dankmar Böhning

University of Southampton

John Bunge

Cornell University

Peter G.M. van der Heijden

Universities of Utrecht and Southampton

CONTENTS

26.1 Introduction and background

Let us consider a simplified situation to start with. N is the unknown population size, f_0 the frequency of uncaptured units, n are the observed units and p_0 denotes the probability of not capturing a unit of the target population. It follows that $E(f_0) = Np_0$ and estimates for f_0 and are readily available as $\hat{f}_0 = n\frac{p_0}{1-p_0}$ and $\hat{N} = n/(1-p_0)$, respectively. We point out that

$$\hat{N} = n + \hat{f}_0. \tag{26.1}$$

If we consider the variance of (26.3) we easily see that there are two components. There is the variation due to n and there is the variation due to $\hat{f}_0$. However, one could argue that interest is typically in the uncertainty attached to $\hat{f}_0$ as there is nothing uncertain about n since this has been observed. What is really of interest is the variance of f_0 as this quantity is unobserved and is *predicted*. Hence there is a fundamental difference between the observed random quantity n and the predicted random quantity $\hat{f}_0$. Hence interest is on the variance $var(\hat{f}_0)$, which is easily computed as

$$var(\hat{f}_0) = \left[\frac{p_0}{1-p_0}\right]^2 var(n) = \left[\frac{p_0}{1-p_0}\right]^2 Np_0(1-p_0), \tag{26.2}$$

as $n \sim Bin(N, 1-p_0)$ and has variance $Np_0(1-p_0)$. The variance (26.2) is neither conditional on n (as this would lead to a zero variance), nor is it an unconditional variance as this would require involving the first part of (26.3). It is more appropriate to view this as a *partial variance* acknowledging the different characteristics involved in (26.2).

Note that the prediction variance $var(\hat{f}_0) = \left[\frac{p_0}{1-p_0}\right]^2 Np_0(1-p_0)$ is smaller than the *total variance* $var(\hat{N}) = \left[\frac{1}{1-p_0}\right]^2 Np_0(1-p_0)$ which is easily computed using that $\hat{N} = n + \hat{f}_0 = n + n\frac{p_0}{1-p_0} = n/(1-p_0)$. Prediction variance and total variance are related by

$$var(\hat{f}_0) = p_0^2 var(\hat{N}). \tag{26.3}$$

We find the prediction variance more appropriate for capture-recapture experiments. Note that in this simple case the prediction variance can be estimated as $\left[\frac{p_0}{1-p_0}\right]^2 \hat{N}p_0(1-p_0) = \left[\frac{p_0}{1-p_0}\right]^2 np_0$.

As a consequence we can calculate a 95% prediction interval based upon prediction variance as $\hat{f}_0 \pm 1.96\sqrt{var(\hat{f}_0)}$ for f_0 and as $n + \hat{f}_0 \pm 1.96\sqrt{var(\hat{f}_0)}$ for N. In contrast, the 95% confidence interval is based upon the total variance and is provided as $\hat{N} \pm 1.96\sqrt{var(\hat{N})}$.

26.2   Computing variances using conditional moments

For two arbitrary random variables X and Y with existing second moments, we use the following result:

$$var(X) = E[var(X|Y)] + Var[E(X|Y)], \tag{26.4}$$

where moments inside the square brackets refer to the conditional distribution of X given Y, and moments outside the square brackets refer to the unconditional distribution of Y. For more details see Böhning [40], van der Heijden et al. [280].

We apply this result using $X = \left[\frac{p_0(\hat{\theta})}{1-p_0(\hat{\theta})}\right]$ and $Y = n$ so that we achieve

$$
\begin{aligned}
var(\hat{f}_0) &= var\left[\frac{p_0(\hat{\theta})}{1-p_0(\hat{\theta})}n\right] \\
&= E\left[var\left(\frac{p_0(\hat{\theta})}{1-p_0(\hat{\theta})}n\right)\Big|n\right] + var\left[E\left(\frac{p_0(\hat{\theta})}{1-p_0(\hat{\theta})}n\Big|n\right)\right].
\end{aligned} \tag{26.5}
$$

The *first term* in (26.5) can be estimated by

$$var\left(\frac{p_0(\hat{\theta})}{1-p_0(\hat{\theta})}n\right), \tag{26.6}$$

using its moment estimate. Note that this term is zero if p_0 is known, as we are conditioning on n. Otherwise the estimate will depend on the form of $p_0(\theta)$.

It is the *second term* in (26.5) where a more general expression can be achieved:

$$var\left[E\left(\frac{p_0(\hat{\theta})}{1-p_0(\hat{\theta})}n\Big|n\right)\right] \approx var\left[\frac{p_0(\theta)}{1-p_0(\theta)}n\right], \tag{26.7}$$

using again the moment estimator for the expected value. This can be simplified to

$$var\left[\frac{p_0(\theta)}{1-p_0(\theta)}n\right] = \left[\frac{p_0(\theta)}{1-p_0(\theta)}\right]^2 Np_0(\theta)(1-p_0(\theta))$$

$$= \left[\frac{p_0(\theta)N}{(1-p_0(\theta))N}\right]^2 Np_0(\theta)(1-p_0(\theta)) \qquad (26.8)$$

which can be readily estimated by

$$\left[\frac{\hat{f}_0}{n}\right]^2 \hat{f}_0 \frac{n}{n+\hat{f}_0} = \frac{\hat{f}_0^3}{n(n+\hat{f}_0)}. \qquad (26.9)$$

Note the difference between

$$var(\hat{f}_0) = var\left[\frac{p_0(\theta)}{1-p_0(\theta)}n\right] = \left[\frac{p_0(\theta)}{(1-p_0(\theta))}\right]^2 Np_0(\theta)(1-p_0(\theta))$$

and

$$var(f_0) = Np_0(\theta)(1-p_0(\theta)).$$

Both agree if $p_0 = (1-p_0)$, otherwise they are different, with the predicted value $\hat{f}_0$ having a smaller variance than f_0 if $p_0 > 0.5$. In summary, we have:

$$var(\hat{f}_0) \approx var\left(\frac{p_0(\hat{\theta})}{1-p_0(\hat{\theta})}n\right) + \left[\frac{\hat{f}_0}{n}\right]^2 \hat{f}_0 \frac{n}{n+\hat{f}_0} = \frac{\hat{f}_0^3}{n(n+\hat{f}_0)}. \qquad (26.10)$$

26.3 Application to log-linear models

Here we are applying the developed tools to calculate the prediction variance for a log-linear model to model the capture-recapture situation in a multiple list or multiple-source setting. According to Section 26.2 only the first term in (26.5) needs to be found, which is

$$var(\hat{f}_0) = var\left(\frac{p_0(\hat{\theta})}{1-p_0(\hat{\theta})}n\right).$$

Now, for a log-linear model we have that

$$\hat{f}_0 = exp(x_0^T \hat{\beta}), \qquad (26.11)$$

where x_0 represents the covariate combination in the log-linear model that defines the missing zero cell. Typically, $x_0 = (0, \cdots, 0)^T$ where the length of the vector depends on the terms (main effects, interactions, ...) in the log-linear model. Hence $\hat{f}_0 = exp(\hat{\alpha})$ where α is the intercept. Using the δ−method we find that

$$var(\hat{f}_0) \approx exp(x_0^T \hat{\beta})^2 var(x_0^T \hat{\beta}). \qquad (26.12)$$

$x_0^T \hat{\beta}$ is called the linear predictor and its variance is typically available in standard statistical packages.

Let us apply (26.12) to a very simple setting. Amstrup et al. [7] report a capture-recapture study on deer mice (*Peromyscus* sp.) of which we take only 2 occasions. There are $n_{11} = 12$ mice caught at both occasions, $n_{10} = 3$ at the first but not at the second and $n_{01} = 8$ at the second but not at the first.

For this simple two-occasion situation, we write the Lincoln–Petersen estimator as a log-linear model with two main effects:

$$\log E(n_{ij}) = \alpha + \beta_1 s1_i + \beta_2 s2_j, \qquad (26.13)$$

where $s1_i$ and $s2_j$ denote binary variables for $i, j = 0, 1$ with $s1_1 = s2_1 = 1$ and $s1_0 = s2_0 = 0$. The model allowing interaction is not identifiable in this case. Under a Poisson likelihood the maximum likelihood estimates are given by

$$
\begin{aligned}
\hat{\beta}_1 &= & \log n_{11} - \log n_{01} \\
\hat{\beta}_2 &= & \log n_{11} - \log n_{10} \\
\hat{\alpha} &= & \log n_{10} + \log n_{01} - \log n_{11},
\end{aligned}
\qquad (26.14)
$$

the latter being the Lincoln–Petersen estimate. The Chapman estimator is achieved by replacing n_{11} by $n_{11} + 1$ and the estimator arises as the maximum likelihood estimate of α in (26.13) using again the Poisson likelihood. For the study on deer mice, we find an estimate of f_0 of 2 with an estimate of its prediction standard error as 1.48 using (26.12). This would provide a prediction interval of $0 - 4.9$.

26.4   Bootstrap for capture-recapture

To achieve better approximations for variance expression, the bootstrap is usually utilized. However, bootstrapping is more complex in the capture-recapture setting. To illustrate the issues, we assume that $p_0 = 0.05$ is known. Suppose 15 units have been observed out of twenty:

$$1\ 0\ 1\ 1\ 0\ 1\ 1\ 1\ 0\ 1\ 1\ 1\ 1\ 1\ 0\ 1\ 1\ 1\ 0.$$

However, only

$$1 \qquad 1\ 1 \qquad 1\ 1\ 1 \qquad 1\ 1\ 1\ 1\ 1 \qquad 1\ 1\ 1\ 1$$

are observed. Hence a classical bootstrap would always deliver the identical resample.

The suggested solution here is to impute $\hat{f}_0 = \frac{p_0}{1-p_0} n = \frac{p_0}{1-p_0} \times 15 = 5$ zeros leading to the imputed sample

$$1 \qquad 1\ 1 \qquad 1\ 1\ 1 \qquad 1\ 1\ 1\ 1\ 1 \qquad 1\ 1\ 1\ 1 \qquad 0\ 0\ 0\ 0\ 0$$

to which the classical bootstrap is applied. The estimated variance $\left[\frac{p_0}{1-p_0}\right]^2 np_0$ provides a value 0.4167, whereas the imputed bootstrap delivers 0.4095 with a bootstrap replication size of 1000, which is indeed very close.

We point that ideally we would like to resample using the true N and not the estimate $\hat{N}$. Resampling with the true N is called the *true bootstrap*. In Anan et al. [9] the imputed bootstrap is compared with the true bootstrap and concluded that the imputed bootstrap provides a reasonable approximation of the true bootstrap if the underlying estimator $\hat{N}$ is valid. The variance for $\hat{N}$ is 6.67 with bootstrap approximation of 6.16.

We apply the imputed bootstrap to the deer mice example from Section 26.3. We find, using a bootstrap replication of 5000 resamples, a prediction standard error of 1.78 which

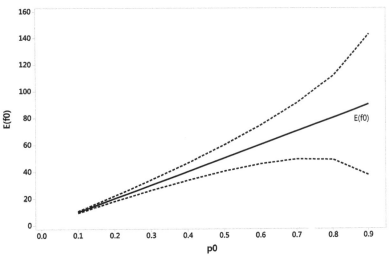

FIGURE 26.1: Expected number of hidden units $E(f_0)$ with 95% pointwise prediction interval.

compares well with the analytical approach taken in 26.3, being slightly on the conservative side. The bootstrap approach also offers a better way to compute prediction intervals, for example a 95% prediction interval based upon the percentile method delivers 0–6.88 which deviates from the normal approximation 95% prediction interval of 0–4.9.

26.5 Choice of sampling effort

According to (26.2) we have that the prediction variance $v(p_0) := var(\hat{f}_0) = [\frac{p_0}{1-p_0}]^2 N p_0 (1 - p_0)$ grows with the population size N. Hence increasing the target population will not help in reducing the prediction variance. However, the prediction variance $v(p_0)$ is strictly increasing as a function of p_0 for $p_0 \in (0, 1)$. Hence we can decrease the prediction variance by choosing p_0 small. This is illustrated in Figure 26.1 in terms of pointwise 95% prediction intervals in dependence on p_0. It seems reasonable to have a value of p_0 not above 0.5. How can this be achieved? Suppose there are T sampling occasions, that sampling takes place independently, and that the identification probability θ remains identical over occasions and sampling units. Then, we have that $p_0 = (1 - \theta)^T$. This can be easily solved for T leading to

$$T = \log(p_0)/\log(1 - \theta) \tag{26.15}$$

sampling occasions. Figure 26.2 shows T in dependence of p_0 and individual identification probability per occasion θ. To illustrate for the deer mice data, we estimate θ as 15/25 (we take the smaller estimate) and need $T = 3$ sampling occasions to have $p_0 = 0.3$.

At times capture-recapture studies do not use a fixed number T of sampling occasions but rather a time window of size T in which repeated identification of the units of the target

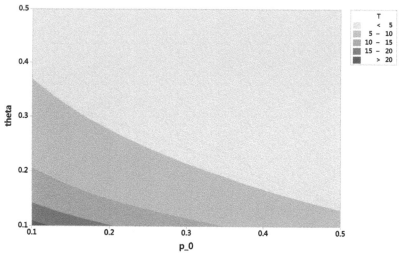

FIGURE 26.2: Contour plot for the choice of sampling effort T in dependence on θ and p_0.

population takes place. In these instances a Poisson model of the type

$$p_x = P(X = x) = \exp(-\theta T)(\theta T)^x/x! \tag{26.10}$$

is often used. Clearly, $p_0 = \exp(-\theta T)$ and the analog equation to (26.15) is

$$T = -\log(p_0)/\theta. \tag{26.17}$$

26.6 Lincoln–Petersen estimation and sampling effort

We consider the problem of estimating the size of an elusive target population with a two occasion approach as it was discussed in Chapter 8 or Chapter 23. A classical Lincoln–Petersen experiment requires sampling the target population on two occasions at which each sampled unit of the target population is marked as identified at occasion 1 or 2 or both, respectively. Let p_{ij} denote the identification probability distribution for both occasions: p_{11} is the probability of identifying a unit at both occasions, p_{10} is the probability of identifying a unit at occasion 1 but not at occasion 2, and so forth. Then, the joint distribution is given under independence in Table 26.1.

The results of a Lincoln–Petersen experiment are displayed in Table 26.2. f_{ij} is the bivariate frequency distribution of units identified at occasion 1 and 2, and f_{11} is the frequency of units identified at both occasions. The frequency f_{00} of units identified at neither occasion is unknown and, hence, the population size $N = f_{11} + f_{10} + f_{01} + f_{00}$ is unknown and the target of the inference.

TABLE 26.1

Joint distribution of identifying a unit at two occasions under independence; $p_{1+} = p_{11} + p_{10}$, $p_{+1} = p_{11} + p_{01}$

		Occasion 2		
		1	0	
	1	$p_{1+}p_{+1}$	$p_{1+}(1 - p_{1+})$	p_{1+}
Occasion 1				
	0	$(1 - p_{1+})p_{+1}$	$(1 - p_{1+})(1 - p_{+1})$	$1 - p_{1+}$
		p_{+1}	$1 - p_{+1}$	

TABLE 26.2

Observed distribution of identifying units at two occasions $(f_{1+} = f_{11} + f_{10}, f_{+1} = f_{11} + f_{01})$

		Occasion 2		
		1	0	
	1	f_{11}	f_{10}	f_{1+}
Occasion 1				
	0	f_{01}	f_{00}	
		f_{+1}		N

The classical Lincoln–Petersen estimator is provided as

$$\ddot{N} = \frac{f_{1+}f_{+1}}{f_{11}},$$

and it is clear from its construction by means of a ratio that its stability largely depends on the size of the denominator m. The smaller f_{11} is the more unreliable the Lincoln–Petersen estimator becomes. This becomes also clear when looking at the associated variance estimator

$$\widehat{Var}(\hat{N}) = \frac{f_{1+}f_{+1}f_{10}f_{01}}{f_{11}^3} = \hat{N}\frac{f_{10}}{f_{11}}\frac{f_{01}}{f_{11}}.$$

Clearly, the variance becomes smaller if f_{11} is increased relative to f_{1+} or f_{+1} or both.

Clearly, f_{11}, on average, cannot be increased if p_{1+} and p_{+1} remain unchanged. Simply enlarging the size N of the target population will not help. However, sometimes experiments are done in such a way that at the first occasion sampling is done repeatedly, for example, live trapping is done over a number of nights. Hence we look at the following scenario. At the first occasion repeated identification is done with T_1 replications and at the second occasion with T_2 replications. Hence, the probability of *not* identifying a unit at the first occasion is $1 - p_{1+} = (1 - \theta_1)^{T_1}$. Similarly, at the second occasion the probability of not identifying a unit is $1 - p_{+1} = (1 - \theta_2)^{T_2}$. Here θ_i is the individual identification probability for each of the T_i replications at occasion i, $i = 1, 2$.

We are interested in maximizing the probability $p_{1+}p_{+1}$ (leading to a maximized frequency of joint identifications f_{11} which determines the variance) keeping the total sampling effort $T = T_1 + T_2$ fixed. With this motivation in mind, we note that

$$p_{1+}p_{+1} = [1 - (1 - \theta_1)^{T_1}][1 - (1 - \theta_2)^{T_2}] = (1 - q_1^{T_1})(1 - q_2^{T_2}),$$

where we used the notation $q_i = (1 - \theta_i)$ for $i = 1, 2$. We introduce

$$f(t; q_1, q_2) = (1 - q_1^t)(1 - q_2^{T-t})$$

for $t = 0, 1, 2, \cdots, T$. Hence the problem of optimizing the sampling effort is to find $\hat{t}$ to maximize $f(t; q_1, q_2)$ in $t = 0, 1, 2, \cdots, T$.

We can achieve a fairly general result if $q_1 = q_2 = q$. Then, $f(t; q) = 1 - q^t - q^{T-t} + q^T$. Hence we only need to consider minimizing $q^t + q^{T-t}$ as a function of t. We have the following result:

The function $f(t; q) = (1 - q^t)(1 - q^{T-t})$ is maximized for $\hat{t} = (T-1)/2$ or $\hat{t} = (T-1)/2+1$ if T is odd. It is maximized for $\hat{t} = T/2$ if T is even.

The result can be used in the following way. If T is even, $T/2$ replications are placed at each of the two occasions. If T is odd, $(T-1)/2 + 1$ replications are placed at one of the occasions, the rest at the remaining occasion. It is also interesting to note that the optimal value does not depend on q or θ.

Proof of the result (sketch). We are minimizing $q^t + q^{T-t}$. Consider $T = 2$. Then, because of symmetry, there are only 2 different values for $q^t + q^{T-t}$: $q^2 + 1$ and $2q$ for $t = 0, 1, 2$. It is clear that $2q \leq q^2 + 1$ since $q^2 + 1 - 2q = (q - 1)^2 \geq 0$. Next, we consider $T = 3$. Then, there are again only 2 different values: $q^3 + 1, q^2 + q$ for $t = 0, 1, 2, 3$. Now, $q^3 + 1 \geq q^2 + q$ since $(1 - q) \geq (1 - q)q^2$. Let us look at $T = 4$. Then, there are 3 different values: $q^4 + 1, q^3 + q, 2q^2$ for $t = 0, 1, 2, 3, 4$. Now, $q^4 + 1 \geq q^3 + q$ since $1 - q \geq q^3(1 - q)$ and $q^3 + q \geq 2q^2$ since $q^2 + 1 \geq q$. We consider $T = 5$. We have 3 different values: $q^5 + 1, q^4 + q, q^3 + q^2$. We have that $q^5 + 1 \geq q^4 + q$ as $(1 - q) \geq q^4(1 - q)$. Also, $q^4 + q \geq q^3 + q^2$ as $1 - q \geq q^2(1 - q)$. The argument continues for increasing T.

To use this result in terms of determining the sampling effort at each occasion, one can proceed as follows. Let $T' = T/2$ (or $(T-1)/2 + 1$ in the odd case although this will be rarely a choice in practice) and then use the results from Section 26.5 to determine T'.

References

[1] AGRESTI, A. and LANG, J. B. (1993). Quasi-symmetric Latent Class models, with application to rater agreement. *Biometrics* **49** 131–139.

[2] AGRESTI, A. (1994). Simple capture-recapture models permitting unequal catchability and variable sampling effort. *Biometrics* **50** 494–500.

[3] ALBERTS, D. S., MARTINEZ, M. E., ROE, D. J., GUILLEN-RODRIGUEZ, J. M., MARSHALL, J. R., VAN LEEUWEN, B., REID, M. E., REITENBAUGH, C., VARGAS, P. A., BHATTACHARYYA, E. D. L., SAMPLINER, R., and THE PHOENIX COLON CANCER PREVENTION PHYSICIAN'S NETWORK (2000). Lack of effect of a high-fiber cereal supplement on the recurrence of colorectal adenomas. *New England Journal of Medicine* **342** 1156–1162.

[4] ALHO, J. (1990). Logistic regression in capture-recapture models. *Biometrics*, **46** 623–635.

[5] ALLMAN, E. S., MATIAS, C., and RHODES, J. A. (2009). Identifiability parameters in latent structure models with many observed variables. *The Annals of Statistics* **37** 3099–3132.

[6] ALLMAN, E. S., RHODES, J. A., STANGHELLINI, E., and VALTORTA, M. (2015). Parameter identification of discrete Bayesian network with hidden variables. *Journal of Causal Inference* **3** 189–205.

[7] AMSTRUP, S. C., MCDONALD, T. L., and MANLY, B.F.J. (2005). *Handbook of Capture-Recapture Analysis*. Princeton: Princeton University Press.

[8] ANAN, O., BÖHNING, D., and MARUOTTI, A. (2017a). Population size estimation and heterogeneity in capture-recapture data: A linear regression estimator based on the Conway–Maxwell–Poisson distribution. *Statistical Methods & Applications* **26** 49–79.

[9] ANAN, O., BÖHNING, D. and MARUOTTI, A. (2017b). Uncertainty estimation in heterogeneous capture-recapture count data. *Journal of Statistical Computation and Simulation* **87** 2094–2114.

[10] ARIEL, A., DE GROOT, M., VAN GROOTHEEST, G., VAN DER LAAN, J., SMIT, J., VERKERK, B., and BAKKER, B.F.M. (2014). Record linkage in health data: A simulation study. *The Hague/Heerlen: Statistics Netherlands*.

[11] ARNOLD, M. E., MARTELLI, F., MCLAREN, I., and DAVIES, R. H. (2014a). Estimation of the rate of egg contamination from *Salmonella*-infected chickens. *Zoonoses and Public Health* **61** 18–27.

[12] ARNOLD, M. E., MARTELLI, F., MCLAREN, I., and DAVIES, R. H. (2014b). Estimation of the sensitivity of environmental sampling for detection of *Salmonella* in commercial layer flocks post-introduction of national control programmes. *Epidemiology and Infection* **142** 1061–1069.

[13] ARNOLD, M. E., PAPADOPOULOU, C., DAVIES, R. H., CARRIQUE-MAS, J. J., EVANS, S. J., HOINVILLE, L. J., COOK, A. J. C., and EVANS, S. J. (2010). Estimation of *Salmonella* prevalence in UK egg-laying holdings. *Preventive Veterinary Medicine* **94** 306–309.

[14] ARNOLD, M. E., CARRIQUE-MAS, J. J., and DAVIES, R. H. (2010). Sensitivity of environmental sampling methods for detecting *Salmonella* Enteritidis in commercial laying flocks relative to the within-flock prevalence. *Epidemiology and Infection* **138** 330–339.

[15] ASMUSSEN, S. and EDWARDS, D. (1983). Collapsibility and response variables in contingency tables. *Biometrika* **70** 567–578.

[16] BAFFOUR, B., BROWN, J.J., and SMITH, P.W.F. (2013). An investigation of triple system estimators in censuses. *Statistical Journal of the International Association for Official Statistics* **29** 53–68.

[17] BAKER, S. G. (1990). A simple EM algorithm for capture-recapture data with categorical covariates (with discussion). *Biometrics* **46** 1193–1197.

[18] BAKKER, B. F. M. (2009). *Trek alle registers open!* Amsterdam: VU University Press.

[19] BAKKER, B. F. M. and DAAS, P. (2012). Some methodological issues of register based research. *Statistica Neerlandica* **66** 2–7.

[20] BAKKER, B. F. M., VAN ROOIJEN, J., and VAN TOOR, L. (2014). The system of social statistical datasets of Statistics Netherlands: An integral approach to the production of register-based social statistics. In: *Statistical Journal of the International Association of Official Statistics* **30** 411–424.

[21] BALABDAOUI, F., DUROT, C., and KOLADJO, F. (2014). On asymptotics of the discrete ... of a pmf. *arXiv preprint arXiv:1404.3094*.

[22] BALABDAOUI, F., DUROT, C., AND KOLADJO, F. (2015). Testing convexity of a discrete distribution. *arXiv preprint arXiv:1701.04367*.

[23] BARGER, K. and BUNGE, J. (2010). Objective Bayesian estimation for the number of species. *Journal of Bayesian Analysis* **5** 765–786.

[24] BARTOLUCCI, F. and FORCINA, A. (2001). Analysis of capture-recapture data with a Rasch-type model allowing for conditional dependence and multidimensionality. *Biometrics* **57** 714–719.

[25] BARTOLUCCI, F. and FORCINA, A. (2006). A class of latent marginal models for capture-recapture data with continuous covariates. *Journal of the American Statistical Association* **101** 786–794.

[26] BARTOLUCCI, F., COLOMBI, R., and FORCINA, A. (2001). An extended class of marginal link functions for modelling contingency tables by equality and inequality constraints. *Statistica Sinica* **17** 691–711.

[27] BÉGUINOT, J. (2014). An algebraic derivation of Chao's estimator of the number of species in a community highlights the condition allowing Chao to deliver centered estimates. *International Scholarly Research Notices* 2014.

[28] BERGER, J. O., BERNARDO, J. M., and SUN, D. (2012). Objective priors for discrete parameter spaces. *Journal of the American Statistical Association* **107** 636–648.

[29] BERGSMA, W., CROON, M. A., and HAGENAARS, J. A. (2009). *Marginal Models: For Dependent, Clustered, and Longitudinal Categorical Data.* New York: Springer Science & Business Media.

[30] BERNARDO, J. M. and SMITH, A. F. M. (2001). *Bayesian Theory.* Chichester: John Wiley & Sons.

[31] BIGGERI, A., STANGHELLINI, E., MERLETTI, F., and MARCHI, M. (1999). Latent class models for varying catchability and correlation among sources in capture-recapture estimation of the size of a human population. *Statistica Applicata* **11** 563–576.

[32] BISHOP, Y.M.M., FIENBERG, S.E., and HOLLAND, P.W. (1975). *Discrete Multivariate Analysis.* Cambridge: M.I.T. Press.

[33] BÖHNING, D., DIETZ, E., SCHAUB, R., SCHLATTMANN, P., and LINDSAY, B. G. (1994). The distribution of the likelihood ratio for mixtures of densities from the one-parameter exponential family. *Annals of the Institute of Statistical Mathematics* **46** 373–388.

[34] BÖHNING, D. (2000). *Computer-Assisted Analysis of Mixtures and Applications. Meta–analysis, Disease Mapping and Others.* Boca Raton: Chapman & Hall/CRC.

[35] BÖHNING, D., SUPPAWATTANABODEE, B., KUSOLVISITKUL, W., and VIWAT-WONGKASEM, C. (2004). Estimating the number of drug users in Bangkok 2001: a capture-recapture approach using repeated entries in one list. *European Journal of Epidemiology* **19** 1075–1083.

[36] BÖHNING, D., DIETZ, E., KUHNERT, R., and SCHÖN, D. (2005). Mixture models for capture-recapture count data. *Statistical Methods and Applications* **14** 29–43.

[37] BÖHNING, D. and SCHÖN, D. (2005) Nonparametric maximum likelihood estimation of population size based on the counting distribution. *Journal of the Royal Statistical Society Series C* **54** 721–737.

[38] BÖHNING, D. and KUHNERT, R. (2006). The equivalence of truncated count mixture distributions and mixtures of truncated count distributions. *Biometrics* **62** 1207–1215.

[39] BÖHNING, D. and DEL RIO VILAS, V. (2008). Estimating the hidden number of Scrapie affected holdings in Great Britain using a simple, truncated count model allowing for heterogeneity. *Journal of Agricultural, Biological and Environmental Statistics* **13** 1–22.

[40] BÖHNING, D. (2008). A simple variance formula for population size estimators by conditioning. *Statistical Methodology* **5** 410–423.

[41] BÖHNING, D., KUHNERT, R., and RATTANASIRI, S. (2008). *Meta-Analysis of Binary Data Using Profile Likelihood.* Boca Raton: Chapman & Hall/CRC.

[42] BÖHNING, D. and VAN DER HEIJDEN, P.G.M. (2009). A covariate adjustment for zero-truncated approaches to estimating the size of hidden and elusive populations. *Annals of Applied Statistics* **3** 595–610.

[43] BÖHNING, D. (2010). Some general comparative points on Chao's and Zelterman's estimators of the population size. *Scandinavian Journal of Statistics* **37** 221–236.

[44] BÖHNING, D., BAKSH, M.F., LERDSUWANSRI, R., and GALLAGHER, J. (2013). The use of the ratio-plot in capture-recapture estimation. *Journal of Computational and Graphical Statistics* **22** 135–155.

[45] BÖHNING, D., LERDSUWANSRI, R., VIDAL-DIEZ, A., VIWATWONGKASEM, C., and ARNOLD, M. (2013). A generalization of Chao's estimator for covariate information. *Biometrics* **69** 1033–1042.

[46] BÖHNING, D. (2015). Power series mixtures and the ratio plot with applications to zero-truncated count distribution modelling. *Metron* **73** 201–216.

[47] BÖHNING, D. (2016). Ratio plot and ratio regression with applications to social and medical sciences. *Statistical Science* **31** 205–218.

[48] BÖHNING, D., ROCCHETTI, I. ALFÓ, M., and HOLLING, H. (2016). A flexible ratio regression approach for zero-truncated capture–recapture counts. *Biometrics* **72** 697–706.

[49] BORCHERS, D. L., BUCKLAND, S. T., and ZUCCHINI, W. (2002). *Estimating Animal Abundance. Closed Populations.* London: Springer.

[50] BORENSTEIN, M., HEDGES, L. V., HIGGINS, J. P. T., and ROTHSTEIN, H. R. (2009) *Introduction to Meta-Analysis.* Chichester: John Wiley & Sons.

[51] BOULANGER, J., WHITE, G. C., McLELLAN, B. N., WOODS, J., PROCTOR, M., and HIMMER, S. (2002). A meta-analysis of grizzly bear DNA mark-recapture projects in British Columbia. *Ursus* **13** 137–152.

[52] BRAZZALE, A. R., DAVISON, A. C., and REID, N. (2007). *Applied Asymptotics. Case Studies in Small-Sample Statistics.* Cambridge: Cambridge Series in Statistical and Probabilistic Mathematics, Cambridge University Press.

[53] BRITTAIN, S. and BÖHNING, D. (2009). Estimators in capture-recapture studies with two sources. *AStA Advances in Statistical Analysis* **93** 23–47.

[54] BROOKMEYER, R. and GAIL, M. H. (1988). A method for obtaining short term projections and lower bound on the size of the Aids epidemic. *Journal of the American Statistical Association* **83** 301–308.

[55] BROWN, J. J., DIAMOND, I. D., CHAMBERS, R. L., BUCKNER, L. J., and TEAGUE, A. D. (1999). A methodological strategy for a one-number census in the UK. *Journal of the Royal Statistical Society Series A* **162** 247–267.

[56] BROWN, J. J., ABBOTT, O., and DIAMOND, I. D. (2006). Dependence in the 2001 one-number census project. *Journal of the Royal Statistical Society Series A* **169** 883–902.

[57] BRUNO, G., LaPORTE, R.E. , BIGGERI, A., McCARTY, D., and PAGANO, G. (1994). National diabetes programs. Application of capture-recapture to count diabetes? *Diabetes Care* **17** 548–556.

[58] BUCKLAND, S. and GARTHWAITE, P. (1991). Quantifying precision of mark-recapture estimates using the bootstrap and related methods. *Biometrics* **47** 255–268.

[59] BUNGE, J. and FITZPATRICK, M. (1993). Estimating the number of species: A review. *Journal of the American Statistical Association* **88** 364–373.

[60] BUNGE, J., WOODARD, L., BÖHNING, D., FOSTER, J. A., CONNOLLY, S., and ALLEN, H. K. (2012). Estimating population diversity with CatchAll. *Bioinformatics* **28** 1045–1047.

[61] BUNGE, J., BÖHNING, D., ALLEN, H., and FOSTER, J. A. (2012). Estimating population diversity with unreliable low frequency counts. *Biocomputing 2012: Proceedings of the Pacific Symposium, Kohala Coast, Hawaii, USA, 2–6 January 2012.* World Scientific Publishing.

[62] BUNGE, J. (2013). A survey of software for fitting capture-recapture models. *Wiley Interdisciplinary Reviews: Computational Statistics* **5** 114–120.

[63] BUNGE, J. WILLIS, A., and WALSH, F. (2014). Estimating the number of species in microbial diversity studies. *Annual Review of Statistics and Its Application* **1** 427–445.

[64] BUNGE, J. (2015). A note on marginal count distributions for diversity estimation. In *Ordered Data Analysis, Modeling and Health Research Methods, Volume 149 of the series Springer Proceedings in Mathematics & Statistics* 147–153. New York: Springer.

[65] BURNHAM, K. P. and OVERTON, W. S. (1978). Estimation of the size of a closed population when capture probabilities vary among animals. *Biometrika* **65** 625–633.

[66] BURNHAM, K. P. and ANDERSON, D. (2002). *Model Selection and Multimodel Inference: A Practical Information-Theoretic Approach.* New York: Springer.

[67] VAN BUUREN, S. (2012). *Flexible Imputation of Missing Data.* Boca Raton: Chapman & Hall/CRC Press.

[68] CAMERON, A.C. and TRIVEDI, P.K. (1998). *Regression Analysis of Count Data.* Cambridge: Cambridge University Press.

[69] CAROTHERS, A.D. (1973). Capture–recapture methods applied to a population with unknown parameters. *Journal of Animal Ecology* **42** 125–146.

[70] CBS, 2015 StatLine *The Hague / Heerlen: CDS.*

[71] CHAO, A. (1984). Nonparametric estimation of the number of classes in a population. *Scandinavian Journal of Statistics* **11** 265–270.

[72] CHAO, A. (1987). Estimating the population size for capture-recapture data with unequal catchability. *Biometrics* **43** 783–791.

[73] CHAO, A. (1989). Estimating population size for sparse data in capture-recapture experiments. *Biometrics* **45** 427–438.

[74] CHAO, A. and LEE, S.-M. (1992). Estimating the number of classes via sample coverage. *Journal of the American Statistical Association* **87** 210–217.

[75] CHAO, A. (2001). An overview of closed capture-recapture models. *Journal of Agricultural, Biological, and Environmental Statistics* **6** 158–175.

[76] CHAO, A. and BUNGE, J. (2002). Estimating the number of species in a stochastic abundance model. *Biometrics* **58** 531–539.

[77] CHAO, A. and HUGGINS, R. M. (2005) Classical closed-population capture-recapture models. In *Handbook of Capture-Recapture Analysis*, Amstrup SC, McDonald TL, Manly BFJ (eds). Princeton: Princeton University Press, pp. 22–35.

[78] CHAO, A., MA, K. H., and HSIEH, T.C. (2015). *SpadeR: Species Prediction and Diversity Estimation with R. R package version 0.1.0.* URL http://chao.stat.nthu.edu.tw/blog/software-download/

[79] CHAO, A., TSAY, P. K., LIN, S.-H., SHAU, W.-Y., and CHAO, D.-Y. (2001). The applications of capture-recapture models to epidemiological data. *Statistics in Medicine* **20** 3123–3157.

[80] CHEN M.-H. and SHAO Q.-M. (1997). On Monte Carlo methods for estimating ratios of normalizing constants. *Annals of Statistics* **25** 1563–1594.

[81] CHEN, Z., ZHANG, G., and LI, J. (2015). Goodness-of-fit test for meta-analysis. *Scientific Reports* **5** 16983.

[82] CHIB, S. (1995). Marginal likelihood from the Gibbs output. *Journal of the American Statistical Association* **90** 1313–1321.

[83] CHUN, Y. H. (2006). Estimating the number of undetected software errors via the correlated capture–recapture model. *European Journal of Operational Research* **175** 1180–1192.

[84] CONWAY, R. W. and MAXWELL, W. L. (1962). A queuing model with state dependent service rates. *Journal of Industrial Engineering* **12** 132–136.

[85] COOK, R. J. and LAWLESS, J. F. (2007). *The Statistical Analysis of Recurrent Events.* London: Springer.

[86] CORMACK, R. M. (1989). Log-linear models for capture-recapture. *Biometrics* **45** 395–413.

[87] CORMACK, R. M. (1992). Interval estimation for mark-recapture studies of closed populations. *Biometrics* **48** 567- 576.

[88] COULL, B. A. and AGRESTI, A. (1999). The use of mixed logit models to reflect heterogeneity in capture-recapture studies. *Biometrics* **55** 294–301.

[89] CRUYFF, M. J. L. F. and VAN DER HEIJDEN, P. G. M. (2008). Point and interval estimation of the population size using a zero truncated negative binomial regression model. *Biometrical Journal* **50** 1035–1050.

[90] CRUYFF, M. J. L. F., VAN DER HEIJDEN, P. G. M., and VAN GILS, G. (2013). *Simulatie Recurrent Events.* Utrecht, Utrecht University, BOA.

[91] CSO (2015). Population and Migration Estimates. 2014 http://www.cso.ie/en/releasesandpublications/-er/pme/populationandmigrationestimatesapril2014/, accessed on 20th Dec 2015.

[92] CULLEN, M. J., WALSH, J., NICHOLSON, L.V., and HARRIS, J. B. (1990). Ultrastructural localization of dystrophin in human muscle by using gold immunolabelling. *Proceedings of the Royal Society of London Series B* **20** 197–210.

[93] DACEY, M. F. (1972). A family of discrete probability distributions defined by the generalized hypergeometric series. *Sankhyā Series B* **34** 243–250.

[94] DALEY, T. and SMITH, A. D. (2013). Predicting the molecular complexity of sequencing libraries. *Nature Methods* **10** 325–327.

[95] DARROCH, J. N. (1958). The multiple-recapture census. I. Estimation of a closed population. *Biometrika* **45** 343–359.

[96] DARROCH, J. N. AND RATCLIFF, D. (1980). A note on capture-recapture estimation. *Biometrics* **36** 149–153.

[97] DARROCH, J. N. and SPEED, T. P. (1983). Additive and multiplicative models and interaction. *Annals of Statistics* **11** 724–738.

[98] DARROCH, J. N., FIENBERG, S. E., GLONEK, G. F. V., and JUNKER, B. W. (1993). A three-sample multiple-recapture approach to census population estimation with heterogeneity catchability. *Journal of American Statistical Association* **88** 1137–1148.

[99] DEMIDENKO, E. (2004). *Mixed Models Theory and Applications*. Hoboken, N.J. : Wiley-Interscience.

[100] DEMPSTER, A. P., LAIRD, N. M., and RUBIN, D. B. (1977). Maximum likelihood estimation from incomplete data via the EM algorithm (with discussion). *Journal of the Royal Statistical Society B* **39** 1–38.

[101] DOBRA, A. and FIENBERG, S. E. (2004). How large is the world wide web? In: *Web Dynamics*, pages 23–43, Springer.

[102] DORAZIO, R. M. and ROYLE, A. J. (2003). Mixture models for estimating the size of a closed population when capture rates vary among individuals. *Biometrics* **59** 351–364.

[103] DUROT, C., HUET, S., KOLADJO, F., and ROBIN, S. (2013). Least-squares estimation of a convex discrete distribution. *Computational Statistics & Data Analysis* **67** 282–298.

[104] DUROT, C., HUET, S., KOLADJO, F., and ROBIN, S. (2015). Nonparametric species richness estimation under convexity constraint. *Environmetrics* **26** 502–513.

[105] DRTON, M. (2009). Likelihood ratio tests and singularities. *Annals of Statistics* **27** 979–1012.

[106] DUNNE, J. (2015). The Irish Statistical System and the emerging census opportunity. *Statistical Journal of the IAOS* **31** 391–400.

[107] DUVAL, S. and TWEEDIE, R. (2000). A simple funnel-plot-based method of testing and adjusting for publication bias in meta-analysis. *Biometrics* **56** 455–463.

[108] EDWARDS, D. (2000). *Introduction to Graphical Modelling, 2nd Ed.* New York: Springer.

[109] EFRON, B. and THISTED, R. (1976). Estimating the number of unseen species: How many words did Shakespeare know? *Biometrika* **63** 435–447.

[110] EGGER, M., SMITH, G. D., and ALTMAN, D. G. (2001). *Systematic Reviews in Health Care: Meta-Analysis in Context.* London: BMJ Publishing Group.

[111] ENGBERSEN, G. B. M., STARING, R., BOOM, J., VAN DER HEIJDEN, P. G. M., and CRUYFF, M. J. L. F. (2002). *Illegale vreemdelingen in Nederland. Omvang, overkomst, verblijf en uitzetting.* Rotterdam: Erasmus Universiteit: RISBO.

[112] EUROPEAN PARLIAMENT, 2008, REGULATION (EC) NO 763/2008 OF THE EUROPEAN PARLIAMENT AND OF THE COUNCIL OF 9 JULY 2008 ON POPULATION AND HOUSING CENSUSES, (13.8.2008). *Official Journal of the European Union*, pp. L 218/14-L 218/20.

[113] FARCOMENI, A. (2012). Quantile regression for longitudinal data based on latent Markov subject-specific parameters. *Statistics and Computing* **22** 141–152.

[114] FARCOMENI, A. (2016). A general class of recapture models based on the conditional capture probabilities. *Biometrics* **72** 116–124.

[115] FARCOMENI, A. and TARDELLA, L. (2010). Reference Bayesian methods for alternative recapture models with heterogeneity. *TEST* **19** 187–208.

[116] FARCOMENI, A. and TARDELLA, L. (2012). Identifiability and inferential issues in capture-recapture experiments with heterogeneous detection probabilities. *Electronic Journal of Statistics* **6** 2602–2626.

[117] FARCOMENI, A. and SCACCIATELLI, D. (2013). Heterogeneity and behavioural response in continuous time capture-recapture, with application to street cannabis use in Italy. *Annals of Applied Statistics* **7** 2293–2314.

[118] FARRINGTON, C. P. (2002) Interval estimation for Poisson capture-recapture models in epidemiology. *Statistics in Medicine* **21** 3079–3092.

[119] FEGATELLI, D. A. and TARDELLA, L. (2016). Flexible behavioral capture-recapture modeling. *Biometrics* **72** 125–135.

[120] FELLEGI, I. P. and SUNTER, A. B. (1969). A theory for record linkage. *Journal of the American Statistical Association* **64** 1183–1210.

[121] FIENBERG, S. E. (1972). The multiple recapture census for closed populations and incomplete 2^k contingency tables. *Biometrika* **59** 591–603.

[122] FIENBERG, S. E., JOHNSON, M. S., and JUNKER, B. W. (1999). Classical multilevel and Bayesian approaches to population size estimation using multiple lists. *Journal of Royal Statistical Society, Series A* **162** 383–405.

[123] FISHER, R., STEVEN-CORBET, A., and WILLIAMS, C. B. (1943). The relation between the number of species and the number of individuals in a random sample of an animal population. *The Journal of Animal Ecology* **12** 42–58.

[124] GALLAY, A., VAILLANT, V., BOUVET, P., GRIMONT, P., and DESENCLOS, J. C. (2000). How many foodborne outbreaks of *Salmonella* infection occurred in France in 1995? Application of the capture-recapture method to three surveillance systems. *American Journal of Epidemiology* **152** 171–177.

[125] GERRITSE, S. C., VAN DER HEIJDEN, P. G. M., and BAKKER, B. F. M. (2015). Sensitivity of population size estimation for violating parametric assumptions in log-linear models. *Journal of Official Statistics* **31** 357–379.

[126] GERRITSE, S. C., BAKKER, B. F. M., and VAN DER HEIJDEN, P. G. M. (2015). Different methods to complete datasets used for capture-recapture estimation. *Statistical Journal of the IAOS* **31**, 613–627.

[127] GERRITSE, S. C., BAKKER, B. F. M., ZULT, D. B., and VAN DER HEIJDEN, P. G. M. (2016). *The Effects of Linkage Errors and Erroneous Captures on the Population Size Estimation.* Statistics Netherland Discussion Paper.

[128] GILLESPIE, I. A., O'BRIEN, S. J., ADAK, G. K., WARD, L. R., and SMITH, H. R. (2005). Foodborne general outbreaks of *Salmonella* Enteritidis phage type 4 infection, England and Wales, 1992–2002: Where are the risks? *Epidemiology and Infection* **133** 795–801.

[129] GILLISPIE, S. B. and GREEN, C. G. (2015). Approximating the Conway–Maxwell–Poisson distribution normalization constant. *Statistics* **49** 1062–1073.

[130] GODWIN, R. (2017). One-inflation and unobserved heterogeneity in population size estimation. *Biometrical Journal* **59** 79–93.

[131] GOOD, I. J. (1953). The population frequencies of species and the estimation of population parameters. *Biometrika* **40** 237–264.

[132] GOODMAN, L. A. (1974). Exploratory latent structure analysis using both identifiable and unidentifiable models. *Biometrika* **61** 215–231.

[133] GIORGI ROSSI, P., MANTOVANI, J., FERRONI, E., FORCINA, A., STANGHELLINI, E., CURTALE, F., and BORGIA, P. (2009) Incidence of bacterial meningitis (2001–2005) in Lazio, Italy: The results of an integrated surveillance system. *BMC Infectious Diseases* **9**:13.

[134] GHOSH, S. K. and NORRIS, J. L. (2005). Bayesian capture-recapture analysis and model selection allowing for heterogeneity and behavioral effects. *Journal of Agricultural, Biological, and Environmental Statistics* **10** 35–49.

[135] GOSKY, R. M. and GHOSH, S. K. (2009). A comparative study of Bayesian model selection criteria for capture-recapture models for closed populations. *Journal of Modern Applied Statistical Methods* **8**:6.

[136] GROENEBOOM, P., JONGBLOED, G., and WELLNER, J. (2008). The support reduction algorithm for computing non-parametric function estimates in mixture models. *Scandinavian Journal of Statistics* **35** 385–399.

[137] GUPTA, A. and NADARAJAH, S. (2004). Mathematical properties of the Beta distribution, in: Gupta, A. and Nadarajah S. (eds.) *Handbook of Beta Distribution and Its Applications*, New York: Marcel Dekker.

[138] HAMBLETON, R. K. and SWAMINATHAN, H. (1985). *Item Response Theory: Principles and Applications.* Boston: Kluwer Nijhoff.

[139] HASSEL, M., ASBJØRNSLETT, B. E., and HOLE, L. P. (2011). Underreporting of maritime accidents to vessel accident databases. *Accident Analysis and Prevention* **43** 2053–2063.

[140] HAY, G. (1997). The selection from multiple data sources in epidemiological capture-recapture studies. *The Statistician* **46** 515–520.

[141] HESSEN, D. J. (2012). Fitting and testing conditional multinormal partial credit models. *Psychometrika* **77** 693–709.

[142] HOAGLIN, D. C. (1980). A Poissonness plot. *The American Statistician* **34** 146–149.

[143] HOGAN, H. (1993). The post-enumeration survey: operations and results. *Journal of the American Statistical Association* **88** 1047–1060.

[144] HOLLAND, P. W. (1990). The Dutch Identity: A new tool for the study of item response model. *Psychometrika* **55** 5–18.

[145] HOLLING, H., BÖHNING, W., BÖHNING, D., and FORMANN, A. K. (2016). The covariate-adjusted frequency plot. *Statistical Methods in Medical Research* **25** 902–916.

[146] HOLZMANN, H., MUNK, A., and ZUCCHINI, W. (2006). On identifiability in capture-recapture models. *Biometrics* **62** 934–936.

[147] HOOGTEIJLING, E.M.J. (2002). Raming van het aantal niet in de GBA geregistreerden, Rapport 177-02-SO. *Voorburg: Centraal Bureau voor de Statistiek*

[148] HOOK, E. B. and REGAL, R. R. (1995). Capture-recapture methods in epidemiology: Methods and limitations. *Epidemiologic Reviews* **17** 243–264.

[149] HSER, Y.-I. (2001). Population estimation of illicit drug users in Los Angeles county. *The Journal of Drug Issues* **23** 323–334.

[150] HWANG, W.-H. and SHEN, T.-J. (2010). Small-sample estimation of species richness applied to forest communities. *Biometrics* **66** 1052–1060.

[151] HUGGINS, R. M. (1989). On the statistical analysis of capture experiments. *Biometrika* **76** 133–140.

[152] HU, J. H. and LAWLESS, J. F. (1996). Estimation of rate and mean functions from truncated recurrent event data. *Journal of the American Statistical Association* **91** 300–310.

[153] HWANG, W. H. and HUGGINS, R. M. (2005). An examination of the effect of heterogeneity on the estimation of population size using capture-recapture data. *Biometrika* **92** 229–233.

[154] IWGDMF — INTERNATIONAL WORKING GROUP FOR DISEASE MONITORING AND FORECASTING (1995a). Capture-recapture and multiple-record systems estimation I: History and theoretical development. *American Journal of Epidemiology* **142** 1047–1058.

[155] IWGDMF — INTERNATIONAL WORKING GROUP FOR DISEASE MONITORING AND FORECASTING (1995b). Capture-recapture and multiple-record systems estimation 2: Applications. *American Journal of Epidemiology* **142** 1059–1068.

[156] JIMÉNEZ-GAMERO, M. D., ALBA-FERNÁNDEZ, M. V., JODRÁ, P., and BARRANCO-CHAMORRO, I. (2015). An approximation to the null distribution of a class of Cramér-von Mises statistics. *Mathematics and Computers in Simulation* **118** 258–272.

[157] JIMÉNEZ-GAMERO, M. D. and KIM, H.-M. (2015). Fast goodness-of-fit tests based on the characteristic function. *Computational Statistics and Data Analysis* **89** 172–191.

[158] JOHNSON, N. L., KEMP, A. W., and KOTZ, S. (2005), *Univariate Discrete Distributions*. New York: Wiley.

[159] KASS, R. E, CARLIN, B. P., GELMAN, A., and NEAL, R. (1998). Markov Chain Monte Carlo in practice: A roundtable discussion. *American Statistician* **52** 93–100.

[160] KATZ, L. (1965). Unified treatment of a broad class of discrete probability distributions. *Classical and Contagious Discrete Distributions.* **1** 175–182.

[161] KEMP, A. W. (1968). A wide class of discrete distributions and the associated differential equations. *Sankhyā Series A* **30** 401–410.

[162] KEATING, K. A., SCHWARTZ, C. C., HAROLDSON, M. A., and MOOD, D. (2002). Estimating numbers of females with cubs-of-the-year in the Yellowstone grizzly bear population. *Ursus* **13** 161–174.

[163] KEMP, A. W. and KEMP C. D. (1966). An alternative derivation of the Hermite distribution. *Biometrika* **53** 627–628.

[164] KEMP, A. W. (2010). Families of power series distributions, with particular reference to the Lerch family. *Journal of Statistical Planning and Inference* **140** 2255–2259.

[165] KIM, S.-H. and KIM, S.-H. (2006). A note on collapsibility in DAG models of contingency tables. *Scandinavian Journal of Statistics* **33** 575–590.

[166] KING, R., MORGAN, B. J. T., GIMENEZ, O., and BROOKS, S. P. (2010). *Bayesian Analysis for Population Ecology.* Boca Raton: Chapman & Hall/CRC.

[167] KÖSE ,T., ORMAN, M., IKIZ, F., BAKSH, F. M., GALLAGHER, J., and BÖHNING, D. (2014). Extending the Lincoln–Petersen estimator for multiple identifications in one source. *Statistics in Medicine* **33** 4237–4249.

[168] KORICHEVA, J. and GUREVITCH, J. (2014). Uses and misuses of meta-analysis in plant ecology. *Journal of Ecology* **102** 828–844.

[169] KULINSKAYA, E., MORGENTHALER, S., and STAUDTE, R. G. (2008). *Meta Analysis: A Guide to Calibrating and Combining Statistical Evidence.* Chichester: John Wiley & Sons.

[170] KUHNERT, R. and BÖHNING, D. (2009). CAMCR: Computer-assisted mixture model analysis for capture-recapture count data. *AStA Advances in Statistical Analysis* **93** 61–71.

[171] LAIRD, N. (1978). Nonparametric maximum likelihood estimation of a mixing distribution. *Journal of the American Statistical Association* **73** 805–811.

[172] LANUMTEANG, K. (2010). *Estimation of the Size of a Target Population Using Capture-Recapture Methods Based upon Multiple Sources and Continuous Time Experiments.* PhD thesis, University of Reading.

[173] LANUMTEANG, K. and BÖHNING, D. (2011). An extension of Chao's estimator of population size based on the first three capture frequency counts. *Computational Statistics and Data Analysis* **55** 2302–2311.

[174] LAZARSFELD, P. F. and HENRY, N. W. (1968). *Latent Structure Analysis.* Boston: Houghton Mifflin.

[175] LAURITZEN, S. L. (1996). *Graphical Models.* Oxford: Oxford University Press.

[176] LEERKES, A., VAN SAN, M., ENGBERSEN, G. B. M., CRUYFF, M. J. L. F., and VAN DER HEIJDEN, P. G. M. (2004). *Wijken voor illegalen. Over ruimtelijke spreiding, huisvesting en leefbaarheid.* Den Haag: SDU.

[177] LERDSUWANSRI,R. (2012). *Generalisation of the Lincoln–Petersen Approach to Non-binary Source Variables.* PhD thesis, University of Reading.

[178] LINDSAY, B. (1995). Mixture Models: Theory, Geometry and Applications. In *NSF-CBMS Regional Conference Series in Probability and Statistics.* Hayward: Institute of Mathematical Statistics.

[179] LINDSAY, B., CLOGG, C., and GREGO, J. (1991). Semiparametric estimation in the Rasch model and related exponential response models, including a simple latent class model for item analysis. *Journal of the American Statistical Association* **86** 96–107.

[180] LINK, W. A. (2003). Nonidentifiability of population size from capture-recapture data with heterogeneous detection probabilities. *Biometrics* **59** 1123–1130.

[181] LINK, W. A., J. YOSHIZAKI, L. L. BAILEY, and POLLOCK, K. H. (2010). Uncovering a latent multinomial: Analysis of mark-recapture data with misidentification. *Biometrics* **66** 178–185.

[182] LITTELL, R. C., MILLIKEN, G. A., STROUP, W. W., WOLFINGER, R. D., and SCHABENBERGER, O. (2006). *SAS® for Mixed Models Second Edition.* Cary, NC: SAS Institute Inc.

[183] LITTLE, R. J. and RUBIN, D. B. (2002). *Statistical Analysis with Missing Data.* New York: John Wiley & Sons.

[184] LIU, G., RONG, G., ZHANG, H., and SHAN, Q. (2015). The adoption of capture-recapture in software engineering: A systematic literature review. *EASE '15 Proceedings of the 19th International Conference on Evaluation and Assessment in Software Engineering,* Article No. 15.

[185] LLOYD, C. J. and FROMMER, D. (2004a). Estimating the false negative fraction for a multiple screening test for bowel cancer when negatives are not verified. *Australian and New Zealand Journal of Statistics* **46** 531–542.

[186] LLOYD, C. J. and FROMMER, D. (2004b). Regression based estimation of the false negative fraction when multiple negatives are unverified. *Journal of the Royal Statistical Society Series C* **53** 619–631.

[187] LLOYD, C. J. and FROMMER, D. (2008). An application of multinomial logistic regression to estimating performance of a multiple-screening test with incomplete verification. *Journal of the Royal Statistical Society Series C* **57** 89–102.

[188] LUM, K., PRICE, M. E., and BANKS, D. (2013). Applications of multiple systems estimation in human rights research. *The American Statistician* **67** 191–200.

[189] MA, Z., MAO, C. X., and YANG, Y. (2014). On log-linear representations of the Rasch model in capture-recapture studies. Technical Report, The Data Engineering Center, Shanghai University of Finance and Economics.

[190] MA, L., SUN, N., LIU, X., JIAO, Y., ZHAO, H., and DENG, X. W. (2005). Organ-specific expression of Arabidopsis genome during development. *Plant Physiology* **138** 80–91.

[191] MADIGAN, D. and YORK, J. C. (1997). Bayesian methods for estimating the size of a closed population. *Biometrika* **84** 19–31.

[192] MAO, C. X. (2006). Inference on the number of species through geometric lower bounds. *Journal of the American Statistical Association* **101** 1663–1670.

[193] MAO, C. X. (2007). Estimating population sizes for capture–recapture sampling with binomial mixtures. *Computational Statistics & Data Analysis* **51** 5211–5219.

[194] MAO, C. X. (2008). On the nonidentifiability of population sizes. *Biometrics* **64** 977–979.

[195] MAO, C. X. AND LINDSAY, B. (2007). Estimating the number of classes. *The Annals of Statistics* **35** 917–930.

[196] MAO, C. X., YANG, N., and ZHONG, J. (2013). On population size estimators in the Poisson mixture model. *Biometrics* **69** 758–765.

[197] MAO, C. X., YANG, Y., and YOU, N. (2014). Appraisal of log-linear models in capture-recapture studies. Technical Report, The Data Engineering Center, Shanghai University of Finance and Economics.

[198] MAO, C. X., YANG, C., YANG, Y., and ZHUANG W. (2017). Estimating population sizes with the Rasch model. *Annals of the Institute of Statistical Mathematics* **69** 705–716.

[199] MAO, C. X. and YOU, N. (2010). Estimate the initial population size from removal data. In: *Biometrics: Methods, Applications and Analyses*, ed. Schuster, H. and Metzger, W., Chapter 10, pages 185–195, Nova Science.

[200] Maple 12.0. Maplesoft, a division of Waterloo Maple Inc., Waterloo, Ontario.

[201] MATTHEWS, J. N. S. and APPLETON, D. R. (1993). An application of the truncated Poisson distribution to immunogold assay. *Biometrics* **49** 617–621.

[202] McCREA, R. S. and MORGAN, B. J. T. (2015). *Analysis of Capture-Recapture Data*. Boca Raton: Chapman & Hall/CRC.

[203] McCULLAGH, P. and NELDER, J. A. (1989). *Generalized Linear Models*. London: Chapman & Hall.

[204] McHUGH, R. B. (1956) Efficient estimation and local identification in latent class analysis. *Psychometrika* **21** 331–347.

[205] McKENDRICK, A. G. (1926). Application of mathematics to medical problems. *Proc. Edinb. Math. Soc.* **44** 98–130.

[206] McLACHLAN, G. and KRISHNAN, T. (1997). *The EM Algorithm and Extensions*. Wiley: New York.

[207] McLACHLAN, G. and PEEL, D. (2000). *Finite Mixture Models*. Wiley: New York.

[208] MEINTANIS, S. and SWANEPOEL, J. (2007). Bootstrap goodness-of-fit tests with estimated parameters based on empirical transforms. *Statistics and Probability Letters* **77** 1004–1013.

[209] MEINTANIS, S. G., JIMÉNEZ GAMERO, M. D., and ALBA-FERNÁNDEZ, V. (2014). A class of goodness-of-fit tests based on transformation. *Communications in Statistics: Theory and Methods* **43** 1708–1735.

[210] MEINTANIS, S. G., SWANEPOEL, J. and ALLISON, J. (2014). The probability weighted characteristic function and goodness-of-fit testing. *Journal of Statistical Planning and Inference* **146** 122–132.

[211] MORGAN, B. J. T and RIDOUT, M. S. (2008). A new mixture model for capture heterogeneity. *Journal of the Royal Statistical Society, Series C* **57** 433–446.

[212] MORIÑA, D., HIGUERAS, M., PUIG, P., and OLIVEIRA, M. (2015). Hermite: Generalized Hermite Distribution (R package). http://CRAN.R-project.org/package=hermite.

[213] MORIÑA, D., HIGUERAS, M., PUIG, P., and OLIVEIRA, M. (2015). Hermite: Generalized Hermite Distribution. Modelling with the R Package hermite. *The R Journal* **7** 263–274.

[214] NADARAJAH, S. (2007). Useful moment and CDF formulations for the COM–Poisson distribution. *Statistical Papers* **50** 617–622.

[215] NAKAMURA, M. and PEREZ-ABREU, V. (1993). Empirical probability generating function: An overview. *Insurance: Mathematics and Economics* **12** 287–295.

[216] NG, C. M., ONH, S.-H., and SRIVASTAVA, H. M. (2013). Parameter estimation by Hellinger type distance for multivariate distributions based upon probability generating functions. *Applied Mathematical Modelling* **37** 7374–7385.

[217] NIREL, R. and GLICKMAN, H. (2009). Sample surveys and censuses. In: D. Pfeffermann and C.R. Rao (eds.) *Sample Surveys: Design, Methods and Applications*, Vol 29A, Chapter 21, 539–565.

[218] NIWITPONG, S. A., BÖHNING, D., VAN DER HEIJDEN, P. G., and HOLLING, H, (2013). Capture-recapture estimation based upon the geometric distribution allowing for heterogeneity. *Metrika* **76** 495–519.

[219] NOACK, A. (1950). On a class of discrete random variables. *Annals of Mathematical Statistics* **21** 127–132.

[220] NORRIS, J. L. and POLLOCK, K. H. (1996). Nonparametric MLE under two closed capture-recapture models with heterogeneity. *Biometrics* **52** 639–649.

[221] NORRIS, J. L. and POLLOCK, K. H. (1998). Non-parametric MLE for Poisson species abundance models allowing for heterogeneity between species. *Environmental and Ecological Statistics* **5** 391–402.

[222] ONS — OFFICE FOR NATIONAL STATISTICS (2013). *Beyond 2011: Producing Population Estimates Using Administrative Data: In Practice*. ONS Internal Report, available at: http://www.ons.gov.uk/ons/about-ons/who-ons-are/programmes-and-projects/beyond-2011/reports-and-publications/index.html

[223] OREMUS, M. (2005). Pers. Communic.

[224] OLIVEIRA, M., EINBECK, J., HIGUERAS, M., AINSBURY, E., PUIG, P., and ROTHKAMM, K.(2015). Zero-inflated regression models for radiation-induced chromosome aberration data: A comparative study. *Biometrical Journal* **58** 259–279.

[225] ORD, J. K. (1967) Graphical methods for a class of discrete distributions. *Journal of the Royal Statistical Society, Series A* **130** 232–238.

[226] OTIS, D.L., BURNHAM, K.P., WHITE, G.C., and ANDERSON, D.R. (1978). Statistical inference from capture data on closed animal populations. *Wildlife Monographs* **62**, The Wildlife Society.

[227] OVERSTALL, A. M., KING, R., BIRD, S. M., HUCHINSON, S. J., and HAY, G. (2014). Incomplete contingency tables with censored cells with application to estimating the number of people who inject drugs in Scotland. *Statistics in Medicine* **33** 1564–1579.

[228] OVERSTALL, A. M. and KING, R. (2014). conting: An R package for Bayesian analysis of complete and incomplete contingency tables. *Journal of Statistical Software* **58** 1–27.

[229] PAWITAN, Y. (2001). *In All Likelihood: Statistical Modelling and Inference Using Likelihood.* Oxford: Oxford University Press.

[230] PELLE, E., HESSEN, D. J., and VAN DER HEIJDEN, P. G. M. (2015). A log-linear multidimensional Rasch model for capture-recapture. *Statistics in Medicine* **35** 622–634.

[231] PETITTI, D. B. (1994). *Meta-Analysis Decision Analysis and Cost-Effectiveness Analysis.* New York: Oxford University Press.

[232] PLEDGER, S. A. (2000). Unified maximum likelihood estimates for closed capture-recapture studies. *Biometrics* **56** 434–442.

[233] PLEDGER, S. A. (2005). The performance of mixture models in heterogeneous closed population capture-recapture. *Biometrics* **61** 868–876.

[234] POLLOCK, K. H., HINES, J. E., and NICHOLS, J. E. (1984). The use of auxilliary variables in capture-recapture and removal experiments. *Biometrics* **40** 329–340.

[235] POLLOCK, K. H. (2002). The use of auxiliary variables in capture-recapture modelling: An overview *Journal of Applied Statistics* **29** 85–102.

[236] PUIG, P. and BARQUINERO, J. F. (2011). An application of compound Poisson modelling to biological dosimetry. *Proceedings of the Royal Society of London, Series A. Mathematical, Physical and Engineering Sciences* **467** 897–910.

[237] PUIG, P. and KOKONENDJI, C. (2017). Nonparametric estimation of the number of zeros in truncated count distributions. *Scandinavian Journal of Statistcs* (to appear).

[238] PUJOL, M., BARQUINERO, J. F., PUIG, P., PUIG, R., CABALLÍN, M. R., and BARRIOS, L. (2014). A new model of biodosimetry to integrate low and high doses. *PlosOne*, 1–19.

[239] RAMSEY, F. and SEVERNS, P. (2010). Persistent models for mark-recapture. *Environmental and Ecological Statistics* **17** 97–109.

[240] RAO, J. N. K. (2003). *Small Area Estimation.* New York: Wiley & Sons.

[241] RASCH, G. (1960). *Probabilistic Models for Some Intelligence and Attainment Tests.* Copenhagen: Danish Institute for Educational Research.

[242] RASCH, G. (1961). On general laws and the meaning of measurement in psychology. In: *Proceedings of the Fourth Berkeley Symposium on Mathematical Statistics and Probability*, Berkeley, CA, University of California Press **4** 321–333.

[243] RENAUD, A. (2007). Estimation of the coverage of the 2000 census of population in Switzerland: Methods and results. *Survey Methodology* **33** 199–210.

[244] RICE J. A. (1995) *Mathematical Statistics and Data Analysis*. California: Duxbury Press.

[245] RISSANEN, J.(1983). A universal prior for integers and estimation by minimum description length. *Annals of Statistics* **11** 416–431.

[246] RIVEST, L.-P. and BAILLARGEON, S. (2014). Capture-recapture methods for estimating the size of a population: Dealing with variable capture probabilities. In: Lawless, J.F. (ed.) *Statistics in Action: A Canadian Outlook*, Boca Raton: CRC Press.

[247] ROCCHETTI, I., BUNGE, J., and BÖHNING, D. (2011). Population size estimation based upon ratios of recapture probabilities. *Annals of Applied Statistics* **5** 1512–1533.

[248] ROCCHETTI, I., ALFÓ, M., and BÖHNING, D. (2014). A regression estimator for mixed binomial capture-recapture data. *Journal of Statistical Planning and Inference* **145** 165–178.

[249] ROSS, S. M. (1985) *Introduction to Probability Models*. Orlando: Academic Press.

[250] RUBIN, D. B. (1976). Inference and missing data. *Biometrika* **63** 581–592.

[251] RÜCKER, G., REISER, V., MOTSCHALL, E., BINDER, H., MEERPOHL, J. J., ANTES, G., and SCHUMACHER, M. (2011). Boosting qualifies capture-recapture methods for estimating the comprehensiveness of literature searches for systematic reviews. *Journal of Clinical Epidemiology* **64** 1364–1372.

[252] SANATHANAN, L. (1977). Estimating the size of a truncated sample. *Journal of the American Statistical Association* **72** 669–672.

[253] SANATHANAN, L. (1972). Estimating the size of a multinomial population. *The Annals of Mathematical Statistics* **43** 142–152.

[254] SCHAFER, J. (1997). *Analysis of Incomplete Multivariate Data*. Boca Raton: Chapman & Hall/CRC.

[255] SCHAFER, J. (1997). *Imputation of Missing Covariates under a General Linear Mixed Model*. Pennsylvania: PennState University, Department of Statistics.

[256] SCHWARZ, G. E. (1978). Estimating the dimension of a model. *Annals of Statistics* **6** 461–464.

[257] SCHULZE, R., HOLLING, H., and BÖHNING, D. (2003). *Meta-Analysis: New Developments and Applications in Medical and Social Sciences*. Göttingen: Hogrefe & Huber.

[258] SCHWARZ, C. and SEBER, G. A. F. (1999). A review of estimating animal abundance III. *Statistical Science* **14** 427–456.

[259] SEBER, G .A. F. (2002). *The Estimation of Animal Abundance and Related Parameters, 2nd Edition*. London: Griffin.

[260] SELF, S. and LIANG, K.(1987). Asymptotic properties of maximum likelihood estimators and likelihood ratio tests under nonstandard conditions. *Journal of the American Statistical Association* **82** 605–610.

[261] SHARIFDOUST, M., NG, C. M., and ONG, S. H. (2016). Probability generating function based Jeffrey's divergence for statistical inference. *Communications in Statistics – Simulation and Computation* **45** 2445–2458.

[262] SHMUELI, G., MINKA, T. P., KADANE, J. B., BORLE, S., and BOATWRIGHT, P. (2005). A useful distribution for fitting discrete data: Revival of the Conway-Maxwell-Poisson distribution. *Journal of the Royal Statistical Society, Series C* **54** 127–142.

[263] SNOW, L. C., DAVIES, R. H., CHRISTIANSEN, K. H., CARRIQUE-MAS, J. J., WALES, A. D., O'CONNOR, J. L., COOK, A. J. C., and EVANS, S. J. (2007). Papers & Articles. *The Veterinary Record* **161** 471–476.

[264] SPANIER, J. and OLDHAM, K. B. (1987). *An Atlas of Functions*. Washington, DC: Hemisphere Publishing Corporation.

[265] SPEVACK, M. (1968). *A Complete and Systematic Concordance to the Works of Shakespeare*. Vols. 1-6. Hildesheim: George Olms.

[266] STANGL, D. K. and BERRY, D. A. (2000). *Meta-Analysis in Medicine and Health Policy*. Basel: Marcel Dekker.

[267] STANGHELLINI, E. and VAN DER HEIJDEN, P. G. M. (2004). A multiple-record systems estimation method that takes observed and unobserved heterogeneity into account. *Biometrics* **60** 510 516.

[268] STANGHELLINI, E. and VANTAGGI, B. (2013). On the identification of discrete concentration graph models with one hidden binary variable. *Bernoulli* **19** 1920–1937.

[269] STANLEY, T. and BURNHAM, K. (1998). Information-theoretic model selection and model averaging for closed-population capture-recapture studies. *Biometrical Journal* **40** 475–494.

[270] Sterne, J. (2009). *Meta-Analysis in Stata: An Updated Collection from the Stata Journal*. Texas. Stata Press.

[271] STOCK, A., JÜRGENS, K., BUNGE, J., and STOECK, T. (2009). Protistan diversity in the suboxic and anoxic waters of the Gotland Deep (Baltic Sea) as revealed by 18S rRNA clone libraries. *Aquatic Microbial Ecology* **55** 267–284.

[272] SUTHERLAND, J.M., SCHWARZ, C. J., and RIVEST, L.-P. (2007). Multilist population estimation with incomplete and partial stratification. *Biometrics* **63** 910–916.

[273] SUTTON, A. J., ABRAMS, K. R., JONES, D. R., SHELDON, T. A., and SONG, F. (2000). *Methods for Meta-Analysis in Medical Research*. New York: Wiley & Sons.

[274] TARDELLA, L. (2002). A new Bayesian method for nonparametric capture-recapture models in presence of heterogeneity. *Biometrika* **89** 807–817.

[275] TENZIN, T., MCKENZIE,J.S., VANDERSTICHEL, R., RAI, B. D., RINZIN, K., TSHERING, Y., PEM, R., TSHERING, C., DAHAL, N., DUKPA, K., DORJEE, S., WANGCHUK, S., JOLLY, P. D., MORRIS, R., and WARD, M. P. (2015). Comparison of mark-resight methods to estimate abundance and rabies vaccination coverage of free-roaming dogs in two urban areas of south Bhutan. *Preventive Veterinary Medicine* **118** 436–448.

[276] TOUNKARA, F. and RIVEST, L. P. (2015). Mixture regression models for closed population capture-recapture data. *Biometrics* **71** 721–730.

[277] Tsay, P. K. and Chao, A. (2001). Population size estimation for capture-recapture models with applications to epidemiological data. *Journal of Applied Statistics* **28** 25–36.

[278] UNECE (2014). Measuring population and housing. Practices of UNECE countries in the 2010 round of censuses. *United Nations Economic Commission for Europe* http://www.unece.org.

[279] Van der Heijden, P. G. M., Cruyff, M.J.L.F., and Van Houwelingen, H. C. (2003). Estimating the size of a criminal population from police records using the truncated Poisson regression model. *Statistica Neerlandica* **57** 1–16.

[280] Van der Heijden, P. G. M., Bustami, R., Cruyff, M. J. L. F., Engbersen, G., and Van Houwelingen, H. C. (2003). Point and interval estimation of the population size using the truncated Poisson regression model. *Statistical Modelling* **3** 305–322.

[281] Van der Heijden, P. G. M., Zwane, E., and Hessen, D. (2009). Structurally missing data problems in multiple list capture-recapture data. *Advances of Statistical Analysis* **93** 5–21.

[282] Van der Heijden, P. G. M., Cruyff, M. J. L. F., and Van Gils, G. (2011). *Schattingen illegaal in Nederland verblijvende vreemdelingen 2009.* Den Haag: WODC.

[283] van der Heijden, P. G. M., Whittaker, J., Cruyff, M. J. L. F., Bakker, B., and van der Vliet, R. (2012). People born in the Middle East but residing in the Netherlands: Invariant population size estimates and the role of active and passive covariates. *Annals of Applied Statistics* **6** 831–852.

[284] Van der Heijden, P. G. M., Cruyff, M. J. L. F., and Böhning, D. (2014). *Analyses daklozen [Utrecht WODC.* Universiteit Utrecht en University of Southampton. Utrecht, 28 januari 2014.

[285] Van der Heijden, P. G. M., Cruyff, M. J. L. F., and van Gils, G. (2015). *Schattingen illegaal in Nederland verblijvende vreemdelingen 2012–2013.* Utrecht: Universiteit Utrecht, Faculteit Sociale Wetenschappen, Afdeling Methoden en Statistiek.

[286] Valente, P. (2010). *Main Results of the UNECE / UNSD Survey on the 2010 / 2011 Round of Censuses in the UNECE Region.* Luxembourg: Eurostat.

[287] Van der Pal, K., Van der Heijden, P. G. M., Buitendijk, S., and Den Ouden, A. (2003). Periconceptional folic acid use and the prevalence of neural tube defects in the Netherlands. *European Journal of Obstetrics Gynecology and Reproductive Biology,* **108** 33–39.

[288] Van der Vaart, A. (1998). *Asymptotic Statistics.* Cambridge: Cambridge University Press.

[289] Vergne, T., Calavas, D., Cazeau, G., Durand, B., Dufour, B., and Grosbois, V. (2012). A Bayesian zero-truncated approach for analysis capture-recapture count data from classical scrapie surveillance in France. *Preventive Veterinary Medicine* **105** 127–135.

[290] Wang, J.-P. and Lindsay, B. G. (2005). A penalized nonparametric maximum likelihood approach to species richness estimation. *Journal of the American Statistical Association* **100** 942–959.

[291] WANG, J.-P. and LINDSAY, B. G. (2008). An exponential partial prior for improving nonparametric maximum likelihood estimation in mixture models. *Statistical Methodology* **5** 30–45.

[292] WANG, J. P (2010). Estimating species richness by a Poisson–compound gamma model. *Biometrika* **97** 727–740.

[293] WANG, X., HE, C. Z., and SUN, D. (2007). Population estimation for small sample capture-recapture data using noninformative priors. *Journal of Statistical Planning and Inference* **137** 1099–1118.

[294] WHITE, G. C. and BENNETTS, R. E. (1996). Analysis of frequency count data using the negative binomial distribution. *Ecology* **77** 2549–2557.

[295] WHITTAKER, J. (1990). *Graphical Models in Applied Multivariate Statistics.* New York: John Wiley & Sons.

[296] WHO (2010). Guidelines on estimating the size of populations most at risk to HIV. WHO Publications.

[297] WILLIS, A. and BUNGE, J. (2015). Estimating diversity via frequency ratios. *Biometrics* **71** 1042–1049.

[298] WILLIS, A., BUNGE, J., and WHITMAN, T. (2015). Inference for changes in biodiversity. arXiv:1506.05710 [stat.ME].

[299] WILSON, R.M. and COLLINS, M.F. (1992). Capture-recapture estimation with samples of size one using frequency data. *Biometrika* **79** 543–553.

[300] WITTES J. T., COLTON T., and SIDEL V. W. (1974). Capture-recapture methods for assessing the completeness of cases ascertainment when using multiple information sources. *Journal of Chronic Diseases* **27** 25–36.

[301] WOLTER, K. (1986). Some coverage error models for census data. *Journal of the American Statistical Association* **81** 338–346.

[302] WRIGHT, J. A., BARKER, R. J., SCHOFIELD, M. R., FRANTZ, A. C., BYROM, A. E., and GLEESON, D. M. (2009). Incorporating genotype uncertainty into mark-recapture-type models for estimating abundance using DNA samples. *Biometrics* **65** 833–840.

[303] XU, C., SUN, D., and HE, C. (2014). Objective Bayesian analysis for a capture–recapture model. *Annals of the Institute of Statistical Mathematics* **66** 245–278.

[304] YEE, T. W. (2010). The VGAM package for categorical data analysis. *Journal of Statistical Software* **32** 1–34.

[305] YEE, T.W., STOKLOSA, J., and HUGGINS, R. M. (2015). The VGAM package for capture-recapture data using the conditional likelihood. *Journal of Statistical Software* **65** 1–33.

[306] ZELTERMAN, D. (1988). Robust estimation in truncated discrete distributions with applications to capture-recapture experiments. *Journal of Statistical Planning and Inference* **18** 225–237.

[307] ZHANG, L.-C. (2012). Topics of statistical theory for register-based statistics and data integration. *Statistica Neerlandica* **66** 41–63.

[308] ZHANG, L.-C. (2015). On modelling register coverage errors. *Journal of Official Statistics* **31** 381–396.

[309] ZWANE, E. N. and VAN DER HEIJDEN, P. G. M. (2004). Semiparametric models for capture-recapture studies with covariates. *Computational Statistics and Data Analysis* **47** 729–743.

[310] ZWANE, E. N. and VAN DER HEIJDEN, P. G. M. (2003). Implementing the parametric bootstrap in capture–recapture models with continuous covariates. *Statistics & Probability Letters* **65** 121–125.

[311] ZWANE, E. N. and VAN DER HEIJDEN, P. G. M. (2007). Analysing capture-recapture data when some variables of heterogeneous catchability are not collected or asked in all registrations. *Statistics in Medicine* **26** 1069–1089.

[312] ZWANE, E. N. and VAN DER HEIJDEN, P. G. M. (2005). Population estimation using the multiple system estimator in the presence of continuous covariates. *Statistical Modelling* **5** 39–52.

[313] ZWANE, E. N., VAN DER PAL-DE BRUIN, K., and VAN DER HEIJDEN, P. G. M. (2004). The multiple-record systems estimator when registrations refer to different but overlapping populations. *Statistics in Medicine* **23** 2267–2281.

Index

9 781032 096698